Lecture Notes in Physics

Volume 862

For further volumes:
http://www.springer.com/series/5304

The Lecture Notes in Physics

The series Lecture Notes in Physics (LNP), founded in 1969, reports new developments in physics research and teaching—quickly and informally, but with a high quality and the explicit aim to summarize and communicate current knowledge in an accessible way. Books published in this series are conceived as bridging material between advanced graduate textbooks and the forefront of research and to serve three purposes:

- to be a compact and modern up-to-date source of reference on a well-defined topic
- to serve as an accessible introduction to the field to postgraduate students and nonspecialist researchers from related areas
- to be a source of advanced teaching material for specialized seminars, courses and schools

Both monographs and multi-author volumes will be considered for publication. Edited volumes should, however, consist of a very limited number of contributions only. Proceedings will not be considered for LNP.

Volumes published in LNP are disseminated both in print and in electronic formats, the electronic archive being available at springerlink.com. The series content is indexed, abstracted and referenced by many abstracting and information services, bibliographic networks, subscription agencies, library networks, and consortia.

Proposals should be sent to a member of the Editorial Board, or directly to the managing editor at Springer:

Christian Caron
Springer Heidelberg
Physics Editorial Department I
Tiergartenstrasse 17
69121 Heidelberg/Germany
christian.caron@springer.com

Sei Suzuki • Jun-ichi Inoue • Bikas K. Chakrabarti

Quantum Ising Phases and Transitions in Transverse Ising Models

Second Edition

Sei Suzuki
Dept. of Physics and Mathematics
Aoyama Gakuin University
Sagamihara, Japan

Bikas K. Chakrabarti
Saha Institute of Nuclear Physics
Kolkata, India

Jun-ichi Inoue
Graduate School of Information Science
and Technology
Hokkaido University
Sapporo, Japan

ISSN 0075-8450
ISSN 1616-6361 (electronic)
Lecture Notes in Physics
ISBN 978-3-642-33038-4
ISBN 978-3-642-33039-1 (eBook)
DOI 10.1007/978-3-642-33039-1
Springer Heidelberg New York Dordrecht London

Library of Congress Control Number: 2012953284

Printed on acid-free paper

Springer is part of Springer Science+Business Media (www.springer.com)

Preface

Quantum phase transitions, driven by quantum fluctuations, have got intriguing features with potentially new application possibilities including those in quantum computations. A significant amount of research in physics today is therefore directed towards the investigations on the nature and behaviour of such quantum phases and transitions in cooperatively interacting many-body quantum systems. Major advances have been made in both theoretical and experimental studies on such systems.

For modelling purposes, although the Heisenberg model and its variants were introduced much earlier (in 1920s), most of the innovative and successful researches today in this field have been obtained by employing, or comparing with, the results of quantum, or transverse field, Ising models (introduced in 1960s). This is because of the advantage in the separability of the cooperative interaction from the (tunable) transverse field or tunnelling term in the Hamiltonian of the transverse field Ising model (in contrast to that in the Heisenberg and other models where the cooperativity and non-commutability are intertwined). Also, a number of condensed matter systems can be modelled accurately by such transverse field Ising models. Because of these, many of the intriguing features observed in the statics or dynamics of quantum phase transitions are mostly limited today to either the theoretical investigations in such quantum Ising models or to the experimental results which can be checked and compared for such models!

This book introduces these quantum Ising models and their theoretical analysis, including numerical ones, at length. Attempts have been made to bring the reader to the research frontiers today.

In an earlier incarnation of this book, it was published in the Lecture Notes in Physics Series of Springer with the same title in 1996 and was authored by Bikas K. Chakrabarti, Amit Dutta & Parangama Sen. Since then, many important developments have taken place, in particular in the dynamic studies like in quantum quenching, annealing etc. Also, the book went out of print rather quickly and Dr. Christian Caron, Executive Publishing Editor of the Springer, requested several times to bring out a second updated edition of the book. However, because of the prolonged nature of the involvements in the respective researches, both Profs. Dutta and Sen could not agree to join the effort. We therefore decided, with their kind consent, to revise this book thoroughly and upgrade it extensively. The resulting book, though seeded in the earlier one, is also considerably modified and upgraded. Indeed, when this revised book manuscript was given by Dr. Caron to one of their reviewers for an opinion, the reviewer wrote "... authors present an excellent overview on the state of the art of Quantum Ising Phases under various situations ... this reviewer is highly impressed about the excellent quality, the comprehensive and thorough presentation of the material and its expert discussion ... this book is a masterpiece which belongs to the shelves (if not on the desks!) of all researchers working in quantum phase transitions and their equilibrium properties and non-equilibrium dynamics ... authors possess a deep knowledge of the field and are able to account for recent modern developments ... much of the technical (but important) details have wisely been organized by deferring such material to concise appendices at the end of the corresponding chapter ... I have only minor suggestions for improving this 'gem' further ...". These words from the anonymous reviewer clearly offer us *a posteriori* justification of our renewed effort to revise and upgrade the book!

We are also extremely happy to receive many helpful comments and suggestions on the draft version of this book from Profs. Subinay Dasgupta, Amit Dutta and Parongama Sen.

We hope, the book will be useful to the young researchers in exploring this exciting field.

Sagamihara, Japan — Sei Suzuki
Sapporo, Japan — Jun-ichi Inoue
Kolkata, India — Bikas K. Chakrabarti

Contents

Chapter 1
Introduction

1.1 The Transverse Ising Models

The study of the transverse Ising Hamiltonian dates back to the early 1960s. de Gennes [98], in 1963, studied this system to theoretically model the order-disorder transition in some double-well ferroelectric systems [40], such as potassium dihydrogen phosphate (KDP or KH_2PO_4) crystals. The mean field phase diagram [48, 379] and the study of susceptibilities (correlations) gave good qualitative agreement with the experimental results for such simple hydrogen-bonded ferroelectric samples. Katsura [216] had already studied this model as a special case of the anisotropic Heisenberg Hamiltonian in a magnetic field and calculated the exact free energy and the dispersion relation obeyed by the elementary excitations for the one-dimensional transverse Ising system. Later, this was re-derived [312] using Jordan-Wigner transformation of spins to fermions and extended for the estimation of correlations. It was also studied on a chain and on a Bethe lattice [142] and its transverse susceptibility was perturbatively calculated, by employing high-temperature and low-temperature series expansions. In fact, the one-dimensional transverse Ising Hamiltonian appeared as a limit in the partition function of the two-dimensional classical Ising model on a square lattice [231, 349]. It was soon established that the transverse Ising system is the simplest one to exhibit zero-temperature quantum phase transition (driven by quantum fluctuations arising due to the transverse field or the tunnelling term) [339] and the zero-temperature quantum phase transition in dimension d belongs to the same "universality" class as that of the $(d + 1)$-dimensional classical Ising model [126, 173, 385–387, 438]. In fact, the pseudo-spin mapping of the BCS Hamiltonian of superconductivity also reduces it (in the low-lying excited-state space) to an XY model in a transverse field, and in its mean field treatment it becomes exactly like that of a transverse Ising system and gives the BCS gap equation [16]. Lately, various transverse Ising systems have been studied intensively in the context of investigations on quantum glasses (using Ising spin glass in a transverse field) [14, 62, 321, 427, 428], on the nature of the quantum-fluctuation-driven ground states of antiferromagnetic or other regularly frustrated systems (using an axial next-nearest-neighbour Ising model in a

S. Suzuki et al., *Quantum Ising Phases and Transitions in Transverse Ising Models*, Lecture Notes in Physics 862, DOI 10.1007/978-3-642-33039-1_1,

transverse field) [9, 24, 67, 102, 115, 291, 336, 361, 412], on quantum dynamics following a quantum quench (with discontinuous or slow change of a transverse field) [28, 56, 118, 334, 335, 365, 445] and quantum hysteresis (using the Ising model in periodically varying or pulsed transverse fields) [3, 4, 22, 38, 92]. In addition, random Ising models representing combinatorial optimisation problems and information processing in a transverse field [130, 192, 211, 298] have joined in the subject of active study, following an increasing interest in computation and information processing using quantum mechanics. These studies on the static and dynamic properties of the random (and frustrated) transverse Ising models have recently inspired intensive research on the nature of the various fluctuation-induced (and stabilised) "solid", "liquid" or "glassy" phases in quantum many-body systems. Such studies, and the progress already made with established results, imply again that the Ising model (with disorder, etc.) in a transverse field might be the simplest nontrivial quantum many-body system (compared to the Heisenberg system, etc.), having many intriguing and rich properties. In fact, the transverse Ising system shows the simplest nontrivial zero-temperature quantum transition, driven by the tunable (often directly in experiments) tunnelling or transverse field. The other zero-temperature quantum transitions such as the Anderson metal-insulator transition [244], the Bose glass [146], and other [82] transitions (mostly driven by disorder) have, so far, been handicapped by various subtle difficulties (such as the lack of identifiable suitable order parameter in the Anderson transition, etc.).

1.2 A Simple Version of the Model and Mean Field Phase Diagram

As mentioned earlier, de Gennes [98] introduced this model to investigate theoretically the ordered configuration of protons and elementary excitations above this ordered configuration, in the ferroelectric phase of KDP. Since each proton of the hydrogen bond of KDP can occupy one of the two minima of the double-well created by oxygen atoms [40], one can associate a double-well potential with each proton site. In the pseudo spin picture, one can ascribe a pseudo-spin $S_i = 1/2$ with the i-th proton, such that $S_i^z = 1/2$ corresponds to one of the well states at the site i and $S_i^z = -1/2$ corresponds to the other. In this pseudo-spin representation, the Hamiltonian of the proton system can be written as

$$H = -\Gamma \sum_i S_i^x - \sum_{\langle ij \rangle} J_{ij} S_i^z S_j^z \tag{1.2.1}$$

where S^α are the Pauli spin operators, J_{ij} is the electrostatic dipolar interaction between the (neighbouring) protons, and Γ is the tunnelling integral, which determines the rate of tunnelling of protons from one potential minima to the other [40] (S_i^x acting on the state $S_i^z = 1/2(-1/2)$ changes it to the state $S_i^z = -1/2(1/2)$). The sum $\sum_{\langle ij \rangle}(\cdots)$ should be taken for all the nearest neighbouring pairs (i, j). In this pseudo-spin picture, one can explain the phenomenon of ferroelectricity in the following way. In absence of any dipolar interaction between protons, the ground state of the protonic system is the symmetric combination of the states corresponding to the minima of a double well (in the pseudo-spin language the net magnetisation in the z direction vanishes; $\langle S^z \rangle = 0$). Due to the presence of dipole-dipole interaction (J_{ij}) because of the asymmetry of the proton position, the ground state will no longer be symmetric (resulting in the ferroelectric order represented by the nonzero value of $\langle S^z \rangle$). Consequently, there will be a competition between the ordering term (J_{ij}) and delocalising or tunnelling term (Γ), arising due to the fact that the barrier separating the double-well is finite in width and height. The ordering term (J_{ij}) therefore helps the stabilisation of order (nonzero $\langle S^z \rangle$) by localising the proton in one side of the double well, and the disordering tunnelling term (Γ) competes with it and tries to delocalise the protons from any specific well. Thus, even at zero temperature, a (quantum) order-disorder transition occurs when the tunnelling term, which can be tuned by deuteration (replacing the proton mass of hydrogen by that of deuterium) or by increasing the pressure on the sample (i.e., increasing double-well overlap integral, etc.), is changed.

The ground state of the Hamiltonian (1.2.1) is obtained by using a semi-classical approximation, where one puts $S_i^z = S\cos\theta$ and $S_i^x = S\sin\theta$, so that the energy per site from the above Hamiltonian (1.2.1) is given by

$$E = -S\Gamma\sin\theta - S^2 J(0)\cos^2\theta \tag{1.2.2}$$

where $J(0) = \sum_j J_{ij}$. With the magnitude $S = 1/2$ here (for the two-spin state of the double well), the energy is minimised for the condition

$$\sin\theta = \Gamma/J(0) \tag{1.2.3}$$

which suggests that if $\Gamma < J(0)$, the ground state is partially polarised (both $\langle S^z \rangle$ and $\langle S^x \rangle$ are nonzero and $\langle S^z \rangle/S = 1$ and $\langle S^x \rangle = 0$ for $\Gamma = 0$), whereas for $\Gamma \geq J(0)$, the ground state is polarised in the x direction ($\langle S^x \rangle/S = 1$; $\langle S^z \rangle = 0$). Hence, as Γ increases from 0 to $J(0)$, the system undergoes a transition from a ferroelectric phase (with order parameter $\langle S^z \rangle \neq 0$) to a para-electric phase (with $\langle S^z \rangle = 0$). Henceforth we use $S_i^z = \pm 1$ instead of $S_i^z = \pm 1/2$ for convenience.

In general, the finite-temperature behaviour of the above pseudo-spin model of ferroelectricity is obtained by using the mean field theory [48, 379]. Using this approximation, one can reduce the many-body Hamiltonian (1.2.1) to a collection of effective (single) spins with Hamiltonian

$$H = -\sum_i \boldsymbol{h}_i \cdot \boldsymbol{S}_i \tag{1.2.4}$$

where h_i is the effective molecular field (Fig. 1.1) at the site i, which is a vector in the pseudo-spin space given by $\mathbf{h}_i = \Gamma\hat{x} + \sum_j J_{ij}\langle S_j^z \rangle \hat{z}$, and $\mathbf{S} = S^z\hat{z} + S^x\hat{x}$. Here, $\hat{x}$ and $\hat{z}$ denote respectively unit vectors in the pseudo-spin space along x and z directions.

For the non-random case (when J_{ij}s are not site or configuration dependent), the mean field equation for the spontaneous magnetisation can be readily written as

$$\langle \mathbf{S} \rangle = \tanh(\beta|h|)\frac{\mathbf{h}}{|h|}; \quad |h| = \sqrt{\Gamma^2 + \left(J(0)\langle S^z \rangle\right)^2}, \tag{1.2.5}$$

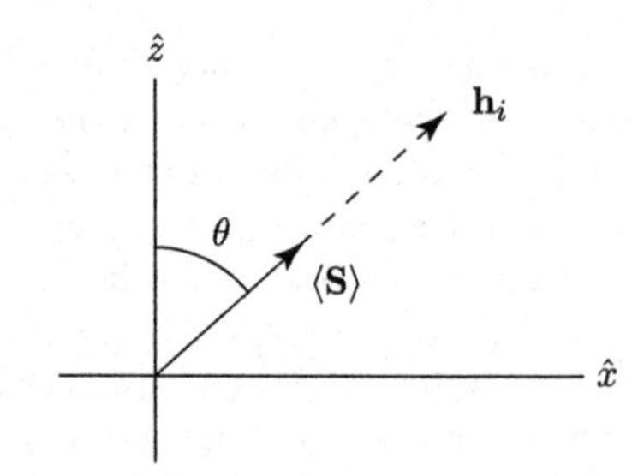

Fig. 1.1 The effective molecular field h_i at the site i

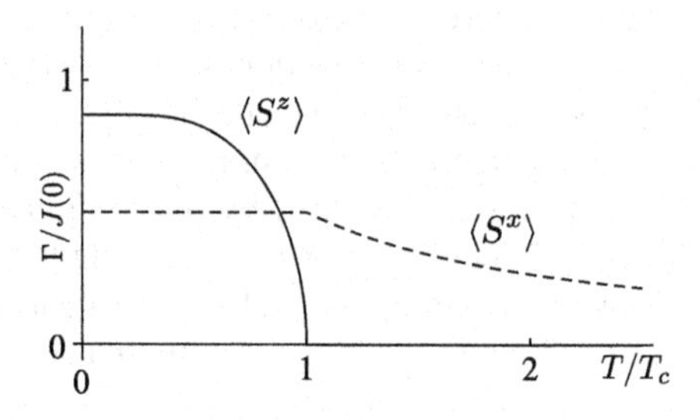

Fig. 1.2 Temperature dependence of $\langle S_z \rangle$ and $\langle S_x \rangle$ in the molecular field approximation

which gives for the components along the $\hat{z}$ and $\hat{x}$ directions as

$$\langle S^z \rangle = \tanh(\beta|h|)\left(\frac{J(0)\langle S^z \rangle}{|h|}\right), \qquad \langle S^x \rangle = \tanh(\beta|h|)\frac{\Gamma}{|h|} \tag{1.2.6}$$

where $\beta = (1/k_B T)$ denotes the inverse temperature and $J(0) = \sum_j J_{ij}$. From the self-consistent solution of the above equations, one can get the mean field phase diagram (Fig. 1.2). At $T = 0$, one gets ferroelectric order (i.e., $\langle S^z \rangle \neq 0$) for $J(0) > \Gamma$, while it disappears ($\langle S^z \rangle = 0$) for $\Gamma \geq J(0)$. The transition point is given by

$$\tanh(\beta_c \Gamma) = \Gamma / J(0). \tag{1.2.7}$$

Also, as $(\tanh \beta|h|)/|h|$ is a constant $(= 1/J(0))$ for $\langle S^z \rangle \neq 0$, $\langle S^x \rangle =$ constant for $T < T_c$ and $\langle S^x \rangle = \Gamma / T$ for higher temperatures (see Fig. 1.2).

In fact, one can also study the elementary excitations of the system, considered as spin waves, using Heisenberg equation of motion obtained from Hamiltonian (1.2.1) (for $\hbar = 1$)

$$\ddot{S}_i^z = 2i\Gamma \dot{S}_i^y = 4\Gamma \sum_j J_{ij} S_i^z S_j^x - 4\Gamma^2 S_i^z. \tag{1.2.8}$$

Using now Fourier transformation ($J(\mathbf{q}) = \sum_j J_{ij} e^{i\mathbf{q}\cdot\mathbf{R}_j}$) and the random phase approximation ($S_i^x S_j^z = S_i^x \langle S_j^z \rangle + S_j^z \langle S_i^x \rangle$, with $\langle S^z \rangle = 0$ in the para-phase) [48, 379] one gets

$$\omega_{\mathbf{q}}^2 = 4\Gamma\left(\Gamma - J(\mathbf{q})\langle S^x \rangle\right) \tag{1.2.9}$$

for the elementary excitations. The largest wavelength mode ($q = 0$) softens ($\omega_0 = 0$) at the same phase boundary (given by (1.2.7)).

1.3 Properties of Ising Models in a Transverse Field: A Summary

Extensive studies have been made in the last six decades or so, to investigate the thermodynamic properties of classical many-body systems with cooperative interactions. Many novel properties of the various thermodynamic phases of such systems with competing and randomly frustrated interactions (as for example in the spin glass) have been studied and established in the last three decades or so. The most extensively studied model system has been the Ising system. In all such systems, the thermodynamic order (coming from the cooperative interactions), which may either be in space (as in ferromagnetic systems) or in time (as in spin glasses), are essentially destroyed by the increasing thermal fluctuation and various kinds of phase transition occur.

Quantum fluctuations (observable at very low temperatures and in high frequency measurements) differ in nature from the thermal fluctuations. However, both can drive the order-disorder phase transitions. In fact, often there is a one-to-one mapping of the zero temperature quantum phase transition (in d dimension) to the thermally driven phase transition in the equivalent classical ($(d+z)$-dimensional; where z is the dynamical exponent [82]) systems (see Sect. 3.1). There has been intense interest in studying the quantum effects on the nature of the ordered phases and the transition between them, in such cooperative systems. Here, in such quantum models, the statics and the dynamics get mixed up, and, due to the linear scaling of energy with the length, the dynamic exponent z is always unity (for pure system). This is responsible for the equivalence of d-dimensional quantum to $(d+1)$-dimensional (pure) classical system [385–387]. It turns out that most often the simplest and the most appropriate model (for comparison with established classical model results) to study quantum effects on cooperative systems (phase transitions) is the Ising model in a transverse field. Here, unlike in other quantum models (e.g., the Heisenberg model, etc.), the cooperative interactions are confined to one spin component only, and this cooperative system is placed in a non-commuting external field, which can be tuned to increase or decrease the tunnelling between the various eigenstates of the (one component) cooperative Hamiltonian. As mentioned earlier, the one dimensional transverse Ising system comes as a limit in the effective transfer matrix of the (classical) isotropic Ising model on square lattice. However, the major interest in the transverse Ising model is independent of this mapping. It comes from the fact that there are many experimental systems where the tunnelling between the cooperatively interacting localised (double-well) states can be accurately expressed by such transverse Ising Hamiltonian. The order-disorder ferroelectrics, discussed earlier (in Sect. 1.1), are such systems [40, 98]. The quantum glass properties of various proton and other glassy systems [8, 14, 89, 96, 317, 321, 427, 428] have been established to be accurately modelled by Ising spin glass system in a transverse field. In fact the discovery of mixed ferromagnetic and antiferromagnetic $LiHo_xY_{1-x}F_4$ systems [14, 321, 427, 428], which can be accurately modelled by the short range interacting transverse Ising spin glass, has led to a recent major upsurge in the interest in quantum glasses and other quantum phases in transverse Ising models in general. Very recently, a precise coincidence between the theory and the experiment has been observed in one dimensional Ising ferromagnet $CoNb_2O_6$ [81]. A comprehensive (but not exhaustive) list of systems, represented by transverse Ising models, are given in Table 1.1 (this table has been partially compiled following Stinchcombe [379]).

In fact, due to the tunability of quantum fluctuations through the transverse field or the tunnelling term, the nature of various quantum phases in many-body systems with random, competing and frustrated interactions, are being intensively studied in the transverse Ising models with appropriate kind of (cooperative) interactions. Because of the extensive studies, the nature of the quantum phases and transitions are getting better understood and established for the quantum Ising or transverse Ising models, rather than in other simple quantum cooperative models (like the Heisenberg model).

As mentioned before, apart from the appearance of a limiting transverse Ising chain Hamiltonian in the effective partition function for the (classical) isotropic Ising model (without any external field) on a square lattice [231, 349], the Ising model in the presence of a transverse field can generally describe accurately the (KDP-type) order-disorder ferroelectrics [40, 98]. Here, each lattice site has a (degenerate) double-well potential (for the protons in the KDP) in which the ionic localisation in any well on any site (represented by two Ising states at that site), gives rise to dipole-dipole (cooperative) interaction. The barrier height and width being finite, tunnelling can occur from one well to another (tunnelling from one Ising state to another, represented by the action of transverse field) at zero (or very low) temperature. As shown in Sect. 1.2, the mean field approximations can easily give an elegant and qualitative picture of the phase transition behaviour of such models. These mean field studies for the pure transverse Ising model have already been discussed earlier in this chapter.

A pure transverse Ising chain Hamiltonian can be exactly diagonalised (e.g., the spin Hamiltonian can be reduced to noninteracting fermion Hamiltonian, using Jordan-Wigner transformation) [216, 312]. One can thus find the nature of the ground state and the excited states and can also formulate the finite temperature thermodynamics. Because of the possibility of comparing with exact results, various approximate real space renormalisation etc. techniques have been developed and tested for such Hamiltonian [314]. Similarly, with exact numerical diagonalisation of finite-size system Hamiltonian, and using finite-size scaling techniques, accurate studies of the quantum phase transition (at zero temperature; with the change of transverse field) have been made [166, 167]. Remarkably, a theory on the pure transverse Ising chain has been verified by a neutron scattering experiment of $CoNb_2O_6$ [81]. These are discussed in Chap. 2.

In the next chapter (Chap. 3), we describe mapping of transverse Ising Hamiltonian (pure d-dimensional) to an effective $(d + 1)$-dimensional classical Hamiltonian, using the Suzuki-Trotter formalism [385–387]. This mapping of the finite temperature partition function of the quantum Hamiltonian (of the transverse Ising system) to the Suzuki-Trotter classical Hamiltonian (at nonzero temperature) helps utilisation of very powerful techniques, like the Monte Carlo methods, to extract

Table 1.1 Systems related to Ising models in a transverse field

Category	Some specific examples	Notes on applicability of		References
		(a) TIM	(b) MFA	
Order disorder ferroelectrics		1		[80, 202, 213]
(a) With tunnelling		2		[98, 230]
	KH_2PO_4	3, 4	6	[212, 340]
	KD_2PO_4	3, 4	6	[53, 340, 371]
	KH_2AsO_4			
	RbH_2AsO_4	4		
	RbD_2AsO_4	4		
	$NH_4H_2AsO_4$			
	$LiH_3(SeO_3)i_2$	3, 4, 7		[340]
	$LiD_3(SeO_3)i_2$	3, 4, 7		[340]
(b) Without tunnelling	TGS	5	6	[431]
	Rochelle Salt			
	$NaNO_3$	5, 8		[431]
	$NaNO_2$	5	6	[429, 430]
	NH_4Cl	5		[199]
	$KNbO_3$			
Simple ferromagnets with uniaxial symmetry	$Dy(C_2H_5SO_4)_39H_2O$	9	11, 12	[83, 424]
	$FeCl_22H_2O$ and related compounds	9	11	[294]
	$CoCl_2$	9	13	[383]
	EuS	10	14	[284]
	$Cu(NH_4)_2Br_42H_2O$	10	14	[420]
	$CoNb_2O_6$	23		[81]
Rare earth compounds with single crystal-field ground state	Rare earth-group V elements with NaCl structure	15	14	[88, 402, 419]
	Rare earth-group VI anion compounds with NaCl structure			
Simple Jahn-Teller systems	$DyVO_4$	16	17	[84, 85, 154]
	$TbVO_4$	16	18	[128]
			18	[123, 127, 168]
	$TmVO_4$			[86, 257]
	$TmAsO_4$			

Table 1.1 (continued)

Category	Some specific examples	Notes on applicability of		References
		(a) TIM	(b) MFA	
Other systems with 'pseudo-spin'-phonon interaction	$CeEthylSO_4$	19	20	[124, 148, 378]
Mixed hydrogen bonded ferroelectrics (proton glasses)	$Rb_{1-x}(NH_4)_xH_2PO_4$		21	[316]
Dipolar magnet	$LiHo_xY_{1-x}F_4$	22		[14, 321, 427, 428]

TIM: Transverse Ising Models; MFA: Mean Field Approximation

1. Evidence for order-disorder character from greater entropy difference between ferroelectric and paraelectric phases and small Curie constant than in displacive ferroelectrics, such as double oxides with perovskite structure [41, 42]

2. Proton-lattice coupling, which gives rise to the spontaneous polarisation, is neglected in transverse Ising model. Kobayashi [230] has extended the model to include the phonons

3. Importance in tunnelling is indicated by decrease of T_c on application of pressure, due to increased tunnelling through lower and narrower barrier [53, 371]

4. Isotope effects in T_c indicates importance of tunnelling (which decreases on deuteration), though isotope effect due in part to change in exchange interaction J [40]

5. Ising model without tunnelling seems to apply: limiting case of tunnelling model

6. Use of molecular field theory as leading approximation supported by largely dipolar character of exchange

7. High T_c indicates large J, implying tunnelling effects less important than in other hydrogen-bonded ferroelectrics

8. Not ferroelectric: NO_3^- carries no dipole moment

9. Very strong uniaxial anisotropy

10. Weak anisotropy

11. One dimensional magnetic character (weak interaction between chains): small number of nearest neighbours

12. Interaction predominantly dipolar

13. Two-dimensional magnetic character (small antiferromagnetic coupling between planes): small coordination number z

14. Nearest neighbour coupling with $\delta_0 \sim 10$: molecular field theory reasonable leading approximation

15. Spin-1/2 transverse field applies well if first excited state is also a singlet

16. Transverse field small

17. Predominance in optical phonon exchange as source of coupling suggests molecular field theory only a crude approximation for $DyVO_4$

18. Use of molecular field theory supported by moderately strong strain coupling in $TbVO_4$, and by shape of specific heat curve in $TmVO_4$

19. Weak coupling to phonons and consequent weak exchange

20. Molecular field theory can account for shape of Schottky anomaly [148]

21. Possibility of tunnelling of the protons between the two potential minima in each hydrogen bond. Infinite range interaction justifiable in view of the dipolar nature of the interbond forces

22. Short range interactions

23. Very strong uniaxial anisotropy and one dimensional magnetic character. Weak effective longitudinal field induced by inter-chain couplings is present

information about various phases in the system and about the transitions between them. However, these numerical techniques require tricky extrapolation schemes [290, 421] or cluster updating schemes [222, 292, 391], which have been well developed for the pure transverse Ising system. These have been discussed in this chapter, where we also discuss the effective field theoretic formulation and renormalisation group results.

The nature of quantum-fluctuation-driven phases in many-body systems (with regular frustration, either due to the lattice structure, or due to the nature of interactions) like in the Heisenberg antiferromagnet, etc., have been studied for a long time [20, 255, 256, 263]. In this context, there have been some studies on the effect of the tunnelling term (the transverse field) on various modulated (commensurate and incommensurate) phases of the Axial Next Nearest Neighbour Ising (ANNNI)-like models. Various appropriate analytical (e.g., exact solution along the special Peschel-Emery line [311], interacting fermions in one dimension [359–362], perturbative analysis [67], renormalisation group analysis [9, 115], large spin analysis [171, 355, 356] etc.) and numerical (e.g., exact diagonalisation for finite sizes [359–362], quantum Monte Carlo [18], density matrix renormalisation group [31, 32], entanglement scaling of matrix product states [291] etc.) techniques have been employed to study the nature of the (quantum) fluctuation-stabilised phases in such (regularly frustrated) systems. These results (for the phase diagram etc.) are discussed in Chap. 4.

There has also been considerable effort to study the phases and the transitions in various randomly disordered transverse Ising models, like the dilute transverse Ising models [37, 169, 189, 380, 382], random exchange and transverse field Ising models [138, 139, 439] and etc. Here the phase diagram contains an interesting discontinuity at the percolation threshold for the dilute Ising models [37, 170, 381, 382]. The intriguing and prominent manifestation [138–141, 188, 235, 236, 315, 415, 439] of the Griffiths phase in such randomly disordered (quantum) transverse Ising chain (without frustration) have been discussed here in Chap. 5.

As mentioned before, there has been a spectacular upsurge in the interest in the zero-temperature quantum glass phase (and in their transition behaviour) of the Ising spin glasses in transverse field. These studies help us to see the effect of tunnelling between the (classically degenerate) localised spin glass states. The recent discovery [14, 321, 427, 428] of the compound material $LiHo_xY_{1-x}F_4$, with the magnetic Holmium ion concentration x, represented by the transverse Ising spin glass, has led to a renewed interest. Here, the strong spin-orbit coupling between the spins and the host crystal restricts the ("Ising") spins to be either parallel or antiparallel to a specific crystalline axis. An applied field, transverse to the preferred axis, flips these "Ising" spins. This, together with the randomness in the interaction, makes it a unique transverse Ising spin glass system.

The transverse Ising spin glass model was first introduced in 1981 [62], when the (unfrustrated) Mattis model of random magnets was studied in the presence of a transverse field. Here, the disorder could be transformed away (even in the presence of the tunnelling term) and the (mapped) phases and transition remain exactly identical to that of a pure transverse Ising system. For the Edwards-Anderson model in a transverse field (considered again in [62]), the effective Suzuki-Trotter Hamiltonian has the original disordered interaction in d-dimensions, whereas the interaction becomes ferromagnetic in the additional Trotter dimension. It seems that this correlation in disorder in the Trotter (time) dimension gives the dynamic exponent (z) value different from unity (unlike pure transverse Ising system, where energy always scales linearly with the length). The effective correspondence therefore shifts to the $(d+z)$-dimensional classical spin glass (for a d-dimensional transverse Ising spin glass) with $z=2$, presumably, at the upper critical dimension $d_u=8$ ([325]; see also [163, 332]), where the quantum glass transition becomes mean-field-like. In fact, the mean field Sherrington-Kirkpatrick model (with infinite range interaction), put in a transverse field, has also been studied for the determination of its phase diagram [195, 417, 418, 433] and to study the question of replica symmetry restoration (due to the quantum tunnelling in the zero-temperature transition) [159, 323, 325, 401]. It appears that for the zero-temperature transition (driven by the transverse field), the replica symmetry breaking is absent due to quantum tunnelling (see e.g., [159, 323, 325, 401]). For finite range interaction numerical studies in $d=2$ and 3 [163, 332], together with finite size scaling, gives estimates of various exponents (including the dynamical exponents). It also appears that, unlike in the classical case where the linear susceptibility gives a cusp and the nonlinear susceptibility diverges (for the systems above the lower critical dimension $2 \ll 3$), the linear susceptibility is found to diverge at the critical tunnelling field for $d=2$ but not for $d=3$ [195, 417, 418, 433]. The p-spin interacting Ising glass models in a transverse field with the $p \to \infty$ limit [61, 158] has been known as an exactly solvable model. It has been shown that the spin glass phase where the replica symmetry breaks exists at low temperatures and low transverse fields [158]. The "valleyed" structure in the free energy landscape of the random (longitudinal) field Ising system has been studied extensively [33]. We have also discussed studies of the effect of quantum fluctuations, induced by the transverse field, on the transition behaviour of random (longitudinal) field Ising system. These have been discussed extensively in this chapter (Chap. 6).

The dynamics of transverse Ising models, in particular when the Hamiltonian (the transverse field) has explicit time dependence, poses very interesting questions. In particular, the questions of quantum relaxation after a sudden quench of a transverse field [28, 334, 335, 365], non-adiabatic time evolution involved by a slow quench of a transverse field across a quantum critical point [118, 318, 445], and hysteresis when the transverse field varies periodically with time (due to, for example, the pressure modulation in KDP-type systems) [3, 4, 22, 38, 92] have been received a growing amount of interests in the last decade. Studies of such time-dependent systems have been developed mainly in a transverse Ising chain and by solving the time-dependent Schrödinger equations analytically and numerically. These studies are discussed in Chap. 7.

One of the most challenging problems with respect to transverse Ising models might be the dynamics of transverse Ising spin glasses in the presence of a time dependent transverse field. Regarding this subject, the study of quantum annealing has been remarkably developed in the last decade. Like simulated annealing for combinatorial optimisation problems, quantum annealing aims to minimise an energy cost function given by a randomly disordered Ising model representing an optimisation problem by lowering the quantum fluctuation [130, 211]. The random Ising model here shares several properties with Ising spin glasses [275]. Hence an understanding of the performance of quantum annealing is very closely related to an understanding of the dynamics of transverse Ising spin glasses. The essence of quantum annealing is an adiabatic time evolution from a trivial ground state of an initial Hamiltonian to the ground state of the random Ising model. The adiabatic time evolution entirely depends on the energy gap which separates the ground state and the excited states. If the energy gap vanishes, then the adiabaticity breaks down and consequently quantum annealing fails. Usually, the energy gap closes at the quantum critical point in the thermodynamic limit. Hence the size scaling of the energy gap at the quantum critical point determines the complexity of the problem with quantum annealing. From this point of view, a lot of efforts have been made to reveal the size scaling of the energy gap, in particular, in systems with an infinite randomness critical point [141] and with a first order quantum phase transition [204, 205, 441]. The scaling of an energy gap has been also analysed using the concept of the Anderson localisation [10]. Apart from the size scaling, the scaling of excitation density with respect to runtime gives another estimation on the adiabaticity. This has been studied using the random transverse Ising chain and shown that errors of quantum annealing decays faster than simulated annealing [58, 119, 389]. In addition to these studies on the performance of quantum annealing, the sufficient condition on the time dependence of the Hamiltonian to assure the success of quantum annealing has been studied and several theorems have been obtained [280, 281]. A comprehensive review on quantum annealing including these studies is given in Chap. 8.

In Chap. 9, we review several applications of the transverse Ising model to various research fields such as neural networks or Bayesian information processing [273, 297]. We extend the Hopfield model in a transverse field [250, 251, 253, 298] and consider both static [298] and dynamical properties [192]. We also investigate the quantum effects in image restoration [190, 397] and error-correcting codes [191].

There have also been considerable studies in related models like the XY model in the transverse field. In particular, using pseudo-spin formalism, the BCS Hamiltonian in the low-lying paired state could be written as a XY model in a transverse field, and its mean field theory gives the BCS gap equation [16]. The (pure) model is also diagonalisable in one dimension [322, 385–387]. A recent mapping of the Harper chain to such transverse XY model [344–346] has indicated the possibility of comparing the metal-insulator transition in such quasiperiodically disordered chains to the correlations in the transverse XY chain. These are discussed in Chap. 10. We have also discussed in this chapter, the studies on the XY spin glass models in transverse field [51]. Recently, the Kitaev model joined in the family of the models which can be mapped to free fermion models by the Jordan-Wigner transformation [229]. Brief discussions on this model including slow quench dynamics across a quantum critical line [174, 366] have been made in Chap. 10.

Chapter 2
Transverse Ising Chain (Pure System)

2.1 Symmetries and the Critical Point

2.1.1 Duality Symmetry of the Transverse Ising Model

Following the duality of the two dimensional Ising model on square lattice [237], one can show [231] the self-duality, and thereby make exact estimate of the critical tunnelling (transverse) field of the one-dimensional spin-$1/2$ transverse Ising model. Before considering the exact solution of the one dimensional quantum mechanical Hamiltonian, we shall study the duality symmetry hidden in the model, and identify the critical value of the transverse field, for which there exists a second order zero-temperature phase transition from a ferromagnetic phase to a paramagnetic phase.

Let us first consider the quantum Hamiltonian of the transverse Ising chain

$$H = -\sum_i \left[\Gamma S_i^x + J S_i^z S_{i+1}^z\right] = -\sum_i \left[S_i^x + \overline{\lambda} S_i^z S_{i+1}^z\right] \tag{2.1.1}$$

where S^αs are the usual the Pauli spin matrices, $\overline{\lambda} = J/\Gamma$, and the Hamiltonian is scaled with the transverse field ($\Gamma = 1$). The spin operators satisfy the usual commutation relations given by (with $\hbar = 1$)

$$\left[S_i^\alpha, S_j^\beta\right] = 2i\delta_{ij}\epsilon_{\alpha\beta\gamma} S_i^\gamma; \quad \alpha, \beta, \gamma = x, y, z. \tag{2.1.2}$$

It may be noted that the Pauli operators commute on different sites and satisfy the anti-commutation relation

$$\left[S_i^\alpha, S_i^\beta\right]_+ = S_i^\alpha S_i^\beta + S_i^\beta S_i^\alpha = 0, \tag{2.1.3}$$

and

$$\left(S_i^\alpha\right)^2 = 1 \tag{2.1.4}$$

on the same site.

To investigate the duality symmetry involved in the model we associate a dual lattice with our original spatial lattice (chain) such that sites in the original chain

S. Suzuki et al., *Quantum Ising Phases and Transitions in Transverse Ising Models*, Lecture Notes in Physics 862, DOI 10.1007/978-3-642-33039-1_2,

are associated with the bonds of the dual chain and vice versa. The operators on the dual lattice are defined as

$$\tau_j^x = S_j^z S_{j+1}^z \quad \text{and} \tag{2.1.5a}$$

$$\tau_j^z = \prod_{k \le j} S_k^x. \tag{2.1.5b}$$

The non-local mapping (2.1.5a), (2.1.5b) represents the "duality" transformation. As mentioned earlier, the new operators are defined on links of the original chain which has one-to-one correspondence with the sites of the "dual" lattice. From (2.1.5a), (2.1.5b) we can easily see that senses whether spins on the original lattice are aligned or not; whereas flips all the spins left to the site i. One can now easily check that the operators τ_i^α's satisfy the same set of commutation relation as the operator i.e., they commute on different sites and anti-commute on the same site

$$\left[\tau_i^x, \tau_i^z\right] = 0 \text{ for } i \neq j \quad \text{and} \quad \left[\tau_i^x, \tau_z\right]_+ = 0. \tag{2.1.6}$$

One can thus readily rewrite the original transverse Ising Hamiltonian in terms of the dual operators τ^α

$$\begin{aligned} H &= -\sum_i \left[\tau_i^z \tau_{i+1}^z + \overline{\lambda} \tau_i^x\right] \\ &= \overline{\lambda}\left[-\sum_i \tau_i^x - (\overline{\lambda})^{-1} \sum_i \tau_i^z \tau_{i+1}^z\right] \end{aligned} \tag{2.1.7}$$

which immediately suggests the (duality) scaling property of the Hamiltonian:

$$H(S; \overline{\lambda}) = \overline{\lambda} H\left(\tau; \overline{\lambda}^{-1}\right). \tag{2.1.8}$$

Since both S and τ operators satisfy the same algebra, the above symmetry implies, for the equivalent classical two dimensional Ising model (where Γ corresponds to inverse temperature), one-to-one correspondence of the high and low temperature phases. Here for the quantum model, it implies that each eigenvalue E of H satisfies the relation,

$$E(\overline{\lambda}) = \overline{\lambda} E(1/\overline{\lambda}). \tag{2.1.9}$$

Equation (2.1.9) has important significance from the point of quantum phase transition in the model. For the quantum phase transition, at the critical point the mass gap (the gap between the ground state and the first excited state of the quantum Hamiltonian) vanishes and the correlation length $\xi(\overline{\lambda})$ (which is the inverse of the mass gap $\Delta(\overline{\lambda})$) diverges. For the present model, if the mass gap vanishes for some nonzero value of then (2.1.9) suggests that it will also have to vanish at. If one assumes that is unique, the self-duality of the model implies that

$$\overline{\lambda}_c = 1, \quad \text{or} \quad \Gamma_c = J \tag{2.1.10}$$

and, as we will show in the next subsection,

$$\xi(\overline{\lambda}) \sim \Delta^{-1} = 2|1 - \overline{\lambda}|^{-1}. \tag{2.1.11}$$

The self-duality of the model thus yields the exact critical value if we assume that the critical value is unique (valid for simple Ising model, but is not true in general).

2.1.2 Perturbative Approach

In this section we shall try to estimate the mass gap associated with the above quantum Hamiltonian (1.2.1) for the transverse Ising chain, using a perturbative approach. We first rewrite the transverse Ising Hamiltonian in the following form

$$H = H_0 + V = \sum_i [1 - S_i^x] - \bar{\lambda} \sum_i S_i^z S_{i+1}^z, \tag{2.1.12}$$

with

$$H_0 = \sum_i [1 - S_i^x] \tag{2.1.13a}$$

$$V = -\sum_i S_i^z S_{i+1}^z, \tag{2.1.13b}$$

and write a perturbation series in powers of for any eigenvalue of the total Hamiltonian:

$$E(\bar{\lambda}) = E^{(0)} + \bar{\lambda} E^{(1)} + (\bar{\lambda})^2 E^{(2)} + \cdots. \tag{2.1.14}$$

The Hamiltonian H_0 represents a one dimensional chain of non-interacting spins in a magnetic field, and one can now trivially show that the lowest eigenvalue E_0 of H_0 is zero:

$$H_0|0\rangle = E_0^{(0)}|0\rangle = 0. \tag{2.1.15}$$

The corresponding bare vacuum $|0\rangle$ is given by

$$|0\rangle = \prod_{i=1}^{N} |+\rangle_i \tag{2.1.16}$$

where $|+\rangle_i$ is the eigenstate of the operator S_i^x with eigenvalue $= +1$. Similarly, one can easily see that the first excited state of the Hamiltonian H_0 consists of just one flipped spin $|0\rangle$ at the site i. But it is clearly seen that this state is N-fold degenerate, since the flipped spin can occur on any site of the one-dimensional lattice. Keeping in mind the translational invariance of the model (any eigenstate has to have a definite momentum) we can choose an appropriate linear combination for the first excited state of the Hamiltonian H_0 given by

$$|1\rangle = N^{1/2} \sum_i S_i^z|0\rangle, \tag{2.1.17}$$

the corresponding eigenvalue equation can be written as

$$H_0|1\rangle = E_1^{(0)}|1\rangle = 2|1\rangle. \tag{2.1.18}$$

Hence considering the unperturbed Hamiltonian ($\bar{\lambda} = 0$), we get the mass gap (at the zeroth order) of the system, given by

$$\Delta_0 \equiv E_1^{(0)} - E_0^{(0)} = 2. \tag{2.1.19}$$

We shall now incorporate the effect of V in a perturbative manner. To do this we must recall that the operator acts as a spin-flip operator on state $|0\rangle$ or for that matter, $|1\rangle$. We can thus immediately get for the first order corrections

$$E_0^{(1)} = \langle 0|V|0\rangle = 0, \tag{2.1.20a}$$

$$E_1^{(1)} = \langle 1|V|1\rangle = -2. \tag{2.1.20b}$$

Putting (2.1.20a), (2.1.20b) in the (2.1.14) we get

$$\begin{aligned} \Delta(\bar{\lambda}) &\equiv E_1(\bar{\lambda}) - E_0(\bar{\lambda}) = \left(E_1^{(0)} + \bar{\lambda}E_1^{(1)}\right) - \left(E_0^{(0)} + \bar{\lambda}E_0^{(1)}\right) \\ &= 2(1 - \bar{\lambda}), \end{aligned} \tag{2.1.21}$$

up to first order in $\bar{\lambda}$.

Equation (2.1.21) suggests that as $\bar{\lambda}$ increases from zero, the mass gap decreases and vanishes at the critical value $\bar{\lambda}_c$, which is found to be unity here from the first order perturbation theory. Near the critical point, however, all higher order contributions to the mass gap are expected to become relevant. But Kogut [231], using Raleigh-Schrödinger perturbation theory, estimated the higher order corrections in $\bar{\lambda}$ to the mass gap $\Delta(\bar{\lambda})$ and found that all the higher order contributions vanish identically in $(1+1)$-dimension. Note that "+1" in the dimension comes from the imaginary-time Trotter direction. We shall discuss in the next chapter the equivalence between a d-dimensional transverse Ising model and a $(d+1)$-dimensional Ising model. In dimensions higher than $d = 1$, of course, all the beneficial properties of the perturbation scheme are lost, the perturbation series does not truncate in the first order.

We have just seen that the mass gap of the elementary excitations vanishes as $\bar{\lambda}$ approaches 1 from below. Although this perturbative estimate is valid for $\bar{\lambda} < 1$, one can always extend it for $\bar{\lambda} > 1$ using the duality symmetry of the Hamiltonian. As the mass gap also satisfies the duality condition (cf. (2.1.9))

$$\Delta(\bar{\lambda}) = \bar{\lambda}\Delta\left(\bar{\lambda}^{-1}\right), \tag{2.1.22}$$

we can write the general form for mass gap as

$$\Delta(\bar{\lambda}) = 2|1 - \bar{\lambda}|. \tag{2.1.23}$$

This suggests that the gap vanishes both from below and above the critical point $\bar{\lambda} = \bar{\lambda}_c = 1$. This also gives the estimate of correlation length $\xi(\bar{\lambda}) \sim |\bar{\lambda} - \bar{\lambda}_c|^{-\nu} \sim \Delta^{-1}(\bar{\lambda})$ where ν represents correlation length exponent. From (2.1.23) we get

$$\xi(\bar{\lambda}) = (1/2)|1 - \bar{\lambda}|^{-1}, \tag{2.1.24}$$

giving $\nu = 1$ for the transverse Ising chain.

We now conclude this section with the note that the hidden duality symmetry makes our task simpler in the case of the spin-$1/2$ transverse Ising system in $1+1$-dimension. Employing duality and perturbation technique we can extract a number of information about the associated quantum phase transition. Unfortunately, this symmetry does not hold for higher dimensional models.

2.2 Eigenvalue Spectrum: Fermionic Representation

The above spin-$1/2$ transverse Ising chain Hamiltonian can be exactly diagonalised, and the entire eigenvalue spectrum and eigenfunctions can be obtained by employing Jordan-Wigner transformation of the spin operators to spinless fermions [245, 312] (see also [216]).

For this, we consider again the Hamiltonian (2.1.1). Using a canonical transformation

$$S^x \to S^z, \qquad S^z \to -S^x,$$

we rewrite the Hamiltonian (2.1.1) as

$$H = -\sum_i S_i^z - \bar{\lambda} \sum_i S_i^x S_{i+1}^x. \tag{2.2.1}$$

We can now express the Hamiltonian H in terms of the raising and lowering operators S_i^+ and S_i^- at every site, where

$$S_i^+ = (1/2)\left[S_i^x + i S_i^y\right]$$
$$S_i^- = (1/2)\left[S_i^x - i S_i^y\right]$$

which satisfy a mixed set of commutation relations, i.e., commute on different sites and satisfy the fermionic anti-commutation relations on the same site:

$$\left[S_i^-, S_j^+\right] = 0, \quad \left[S_i^+, S_j^-\right] = 0; \quad i \neq j$$

and

$$\left[S_i^-, S_i^+\right]_+ = 1$$
$$\left[S_i^-, S_i^-\right]_+ = \left[S_i^+, S_i^+\right]_+ = 0.$$

The last condition implies that if an arbitrary state $|F\rangle$ is not annihilated by , then it is annihilated by $(S_i^+)^2$ (a spin can be flipped only once):

$$S_i^+\left[S_i^+|F\rangle\right] = 0; \qquad \left(S_i^+\right)^2 = 0.$$

The above equation implies that, although the operator S_i^+ creates a bosonic excitation at the site i, it is impossible to have two such excitations at the same site. This is the hard-sphere condition, and the raising and lowering operators should appropriately be treated as hard-core bosons.

The standard procedure to treat one dimensional hard-core bosons is to transform the spin operators into fermions by using the Jordan-Wigner transformation

$$c_1 = S_1^-, \qquad c_i = \prod_{j=1}^{i-1} \exp\left[i\pi S_j^+ S_j^-\right] S_i^- \quad (i = 2, \ldots, N) \tag{2.2.2a}$$

$$c_1^\dagger = S_1^+, \qquad c_i^\dagger = S_i^+ \prod_{j=1}^{i-1} \exp\left[-i\pi S_j^+ S_j^-\right] \quad (i = 2, \ldots, N). \tag{2.2.2b}$$

One can easily check that (see Sect. 2.A.1) the operators c_i and $c_i^\dagger$ are fermionic operators satisfying

$$\left[c_i, c_j^\dagger\right]_+ = \delta_{ij}; \qquad [c_i, c_i]_+ = 0. \tag{2.2.3}$$

Unlike the spin operators, the fermion operators anti-commute even at different sites and this is due to the presence of the non-local factor

$$K_i = \exp\left[i\pi \sum_{j=1}^{i-1} S_j^+ S_j^-\right] \quad (i = 2, \ldots, N), \tag{2.2.4}$$

which is called the disorder or "soliton" term which provides an extra minus sign to convert a commutator to an anti-commutator at different sites. This term is a unitary operator which rotates (up to a phase factor) the spin configuration of all sites left to the i-th site by an angle π about the z-axis. This term is called the disorder term since it cannot have a nonzero expectation value in a state having a long-range order. On the other hand, it may have a nonzero expectation value in a state having no long-range order. We can therefore transform the spin system into a system of spinless fermions with chemical potential zero, defined on a one dimensional lattice.

To write the Hamiltonian (2.2.1), in terms of Jordan-Wigner fermions, we need to worry about the boundary condition. If the spin chain has a periodic boundary condition, i.e.,

$$S_1^\alpha = S_{N+1}^\alpha; \quad \alpha = x, y, z \tag{2.2.5}$$

then we can recast the transverse Ising Hamiltonian in the following form (see Sect. 2.A.1)

$$H = N - 2\sum_i c_i^\dagger c_i - \bar{\lambda} \sum_i \left[c_i^\dagger - c_i\right]\left[c_{i+1}^\dagger + c_{i+1}\right], \tag{2.2.6}$$

where we have neglected the correction term $[(c_1^\dagger + c_1)(c_N^\dagger - c_N)\{\exp(i\pi L) + 1\}]$, where $L = \sum_{i=1}^{N} c_i^\dagger c_i$, arising from the periodic boundary condition of spins for large systems. One must note here that although the number of fermions $\sum_i c_i^\dagger c_i$ is not a constant of motion, its parity is conserved and hence $\exp(i\pi L)$ is a constant of motion having the value $+1$ or -1. Hence, the fermion problem must have an anti-periodic boundary condition if there is an even number of fermions and periodic boundary condition if there is an odd number of fermions. The correction term can be neglected for a thermodynamically large system in which case we call it the "c-cyclic" problem (the original problem being the "a-cyclic" one). The above transformed Hamiltonian is already quadratic in the fermion operators and it is obviously diagonalisable. To do so, let us consider fermions in momentum space

$$c_q = (1/N)^{1/2} \sum_{j=1}^{N} c_j \exp(iqR_j) \tag{2.2.7a}$$

$$c_q^\dagger = (1/N)^{1/2} \sum_{j=1}^{N} c_j^\dagger \exp(-iqR_j) \tag{2.2.7b}$$

where the complete set of wavevector is $q = 2\pi m/N$,

$$m = -(N-1)/2, \dots, -1/2, 1/2, \dots, (N-1)/2 \quad (\text{for } N \text{ even})$$

$$m = N/2, \dots, 0, \dots, N/2 \quad (\text{for } N \text{ odd})$$

and the final form of the transverse Ising chain Hamiltonian becomes

$$H = N - 2\sum_q (1 + \bar{\lambda}\cos q) c_q^\dagger c_q - \bar{\lambda}\sum_q \left(\mathrm{e}^{-iq} c_q^\dagger c_{-q}^\dagger - \mathrm{e}^{iq} c_q c_{-q}\right). \tag{2.2.8}$$

To diagonalise the Hamiltonian, we employ the Bogoliubov transformation in which new fermion creation operators $\eta_q^\dagger$ are formed as a linear combination of $c_q^\dagger$ and c_q in order to remove terms in Hamiltonian which do not conserve the particle number.

For this, it is convenient to sum over modes with $q > 0$ and include the others by simply writing them out. Then above Hamiltonian (2.2.8) becomes

$$\begin{aligned} H &= -2\sum_{q>0} (1 + \bar{\lambda}\cos q)\left(c_q^\dagger c_q - c_{-q} c_{-q}^\dagger\right) + 2i\bar{\lambda}\sum_{q>0} \sin q \left(c_q^\dagger c_{-q}^\dagger - c_{-q} c_q\right) \\ &= -2\begin{pmatrix} c_q^\dagger & c_{-q} \end{pmatrix} \begin{pmatrix} 1 + \bar{\lambda}\cos q & -i\bar{\lambda}\sin q \\ i\bar{\lambda}\sin q & -1 - \bar{\lambda}\cos q \end{pmatrix} \begin{pmatrix} c_q \\ c_{-q}^\dagger \end{pmatrix}, \end{aligned} \tag{2.2.9}$$

where the constant N in (2.2.8) is cancelled by $2\sum_{q>0} 1$. Note that the vacuum of H is not the vacuum of the operator c_q because of the presence of the off-diagonal term $(c_q^\dagger c_{-q}^\dagger + c_q c_{-q})$ in the Hamiltonian. We wish to write the Hamiltonian in the form

$$H = \sum_q \omega_q \eta_q^\dagger \eta_q + \text{const}, \tag{2.2.10}$$

so that the single particle excitations are identifiable above the vacuum state $(\eta_q|0\rangle = 0)$. This is achieved by making a canonical transformation from the operators c_q, $c_q^\dagger$ to the operators η_q, $\eta_q^\dagger$ (Bogoliubov transformations)

$$\begin{pmatrix} \eta_q \\ \eta_{-q}^\dagger \end{pmatrix} = \begin{pmatrix} u_q & iv_q \\ iv_q & u_q \end{pmatrix} \begin{pmatrix} c_q \\ c_{-q}^\dagger \end{pmatrix} = \begin{pmatrix} u_q c_q + iv_q c_{-q}^\dagger \\ iv_q c_q + u_q c_{-q}^\dagger \end{pmatrix} \tag{2.2.11}$$

where $q > 0$ everywhere. The functions u_q, v_q are determined by two criteria: (1) These new set of operators η_q, $\eta_q^\dagger$, are fermionic operators. (2) The Hamiltonian (2.2.9) is diagonalised when written in terms of these operators. The choice of the transformation is made in the way such that u_q, v_q are real. The first criterion

$$[\eta_{q'}, \eta_q^\dagger]_+ = \delta_{q',q}, \qquad [\eta_{q'}, \eta_q]_+ = 0 = [\eta_{q'}^\dagger, \eta_q^\dagger]_+,$$

leads to the relation

$$u_q^2 + v_q^2 = 1. \tag{2.2.12}$$

Using now the inverse transformation of (2.2.11)

$$\begin{pmatrix} c_q \\ c_{-q}^\dagger \end{pmatrix} = \begin{pmatrix} u_q & -iv_q \\ -iv_q & u_q \end{pmatrix} \begin{pmatrix} \eta_q \\ \eta_{-q}^\dagger \end{pmatrix}, \tag{2.2.13}$$

we can rewrite the Hamiltonian H in terms of the Bogoliubov fermions

$$\begin{aligned} H &= \sum_{q>0} H_q \\ H_q &= -2\begin{pmatrix} \eta_q^\dagger & \eta_{-q} \end{pmatrix} \begin{pmatrix} u_q & iv_q \\ iv_q & u_q \end{pmatrix} \begin{pmatrix} 1+\bar{\lambda}\cos q & -i\bar{\lambda}\sin q \\ i\bar{\lambda}\sin q & -1-\bar{\lambda}\cos q \end{pmatrix} \\ &\quad \times \begin{pmatrix} u_q & -iv_q \\ -iv_q & u_q \end{pmatrix} \begin{pmatrix} \eta_q \\ \eta_{-q}^\dagger \end{pmatrix}. \end{aligned} \tag{2.2.14}$$

To recast the Hamiltonian in the diagonal form we demand that $(u_q, -iv_q)^T$ is an eigenvector of the Hamiltonian matrix

$$\begin{pmatrix} 1+\bar{\lambda}\cos q & -i\bar{\lambda}\sin q \\ i\bar{\lambda}\sin q & -1-\bar{\lambda}\cos q \end{pmatrix} \begin{pmatrix} u_q \\ -iv_q \end{pmatrix} = -\omega_q \begin{pmatrix} u_q \\ -iv_q \end{pmatrix}. \tag{2.2.15}$$

This 2×2 eigenvalue problem is easily solved to yield

$$\omega_q = \left(1 + 2\bar{\lambda}\cos q + \bar{\lambda}^2\right)^{1/2}, \tag{2.2.16}$$

and

$$u_q = \frac{\bar{\lambda}\sin q}{\sqrt{2\omega_q(\omega_q + 1 + \bar{\lambda}\cos q)}}, \qquad v_q = \frac{\omega_q + 1 + \bar{\lambda}\cos q}{\sqrt{2\omega_q(\omega_q + 1 + \bar{\lambda}\cos q)}}, \tag{2.2.17}$$

where we have chosen the sign of u_q and v_q such that $u_q > 0$ for $0 < q < \pi$. Having obtained the eigenvalue ω_q, one can write the Hamiltonian in a diagonal form

$$H = 2\sum_{q>0} \omega_q \left(\eta_q^\dagger \eta_q - \eta_{-q}\eta_{-q}^\dagger\right) = 2\sum_q \omega_q \eta_q^\dagger \eta_q + E_0. \tag{2.2.18}$$

The zero-point (ground-state) energy of the spinless fermion system is given by

$$E_0 = -\sum_q \omega_q. \tag{2.2.19}$$

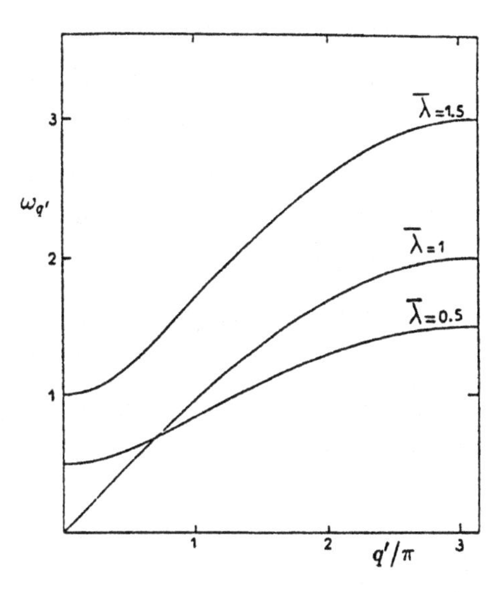

Fig. 2.1 Elementary excitation energy as a function of q'/π for different $\overline{\lambda}$ [312]

The Bogoliubov fermion operators are written as the linear combination of the original Jordan-Wigner fermions

$$\eta_q = \sum_i \left[\left(\frac{\phi_{qi} + \psi_{qi}}{2} \right) c_i + \left(\frac{\phi_{qi} - \psi_{qi}}{2} \right) c_i^\dagger \right], \tag{2.2.20}$$

where for $\overline{\lambda} \neq 1$ the normal modes [245, 312] are given by

$$\phi_{q,j} = \left(\frac{2}{N} \right)^{1/2} \sin(qR_j); \quad q > 0 \tag{2.2.21a}$$

$$= \left(\frac{2}{N} \right)^{1/2} \cos(qR_j); \quad q \leq 0 \tag{2.2.21b}$$

$$\psi_{qj} = -\omega_q^{-1} \left[(1 + \overline{\lambda} \cos q) \phi_{qj} + (\overline{\lambda} \sin q) \phi_{-qj} \right]. \tag{2.2.21c}$$

For $\overline{\lambda} = 1$ and $m = -1/2$ (i.e., $q = -\pi$),

$$\omega_q = 0, \qquad \phi_{qj} = N^{-1/2}, \qquad \psi_{qj} = \pm N^{-1/2}. \tag{2.2.22}$$

One can thus transform the spin Hamiltonian into a non-interacting set of fermions obeying the dispersion relation given by (2.2.16).

The energy of the elementary excitations as a function of wave vector $q' = \pi - q$ for different values of $\overline{\lambda}$ is shown in Fig. 2.1. There is an energy gap in the excitation spectrum of the system which goes to zero at $q' = 0$ for $\overline{\lambda} \equiv \overline{\lambda}_c = 1$ as

$$\Delta(\overline{\lambda}) \equiv E_1 - E_0 = 2|1 - \overline{\lambda}|, \tag{2.2.23}$$

indicating the divergence of the correlation length and a quantum phase transition at $\overline{\lambda} = \overline{\lambda}_c = 1$ from an ordered state ($\langle S^x \rangle = 1$) to a disordered state ($\langle S^x \rangle = 0$). One should mention at this point that the critical value of the transverse (tunnelling) field, obtained from the exact solution of the one-dimensional chain, is $\Gamma_c = J$, whereas the mean field theory employed by Brout et al. (see Sect. 1.2), overestimates for the one-dimensional nearest neighbour chain and yields $\Gamma_c = 2J$.

At this point we remark on the simplification resulting from the consideration of the "c-cyclic" problem rather than the "a-cyclic" problem. The Hamiltonian for the "a-cyclic" problem is complicated by the presence of the term

$$\left[\left(c_1^\dagger + c_1\right)\left(c_N^\dagger - c_N\right)\right]\left[\exp(i\pi L) + 1\right].$$

As mentioned earlier, that although L is not a constant of motion, $\exp(i\pi L)$ is. Now in the ground state of the "c-cyclic" problem, in all states with even number of excitations, the number of fermions is odd (the q's are assumed to be occupied symmetrically around $q = 0$, except that $q = \pi$ but not $q = -\pi$ is occupied). Therefore, the additional term gives zero acting on such states and they remain eigenstates of the "a-cyclic" problem. States with odd number of excitation, on the other hand, have L even, giving an additional term in the Hamiltonian. This has the effect of making changes of the order of $1/N$ in the q's, ϕ_q's etc., which one can ignore for a thermodynamically large system. Strictly speaking the elementary excitations are not independent in the "a-cyclic" problems [245].

One must note that the above method of diagonalising a spin Hamiltonian in terms of free fermions is exact only in the case of pure transverse Ising chain in absence of any longitudinal field. One cannot generalise it to higher dimensional systems. For spin chains incorporating frustrations the resulting fermions are interacting (see Chap. 4) and hence can not be diagonalised exactly. For random interaction and transverse field Ising chain (see Chap. 5), one can map the problem to free fermion problem but as both J_i's and Γ_i's are random one has to adopt numerical diagonalisation technique.

2.2.1 The Ground State Energy, Correlations and Exponents

The ground state energy of the fermion system, given by the expression (2.2.19), can be reduced to the elliptic integral of second kind

$$-\frac{E_0}{NJ} = \frac{2}{\pi}(1+\overline{\lambda})E\left(\frac{\pi}{2}, \theta\right); \quad \theta^2 = \frac{4\overline{\lambda}}{(1+i\overline{\lambda})^2}. \tag{2.2.24}$$

The ground state energy decreases monotonically with the transverse field and is non-analytic at the point $\overline{\lambda} = 1$ (see Fig. 2.2).

Defining the divergence of correlation length ($\xi(\overline{\lambda}) = \Delta^{-1}$) in the neighbourhood of the critical point $\overline{\lambda}_c = 1$ as $\xi(\overline{\lambda}) \sim (\overline{\lambda}_c - \overline{\lambda})^{-\nu}$, one can readily obtain from (2.2.23) the value of exponent $\nu = 1$. If one defines the variation of the energy gap as $\Delta(\overline{\lambda}) \sim (\overline{\lambda}_c - \overline{\lambda})^s$, the same equation yields $s = 1$. To evaluate the other exponents of the transition one has to calculate magnetisations and the different correlation functions using "Wick's Theorem" to evaluate the vacuum expectation value of many fermion operators [312]. The longitudinal magnetisation $\langle S^x \rangle$ (the order parameter

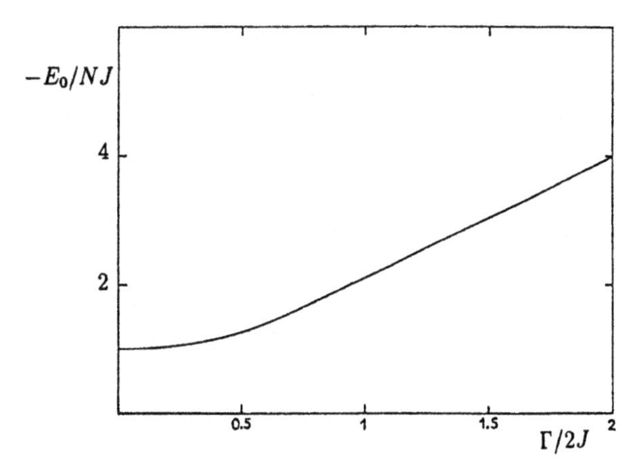

Fig. 2.2 Ground state energy as a function of $\Gamma/2J$ [312]

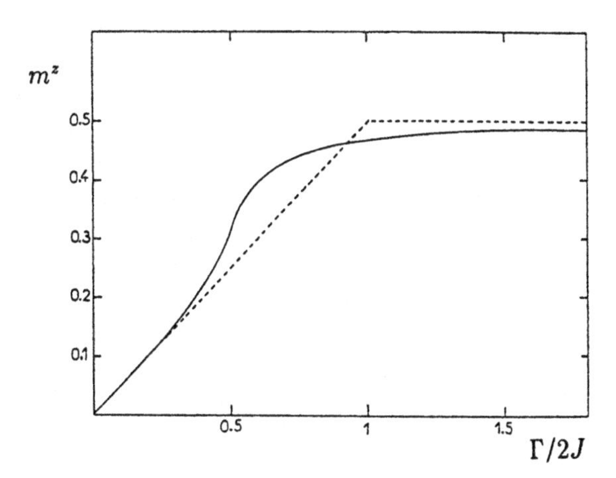

Fig. 2.3 Variation of the transverse magnetisation $m^z = \langle S^z \rangle$ with $\Gamma/2J$. *Dashed curve* is the mean field result [312]

in the present problem) cannot be evaluated directly because of the up-down symmetry of the Hamiltonian unless one applies a symmetry breaking longitudinal field. The transverse magnetisation can be written as [312] (see Sect. 2.A.3)

$$\begin{aligned} m^z = G_0 &= \langle \psi_0 | S_i^z | \psi_0 \rangle \\ &= \frac{1}{\pi} \int_0^{\pi} dq\, \omega_q^{-1} + \frac{1}{\pi} \int_0^{\pi} dq\, \omega_q^{-1} \cos q, \end{aligned} \tag{2.2.25}$$

which gives an elliptic integral of the first kind, and is non-analytic at $\overline{\lambda} = \overline{\lambda}_c = 1$. (Fig. 2.3).

The correlation functions given in Sect. 2.A.3 depend only on G_r. As a function of $\overline{\lambda}$, G_r is non-analytic for $\overline{\lambda} = 1$ and so are the correlation functions. The critical case $\overline{\lambda} = 1$ can be studied more easily because of the simple form of G_r (see Sect. 2.A.3). We get the correlation functions given as [265, 312]

$$C_r^z = \langle\psi_0|S_i^z S_{i+r}^z|\psi_0\rangle = \frac{4}{\pi^2}\frac{1}{4r^2-1}$$

$$C_r^x = \langle\psi_0|S_i^x S_{i+r}^x|\psi_0\rangle = \left(\frac{2}{\pi}\right)^r 2^{2r(r-1)}\frac{H(r)^4}{H(2r)}$$

where $H(r) = 1^{r-1}2^{r-1}\cdots(r-1)$. C_r^y and C_r^z behave in the analogous way and all the correlations at $\overline{\lambda} = 1$ go to zero as $r \to \infty$.

The spin-spin correlation in the longitudinal direction C_r^x vanishes for large r ($\lim_{r\to\infty} C_r^x = 0$) when $\overline{\lambda} \leq 1$, indicating the absence of any long-range correlation in the disordered phase. For $\overline{\lambda} > 1$ [265, 312]

$$\lim_{r\to\infty} C_r^x = \left(1 - \overline{\lambda}^{-2}\right)^{1/4}, \tag{2.2.26}$$

and hence the longitudinal magnetisation (as $r \to \infty$) can be written as

$$m^x = \langle\psi_0|S_i^x|\psi_0\rangle = 0; \quad \overline{\lambda} \leq 1, \tag{2.2.27a}$$

$$= \left(1 - \overline{\lambda}^{-2}\right)^{1/8} \quad \overline{\lambda} > 1. \tag{2.2.27b}$$

Equation (2.2.27b) gives the value of the magnetisation exponent $\beta = 1/8$. All the other exponents can be obtained in a similar fashion and it is readily observed that the value of these exponents are identically same with the exponents associated with the thermal phase transition in classical two-dimensional Ising model. One should mention that these correlation functions can be derived alternatively using "bosonisation" technique [367].

Following the development in the study of quantum computation and quantum information, the quantum entanglement has drawn increasing attention [295]. In the context of the statistical physics, the behaviour of the entanglement near a quantum critical point has been often discussed. It has been known in general that the entanglement increases when the system approaches the quantum phase transition.

The entanglement is basically measured by the entanglement entropy. To define the entanglement entropy, one considers a partition of a system into subsystems A and B. Let $\rho_A = \mathrm{Tr}_B(\rho)$ be the density matrix of the subsystem A, which is given by the trace of the density matrix of the whole system ρ with respect to the subsystem B. The entanglement entropy between A and B is defined by the von Neumann entropy of ρ_A as $S(\rho_A) = \mathrm{Tr}_A(-\rho_A \log \rho_A)$. One can readily see that the entanglement entropy vanishes if the subsystem A is decoupled from B such that $\rho = \rho_A\rho_B$ and the rank of ρ_A is 1, whereas it has the maximum when all the eigenvalues of ρ_A are equivalent. Let us now consider an infinite transverse Ising chain and focus on a subsystem of size l. The entanglement entropy S_l of this block is obtained by considering the matrix [12, 410]

$$\Lambda = \begin{pmatrix} \Pi_0 & \Pi_1 & \cdots & \Pi_{l-1} \\ \Pi_{-1} & \Pi_0 & \cdots & \Pi_{l-2} \\ \vdots & \vdots & \ddots & \vdots \\ \Pi_{-l+1} & \Pi_{-l+2} & \cdots & \Pi_0 \end{pmatrix}, \tag{2.2.28}$$

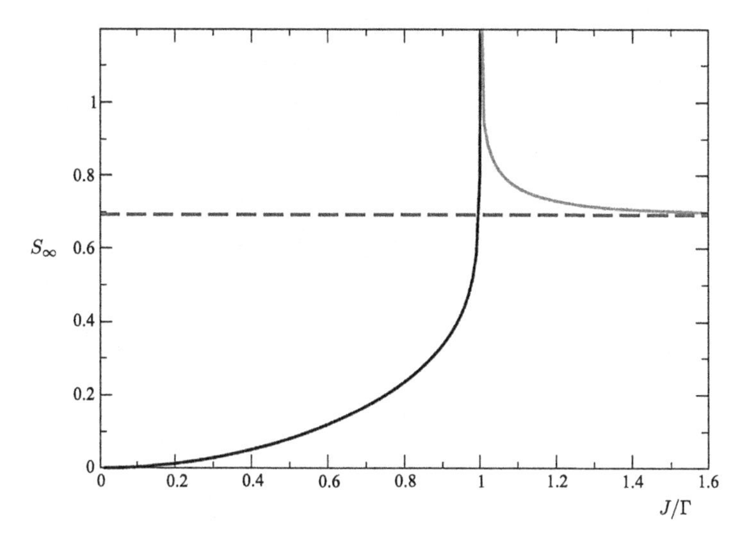

Fig. 2.4 Entanglement entropy $S_{l=\infty}$ versus J/Γ. $S_{l=\infty}$ has a singularity at $\Gamma = J$, namely, at $\overline{\lambda} = 1$. Note that $S_\infty = \log 2$ when $\Gamma = 0$. This comes from doubly degenerated ground states of the Ising chain. (After Calabrese and Cardy [54])

with

$$\Pi_n = \begin{pmatrix} 0 & g_n \\ -g_n & 0 \end{pmatrix}; \quad g_n = \int_0^{2\pi} \frac{d\phi}{2\pi} e^{-in\phi} \frac{e^{-i\phi} - \overline{\lambda}}{|e^{-i\phi} - \overline{\lambda}|}.$$

Letting λ_n's denote positive eigenvalues of the matrix (2.2.28) and defining $h(x) = -x\log(x) - (1-x)\log(1-x)$, S_l is given by

$$S_l = \sum_{n=1}^{l} h\left(\frac{1+\lambda_n}{2}\right). \tag{2.2.29}$$

The behaviour of (2.2.29) has been studied numerically [410] and analytically [54, 200, 310]. Figure 2.4 shows $\overline{\lambda}$ dependence of $S_{l=\infty}$ [54]. As shown in this figure, the entanglement entropy $S_{l=\infty}$ diverges at the critical point $\overline{\lambda} = 1$. Focusing on the critical point, S_l scales as $S_l \sim \frac{1}{6}\log l$ [410]. This logarithmic scaling of the entanglement entropy is universal in one-dimensional quantum critical systems. It has been known from the conformal field theory that, in general, the entanglement entropy of a one-dimensional quantum critical system with the central charge c behaves as $S_l \sim \frac{c}{3}\log l$ [54, 176]. The prefactor $\frac{1}{6}$ in S_l of the transverse Ising chain $\overline{\lambda} = 1$ is consistent with $c = \frac{1}{2}$ of the same system.

Another measure of the entanglement is the concurrence [425]. We consider a density matrix with respect to sites i and $i+r$, $\rho_{i,r} = \mathrm{Tr}'(\rho)$, where Tr' stands for the partial trace with respect to the degree of freedom on sites except i and $i+r$. We define the time-reversal of $\rho_{i,r}$ as $\tilde{\rho}_{i,r} = (S_i^y S_{i+r}^y)\rho_{i,r}^*(S_i^y S_{i+r}^y)$, where $\rho_{i,r}^*$ is the complex conjugate of $\rho_{i,r}$. Let λ_k ($k = 1, \ldots, 4$) be the eigenvalues of $\rho_{i,r}\tilde{\rho}_{i,r}$. Note that λ_k is a non-negative number. We fix k such that $\lambda_1 \geq \cdots \geq \lambda_4$. Then

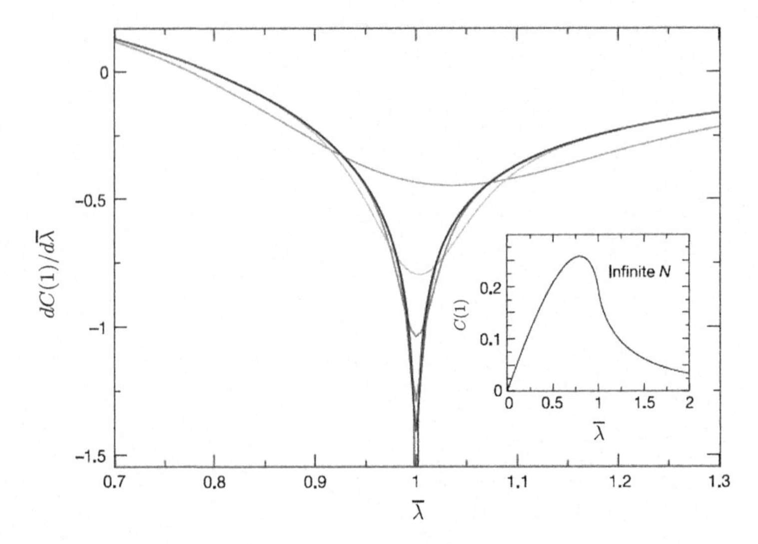

Fig. 2.5 Nearest neighbour concurrence versus $\overline{\lambda} = J/\Gamma$. The main panel shows the first derivative $dC(1)/d\overline{\lambda}$ for several sizes, while the *inset* shows $C(1)$ of the infinite size system. The dip of $dC(1)/d\overline{\lambda}$ at the critical point $\overline{\lambda} = 1$ becomes deeper with increasing size. In the infinite size system, $dC(1)d\overline{\lambda}$ behaves as $(8/3\pi^2)\ln|\overline{\lambda} - 1| + \text{const}$ and diverges at $\overline{\lambda} = 1$. (After Osterloh et al. [303])

the concurrence is defined by $C(r) = \max\{0, \sqrt{\lambda_1} - \sqrt{\lambda_2} - \sqrt{\lambda_3} - \sqrt{\lambda_4}\}$. It has been shown that this quantity gives the so-called entanglement of formation, which is an elementary measure of the entanglement, and thus can be regarded itself as an entanglement measure [425]. The concurrence in the transverse Ising chain has been studied in Refs. [303, 392]. It has been shown that $C(r)$ vanishes for $r \geq 3$ even at the critical point $\overline{\lambda} = 1$. Confining our attention to the concurrence between the nearest neighbour $C(1)$, it has a peak near the critical point and is a smooth function of $\overline{\lambda}$. The singularity at the critical point is clearer in the first derivative of $C(1)$. Osterloh et al. [303] have shown that, as $\overline{\lambda}$ approaches its critical point $\overline{\lambda}_c = 1$, $dC(1)/d\overline{\lambda}$ behaves as $dC(1)/d\overline{\lambda} = (8/3\pi^2)\ln|\overline{\lambda} - 1| + \text{const}$ in the infinite size system. Figure 2.5 shows $C(1)$ and $dC(1)/d\overline{\lambda}$ as functions of $\overline{\lambda}$ [303].

2.3 Diagonalisation Techniques for Finite Transverse Ising Chain

Although, as discussed in the previous section, the spin-1/2 transverse Ising chain (and its classical analogue i.e., classical Ising system on a square lattice) is exactly solvable, the subtlety for the condition of exact solubility is very unique. With slightest change in the conditions (like presence of next nearest neighbour interaction or disorder, or in higher dimensions etc.), the above tricks fail and one requires accurate approximate methods. One such method, which can be (and has been) utilised to extract accurate information in such cases (see Chaps. 4 and 6), is exact diagonalisation of finite-size quantum systems and employment of finite size scaling. In order to test the accuracy of such methods, we can employ it here for a spin-1/2 transverse Ising chain (where we can compare with the exact results).

2.3.1 *Finite-Size Scaling*

The idea of finite-size scaling was introduced by Fisher and Barber [143] to explain the effect of the thermodynamic singularities on finite size variations. Let $\psi(\lambda)$ be some quantity which diverges in the thermodynamic limit at a critical value (λ_c)

$$\psi(\lambda) \sim A|\Delta|^{-\psi}; \qquad \Delta\lambda = \frac{(\lambda - \lambda_c)}{\lambda_c} \to 0, \tag{2.3.1}$$

where the correlation length $\xi \sim |\Delta\lambda|^{-\nu}$ also diverges with the correlation length exponent λ. For a finite system of linear dimension L, the behaviour of $\psi(\lambda, L)$ is given by the finite size scaling ansatz

$$\psi(\lambda, L) = L^{\psi/L} f\big(L/\xi(\lambda)\big) \tag{2.3.2}$$

where the scaling function $f(x)$ is asymptotically defined with a power law: $f(x) \sim x^{-\psi/\nu}$, as $x \to \infty$. One thus recovers (2.3.1) from (2.3.2) when $L \to \infty$.

The above ansatz (2.3.2) is helpful in the following ways: (1) If the critical point λ_c is exactly known, then the finite size variation there is precisely given by

$$\psi(\lambda_c, L) = L^{\psi/\nu}. \tag{2.3.3}$$

Using thus the data of finite systems one can estimate the exponent relations for the transition in infinite system. (2) If the critical point for the infinite system is not exactly known, then $\psi(\lambda, L)$ is evaluated at the effective $\lambda_c(L)$ (at which the effective order parameter vanishes). This $\lambda_c(L)$ approaches the true critical value as $L \to \infty$. Since the effective $\lambda_c(L)$ is the point where $\xi(\lambda_c(L)) \sim L$, one immediately gets

$$\lambda_c(L) = \lambda_c + AL^{-1/\nu} \tag{2.3.4}$$

(where A is a constant) and can extract the value of ν.

For a quantum phase transition, the inverse of the mass gap of the quantum Hamiltonian gives the correlation length. At the transition point, the mass gap $\Delta(\lambda)$ vanishes inversely as the correlation length and for finite sizes the mass gap variation is given by

$$\Delta(\lambda, L) \sim L^{-1} g\big(\Delta\lambda L^{1/\mu}\big),$$

where $g(x) \sim x^{\nu}$ as $x \to 0$. Hence at the critical point

$$L\Delta(\lambda_c, L) = (L+1)\Delta(\lambda_c, L+1). \tag{2.3.5}$$

Using the above relation one can estimate the critical point for an infinite system from the mass gap obtained from the diagonalisation of finite systems. It has been analytically established by Hamer and Barber [167] that finite size scaling is exact for the mass gap of the transverse Ising chain in the limit $L \to \infty$ and $\lambda \to 1$ with $(1 - \lambda)L$ of the order of unity.

2.3.2 The Diagonalisation Techniques

One considers here a system of finite size N and diagonalises $2N \times 2N$ dimensional Hamiltonian matrix. Here, the observation of the symmetries of H_N can help in reducing the size of the matrix. Also the interest in some very specific states (like the ground state and the low lying excited states) may not need the search in the entire Hilbert space and one can restrict to very specific subspace. Let us now consider here a spin-$1/2$ transverse Ising system on an open chain of N spins

$$H_N = -\sum_{i=1}^{N-1} S_i^z S_{i+1}^z - \lambda \sum_{i=1}^{N} S_i^x. \qquad (2.3.6)$$

The above Hamiltonian has discrete spin-flip symmetry. We choose a set of basis vectors in which is diagonal. We can write the 2^N basis vectors, spanning the Hilbert space associated with the Hamiltonian, in the form $|\epsilon_1, \ldots, \epsilon_p, \ldots, \epsilon_N\rangle$ where ϵ_p is given by

$$S_p^x |\epsilon_1, \ldots, \epsilon_p, \ldots, \epsilon_N\rangle = \epsilon_p |\epsilon_1, \ldots, \epsilon_p, \ldots, \epsilon_N\rangle. \qquad (2.3.7)$$

Clearly, ϵ_p can take two values $+1$ and -1. We can now easily check that the Hamiltonian H_N acting on a basis vector does not change the parity of total number of $(+)$ and $(-)$ spins. Hence, we can divide the entire 2^N-dimensional Hilbert space into two 2^{N-1}-dimensional subspaces: one having basis vectors with even number of up spins and the other with basis vectors with odd number of up spins. At this point one can separately diagonalise the Hamiltonian in two subspaces and reduce a 2^N-dimensional problem into two 2^{N-1} problems. In the present case, the ground state of the Hamiltonian in the first subspace gives the ground state in the entire space whereas the ground state in the second subspace gives the first excited state; the difference of the two gives us the required mass gap of the quantum Hamiltonian.

2.3.2.1 Strong Coupling Eigenstate Method

The idea of strong coupling eigenstate method was introduced by Hamer and Barber [167] to estimate approximately (in a very small subspace) the mass gap of a quantum Hamiltonian, for a finite size N. We rewrite the quantum Hamiltonian in the form

$$H_N = H_0 + V \qquad (2.3.8)$$

$$V = -\sum_{i=1}^{N-1} S_i^z S_{i+1}^z$$

$$H_0 = -\lambda \sum_{i=1}^{N} S_i^x.$$

The essential idea is to generate a set of (strong coupling) eigenstate of H_0 by the successive application of the operator V to an unperturbed eigenstate of H_0. The advantage of using this scheme is that the number of strongly coupled eigenstates is much smaller than the total number of basis vectors, and hence the effective Hamiltonian matrix is of much reduced dimension.

Let us consider, for example, a lattice of 5 spins with periodic boundary condition. The ground state of the Hamiltonian H_0 is

$$S_i^x|0\rangle = 1|0\rangle, \quad i = 1, 2, \ldots, 5.$$

Using the notation introduced in the previous section we can write the ground state in the form

$$|0\rangle = |+++++\rangle.$$

Applying the operator V on the ground state we get a state

$$\begin{aligned}|1\rangle = (1/5)^{1/2}\big(&|--+++\rangle + |+--++\rangle + |++--+\rangle + |+++--\rangle \\ &+ |-+++-\rangle\big).\end{aligned} \tag{2.3.9}$$

Considering the translational invariance of the model we simply write the state $|1\rangle = |--\rangle$. Applying V on this state $|1\rangle$ gives

$$V|--\rangle = 2\big(|-+-++\rangle + |----+\rangle + 5|0\rangle\big).$$

We call these states $|-+-++\rangle = |2\rangle$ and $|----+\rangle = |3\rangle$. Application of V on this states does not generate any new state. We have thus effectively generated a complete set of strongly coupled eigenstates of the operator H_0. In the present case the series terminates because the unperturbed Hamiltonian has a finite number of states, whereas in the case of Hamiltonians with continuous symmetry the series does not terminate but converges very rapidly. Using the above four basis vectors we can write the Hamiltonian in the truncated 4-dimensional Hilbert space and now diagonalisation is much easier. To evaluate the first excited one has to repeat the same procedure starting from the state with a single spin flipped. This method gives the exact value of the mass gap for transverse Ising system and is also effective for estimating the phase diagram of other quantum Hamiltonians.

We now give some typical numerical results. The variation of the mass gap $\Delta(\lambda, L)$ with λ of the transverse Ising chain (of size $3 < L < 50$) is shown below (Fig. 2.6). As the size of the system increases, the curve approaches the exact variation of mass gap with λ for an infinite chain ($\Delta(\lambda) \sim |1 - \lambda|$), but for a finite chain the mass gap vanishes only in the limit of asymptotic value of λ, indicating the absence of phase transition in a finite size system. To estimate numerically the critical coupling of an infinite system, using finite size results, one has to consider the "scaled mass gap ratio" R_L given by

$$R_L = \frac{L\Delta(\lambda, L)}{(L-1)\Delta(\lambda, L-1)} \tag{2.3.10}$$

and calculate the effective critical coupling $\lambda_c(L)$ from the equation $R_L(\lambda_c) = 1$. For example, in the case of transverse Ising system, one gets, with $L = 9$, the value of effective critical coupling $\lambda_c(L = 9) = 0.9985$ [143, 167] which converges to the

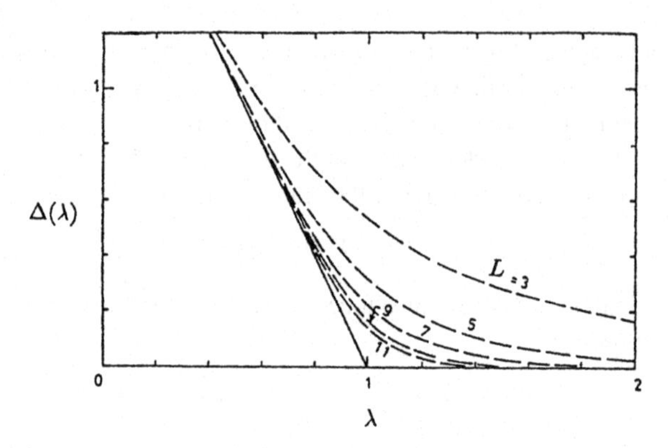

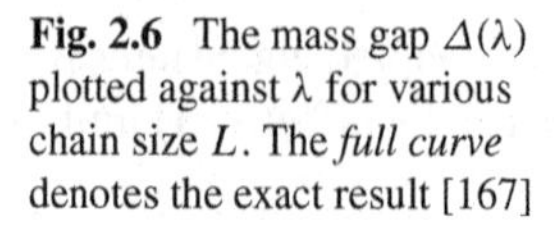
Fig. 2.6 The mass gap $\Delta(\lambda)$ plotted against λ for various chain size L. The *full curve* denotes the exact result [167]

critical value ($\lambda_c = 1$) for an infinite system, as L increases. One can thus extrapolate finite size results to make a very good approximate (exact, for a transverse Ising chain) estimation for the infinite system results.

Using the extrapolated value of critical coupling for an infinite system, as obtained above, we can estimate the value of the correlation length exponent ν using the scaling relation (2.3.4). For a transverse Ising system, using $L = 9$, one gets $\nu = 0.995$, which again extrapolates to the exact value ($\nu = 1$) in the infinite chain limit. To calculate the susceptibility exponent γ one applies a magnetic field h in the longitudinal direction and the zero field susceptibility $\chi = -(\partial^2 E_0/\partial h^2)_{h=0}$ shows a peak for a finite system at the point $\lambda_c(L)$. This peak becomes sharper in the limit of large L (Fig. 2.7), and from the slope of the logarithmic plot of the above curve one can estimate the value of γ for a finite system: for transverse Ising chain of size $L = 9$, $\gamma = 1.758$. One can also use (2.3.3), with the extrapolated approximate value of the coupling constant to estimate the value of γ for an infinite system. The approximate value of γ, thus obtained, is 1.75 ± 0.005, compared to the exact value 1.75. These demonstrate the strength of diagonalisation technique, when implemented along with finite size scaling. This diagonalisation technique has also been applied to other quantum systems with appreciable amount of success [143].

2.4 Real-Space Renormalisation

Real-space renormalisation group (RSRG) techniques have been frequently used to study the phase transition in classical systems. These techniques have also been extended to study quantum systems at $T = 0$ [113, 208] (see also Ref. [209]), where one develops various schemes for generating (rescaling) the ground state and the low-lying excited states of the quantum Hamiltonians at various levels of scaling. In this section, we shall discuss block renormalisation group method, introduced by Drell et al. [113] and extended by Jullien et al. [208], to the problem of the spin-1/2 transverse Ising chain.

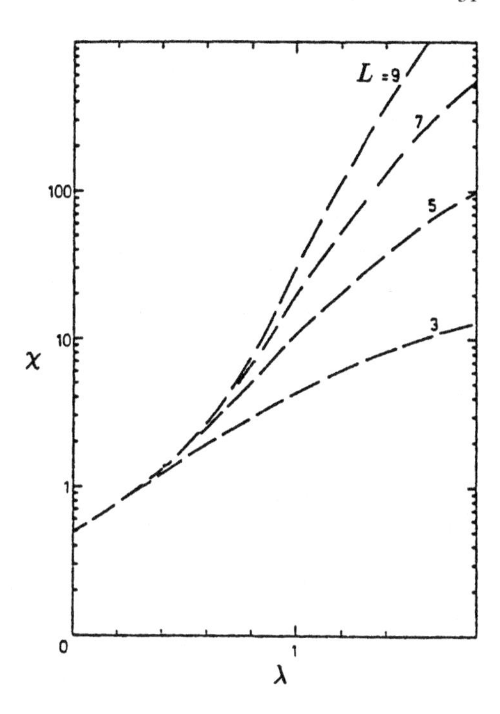

Fig. 2.7 A semi-logarithmic plot of the finite lattice susceptibility against λ [167]

2.4.1 Block Renormalisation Group Method

We shall illustrate the essential idea of the block RSRG method by applying the above to the spin-1/2 transverse Ising Hamiltonian (2.1.1) on an open chain. The idea is to generate an iterative procedure which yields for the Hamiltonian (2.1.1) in the nth iteration

$$H^{(n)} = -\sum_i \left(J^{(n)} S_i^{z(n)} S_{i+1}^{z(n)} + \Gamma^{(n)} S_i^{x(n)}\right) + c^{(n)} \sum_i I_i^{(n)} \tag{2.4.1}$$

where $I_i^{(n)}$ is the 2×2 identity matrix.

To generate an iterative scheme as shown in (2.4.1), the starting point is to divide the entire one dimensional chain of N sites into N/b blocks each having b spins (i.e., changing the length scale of the problem by a factor b) and rewrite the Hamiltonian (2.1.1) as $H = H_B + H_{IB}$ where H_B is the intra-block part and H_{IB} is the inter-block part of the Hamiltonian

$$H_B = \sum_{p=1}^{N/b} H_p, \qquad H_{IB} = \sum_{p=1}^{N/b-1} H_{p,p+1} \tag{2.4.2}$$

with

$$H_p = -\sum_{i=1}^{b-1} J S^z_{i,p} S^z_{i+1,p} + \Gamma \sum_{i=1}^{b} S^x_{i,p} \tag{2.4.3a}$$

$$H_{p,p+1} = -J S^z_{n,p} S^z_{1,p+1}, \tag{2.4.3b}$$

where indices i, p refers to the ith spin in the pth block.

The next task is to diagonalise the Block Hamiltonian H_p using the same tricks as discussed in Sect. 2.3.2. Using the same notations we can write the eigenvectors corresponding to the ground state energy E_0 and first excited state energy E_1 as

$$|0\rangle = \sum^{+} \lambda^{+}_{\epsilon_1,\ldots,\epsilon_b} |\epsilon_1, \epsilon_2, \ldots, \epsilon_b\rangle \tag{2.4.4a}$$

$$|1\rangle = \sum^{-} \lambda^{-}_{\epsilon_1,\ldots,\epsilon_b} |\epsilon_1, \epsilon_2, \ldots, \epsilon_b\rangle \tag{2.4.4b}$$

where $\sum^+ (\sum^-)$ is summation restricted to the Hilbert space consisting of the basis vectors with even (odd) number of down spins.

To perform the renormalisation procedure we just retain the ground state and the first excited state of the block Hamiltonian and introduce a new set of spin operator $S_p^{\alpha(1)}$ associated with each block p, such that the eigenstates of $S_p^{x(1)}$ are precisely the states $|0\rangle$ and $|1\rangle$. Thus, we can rewrite the block Hamiltonian in a renormalised form (in the first iteration) in terms of the new block spins,

$$H_p^{(1)} = -\Gamma^{(1)} S_p^{x(1)} + c^{(1)} I_p^{(1)}, \tag{2.4.5}$$

where

$$\Gamma^{(1)} = (1/2)(E_1 - E_0), \qquad c^{(1)} = (1/2)(E_1 + E_0). \tag{2.4.6}$$

To rewrite the renormalised form of the total Hamiltonian, we include the interblock part of the Hamiltonian in a perturbative way. To the zeroth order $H^{(1)} = H_p^{(1)}$. To the first order in perturbation, obtained by taking the matrix element of old spin operators $S^z_{i,p}$ between the new block states $|0\rangle$ and $|1\rangle$, $H_{p,p+1}$ takes the form

$$H^{(1)}_{p,p+1} = -J^{(1)} \sum_p S_p^{z(1)} S_{p+1}^{z(1)}$$

where

$$J^{(1)} = \left(\sum^{+} \lambda^{+}_{\epsilon_1,\ldots,\epsilon_p,\ldots,\epsilon_b} \lambda^{-}_{\epsilon_{-1},\ldots,\epsilon_p,\ldots,\epsilon_b} \right)^2 J \equiv \left(\eta_1^{(0)}\right)^2 J. \tag{2.4.7}$$

Equations (2.4.6) and (2.4.7) constitute the recursion relation of J and Γ in the first iteration. Thus one obtains for $(n+1)$th iteration

$$\Gamma(n+1) = \frac{1}{2}\left(E_1^{(n)} - E_0^{(n)}\right) \tag{2.4.8a}$$

$$J(n+1) = \left(\eta_1^{(n)}\right)^2 J^{(n)} \tag{2.4.8b}$$

$$c(n+1) = bc^{(n)} + \frac{1}{2}\left(E_1^{(n+1)} + E_0^{(n+1)}\right). \tag{2.4.8c}$$

The above recursion relations along with the initial conditions

$$J(0) = J, \qquad \Gamma(0) = \Gamma \quad \text{and} \quad c(0) = 0,$$

define a renormalisation group transformation, which can be readily exploited to estimate the critical point and exponents for the Hamiltonian (2.1.1).

We now apply the scheme to the transverse Ising chain, using block size $b = 2$ [183, 184, 208]. Let us start from the canonically transformed Hamiltonian (2.2.1)

$$H = -\Gamma \sum_i S_i^z - J \sum_i S_i^x S_{i+1}^x \tag{2.4.9}$$

and write in the form of (2.4.2) with N sites of the chain divided into $N/2$ blocks ($b = 2$), and

$$H_p = -\Gamma\left(S_{1,p}^x + S_{2,p}^x\right) - J S_{1,p}^x S_{2,p}^x$$

and

$$H_{p,p+1} = -J S_{2,p}^x S_{1,p+1}^x.$$

The block Hamiltonian H_p can be diagonalised exactly, with the eigenstates given by

$$|0\rangle = \frac{1}{(1+a^2)^{1/2}}\left(|{+}{+}\rangle + a|{-}{-}\rangle\right)$$

$$|1\rangle = \frac{1}{\sqrt{2}}\left(|{+}{-}\rangle + |{-}{+}\rangle\right)$$

$$|2\rangle = \frac{1}{\sqrt{2}}\left(|{+}{-}\rangle - |{-}{+}\rangle\right)$$

$$|3\rangle = \frac{1}{(1+a^2)^{1/2}}\left(a|{+}{+}\rangle - |{-}{-}\rangle\right)$$

where $a = (1/J)[(4\Gamma^2 + J^2)^{1/2} - 2\Gamma]$, and the respective eigenenergies given by E_0, E_1, $-E_1$ and $-E_0$, where

$$E_0 = -\sqrt{4\Gamma^2 + J^2} \quad \text{and} \quad E_1 = -J. \tag{2.4.10}$$

Retaining now the two lowest energy states $|0\rangle$ and $|1\rangle$ and treating them as the renormalised block spin state $S^{z(1)}$, the inter-block interaction can be written as

$$-J^{(1)} S_p^{x(1)} S_{p+1}^{x(1)}$$

where

$$J^{(1)} = J\frac{(1+a)^2}{2(1+a^2)}. \tag{2.4.11a}$$

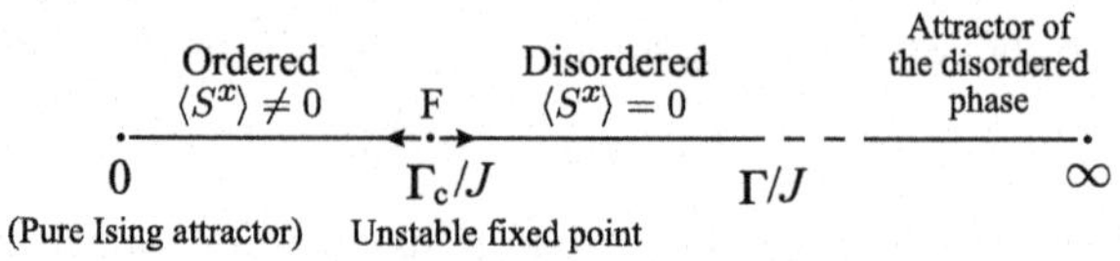

Fig. 2.8 Schematic phase diagram for renormalisation group flow for transverse Ising chain

This is because, $\langle 0|S^x|1\rangle = (1+a)/[2(1+a^2)]^{1/2}$. Since $2\Gamma^{(1)}$ equals the energy difference of states $|0\rangle$ and $|1\rangle$, we get

$$\Gamma^{(1)} = \frac{E_1 - E_0}{2} = \frac{\sqrt{4\Gamma^2 + J^2} - J}{2} \tag{2.4.11b}$$

$$c^{(1)} = \frac{E_1 + E_0}{2} = \frac{\sqrt{4\Gamma^2 + J^2} + J}{2}, \tag{2.4.11c}$$

with the last term as the additive constant appearing due to renormalisation (see (2.4.1)).

One can now write the recursion relation for the variable $\lambda = \gamma/J$ as

$$\lambda^{(1)} = \frac{[\sqrt{4\lambda^2 + 1} - 1](1+a^2)}{(1+a)^2} \tag{2.4.12}$$

where $a = \sqrt{4\lambda^2 + 1} - 2\lambda$. Solving numerically (2.4.12) for the fixed point λ^*, one gets (apart from the trivial fixed points at $\lambda^* = 0$ or ∞) $\lambda^* = 1.277$, and linearising the recursion relation (2.4.12) in the neighbourhood of the fixed point, one gets

$$\lambda' - \lambda = \Omega(\lambda - \lambda^*), \tag{2.4.13}$$

which gives the correlation length exponent $\lambda = \ln \Omega / \ln 2 \simeq 1.47$ (with $b = 2$ here), compared to the exact value $\lambda^* = \lambda_c = 1$ and $\nu = 1$. Also, writing $J^{(1)}/J = \Gamma(1)/\Gamma \sim b^{-z}$ near the fixed point, one gets $z = 0.55$, compared to the exact value $z = 1$. It may be noted here if the mass gap $\Delta(\lambda) \sim |\lambda - \lambda_c|^s$, then $s = \nu z$. The results improve considerably with larger block size [183, 184, 208].

The schematic flow diagram of transverse Ising chain, as obtained from the above block real-space renormalisation group technique, is as shown in Fig. 2.8. The information we derive is the following:

(1) There exists an unstable (critical) fixed point in the one parameter space $\gamma/J = \lambda$ at a point λ_c. If one starts from a point $\lambda \geq \lambda_c$, the system is iterated towards the trivial fixed point at $\lambda = \infty$ (the disordered phase), whereas if started from a point $\lambda \leq (\gamma/J)_c$ the system is iterated towards the point $\lambda = 0$ (classical Ising phase).
(2) The value of λ_c approaches the exact value ($\lambda_c = 1$) as one considers larger and larger block size.
(3) Ground state energy per site is given by

$$(E_0/N)_{N\to\infty} = \lim_{n\to\infty} \left(c^{(n)}/b^{(n)}\right). \tag{2.4.14}$$

The second derivative of E_0 with respect to Γ, $-\partial^2(E_0/N)/\partial\Gamma^2$, shows a non-analytic behaviour at the point λ_c (Fig. 2.9). This singular behaviour approaches the exact behaviour as larger block size is considered.

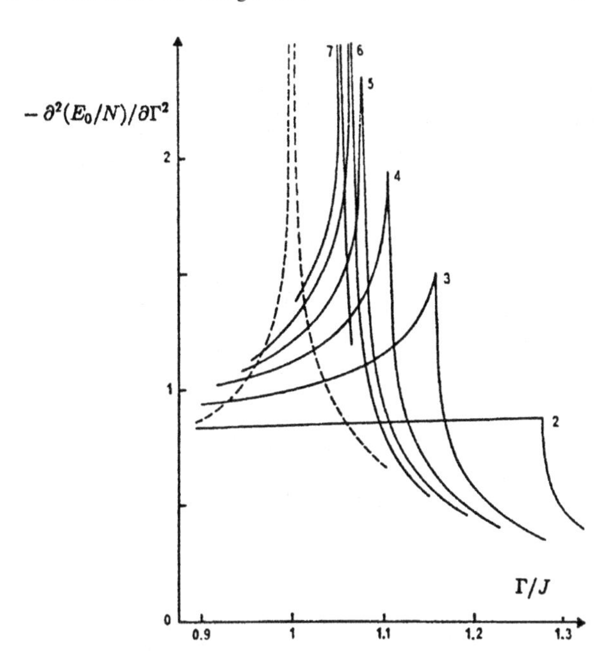

Fig. 2.9 Variation of specific heat ($= -\partial^2(E_0/N)/\partial\Gamma^2$) for different b. The exact result is represented by the *dashed line* [208]

As mentioned before, the values of these exponents, thus obtained, can be made more accurate by using larger block [208]. Hirsch and Mazenko [175], obtained better results for the mass gap and the critical exponents considering the next order in perturbation. Better results can also be derived by increasing the number of energy states of the block Hamiltonian, retained in each iteration [207].

We shall conclude this section with the note that, although we have restricted our discussion to the transverse Ising chain, this model has also been applied to interacting fermions, spin-1/2 XY and Heisenberg Chain, XY chain in 2 and 3 dimension [208] etc. One should note that the block renormalisation group technique has been extended to study quantum spin systems at finite temperature [377]. Recently transverse Ising system has also been studied using density matrix renormalisation group techniques [114].

2.5 Finite Temperature Behaviour of the Transverse Ising Chain

In this section, we shall discuss the finite temperature behaviour of the spin-1/2 transverse Ising chain. Using the dispersion relation of the elementary excitations (2.2.16), one can readily write out the free energy of the system

$$F = -Nk_BT\left[\ln 2 + \int_0^{\pi} dq \ln\cosh\left(\frac{1}{2}\beta\omega_q\right)\right]; \quad \beta = \frac{1}{k_BT}, \qquad (2.5.1)$$

from which all the thermodynamic quantities can be obtained. The free energy does not show any singularity at any finite temperature. Since the model is one-dimensional, even when the transverse field is absent, the long range order and the

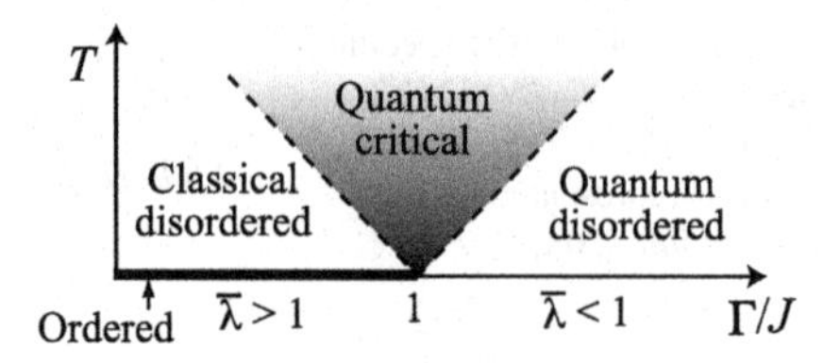

Fig. 2.10 Schematic phase diagram of the transverse Ising chain. The finite temperature phase is divided into three phases. *Dashed lines* stand for crossovers. (After Sachdev [339])

correlations of the system are destroyed with the introduction of infinitesimal thermal fluctuations ($T_c = 0$). Hence, there is no finite temperature transition for the one-dimensional model.

In spite of the absence of the long range order, the finite temperature phase diagram is divided into three phases by crossover lines $T \sim \Delta(\bar{\lambda}) = 2|1 - \bar{\lambda}|$. For $T < \Delta(\bar{\lambda})$ with $\bar{\lambda} < 1$ and with $\bar{\lambda} > 1$ one has the classical disordered phase and the quantum disordered phase respectively. There exists a quantum critical phase between these phases where $T > \Delta(\bar{\lambda})$. This picture is corroborated by analysis of finite temperature correlation functions [338, 339] (see Sect. 2.A.3). We show the schematic phase diagram in Fig. 2.10. In the classical disordered phase, the thermal fluctuation destroys the long-range order as the classical Ising chain at finite temperature is. On the other hand, the quantum fluctuation is responsible for the quantum disordered state. In the quantum critical region, the system is governed by the quantum criticality. It is remarkable that a nature of quantum criticality may be observed even at finite temperatures in the quantum critical region.

2.6 Experimental Studies of the Transverse Ising Chain

In realisation of a transverse Ising chain, we are restricted by dimensionality and the Ising anisotropy. $CoNb_2O_6$ has been perhaps the best candidate so far that provides us with a quasi-one-dimensional spin system with strong Ising anisotropy. Coldea et al. [81] performed neutron scattering experiments for this material in the presence of a transverse field and obtained distinct excitation spectra. Due to an interchain coupling, each Ising-spin chain in this material undergoes a ferromagnetic transition with a critical temperature $T = 2.95$ K. At sufficiently low fields and low temperature (i.e., $\bar{\lambda} > 1$ and $T \ll \Delta(\bar{\lambda})$), the elementary excitations from the ordered ground state is a kink and anti-kink. They form a continuum in the excitation spectrum. On the other hand, when the transverse field is sufficiently high ($\bar{\lambda} < 1$ and $T \ll \Delta(\bar{\lambda})$), the excitation consists of a flip of a single spin from the ground state polarised along the transverse axis. This excitation is signified by a single branch in the excitation spectrum. The results of experiment [81] clearly show the change of excitation spectrum between two regimes (Fig. 2.11). In addition to this observation, Coldea et al. detected a nature of quantum criticality, which is described as follows, in this material. The present system is represented effectively by an Ising chain with a weak longitudinal mean field induced by interchain couplings. This model in the presence of the transverse field cannot be mapped to a free fermion model (see Sect. 2.2).

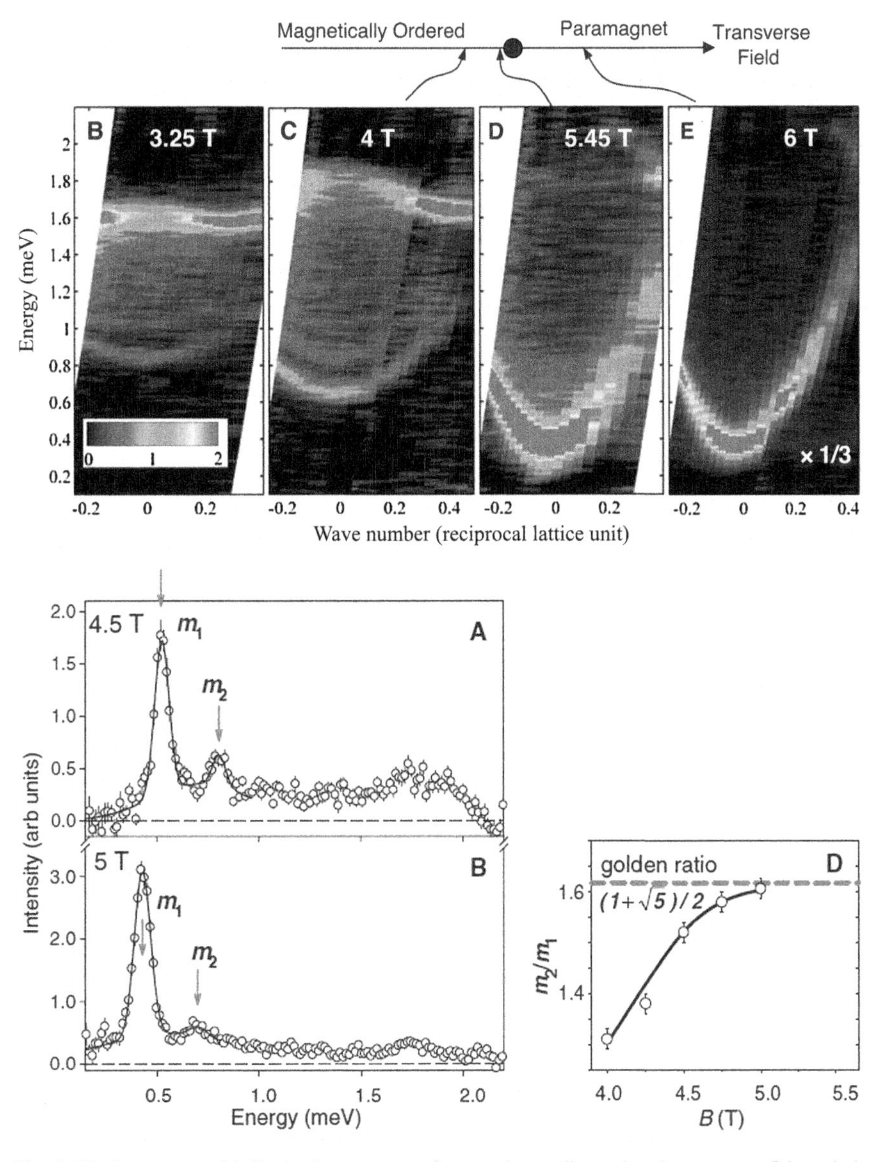

Fig. 2.11 (*upper panels*) Excitation spectra of a quasi-one-dimensional transverse Ising chain observed in $CoNb_2O_6$ by neutron scattering. The *left two panels* **B** and **C** show spectra in the low field regime, where one finds a continuum formed by scattering states of kink-antikink pair. In the *rightmost panel* **E**, the spectrum consists of a single branch which is attributed to the excitation by a single-spin flip from the polarised ground state along the transverse axis. The *panel* **D** is the result near the quantum critical point. (*lower panels*) **A** and **B** show spectral intensities below and near the critical field respectively. m_1 and m_2 denotes the lowest and the second lowest excitation energies. **D** is a plot of the ratio m_1/m_2 as a function of the transverse field. One finds that the ratio approaches the golden ratio with increasing the field toward to the critical value. (After Coldea et al. [81])

However, at the critical transverse field, the system is described by an integrable field theory [289]. Zamolodchikov [442] showed that the system possesses the E_8 symmetry of the exceptional Lie group and eight elementary excitations with different masses should exist. Coldea et al. [81] confirmed that the ratio of the lowest two excitation energies approaches the golden ratio $(1+\sqrt{5})/2$ as one shifts the transverse field to the critical point (Fig. 2.11). The golden ratio is nothing but what Zamolodchikov predicted. This experimental result reveals that the transverse Ising chain is not only a mathematical model but a realisable system which contains essences of quantum critical phenomena.

Appendix 2.A

2.A.1 *Jordan-Wigner Fermions*

To check that the operators c_i and $c_i^\dagger$ satisfy fermionic anti-commutation relations

$$\left[c_i, c_i^\dagger\right]_+ = \delta_{ik}, \qquad [c_i, c_k]_+ = 0 \quad \text{and} \quad \left[c_i^\dagger, c_k^\dagger\right]_+ = 0, \tag{2.A.1}$$

one uses the simple relations obeyed by Pauli spin operators

$$S_i^- S_i^+ = \frac{1}{2}\left(1 - S_i^z\right); \quad S_i^+ S_i^- = \frac{1}{2}\left(1 + S_i^z\right); \quad \exp\left(\frac{i\pi}{2} S_i^z\right) = i S_i^z. \tag{2.A.2}$$

So that, the Jordan-Wigner transformation (2.2.2a), (2.2.2b) can be rewritten as

$$c_i = \prod_{j=1}^{i-1} \left[-S_j^z\right] S_i^- \tag{2.A.3a}$$

$$c_i^\dagger = S_i^+ \prod_{j=1}^{i-1} \left[-S_j^z\right]. \tag{2.A.3b}$$

Using (2.A.3a) and (2.A.3b), one can immediately check that

$$c_i c_i^\dagger + c_i^\dagger c_i = S_i^- S_i^+ + S_i^+ S_i^- = 1. \tag{2.A.4}$$

To prove $[c_i, c_k^\dagger]_+ = 0$, for $k \neq i$, without loss of generality we assume $k < i$. Now

$$c_k c_i^\dagger + c_i^\dagger c_k = S_k^- \prod_{j=k}^{i-1} \left[-S_j^z\right] S_i^+ + S_i^+ \prod_{j=k}^{i-1} \left[-S_j^z\right] S_k^-. \tag{2.A.5}$$

If one now uses the fact

$$S_m^- S_m^z = -S_m^z S_m^-$$

the right hand side of (2.A.5) vanishes identically. Similarly, one can derive the other anti-commutation relations.

To express the spin Hamiltonian in terms of the fermion operators, we use the relation, which can be checked using (2.A.3a) and (2.A.3b). Now from (2.A.2), we get

$$S_i^z = 2S_i^+ S_i^- - 1 = 2c_i^\dagger c_i - 1, \tag{2.A.6}$$

which can be easily derived using (2.A.3a), (2.A.3b). The coupling term

$$S_i^x S_{i+1}^x = \left[S_i^+ + S_i^-\right]\left[S_{i+1}^+ + S_{i+1}^-\right] \tag{2.A.7}$$

is now to be rewritten in terms of fermions. Consider the product

$$c_i^\dagger c_{i+1} = S_i^+\left[-S_i^z\right]S_{i+1}^- \quad \bigl(\text{from (2.A.3a), (2.A.3b)}\bigr). \tag{2.A.8}$$

But $S_i^+ S_i^z = -S_i^+$, so we get

$$c_i^\dagger c_{i+1} = S_i^+ S_{i+1}^-. \tag{2.A.9a}$$

Similarly one can derive the following relations

$$c_i c_{i+1}^\dagger = -S_i^- S_{i+1}^+, \tag{2.A.9b}$$

$$c_i^\dagger c_{i+1}^\dagger = S_i^+ S_{i+1}^+, \tag{2.A.9c}$$

$$c_i c_{i+1} = -S_i^- S_{i+1}^-. \tag{2.A.9d}$$

The coupling term can be written in terms of fermion operators which are also only coupled to nearest neighbours. This result is a consequence of the Ising character $S_i^2 = +1$ of the original degrees of freedom. Collecting all the terms in (2.A.5) and (2.A.9a), (2.A.9b), (2.A.9c), (2.A.9d), one can arrive at the Hamiltonian (2.2.6). One should note here that if the original spin Hamiltonian incorporates next-nearest neighbour interaction, the resulting fermion Hamiltonian includes four-fermion term.

2.A.2 To Diagonalise a General Hamiltonian Quadratic in Fermions

We wish to diagonalise a general quadratic Hamiltonian of the form

$$H = \sum_{ij} c_i^\dagger A_{ij} c_j + \frac{1}{2}\sum_{ij} c_i^\dagger B_{ij} c_j^\dagger + h.c. \tag{2.A.10}$$

where c_i and $c_i^\dagger$ are fermion annihilation and creation operators respectively. For system size N, A and B are both $N \times N$ matrices. Hermiticity of H demands A to be a Hermitian matrix and anti-commutation of fermion operators demands B to be antisymmetric. Both can be chosen to be real.

Equation (2.A.10) is most effectively diagonalised by Bogoliubov transformation employed for fermion operators by Lieb et al. [245]. In particular, the technique is directly applicable for quite general nearest neighbour interactions, including for example quasiperiodic and random ones [77]. One wants to rewrite the Hamiltonian in a diagonal form using variables η_q and $\eta_q^\dagger$ as

$$H = \sum_q \omega_q \eta_q^\dagger \eta_q + \text{const}, \tag{2.A.11}$$

ω_q are one fermion energies. One makes a linear transformation of the form [245]

$$\eta_q = \sum_i (g_{qi} c_i + h_{qi} c_i^\dagger) \tag{2.A.12a}$$

$$\eta_q^\dagger = \sum_i (g_{qi} c_i^\dagger + h_{qi} c_i) \tag{2.A.12b}$$

where g_{qi} and h_{qi} can be chosen to be real. For η_q's to satisfy fermionic anti-commutation relations we require

$$\sum_i (g_{qi} g_{q'i} + h_{qi} h_{q'i}) = \delta_{qq'} \tag{2.A.13a}$$

$$\sum_i (g_{qi} h_{q'i} - g_{q'i} h_{qi}) = 0. \tag{2.A.13b}$$

If (2.A.11) holds, then we must have

$$[\eta_q, H]_+ - \omega_q \eta_q = 0. \tag{2.A.14}$$

Using (2.A.12a), (2.A.12b) in (2.A.14) one finds

$$\omega_q g_{qi} = \sum_j (g_{qj} A_{ji} - h_{qj} B_{ji}) \tag{2.A.15a}$$

$$\omega_q h_{qi} = \sum_j (g_{qj} B_{ji} - h_{qj} A_{ji}). \tag{2.A.15b}$$

The above coupled equations can be written in the following form

$$\Phi_q (A - B) = \omega_q \Psi_q \tag{2.A.16a}$$

$$\Psi_q (A + B) = \omega_q \Phi_q \tag{2.A.16b}$$

where the components of the $2N$ vectors Φ and Ψ are obtained from the matrices g and h as

$$(\Phi_q)_i = g_{qi} + h_{qi} \tag{2.A.17a}$$

$$(\Psi_q)_i = g_{qi} - h_{qi}. \tag{2.A.17b}$$

One fermion energies ω_q are obtained from the eigenvalues of a $N \times N$ matrix

$$(A + B)(A - B)\Phi_q = \omega_q^2 \Phi_q \tag{2.A.18a}$$

$$\Psi_q (A + B)(A - B) = \omega_q^2 \Psi_q. \tag{2.A.18b}$$

For $\omega_q \neq 0$, either (2.A.18a) or (2.A.18b) is solved for Φ_q or Ψ_q and the other vector is obtained from (2.A.17a) or (2.A.17b). For $\omega_q = 0$ these vectors are determined by (2.A.18a), (2.A.18b) or more simply by (2.A.17a), (2.A.17b), their relative sign being arbitrary.

The achievement of the method is obvious. The problem of diagonalising a $2N \times 2N$ matrix has been reduced to the eigenvalue problem of a $N \times N$ matrix [77, 245]. Even when the eigenvalues of $M = (A+B)(A-B) = (A-B)(A+B)^T$ cannot be found analytically, numerical studies can be made for large systems, for example, in one dimensional nearest-neighbour model with random transverse field (see Sect. 5.3). The price to pay is that one only obtains the one particle energies directly. Multiparticle states have to be built up by linear superposition of the one particle ones. Since A is symmetric and B is antisymmetric both $(A-B)(A+B)$ and $(A+B)(A-B)$ are symmetric and at least positive semi-definite. Thus all the λ_q's are real and all the Φ_q's and Ψ_q's can be chosen to be real as well as orthogonal. If Φ_q's are normalised, Ψ_q's are automatically normalised. To evaluate the constant in (2.A.11) we use the trace invariance of the Hamiltonian under the canonical transformation to the variables η. From (2.A.10)

$$\mathrm{Tr}H = 2^{N-1}\sum_i A_{ii}.$$

Again

$$\mathrm{Tr}H = 2^{N-1}\sum_q \omega_q + 2^N \times \text{const},$$

from (2.A.11). The constant is thus

$$\frac{1}{2}\left[\sum_i A_{ii} - \sum_q \omega_q\right]. \tag{2.A.19}$$

The complete diagonalised form of the Hamiltonian (2.A.10) is therefore written as

$$H = \sum_q \omega_q \eta_q^\dagger \eta_q + \frac{1}{2}\left(\sum_i A_{ii} - \sum_q \omega_q\right). \tag{2.A.20}$$

We illustrate the above diagonalisation procedure using the example of the one dimensional anisotropic XY chain in a transverse field described by the Hamiltonian (see Sect. 10.1.2)

$$H = -\frac{J}{2}\sum_i\left[(1+\gamma)S_i^x S_{i+1}^x + (1-\gamma)S_i^y S_{i+1}^y\right] - \Gamma\sum_i S_i^z \tag{2.A.21}$$

where γ is the measure of anisotropy. $\gamma = 0$ corresponds to the isotropic XY chain in a transverse field, $\gamma = 1$ corresponds to the transverse Ising chain. The Hamiltonian (2.A.21) is written in terms of Jordan-Wigner fermions as

$$H = -2\left(\sum_i\left(c_i^\dagger c_i - \frac{1}{2}\right) + \frac{1}{2}\lambda\sum_{i=1}^{N}\left(c_i^\dagger c_{i+1} + c_i c_{i+1}^\dagger + \gamma c_i^\dagger c_i^\dagger + \gamma c_i c_{i+1}\right)\right), \tag{2.A.22}$$

where $\overline{\lambda} = J/\Gamma$. Equation (2.A.22) can be put in the general form

$$H = \left(\sum_{ij} c_i^\dagger A_{ij} c_j + \frac{1}{2} \sum_{ij} (c_i^\dagger B_{ij} c_j^\dagger + h.c.) \right) + N, \qquad (2.A.23)$$

where

$$A_{ii} = -1, \quad A_{i\,i+1} = -\frac{1}{2}\gamma\overline{\lambda} = A_{i+1\,i}, \quad B_{i\,i+1} = -\frac{1}{2}\gamma\overline{\lambda}, \quad B_{i+1\,i} = \frac{1}{2}\gamma\overline{\lambda}. \qquad (2.A.24)$$

The eigenvalue problem (2.A.18a), (2.A.18b) is solved considering an ansatz wave function $\exp(i\boldsymbol{q} \cdot \boldsymbol{R}_j)$ and the excitation spectrum is given as

$$\omega_q = \sqrt{(\gamma\overline{\lambda})^2 \sin^2 q + (1 - \overline{\lambda}\cos q)^2}, \qquad (2.A.25)$$

so that the diagonalised form of the Hamiltonian (with $\sum_i A_{ii} = -N$) is given by

$$H = 2\sum_q \sqrt{(\gamma\overline{\lambda}\sin q)^2 + (1 - \overline{\lambda}\cos q)^2} \eta_q^\dagger \eta_q - \sum_q \omega_q \qquad (2.A.26)$$

which reduces to the transverse Ising Hamiltonian (2.2.18) for $\gamma = 1$.

2.A.3 Calculation of Correlation Functions

Longitudinal spin-spin correlation function is defined as

$$\begin{aligned} C_{ij}^x &= \langle \psi_0 | S_i^x S_j^x | \psi_0 \rangle \\ &= \langle \psi_0 | (S_i^+ + S_i^-)(S_j^+ + S_j^-) | \psi_0 \rangle. \end{aligned} \qquad (2.A.27)$$

In terms of Jordan-Wigner fermions (with $j > i$)

$$C_{ij}^x = \langle \psi_0 | (c_i^\dagger + c_i) \exp\left(-i\pi \sum_{i=0}^{j-1} c_l^\dagger c_l \right) (c_j^\dagger + c_j) | \psi_0 \rangle \qquad (2.A.28)$$

where the averages are calculated over the ground state. One can now verify using a representation in which $c_l^\dagger c_l$ is diagonal, that

$$\begin{aligned} \exp[-i\pi c_l^\dagger c_l] &= -(c_l^\dagger - c_l)(c_l^\dagger + c_l) \\ &= (c_l^\dagger + c_l)(c_l^\dagger - c_l). \end{aligned}$$

Defining $A_l = c_l^\dagger + c_l$ and $B_l = c_l^\dagger - c_l$ and noting that $A_l^2 = 1$, we have

$$C_{ij}^x = \langle \psi_0 | B_i (A_{i+1} B_{i+1} \cdots A_{j-1} B_{j-1}) A_j | \psi_0 \rangle. \qquad (2.A.29)$$

The complicated expression can be simplified using Wick's theorem, and following relations

$$\langle \psi_0 | A_i A_j | \psi_0 \rangle = \langle \psi_0 | (\delta_{ij} - c_j^\dagger c_i + c_i^\dagger c_j) | \psi_0 \rangle = \delta_{ij}, \qquad (2.A.30)$$

and

$$\langle\psi_0|B_i B_j|\psi_0\rangle = -\delta_{ij}. \tag{2.A.31}$$

Only nonzero contractions are $\langle A_j B_i\rangle$ and $\langle B_i A_j\rangle$, since $\langle A_i A_j\rangle$ and $\langle B_i B_j\rangle$ never occur. Defining $\langle B_i A_{i+r}\rangle = -\langle A_{i+r} B_i\rangle = G_{i\,i+r} = G_r = G_{-r}$, the correlation function is given by a determinant

$$\begin{aligned} C^x_{i\,i+r} &= \det\begin{pmatrix} G_{i\,i+1} & G_{i\,i+2} & \cdots & G_{i\,i+r} \\ G_{i+1\,i+1} & G_{i+1\,i+2} & \cdots & G_{i+1\,i+r} \\ \cdots & \cdots & \cdots & \cdots \\ G_{i+r-1\,i+1} & G_{i+r-1\,i+2} & \cdots & G_{i+r-1\,i+r} \end{pmatrix} \\ &= \det\begin{pmatrix} G_{-1} & G_{-2} & \cdots & G_{-r} \\ G_0 & G_{-1} & \cdots & G_{-r+1} \\ \cdots & \cdots & \cdots & \cdots \\ G_{r-2} & G_{r-3} & \cdots & G_{-1} \end{pmatrix} \end{aligned} \tag{2.A.32}$$

which is of size r. Similarly one can evaluate the transverse correlation function defined as

$$\begin{aligned} C^z_{i\,i+r} &= \langle\psi_0|S^z_i S^z_{i+r}|\psi_0\rangle - \left(m^z\right)^2 \\ &= \langle\psi_0|A_i B_i A_{i+r} B_{i+r}|\psi_0\rangle \\ &= (G_{i\,i} G_{i+r\,i+r} - G_{i\,i+r} G_{i+r\,i}) - \left(m^z\right)^2. \end{aligned} \tag{2.A.33}$$

One can now check that,

$$\begin{aligned} m^z &= \langle\psi_0|S^z_i|\psi_0\rangle \\ &= \langle\psi_0|2c^\dagger_i c_i - 1|\psi_0\rangle \\ &= 2\langle\psi_0|c^\dagger_i c_i|\psi_0\rangle - 1 = G_{ii} = G_0 \end{aligned} \tag{2.A.34}$$

so that

$$C^z_{i\,i+r} = -G_{i\,i+r} G_{i+r\,i}. \tag{2.A.35}$$

To evaluate $G_{i\,i+r}$ we consider the inverse Fourier-Bogoliubov transformation (see (2.2.7a), (2.2.7b) and (2.2.13)) given by

$$c^\dagger_i = \sqrt{\frac{1}{N}} \sum_q e^{iqR_i}\left(u_q \eta^\dagger_q + i v_q \eta_{-q}\right) \tag{2.A.36}$$

so that we get

$$c^\dagger_i + c_i = \sqrt{\frac{1}{N}} \sum_q \left(e^{iqR_i} u_q \eta^\dagger_q + e^{-iqR_i} u_q \eta_q - i e^{-iqR_i} v_q \eta^\dagger_{-q} + i e^{iqR_i} v_q \eta_{-q}\right) \tag{2.A.37a}$$

and

$$c_i^\dagger - c_i = \sqrt{\frac{1}{N}} \sum_q \left(e^{iqR_i} u_q \eta_q^\dagger - e^{-iqR_i} u_q \eta_q + i e^{-iqR_i} v_q \eta_{-q}^\dagger + i e^{iqR_i} v_q \eta_{-q} \right). \tag{2.A.37b}$$

Note that u_q and v_q are given by (2.2.17). In the ground state, $G_r = G_{i\,i+r} = \langle \psi_0 | B_i A_{i+r} | \psi_0 \rangle$. At a finite temperature T, $G_r = G_{i\,i+r} = \langle B_i A_{i+r} \rangle_\beta$, where $\langle B_i A_j \rangle_\beta$ denotes an average over the canonical ensemble at temperature $k_B T = 1/\beta$. Thus,

$$\begin{aligned} G_r(\beta) = \frac{1}{N} \sum_q \big[& e^{-iqr} \{ (u_q^2 + i u_q v_q) \langle \eta_q^\dagger \eta_q \rangle + (v_q^2 - i u_q v_q) \langle \eta_{-q} \eta_{-q}^\dagger \rangle \} \\ & - e^{iqr} \{ (u_q^2 - i u_q v_q) \langle \eta_q \eta_q^\dagger \rangle + (v_q^2 + i u_q v_q) \langle \eta_{-q}^\dagger \eta_{-q} \rangle \} \big]. \end{aligned} \tag{2.A.38}$$

The average fermion occupation at temperature T

$$\langle \eta_q^\dagger \eta_q \rangle_\beta = \langle \eta_{-q}^\dagger \eta_{-q} \rangle_\beta = \left(\exp(\beta \cdot 2\omega_q) + 1 \right)^{-1}$$

so that,

$$\begin{aligned} G_r(\beta) &= \frac{1}{N} \sum_q e^{iqr} \left(1 - 2u_q^2 + 2i u_q v_q \right) \tanh(\beta \omega_q) \\ &\stackrel{N\to\infty}{\to} \int_{-\pi}^{\pi} \frac{dq}{2\pi} e^{iqr} \left(\frac{1 + \bar{\lambda} e^{iq}}{1 + \bar{\lambda} e^{-iq}} \right)^{1/2} \tanh(\beta \omega_q). \end{aligned} \tag{2.A.39}$$

For ground state $\tanh(\beta\omega_q/2) = 1$, we have

$$G_r = \int_{-\pi}^{\pi} \frac{dq}{2\pi} e^{iqr} \left(\frac{1 + \bar{\lambda} e^{iq}}{1 + \bar{\lambda} e^{-iq}} \right)^{1/2}. \tag{2.A.40}$$

One can now evaluate the following values G_r for some special values of $\bar{\lambda}$, e.g.,

$$G_r = \begin{cases} \frac{2}{\pi} \frac{(-1)^r}{2r+1} & \text{for } \bar{\lambda} = 1 \\ \delta_{r,-1} & \text{for } \bar{\lambda} = \infty \\ \delta_{r,0} & \text{for } \bar{\lambda} = 0 \end{cases}$$

etc.

In order to evaluate the determinant (2.A.32), let us define

$$\begin{aligned} D_p = G_{i\,i-1+p} &= G_{p-1} \\ &= \int_{-\pi}^{\pi} \frac{dq}{2\pi} e^{iq(p-1)} \left(\frac{1 + \bar{\lambda} e^{iq}}{1 + \bar{\lambda} e^{-iq}} \right)^{1/2} \tanh(\beta \omega_q) \\ &= \int_{-\pi}^{\pi} \frac{dq}{2\pi} e^{iqp} \left(\frac{1 + \bar{\lambda}^{-1} e^{-iq}}{1 + \bar{\lambda}^{-1} e^{iq}} \right)^{1/2} \tanh(\beta \omega_q). \end{aligned} \tag{2.A.41}$$

In terms of this, (2.A.32) is rewritten as

$$C^x_{i\,i+r} = \det\begin{pmatrix} D_0 & D_{-1} & \cdots & D_{-r+1} \\ D_1 & D_0 & \cdots & D_{-r+2} \\ \cdots & \cdots & \cdots & \cdots \\ D_{r-1} & D_{r-2} & \cdots & D_0 \end{pmatrix}. \tag{2.A.42}$$

This Toeplitz determinant can be evaluated by using the Szegö's theorem [393] (see also Ref. [144]), which stated as follows. We refer to Ref. [268] for the proof.

Szegö's Theorem *Assume that* $\hat{D}(e^{iq})$ *and* $\ln\hat{D}(e^{iq})$ *are continuous cyclic functions of* q *for* $q \in [-\pi, \pi]$. *Let*

$$D_p = \int_{-\pi}^{\pi}\frac{dq}{2\pi}\hat{D}\big(e^{iq}\big)e^{-iqp}, \qquad d_p = \int_{-\pi}^{\pi}\frac{dq}{2\pi}\ln\hat{D}\big(e^{iq}\big)e^{-iqp}.$$

The Toeplitz determinant

$$\Delta_r = \det\begin{pmatrix} D_0 & D_{-1} & \cdots & D_{-r+1} \\ D_1 & D_0 & \cdots & D_{-r+2} \\ \cdots & \cdots & \cdots & \cdots \\ D_{r-1} & D_{r-2} & \cdots & D_0 \end{pmatrix}$$

is given by

$$\lim_{r\to\infty}\frac{\Delta_r}{e^{rd_0}} = \exp\Bigg[\sum_{p=1}^{\infty} p d_{-p} d_p\Bigg], \tag{2.A.43}$$

whenever the sum in (2.A.43) *converges.*

Applying this theorem to (2.A.41) and (2.A.42), one finds that the longitudinal correlation function behaves as $C^x_{i\,i+r} \approx Ae^{-r/\xi}$ for $r \gg 1$ and the correlation length ξ is given by

$$\begin{aligned} \frac{1}{\xi} = -d_0 &= -\int_{-\pi}^{\pi}\frac{dq}{2\pi}\frac{1}{2}\ln\left(\frac{1+\overline{\lambda}^{-1}e^{-iq}}{1+\overline{\lambda}^{-1}e^{iq}}\right) - \int_{-\pi}^{\pi}\frac{dq}{2\pi}\ln\tanh(\beta\omega_q) \\ &= \begin{cases} \int_{-\pi}^{\pi}\frac{dq}{2\pi}\ln\coth(\beta\omega_q) & \text{for } \overline{\lambda} > 1 \\ \int_{-\pi}^{\pi}\frac{dq}{2\pi}\ln\coth(\beta\omega_q) - \ln\overline{\lambda} & \text{for } \overline{\lambda} < 1 \end{cases}. \end{aligned} \tag{2.A.44}$$

(Note that, when $\overline{\lambda} < 1$, one needs additional algebras because $\ln\hat{D}(e^{iq})$ is not cyclic with respect to q. See Ref. [265].) Substituting (2.2.16) for ω_q and recovering the lattice constant a and scale of energy Γ, one can write (2.A.44) as

$$\frac{1}{\xi} = \begin{cases} \int_0^{\pi/a}\frac{dq}{\pi}\ln\coth\beta\Gamma\sqrt{1+2\overline{\lambda}\cos qa+\overline{\lambda}^2} & \text{for } \overline{\lambda} > 1 \\ \int_0^{\pi/a}\frac{dq}{\pi}\ln\coth\beta\Gamma\sqrt{1+2\overline{\lambda}\cos qa+\overline{\lambda}^2} - \frac{1}{a}\ln\overline{\lambda} & \text{for } \overline{\lambda} < 1 \end{cases}. \tag{2.A.45}$$

Let us now define $\Delta = |1 - \overline{\lambda}|/2\Gamma a^2$ and $c = 2\Gamma a$. By making $a \to 0$ with fixing β, Δ and c, and defining a new variable of integral $y = \beta c q$, we reach an expression for the continuous limit [338]

$$\frac{1}{\xi} = \frac{1}{\beta c} f\left(\beta \Delta c^2\right), \tag{2.A.46}$$

where

$$f(x) = \begin{cases} \int_0^\infty \frac{dy}{\pi} \ln \coth \frac{\sqrt{y^2+x^2}}{2} & \text{for } x > 0 \\ \int_0^\infty \frac{dy}{\pi} \ln \coth \frac{\sqrt{y^2+x^2}}{2} - x & \text{for } x < 0 \end{cases}. \tag{2.A.47}$$

One can easily show that the function $f(x)$ is not only continuous but also smooth at $x = 0$. This fact shows that the transverse Ising chain has no singularity at finite temperatures. In special cases, $f(x)$ takes following forms:

$$f(x) = \begin{cases} \sqrt{\frac{2x}{\pi}} e^{-x} & (x \to \infty) \\ \frac{\pi}{4} & (x = 0) \\ -x + \sqrt{2|x|\pi}\, e^{-|s|} & (x \to -\infty) \end{cases}. \tag{2.A.48}$$

Substituting this in (2.A.46) and recovering the original parameters, one obtains

$$\frac{1}{\xi} = \begin{cases} \frac{1}{a}\sqrt{\frac{T}{\pi}(\overline{\lambda} - 1)} e^{-2(\overline{\lambda}-1)/T} & (\overline{\lambda} > 1,\ T \to 0) \\ \frac{\pi}{4}\frac{T}{2a} & (\overline{\lambda} = 1) \\ \frac{1-\overline{\lambda}}{a} + \frac{1}{a}\sqrt{\frac{T}{\pi}(1 - \overline{\lambda})} e^{-2(1-\overline{\lambda})/T} & (\overline{\lambda} < 1,\ T \to 0) \end{cases}. \tag{2.A.49}$$

At the zero temperature, one finds $\xi = a/(1 - \overline{\lambda})$ for $\overline{\lambda} < 1$, so that the exponent ν turns out to be 1.

The result (2.A.49) implies that the low temperature phase of the transverse Ising chain is divided into three phases by crossover lines $T \approx |\overline{\lambda} - 1|$. We show a schematic phase diagram in Fig. 2.10.

Chapter 3
Transverse Ising System in Higher Dimensions (Pure Systems)

3.1 Mapping to the Effective Classical Hamiltonian: Suzuki-Trotter Formalism

In this section, we shall discuss the Suzuki-Trotter formalism, to derive the classical analogue of a quantum mechanical model and apply it to the case of pure transverse Ising model. Elliott et al. [126] numerically established, from series studies, the equivalence of the ground state singularities of a d-dimensional transverse Ising model to those of the $(d+1)$-dimensional classical Ising model. Later, Suzuki [385, 386] (see also [387]), using a generalised version of Trotter formula [403], analytically established that the ground state of a d-dimensional quantum spin system is equivalent to a certain $(d+1)$-dimensional classical Ising model with many-body interactions: the exponents associated with the ground state phase transition of the quantum system are the same as the exponents of thermal phase transition in the equivalent $(d+1)$-dimensional classical model, and for $d > 3$ the exponents of the quantum transition assume the mean field values of the exponents of the classical model. The interaction in the classical system is finite-ranged if the original quantum system has finite-range interaction.

The above equivalence can be analytically proved, as mentioned earlier, using the generalised form of Trotter formula [403], which can be written as

$$\exp[\hat{A}_1 + \hat{A}_2] = \lim_{M\to\infty} \left(\exp(\hat{A}_1/M)\exp(\hat{A}_2/M)\right)^M \tag{3.1.1}$$

where $\hat{A}_1$ and $\hat{A}_2$ are quantum mechanical bounded operators not commuting with each other.

One can now employ the above mentioned formalism to the example of one dimensional Ising model in a transverse field. Let us start from the nearest-neighbour interacting Hamiltonian (with periodic boundary condition)

$$H = -J\sum_{j=1}^{N} S_j^z S_{j+1}^z - \Gamma\sum_{j=1}^{N} S_j^x; \quad S_{N+1}^z = S_1^z. \tag{3.1.2}$$

S. Suzuki et al., *Quantum Ising Phases and Transitions in Transverse Ising Models*, Lecture Notes in Physics 862, DOI 10.1007/978-3-642-33039-1_3,

The partition function of the above quantum Hamiltonian can be written as

$$Z = \mathrm{Tr}\,\mathrm{e}^{-\beta H} = \mathrm{Tr}\exp\left[\sum_{j=1}^{N}\left(K S_j^z S_{j+1}^z + \gamma S_j^x\right)\right]; \quad \beta = 1/k_B T, \tag{3.1.3}$$

where $K = J\beta$, $\gamma = \Gamma\beta$. Using Trotter representation (3.1.1), and using the complete set of eigenvectors of the operator, one can transform the partition function of the quantum Hamiltonian as

$$Z = \lim_{M\to\infty} A^{MN}\,\mathrm{Tr}\exp\left[\sum_{j=1}^{N}\sum_{k=1}^{M}\left(\frac{K}{M} S_{j,k} S_{j+1,k} + K_M S_{j,k} S_{j,k+1}\right)\right], \tag{3.1.4}$$

where $A = [(1/2)\sinh(2\gamma/M)]^{1/2}$ and $K_M = (1/2)\ln\coth(\gamma/M)$. In deriving Eq. (3.1.4), one uses the relation

$$\langle S|\mathrm{e}^{\gamma S^x}|S'\rangle = \left[(1/2)\sinh 2\gamma\right]^{1/2}\exp\left[(1/2)(\ln\coth\gamma) S S'\right], \tag{3.1.5}$$

where $|S\rangle$, $|S'\rangle$ are eigenstates of operator S^z (see Sect. 3.A.1). Equation (3.1.4) clearly indicates that by using the Trotter representation of exponential operators, one can transform the spin-1/2 transverse Ising chain to a classical Ising model on a $M \times N$ square lattice, with $M \to \infty$, having anisotropic coupling in the space and Trotter direction. In the limit $M \to \infty$, the strength of interaction in the Trotter direction, K_M, diverges logarithmically whereas the strength in the spatial direction, (K/M), vanishes except when $\beta = \infty$ $(T = 0)$. Hence, the correspondence between the quantum and the equivalent classical model strictly holds in the $T \to 0$ limit. It may be noted that the above example of equivalence of the quantum Hamiltonian (3.1.3) to an effective (classical) Hamiltonian

$$H_{\mathrm{eff}} = -\sum_{j=1}^{N}\sum_{k=1}^{M}\left[\frac{K}{M} S_{j,k} S_{j+1,k} + K_M S_{j,k} S_{j,k+1}\right], \tag{3.1.6}$$

in the sense that $Z = \mathrm{Tr}\exp(-\beta H_{\mathrm{eff}})$ is identical to that of H, is not limited to one dimension. It is valid in general and gives for

$$H = -\sum_{ij} J_{ij} S_i^z S_j^z - \Gamma \sum_i S_i^x \tag{3.1.7}$$

the effective Hamiltonian

$$H_{\mathrm{eff}}(M) = -\sum_{k=1}^{M}\left[\frac{K}{M}\sum_{ij} S_{i,k} S_{j,k} + K_M \sum_i S_{i,k} S_{i,k+1}\right], \tag{3.1.8}$$

where $K_M = (1/2)\ln[\coth(\Gamma/M k_B T)]$. One should also mention here, that starting from the classical model one can (reversely) arrive at the equivalent (low-dimensional) quantum model as well [349] (see Sect. 3.A.2).

3.2 The Quantum Monte Carlo Method

In the previous section, we have seen that the Trotter expansion formula (3.1.1) enables to transform a d-dimensional quantum spin system into the effective $(d+1)$-dimensional classical system. In particular, the finite temperature thermodynamics of a d-dimensional transverse Ising Hamiltonian (3.1.7) can be obtained using the effective classical Ising Hamiltonian given by Eq. (3.1.8). Using now the standard Monte Carlo techniques [241] for the classical spin system, one can obtain physical quantities and study the transitions in such systems [290, 421].

As in the usual Monte Carlo method, one scans through the lattice sequentially and at each site, the possible change in energy ΔE (coming from the neighbouring spin configuration) for the flip of the spin state at the central site is calculated and flipped with the normalised probability (say, $\mathrm{e}^{-\Delta E/k_B T}/(1+\mathrm{e}^{-\Delta E/k_B T})$ which ensures Boltzmann distribution in equilibrium. With the Suzuki-Trotter Hamiltonian (3.1.8), the interaction in d-dimension (spatial) remains the same (K, normalised by the Trotter index M) as in the original quantum system, while in the Trotter direction it becomes non-trivially different (K_M depends on M and temperature) and anisotropic. The singularities of this interaction (K_M) cause practical problem for simulation, especially at very low temperatures. In fact, for the same reason, although one should have taken the limit of Trotter size $M \to \infty$, one has to consider the above mentioned problems of anisotropy and optimal choice of M for the best result. In exactly solved cases (e.g., in one dimension) one can find some optimal choice for the Trotter size M (dependent on temperature), by comparing the result and looking for the best numerical agreement [421]. In other cases, the optimality can be chosen by comparing the result in some limiting cases (e.g., the classical limit) and the anisotropy is kept at minimum. Practically, these considerations give $M \sim O(10)$ [290, 421].

The quantum Monte Carlo results for the susceptibility, ground state energy, transverse susceptibility and the transverse magnetisation of the transverse Ising chain (with Trotter dimension $M = 4, 8, 12$ at $\Gamma/J = 0.5$) are shown in Figs. 3.1 and 3.2 [290]. Here the results are also compared with the exact analytical results. For $M \simeq 12$, the Monte Carlo results are in good agreement with the analytical results in the high temperature region. The results also indicate a (non-vanishing) Γ_c for order-disorder transition at $T = 0$. Similar results for two and three dimensional transverse Ising model (on hypercubic lattices) are shown in the Figs. 3.3, 3.4, where $M = 12$ is used [290]. The variation of specific heat (C) and the longitudinal susceptibility (χ) for the transverse Ising model on a square lattice at $\Gamma/J = 1.0$ is shown in the Fig. 3.2. Here, the lattice size is 24×24 and $M = 12$. From the position of the susceptibility peak, one estimates [290] $k_B T_c(\Gamma = J) \simeq 2.2J$ (compared to $k_B T_c(\Gamma = 0) = 2.269185J$). The average magnetisation $\langle S^z \rangle$ and $\langle S^x \rangle$ for the same lattices, plotted as a function of transverse field, are shown the Fig. 3.4. The expected and compiled phase diagram of the of transverse Ising Hamiltonian on hypercubic lattices (in $d = 2, 3$), obtained from the Monte Carlo studies [290], are shown in the Fig. 3.5. These results agree well with the results obtained from series study [126], except in the low temperature region.

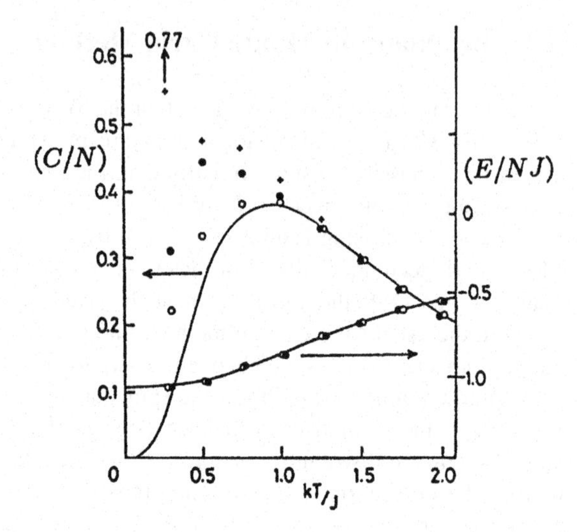

Fig. 3.1 The energy E and specific heat $C = (\partial E/\partial T)$ for transverse Ising chain. The *cross*, *solid circles* and *open circles* denote MC data for $M = 4, 8, 12$ respectively, at $(\Gamma/J = 0.5)$. *Solid lines* denote the exact results [290]

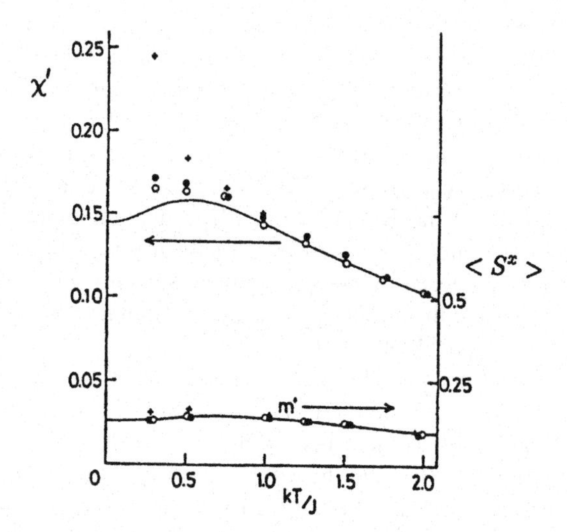

Fig. 3.2 Transverse magnetisation $\langle S^x \rangle$ and the transverse susceptibility $\chi'(= \partial\langle S^x\rangle/\partial\Gamma)$ for the transverse Ising chain [290]

3.2.1 Infinite M Method

The quantum Monte Carlo method has undergone considerable developments so far [222]. In particular, the infinite M method with the aid of cluster updates [292] is worth noting its significance, since it enables to make the Trotter number M infinity (strictly speaking, the largest number allowed by the computer language) with temperature T and transverse field Γ remaining finite. To introduce the infinite M method, we start from the cluster update method proposed by Swendsen and Wang [391].

Let us consider a pure one-dimensional Ising ferromagnet with coupling constant J. Suppose that we have a spin configuration $(S_1, S_2, \ldots)$ denoted symboli-

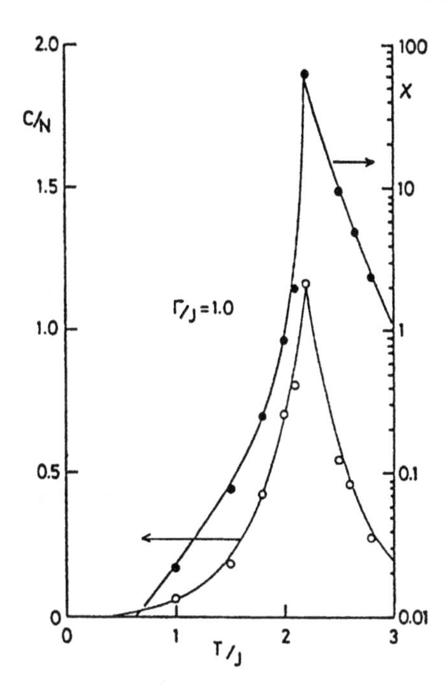

Fig. 3.3 MC data for specific heat C (*open circles*) and longitudinal susceptibility χ (*solid circles*) for 2-d ferromagnetic transverse Ising model. *Solid lines* are guide to the eye [290]

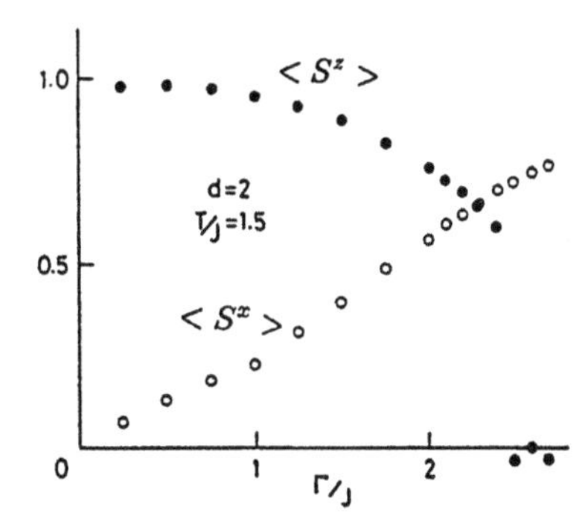

Fig. 3.4 MC data for $\langle S^z \rangle$ and $\langle S^x \rangle$ of 2-d ferromagnetic transverse Ising models

cally by C. For each coupled adjoining spin pair (S_i, S_{i+1}), we define the Swendsen-Wang (SW) bond with the probability

$$p = \begin{cases} p_0 = 1 - e^{-2\beta J} & \text{when } S_i = S_{i+1}, \\ 0 & \text{when } S_i \neq S_{i+1}, \end{cases} \tag{3.2.1}$$

where $\beta = 1/T$. We cluster the spins connected by SW bonds. We then flip all spins included in each cluster with the probability $1/2$. This procedure yields a new configuration C'. It is obvious that this Markov chain satisfies the ergodicity condition since any spin configuration can be reached by a finite series of updates from any state. The satisfaction of the detailed balance condition can be shown as follows. Define $P(C) = \exp(K \sum_i S_i S_{i+1})/Z$ with $Z = \mathrm{Tr} \exp(K \sum_i S_i S_{i+1})$. If a spin configuration C is given, then the number of ferromagnetic bond denoted by

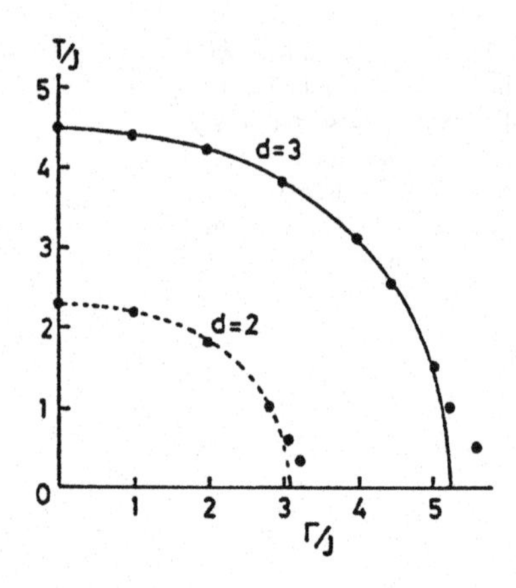

Fig. 3.5 The phase boundaries for 2 and 3 dimensional transverse Ising model [290]. The *solid* and *broken lines* represent series results (see next section)

n_f is also given. We here consider a configuration of SW bonds denoted by B. Note that, once a configuration B is given, the number of SW bonds b is determined. The probability that a configuration of SW bonds B is realised from given spin configuration C is written as

$$P(B|C) = p_0^b(1-p_0)^{n_\mathrm{f}-b}. \tag{3.2.2}$$

Now we consider another spin configuration C' such that it can generate the SW-bond configuration B, namely,

$$P\big(B|C'\big) = p_0^b(1-p_0)^{n'_\mathrm{f}-b}, \tag{3.2.3}$$

where n'_f is the number of ferromagnetic bonds in C'. The probabilities to have C' and C from B are the same:

$$P\big(C'|B\big) = P(C|B). \tag{3.2.4}$$

Using (3.2.2) and (3.2.3), one obtains

$$\frac{P(B|C)P(C)}{P(B|C')P(C')} = \frac{(1-p_0)^{n_\mathrm{f}}P(C)}{(1-p_0)^{n'_\mathrm{f}}P(C')}.$$

Since $P(C) = e^{n_\mathrm{f}\beta J-(N_b-n_\mathrm{f})\beta J} = e^{(2n_\mathrm{f}-N_b)\beta J}$, it turns out that

$$\frac{P(B|C)P(C)}{P(B|C')P(C')} = \frac{e^{-2n_\mathrm{f}\beta J}e^{2n_\mathrm{f}\beta J}}{e^{-2n'_\mathrm{f}\beta J}e^{2n'_\mathrm{f}\beta J}} = 1,$$

namely,

$$P\big(B|C'\big)P\big(C'\big) = P(B|C)P(C). \tag{3.2.5}$$

Let $W(C'|C) = \sum_B P(C'|B)P(B|C)$ be the transition probability from C to C'. Multiplying (3.2.5) by (3.2.4) and summing the resultant equality with respect to B, one obtains the detailed balance condition $W(C|C')P(C') = W(C'|C)P(C)$.

Now we return to the $(d+1)$-dimensional Ising system. We apply the above clustering method to the chain along the Trotter direction. Let us focus on a chain of length M along the Trotter direction. The state of M spins can be identified, if one knows the spin state at an edge site and the position of the domain walls (antiferromagnetic bonds). As we shall show below, the average number of domain walls remains to be a finite number given by $\beta\Gamma$ when $M \to \infty$. Therefore one can represent the spin state of the infinite ($M \to \infty$) chain in terms of one sign and $\beta\Gamma$ rational numbers in between 0 and 1, corresponding to the state of an edge spin and the position of domain walls respectively. This is an essence of the infinite M method. In practice, it is convenient to choose $M = 2^{32}$ for the 4 byte representation of the integer and use integers ranging from 0 to $2^{32} - 1$ to indicate the position of domain walls.

The updating procedure in the infinite M method is described as follows. We suppose that a domain wall configuration and the spin state at an edge is given. A trial configuration of domain walls is generated by creating domain walls with probability $1 - p_0 = e^{-2\beta J}$ at each position i ($0 \le i \le M - 1$) of the bonds. Substituting $\beta J = K_M = (1/2)\ln[\coth(\beta\Gamma/M)]$, this probability reads as $1 - p_0 = \tanh(\beta\Gamma/M) \approx \beta\Gamma/M$. Since $\beta\Gamma \ll M$, the creation of domain walls is the Poisson process. The average number of created domain walls is given by $(1 - p_0)M \to \beta\Gamma$ ($M \to \infty$). Hence one may generate a trial configuration by a Poisson process with mean $\beta\Gamma$. If the position of a generated domain wall happens to coincides with the position of that of the original configuration, we remove the old one at this position. Having obtained a trial configuration, we flip the sign of each cluster between domain walls according to the molecular field from the coupled chains using the Metropolis method. After updating the cluster states, the domain walls between the clusters with the same sign are removed. Thus the update of a chain along the Trotter direction at a real-space site is completed.

The infinite M method presented here has two advantages. At first, the error of Suzuki-Trotter decomposition with finite M is removed. Since we are making $M \gg 1$ with $\beta\Gamma$ being finite, reliable results at finite T should be obtained. The second is on the singularity of K_M with $M \to \infty$. The cluster update method enables to avoid the unphysical slowing down [421] coming from the logarithmic divergence of K_M. We note that the physical dynamics intrinsic in the original system is preserved [292] because the cluster update method is not applied in the real space. Therefore the intrinsic dynamics of a system can be simulated by the present method.

So far, the infinite M quantum Monte Carlo method has been applied to the transverse Ising model in two dimension [292], frustrated transverse Ising models in checkerboard lattice [197, 198], the transverse Ising spin glass in two dimension [283], and so on. The range of application of the present method covers any kind of transverse Ising systems. In addition, the sign problem of the quantum Monte Carlo method does not appear as far as transverse Ising systems are concerned. The present method should be useful in the study of a variety of transverse Ising systems.

3.3 Discretised Path Integral Technique for a Transverse Ising System

In the path integral approach [358, 384], one can study the transverse Ising system in any dimension with arbitrary number of nearest neighbours. The effective Hamiltonian (3.1.8), obtained after the Suzuki-Trotter transformation, is written as

$$\beta H_{\text{eff}} = \beta H_0 + \beta V, \tag{3.3.1}$$

where

$$-\beta H_0 = \sum_{k=1}^{M} \sum_{i=1}^{N} K_M S_{i,k} S_{i,k+1}, \tag{3.3.2}$$

and

$$-\beta V = \sum_{k=1}^{M} \sum_{i,j=1}^{N} \frac{K}{M} S_{i,k} S_{j,k}, \tag{3.3.3}$$

with $K_M = (1/2)\ln[\coth(\beta\Gamma/M)]$. Here k is the Trotter index and M is the Trotter dimension. We have neglected the constant c, where

$$c = \left[(1/2)\sinh(2\beta\Gamma/M)\right]^{NM/2}. \tag{3.3.4}$$

Treating the spins as M-component vector spins with components

$$S_j^k = (\pm 1, \pm 1, \ldots, \pm 1) \tag{3.3.5}$$

where $k = 1, 2, \ldots, M$, one can write the parts of the effective Hamiltonian (3.3.1) as

$$-\beta H_0 = \sum_i \mathbf{S}_i \cdot a \cdot \mathbf{S}_i \tag{3.3.6}$$

and

$$-\beta V = \frac{K}{M} \sum_{\langle ij \rangle} \mathbf{S}_i \cdot \mathbf{S}_j \tag{3.3.7}$$

where the matrix $a_{k,k'} = (1/2)\ln[\coth(\beta\Gamma/M)]\delta_{k,k'}$.

Now the full Hamiltonian can be treated perturbatively such that the free energy $F(= \ln Q;\ Q = \mathrm{Tr}\exp(-\beta H_{\text{eff}}))$ is given by

$$-\beta F = -\beta F_o + \sum_n \frac{1}{n!}(-\beta)^n C_n(V), \tag{3.3.8}$$

with F_o the free energy corresponding to the unperturbed Hamiltonian H_o such that

$$-F_o = \ln Q_o \tag{3.3.9}$$

with

$$Q_o = \sum \exp(-\beta H_o), \tag{3.3.10}$$

Table 3.1 Zero-temperature critical transverse field $\lambda_c (= \Gamma_c/J)$

Lattice[a]	L	Sq	Sc	Fcc
δ_0	2	4	6	12
$(\lambda_c)_{\text{mf}}$	2.000	4.000	6.000	12.00
$(\lambda_c)_{\text{pi}}$	–	3.225[b]	5.291[b]	11.34[b]
$(\lambda_c)_{\text{series}}$	1.00[c]	3.08[d]	5.08 ± 0.04[d]	10.66 ± 0.6[d]

[a]L: linear; Sq: square; Sc: Simple cubic; Fcc: Face centred cubic

[b]Path integral technique [384]

[c]Exact result; see Sect. 2.2

[d]High temperature series results [125, 301]

and the cumulants are given by

$$C_1 = \langle V \rangle_o; \qquad C_2 = \langle V \rangle_o^2, \tag{3.3.11}$$

etc. The above expression can be regarded as an expansion in successively higher order of fluctuations. With classical systems, the first order term gives the mean field estimate ($\Gamma/J\delta_0 = 1$ where δ_0 is the coordination number) and higher orders constitute fluctuation corrections. The critical field where the average magnetisation vanishes, is obtained by performing calculations up to second order (following Kirkwood's prescription of classical spins) and is given by [384]

$$\Gamma_c/J = \left[\delta_0 + \left\{\delta_0(\delta_0 - 5/2)\right\}^{1/2}\right]/2. \tag{3.3.12}$$

Obviously, the mean field results are obtained in the infinite dimension limit ($\delta_0 \to \infty$) of the above equation. The path integral method does not give any result for the one-dimensional Ising system (where the exact result $\Gamma/J = 1$ is available) as the right hand side of (3.3.12) becomes imaginary (here $\delta_0 = 2$). Thus, results are obtained only above a lower critical dimension $d = 1$. However, a comparison of the results for the zero temperature critical transverse fields obtained from different methods (mean field, series expansion and the present method) in different dimensions shows that it gives better results than the mean field estimates for $d \geq 1$ (see Table 3.1). Also, like the mean field theory, estimates from the path integral method improve at higher dimensions (see Fig. 3.6).

3.4 Infinite-Range Models

In this subsection, we shall discuss infinite-range models, where each spin interacts with all other spins with the same strength. For such systems the mean field theory gives exact results. We introduce the Husimi-Temperley-Curie-Weiss model and the p-body model in a transverse field, and study thermal and quantum phase transitions in their models.

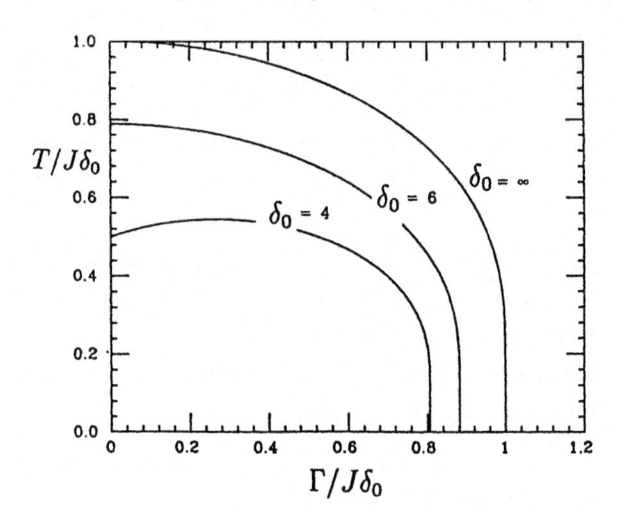

Fig. 3.6 The phase diagram of transverse Ising model as obtained from the path integral method for different δ_0 (coordination number) [384]; $\delta_0 = \infty$ indicates the mean field result (see Sect. 3.4.1)

3.4.1 Husimi-Temperley-Curie-Weiss Model in a Transverse Field

The Husimi-Temperley-Curie-Weiss model in a transverse field is written as [72, 238, 384]

$$\begin{aligned} H &= -\frac{J}{2N}\sum_{i,j} S_i^z S_j^z - \Gamma \sum_{i=1}^{N} S_i^x - h\sum_{i=1}^{N} S_i^z \\ &= -\frac{J}{2N}\Big(\sum_i S_i^z\Big)^2 - \Gamma \sum_{i=1}^{N} S_i^x - h\sum_{i=1}^{N} S_i^z, \end{aligned} \tag{3.4.1}$$

where, in the Ising model part, each spin interacts with all others including itself [186, 398]. Note that the coupling constant includes the factor $1/N$ in order to ensure the thermodynamic limit, where N is the number of spins. In the absence of the transverse field, the Weiss' mean-field treatment provides exact results, so this model is also called the Curie-Weiss model. Now, in the presence of the transverse field, the Suzuki-Trotter transformation leads to the following partition function:

$$\begin{aligned} Z = \lim_{M\to\infty} A^{MN} \operatorname{Tr} \exp\Bigg[\sum_{k=1}^{M}\Bigg\{\frac{\beta J}{2NM}\Big(\sum_i S_{i,k}^z\Big)^2 \\ + \frac{\beta h}{M}\sum_{i=1}^{N} S_{i,k}^z + K_M \sum_{i=1}^{N} S_{i,k}S_{i,k+1}\Bigg\}\Bigg], \end{aligned} \tag{3.4.2}$$

where $A = [(1/2)\sinh(2\beta\Gamma/M)]^{1/2}$, $K_M = (1/2)\ln\coth(\beta\Gamma/M)$, and $S_{i,k}$ is an Ising spin with the real-space index i and the Trotter index k. We assume the periodic boundary condition for k, namely $S_{i,N+1} = S_{i,1}$. We apply the Hubbard-Stratonovich transformation for this partition function using an identity:

$$1 = \sqrt{\frac{N\beta J}{2\pi M}} \int_{-\infty}^{\infty} dm \exp\left[-\frac{N\beta J}{2M} \left(m - \frac{1}{N} \sum_{i=1}^{N} S_{i,k} \right)^2 \right], \tag{3.4.3}$$

to yield

$$Z = \lim_{M\to\infty} A^{MN} \left(\frac{N\beta J}{2\pi M} \right)^{M/2} \int_{-\infty}^{\infty} \left[\prod_{k=1}^{M} dm_k \right] \exp\left(-\frac{N\beta J}{2M} \sum_{k=1}^{N} m_k^2 \right)$$
$$\times \operatorname{Tr} \exp\left[\sum_{i=1}^{N} \left(\frac{\beta J}{M} \sum_{k=1}^{M} m_k S_{i,k} + \frac{\beta h}{M} \sum_{k=1}^{M} S_{i,k} + K_M \sum_{k=1}^{M} S_{i,k} S_{i,k+1} \right) \right]. \tag{3.4.4}$$

Since each real-space site is decoupled from others, the traces with respect to $S_{i,k}$ and $S_{j,k}$ can be performed independently. Hence one can express the trace in (3.4.4) as the power N of a trace with respect to Ising-spin variables S_k ($k = 1, 2, \dots, M$)

$$Z = \lim_{M\to\infty} A^{MN} \left(\frac{N\beta J}{2\pi M} \right)^{M/2} \int_{-\infty}^{\infty} \left[\prod_{k=1}^{M} dm_k \right] \exp\left(-\frac{N\beta J}{2M} \sum_{k=1}^{N} m_k^2 \right)$$
$$\times \left(\operatorname{Tr} \exp\left[\frac{\beta J}{M} \sum_{k=1}^{M} m_k S_k + \frac{\beta h}{M} \sum_{k=1}^{M} S_k + K_M \sum_{k=1}^{M} S_k S_{k+1} \right] \right)^N$$
$$= \lim_{M\to\infty} A^{MN} \left(\frac{N\beta J}{2\pi M} \right)^{M/2} \int_{-\infty}^{\infty} \left[\prod_{k=1}^{M} dm_k \right] \exp\left(-\frac{N\beta J}{2M} \sum_{k=1}^{N} m_k^2 \right.$$
$$\left. + N \ln \operatorname{Tr} \exp\left[\frac{\beta J}{M} \sum_{k=1}^{M} m_k S_k + \frac{\beta h}{M} \sum_{k=1}^{M} S_k + K_M \sum_{k=1}^{M} S_k S_{k+1} \right] \right). \tag{3.4.5}$$

We evaluate the integrals of m_k by the saddle-points. The saddle-point condition for m_k is given by

$$0 = \frac{\partial}{\partial m_k} \left(-\frac{\beta J}{2M} \sum_{k=1}^{N} m_k^2 \right.$$
$$\left. \left. + \ln \operatorname{Tr} \exp\left[\frac{\beta J}{M} \sum_{k=1}^{M} m_k S_k + \frac{\beta h}{M} \sum_{k=1}^{M} S_k + K_M \sum_{k=1}^{M} S_k S_{k+1} \right] \right) \right|_{m_k = m_k^*}, \tag{3.4.6}$$

where m_k^* denotes the saddle-point solution. This equation is arranged into

$$m_k^* = \frac{\operatorname{Tr}[S_k \exp\{\frac{\beta J}{M} \sum_{k=1}^{M} m_k^* S_k + \frac{\beta h}{M} \sum_{k=1}^{M} S_k + K_M \sum_{k=1}^{M} S_k S_{k+1}\}]}{\operatorname{Tr}[\exp\{\frac{\beta J}{M} \sum_{k=1}^{M} m_k^* S_k + \frac{\beta h}{M} \sum_{k=1}^{M} S_k + K_M \sum_{k=1}^{M} S_k S_{k+1}\}]}, \tag{3.4.7}$$

by which it turns out that m_k^* is the longitudinal magnetisation. Since the present effective classical model is translationally invariant along the Trotter direction, the longitudinal magnetisation should be independent of the Trotter index k. One may thus drop the index k from m_k^* such as $m_k^* = m^*$. In terms of this saddle-point solution, the free energy per spin is obtained as

$$f = -\frac{1}{\beta N} \ln Z = -\frac{1}{\beta} \lim_{M\to\infty} \left[M \ln A - \frac{\beta J}{2} m^{*2} \right.$$
$$\left. + \ln \operatorname{Tr} \exp\left\{ \frac{\beta(Jm+h)}{M} \sum_{k=1}^{M} S_k + K_M \sum_{k=1}^{M} S_k S_{k+1} \right\} \right], \quad (3.4.8)$$

where note that we took the limit $N \to \infty$ before $M \to \infty$ and make $M/N \to 0$. The trance is easily computed using a transfer matrix:

$$\hat{T} = \begin{bmatrix} e^{L+K_M} & e^{-K_M} \\ e^{-K_M} & e^{-L+K_M} \end{bmatrix}, \quad (3.4.9)$$

as

$$\operatorname{Tr} \exp\left\{ L \sum_{k=1}^{M} S_k + K_M \sum_{k=1}^{M} S_k S_{k+1} \right\} = \operatorname{Tr}(\hat{T}^M), \quad (3.4.10)$$

where $L = \beta(Jm^* + h)/M$. The eigenvalues of $\hat{T}$ are given by

$$\lambda_\pm = e^{K_M} \cosh L \pm \sqrt{e^{2K_M} \sinh^2 L + e^{-2K_M}}, \quad (3.4.11)$$

which is followed by

$$A\lambda_\pm = \cosh\left(\frac{\beta\Gamma}{M}\right) \cosh L \pm \sqrt{\cosh^2\left(\frac{\beta\Gamma}{M}\right) \cosh^2 L - 1}$$
$$\stackrel{M\to\infty}{\longrightarrow} 1 \pm \frac{\beta}{M} \sqrt{(Jm^* + h)^2 + \Gamma^2} + O\left(\frac{1}{M^2}\right), \quad (3.4.12)$$

and

$$A^M \operatorname{Tr}(\hat{T}^M) = (A\lambda_+)^M + (A\lambda_-)^M \stackrel{M\to\infty}{\longrightarrow} e^{\beta\sqrt{(Jm^*+h)^+\Gamma^2}} + e^{-\beta\sqrt{(Jm^*+h)^+\Gamma^2}}$$
$$= 2\cosh\left(\beta\sqrt{(Jm^* + h)^2 + \Gamma^2}\right). \quad (3.4.13)$$

Thus the free energy per spin results in

$$f = \frac{Jm^{*2}}{2} - \frac{1}{\beta} \ln\left\{ 2\cosh\left(\beta\sqrt{(Jm^* + h)^2 + \Gamma^2}\right) \right\}. \quad (3.4.14)$$

The longitudinal and transverse magnetisations are obtained by differentiating f by h and Γ, respectively, as

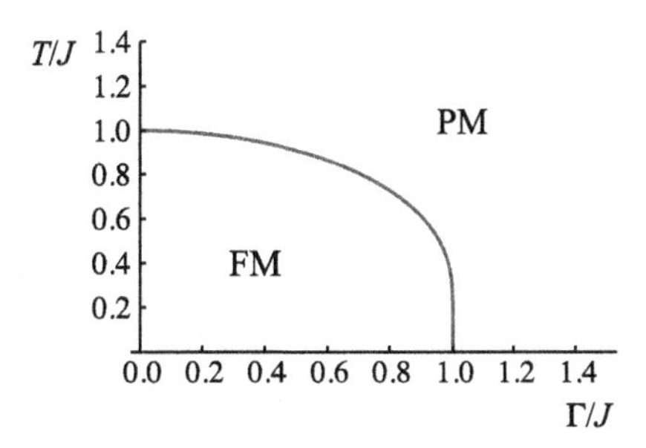

Fig. 3.7 Phase diagram of the Husimi-Temperley-Curie-Weiss model in the transverse field with $h \to +0$. The paramagnetic-ferromagnetic phase boundary is given by $T_c = \Gamma_c / \tanh^{-1}(\Gamma_c/J)$

$$m_z = -\frac{\partial f}{\partial h} = \frac{Jm^* + h}{\sqrt{(Jm^* + h)^2 + \Gamma^2}} \tanh\Big(\beta\sqrt{(Jm^* + h)^2 + \Gamma^2}\Big), \tag{3.4.15}$$

$$m_x = -\frac{\partial f}{\partial \Gamma} = \frac{\Gamma}{\sqrt{(Jm^* + h)^2 + \Gamma^2}} \tanh\Big(\beta\sqrt{(Jm^* + h)^2 + \Gamma^2}\Big). \tag{3.4.16}$$

As mentioned above, the solution of the saddle-point equation is identical to the longitudinal magnetisation. Hence it is obtained by solving (3.4.15) with making $m_z = m^*$.

Having obtained state equations (3.4.15), we move on to derive the spontaneous magnetisation. At first, when the transverse field is absent, (3.4.15) reduces to

$$m_z = \tanh\big(\beta(Jm_z + h)\big). \tag{3.4.17}$$

With $h \to +0$, this equation has a non-zero solution for $\beta > \beta_c = 1/J$. The transition temperature is found to be $T_c = 1/\beta_c = J$. When the transverse field Γ is finite, (3.4.15) with $h \to +0$ is written as

$$m_z = \frac{Jm_z}{\sqrt{(Jm_z)^2 + \Gamma^2}} \tanh\Big(\beta\sqrt{(Jm_z)^2 + \Gamma^2}\Big). \tag{3.4.18}$$

Therefore the ferromagnetic solution ($m_z \neq 0$) satisfies

$$\sqrt{m_z^2 + (\Gamma/J)^2} = \tanh\Big(\beta J\sqrt{m_z^2 + (\Gamma/J)^2}\Big). \tag{3.4.19}$$

The ferromagnetic-paramagnetic phase boundary is determined by making $m_z = 0$ in this equation:

$$\Gamma_c/J = \tanh(\beta_c \Gamma_c), \tag{3.4.20}$$

which yields

$$T_c = 1/\beta_c = \Gamma_c / \tanh^{-1}(\Gamma_c/J). \tag{3.4.21}$$

For $T < T_c$ and $\Gamma < \Gamma_c$ the paramagnetic solution gives a local maximum of f and the ferromagnetic solution is selected as a stable solution. We show the phase diagram in Fig. 3.7. $\Gamma_c = J$ and $T_c = 0$ provides a quantum phase transition point. The spontaneous magnetisation at zero temperature is obtained from (3.4.19) by making $\beta = \infty$ to yield $m_z = \sqrt{1 - (\Gamma/J)^2}$. This implies a continuous quantum phase transition at $\Gamma = J$.

3.4.2 Fully Connected p-Body Model in a Transverse Field

We consider the Hamiltonian [205]

$$H = -\frac{1}{N^{p-1}} \sum_{i_1,\dots,i_p=1}^{N} S^z_{i_1} \cdots S^z_{i_p} - \Gamma \sum_i S^z_i, \tag{3.4.22}$$

where N is the number of spins in the whole system and p denotes the number of coupled spins. Such models with the p-spin interaction are originally discussed in the context of the spin-glass theory [103, 275]. Let us write the longitudinal and transverse magnetisations as

$$m_z(S^z) = \frac{1}{N} \sum_i S^z_i, \qquad m_x(S_x) = \frac{1}{N} \sum_i S^x_i. \tag{3.4.23}$$

The Hamiltonian (3.4.22) is written in terms of magnetisations as

$$H = -N\{(m_z(S^z))^p + \Gamma m_x(S^x)\}. \tag{3.4.24}$$

If p is an even number, then the ground state of the system with $\Gamma = 0$ is the state where all spins are completely aligned up or down. On the other hand, if p is an odd number and $\Gamma = 0$, one has the unique ground state where all spins are up. When p is odd, the classical Hamiltonian with $\Gamma = 0$ is quite simplified in the limit of $p \to \infty$. In such a case, only two states have non-zero energy: The state with only up spins ($m_z(S^z) = 1$) has an energy $E = -N$, while the state with only down spins ($m_z(S^z) = -1$) has $E = N$. The energy of all other states vanishes. The partition function of this system is given by

$$Z_{\Gamma=0} = 2^N - 2 + e^{-\beta N} + e^{\beta N}, \tag{3.4.25}$$

which for $N \gg 1$ reduces to

$$Z_{\Gamma=0} \approx e^{\beta N}\left(1 + e^{N(\ln 2 - \beta)}\right), \tag{3.4.26}$$

where $\beta = 1/T$ is the inverse of the temperature. The free energy per spin is then obtained as

$$f_{\Gamma=0} = -\frac{T}{N} \ln Z \approx -1 - \frac{T}{N} \ln\left(1 + e^{N(\ln 2 - \beta)}\right). \tag{3.4.27}$$

In the thermodynamic limit one finds

$$f_{\Gamma=0} \stackrel{N\to\infty}{\longrightarrow} \begin{cases} -1 & (\beta > \ln 2) \\ -T \ln 2 & (\beta < \ln 2) \end{cases}. \tag{3.4.28}$$

It turns out that a discontinuous phase transition takes place at $T_c = 1/\ln 2$, which separates the high temperature paramagnetic phase (PM) and the low temperature ferromagnetic phase (FM).

We next consider the other limit, namely, $\Gamma \to \infty$. In this case, the classical Hamiltonian is negligible, compared to the transverse-field term, and the quantum

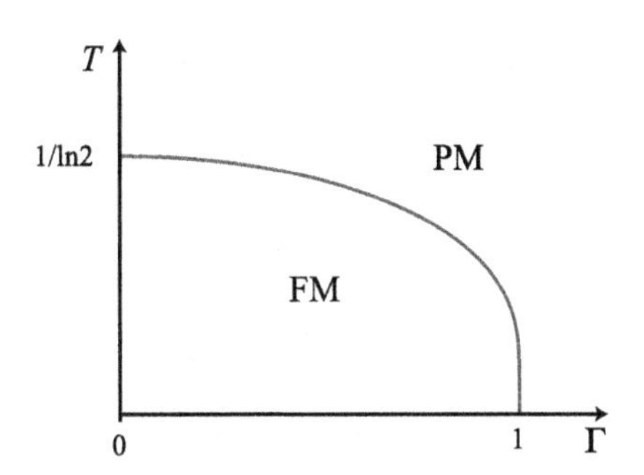

Fig. 3.8 Phase diagram of the fully connected p-body ferromagnetic model in the transverse field. The paramagnetic-ferromagnetic phase boundary is given by $\Gamma_c = T_c \operatorname{arccosh} \frac{e^{1/T_c}}{2}$

fluctuation stabilises the (quantum) paramagnetic state. The partition function is then given by

$$Z_{\Gamma\to\infty} = \left(e^{\beta\Gamma} + e^{-\beta\Gamma}\right)^N = 2^N \cosh^N \beta\Gamma, \tag{3.4.29}$$

which leads to the free energy per spin,

$$f_{\Gamma\to\infty} = -T\ln 2 - T\ln\cosh\beta\Gamma. \tag{3.4.30}$$

Properties in the intermediate case where Γ is finite are described as follows. For $T > 1/\ln 2$, one can expect that the paramagnetic phase extends from $\Gamma = 0$ to ∞. In this high temperature region, the free energy at $\Gamma = 0$ is given by (3.4.28), $f = -T\ln 2$. This implies that zero-energy paramagnetic states are crucial and non-zero-energy states are negligible. Hence the matrix elements of the classical Hamiltonian with respect to quantum paramagnetic states (i.e., eigenstates of the transverse-field term) vanishes, and consequently the free energy (3.4.30) is valid for whole region of Γ. Now in the low temperature case with $T < 1/\ln 2$, two distinct phases, the quantum paramagnetic phase and the ferromagnetic phase, might exist. As we see below, the free energy of the quantum paramagnetic phase is given by (3.4.30), and that of the ferromagnetic phase is by $f = -1$ from (3.4.28) irrespective of Γ. The transition between these two phases is discontinuous and the phase boundary is determined by equating two free energies:

$$-1 = -T\ln 2 - T\ln\cosh\beta\Gamma, \tag{3.4.31}$$

from which one obtains

$$\Gamma_c = T_c \operatorname{arccosh}\frac{e^{1/T_c}}{2}. \tag{3.4.32}$$

In particular, the quantum critical point is found to be $\Gamma_c = 1$. We show the phase diagram in Fig. 3.8.

To corroborate above argument, we introduce the Suzuki-Trotter transformation. The partition function obtained from the Hamiltonian (3.4.22) is written as

$$Z = \lim_{M\to\infty} A^{MN} \operatorname{Tr}\exp\left[\frac{\beta J}{MN^{p-1}}\sum_{k=1}^{M}\left(\sum_{k=1}^{N} S_{i,k}\right)^p + K_M\sum_{k=1}^{M}\sum_{i=1}^{N} S_{i,k}S_{i,k+1}\right], \tag{3.4.33}$$

where $A = [(1/2)\sinh(2\beta\Gamma/M)]^{1/2}$, $K_M = (1/2)\ln\coth(\beta\Gamma/M)$, and $S_{i,k}$ is an Ising spin. i labels the real-space site and k is the Trotter index. We assume the periodic boundary condition, $S_{i,M+1} = S_{i,1}$, in the Trotter direction.

To linearise the interaction terms, we make use of a formula on the delta function and the Lagrange multiplier:

$$f(x) = \int_{-\infty}^{\infty} dm\, \delta(m-x) f(m)$$
$$\approx \int_{-\infty}^{\infty} dm\, d\lambda \exp\left[-\frac{N}{M}\lambda(m-x)\right] f(m), \qquad (3.4.34)$$

where we note that the delta function is recovered from the right hand side by application of the saddle point method to the integral by λ with $N \to \infty$. Applying (3.4.34) with $x = (1/N)\sum_i S_{i,k}$ and $f(x) = \exp[\beta J N x^p/M]$ to (3.4.33), one obtains

$$Z = \lim_{M\to\infty} A^{MN} \int_{-\infty}^{\infty} \left[\prod_{k=1}^{M} dm_k\, d\lambda_k\right] \exp\left(\frac{\beta J N}{M} \sum_{k=1}^{M} m_k^p\right)$$
$$\times \operatorname{Tr} \exp\left[-\frac{N}{M}\sum_{k=1}^{M} \lambda_k \left(m_k - \frac{1}{N}\sum_{i=1}^{N} S_{i,k}\right) + K_M \sum_{k=1}^{M}\sum_{i=1}^{N} S_{i,k}S_{i,k+1}\right]$$
$$= \lim_{M\to\infty} A^{MN} \int_{-\infty}^{\infty} \left[\prod_{k=1}^{M} dm_k\, d\lambda_k\right] \exp\left(-\frac{N}{M}\sum_{k=1}^{M}(\lambda_k m_k - \beta J m_k^p)\right)$$
$$\times \exp\left[N \ln \operatorname{Tr} \exp\left(\frac{1}{M}\sum_{k=1}^{M} \lambda_k S_k + K_M \sum_{k=1}^{M} S_k S_{k+1}\right)\right], \qquad (3.4.35)$$

where we remark that the trace with respect to all spins is replaced by Nth power of the trace with respect to a single spin and hence the site index of the spin variable was dropped in the last line. Now the saddle-point condition for m_k yields

$$\lambda_k = p\beta J m_k^{*p-1}, \qquad (3.4.36)$$

where m_k^* denotes the solution of the saddle-point equation. Note that m_k^* is a function of λ_k. Due to the saddle-point method for the integrals by m_k, we obtain

$$Z = \lim_{M\to\infty} A^{MN} \int_{-\infty}^{\infty} \left[\prod_{k=1}^{M} d\lambda_k\right] \exp\left[-N\frac{(p-1)\beta J}{M} \sum_{k=1}^{M}\left(-\frac{\lambda_k}{p\beta J}\right)^{p/(p-1)}\right]$$
$$\times \exp\left[N \ln \operatorname{Tr} \exp\left(\frac{1}{M}\sum_{k=1}^{M} \lambda_k S_k + K_M \sum_{k=1}^{M} S_k S_{k+1}\right)\right]. \qquad (3.4.37)$$

We furthermore apply the saddle-point approximation for the integrals by λ_k. The saddle-point condition is given by

$$\left(\frac{\lambda_k^*}{p\beta J}\right)^{1/(p-1)} = \frac{\operatorname{Tr}\{S_k \exp(\frac{1}{M}\sum_{k=1}^{M} \lambda_k^* S_k + K_M \sum_{k=1}^{M} S_k S_{k+1})\}}{\operatorname{Tr}\{\exp(\frac{1}{M}\sum_{k=1}^{M} \lambda_k^* S_k + K_M \sum_{k=1}^{M} S_k S_{k+1})\}}, \qquad (3.4.38)$$

which with (3.4.36) is rewritten as

$$m_k^{**} = \frac{\mathrm{Tr}\{S_k \exp(\frac{1}{M}\sum_{k=1}^{M}\lambda_k^* S_k + K_M \sum_{k=1}^{M} S_k S_{k+1})\}}{\mathrm{Tr}\{\exp(\frac{1}{M}\sum_{k=1}^{M}\lambda_k^* S_k + K_M \sum_{k=1}^{M} S_k S_{k+1})\}}, \tag{3.4.39}$$

where m_k^{**} is the solution of Eqs. (3.4.36) and (3.4.39). Equation (3.4.39) implies that m_k^{**} is the longitudinal magnetisation. Since the present system is uniform along the Trotter direction, m_k^{**} must be independent of the imaginary-time index k. Hence we may write $m_k^{**} = m^{**}$ and $\lambda_k^* = \lambda^*$. With this consideration, the trace is computed as

$$\begin{aligned} A^M \,\mathrm{Tr}\exp\left(\frac{\lambda^*}{M}\sum_{k=1}^{M} S_k + K_M \sum_{k=1}^{M} S_k S_{k+1}\right) &\overset{M\to\infty}{\longrightarrow} 2\cosh\left(\beta\sqrt{(\lambda^*/\beta)^2 + \Gamma^2}\right) \\ &= 2\cosh\left(\beta\sqrt{(pJm^{**p-1})^2 + \Gamma^2}\right), \end{aligned} \tag{3.4.40}$$

so that (3.4.39) reduces to

$$m^{**} = \frac{\tanh(\beta\sqrt{(pJm^{**p-1})^2 + \Gamma^2})}{\sqrt{(pJm^{**p-1})^2 + \Gamma^2}} pm^{**p-1}. \tag{3.4.41}$$

The free energy per spin is obtained using the saddle-point solution as

$$f = -\frac{1}{\beta}\frac{1}{N}\ln Z \overset{M\to\infty}{\longrightarrow} (p-1)Jm^{**p} - \frac{1}{\beta}\ln 2\cosh(\beta\sqrt{(pJm^{**p-1})^2 + \Gamma^2}). \tag{3.4.42}$$

Taking $p \to \infty$, (3.4.41) has only the two solutions, $m^{**} = 0$ and $m^{**} = 1$, corresponding to the paramagnetic magnetisation and the ferromagnetic one respectively. The free energy is given by $f = -(1/\beta)\ln 2\cosh(\beta\Gamma)$ for the paramagnetic phase and $f = -1$ for the ferromagnetic phase. The phase boundary is determined by the crossing point of these free energies. The result reproduces (3.4.32).

3.5 Scaling Properties Close to the Critical Point

The scaling properties of a system close to a quantum phase transition point can be derived considering the linearised renormalisation group (see the next section) equations near the unstable fixed point governing the transition. The set of exponents associated with the unstable fixed point determines the universality class of the system. Let us consider the transverse Ising Hamiltonian,

$$H = -J\sum_{\langle ij\rangle} S_i^z S_j^z - \Gamma\sum_i S_i^x - h\sum_i S_i^z \tag{3.5.1}$$

where $\langle ij\rangle$ denotes nearest neighbour interaction and h is an external longitudinal magnetic field. The schematic phase diagram (in the case when $h = 0$) as obtained from the renormalisation group study is shown in the Fig. 3.9 (cf. [208]). For the

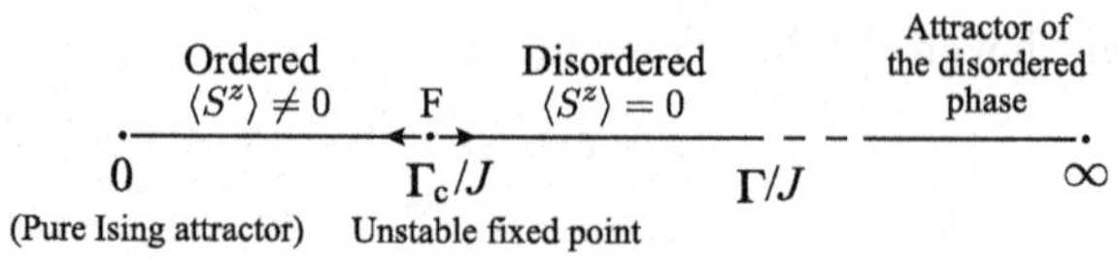

Fig. 3.9 Fixed points and renormalisation group flow in the transverse Ising model

Ordered $\langle S^z \rangle \neq 0$ F Disordered $\langle S^z \rangle = 0$ Attractor of the disordered phase

0 Γ_c/J Γ/J ∞

(Pure Ising attractor) Unstable fixed point

zero-temperature transition, the one-dimensional parameter space consists of two trivial (stable) fixed points: one at $\lambda = 0$ (the pure Ising attractor, the state with long range order in the z-direction) and at $\lambda = \infty$ (attractor of the disordered state), where $\lambda = \Gamma/J(0)$, $J(0) = \delta_0 J$, δ_0 being the number of nearest neighbours. Apart from these two trivial fixed points, there is an unstable critical fixed point at $\lambda = \lambda_c = 1$. This fixed point characterises the zero-temperature phase transition from the state of nonzero order parameter ($\langle S_z \rangle = 0$) to the state with vanishing order parameter. At this critical point, the correlation length (as well as the relaxation time) diverges and associated with this unstable fixed point we have a set of critical exponents which determines the universality class of the quantum phase transition. Let us now study how the parameters (J, Γ, h) of the Hamiltonian (3.5.1) scale under a length scale transformation by a factor b in the proximity of the unstable fixed point λ_c [82]. Let us denote the distance $\Delta\lambda$ of a particular point (λ) in the one-dimensional parameter space from the critical point λ_c by $\Delta\lambda = (\lambda - \lambda_c)/\lambda_c$.

Let us consider a scale transformation by a factor b

$$\Delta\lambda' = b^p \Delta\lambda, \qquad J' = b^{-q} J, \qquad \Gamma' = b^{-q} \Gamma, \qquad h' = b^r h, \tag{3.5.2}$$

where p, q, r are scaling exponents and the prime refers to the renormalised quantities. Since $(\Gamma/J)' = (\Gamma/J)$ at the fixed point, Γ and J scales in identical fashion. The singular part of ground state energy density has the scaling form $E = Jf(\Delta\lambda, h/J)$, where $f(x, y)$ is the scaling function. The renormalised form of the ground state energy density and the correlation length can be written as

$$E' = J' f\big(\Delta\lambda', h'/J'\big) = b^d E, \qquad \xi' = b^{-1}\xi(\Delta\lambda, h/J) = \xi\big(\Delta\lambda', h'/J'\big), \tag{3.5.3}$$

where the lattice dimensionality d enters the energy density as the density of the degrees of freedom scales with the same exponent. Using the scaling relations (3.5.2), we get

$$E(\Delta\lambda, h/J)/J = b^{-(d+q)} f\big(b^p \Delta\lambda, b^{(r+q)}(h/J)\big), \tag{3.5.4a}$$

$$\xi(\Delta\lambda, h/J) = b\xi\big(b^p \Delta\lambda, b^{(r+q)}(h/J)\big). \tag{3.5.4b}$$

As the choice of scaling factor is arbitrary, we set $b^p \Delta\lambda = 1$, which gives

$$E(\Delta\lambda, h/J)/J = (\Delta\lambda)^{(d+q)/p} f\big(1, (h/J)/(\Delta\lambda)^{(r+q)/p}\big), \tag{3.5.5a}$$

$$\xi(\Delta\lambda, h/J) = (\Delta\lambda)^{-(1/p)} \xi\big(1, (h/J)/(\Delta\lambda)^{(r+q)/p}\big). \tag{3.5.5b}$$

Writing $\xi(h = 0) \sim (\Delta\lambda)^{-\nu}$ for the growth of correlation length near the critical point λ_c and $E(h = 0) \sim (\Delta\lambda)^{\varepsilon}$ for the ground state energy density variation with λ (near λ_c), we get

$$\nu = 1/p \quad \text{and} \quad \varepsilon = \nu(d + q). \tag{3.5.6}$$

Writing for the singular part of ground state energy variation as $(\Delta\lambda)^{2-\alpha}$ (expressing the critical variation of the analogue of specific heat, which comes as the second derivative in energy, in thermal phase transition as $(\Delta\lambda)^{-\alpha}$), we arrive at the hyper-scaling relation

$$2-\alpha=\nu(d+q). \tag{3.5.7}$$

Defining magnetisation

$$m=-(\partial E/\partial h)_{h=0}\sim(\Delta\lambda)^{\beta} \tag{3.5.8}$$

and the susceptibility

$$\chi=-\left(\partial^2 E/\partial h^2\right)_{h=0}\sim(\Delta\lambda)^{-\gamma}, \tag{3.5.9}$$

where β and γ denote order parameter and susceptibility exponents respectively, one gets (from (3.5.5a))

$$\beta=\nu(d-r), \qquad \gamma=\nu(2r+q-d). \tag{3.5.10}$$

This leads to the scaling relation

$$\alpha+2\beta+\gamma=2. \tag{3.5.11}$$

Defining $m\sim h^{1/\delta}$ at $\lambda=\lambda_c$, one can get the relation

$$\delta=(\beta+\gamma)/\beta. \tag{3.5.12}$$

This is because, one gets $E\sim h^{(d+q)/(r+q)}$ from (3.5.4a) when one sets $hb^{r+q}=1$. This gives $\delta=(r+q)/(d-r)=(\beta+\gamma)/\beta$.

It may be noted that in quantum transitions, the scaling of energy and time becomes interdependent through the uncertainty relation. If the typical time τ scales as $\tau'=b^z\tau$, where z is the dynamic exponent, dose to the zero temperature transition point λ_c, then expressing the critical fluctuations in J and τ as ΔJ and $\Delta\tau$ respectively, we get

$$\Delta J'\Delta\tau'=b^{z-q}\Delta J\Delta\tau\geq\hbar, \tag{3.5.13}$$

where $\hbar$ is the Planck's constant. This gives $q=z$ and consequently the hyper-scaling relation (3.5.7) can be written as

$$2-\alpha=\nu(d+z). \tag{3.5.14}$$

This hyper-scaling relation suggests that for quantum transitions (at zero temperature) the scaling relations for the exponents become classical-like, when the dimension d is replaced by the effective dimension $d_{\text{eff}}=d+z$, where z is the dynamic exponent. This shift in effective dimension has already been discussed above. In fact, in pure transverse Ising system (in any dimension) time (or energy) scales linearly with length [173, 438], giving $z=1$. This suggests that the zero-temperature transition (critical) behaviour of a pure d-dimensional quantum system is equivalent to that of a $(d+1)$-dimensional classical pure system. It may also be noted here that dynamic exponent z is different from unity for quantum glass transitions (see Chap. 6) and the effective correspondence with the classical system, if any, is expected to be different.

3.6 Real-Space and Field-Theoretic Renormalisation Group

3.6.1 Real-Space Renormalisation Group

As mentioned in Sect. 2.4, for (pure) transverse Ising chain, one can apply the real-space renormalisation group (RSRG) technique to estimate the value of the critical point and exponents for the zero-temperature (quantum) transitions. Here, as discussed in Sect. 2.4, one writes the recursion relation of various quantities like the energy gap $\Delta_n(\lambda)$ (difference between the ground state and first excited state energy $E_{1,n}(\lambda, h=0) - E_{0,n}(\lambda, h=0)$, where h is a longitudinal magnetic field), susceptibilities $\chi_n (= (\partial^2 E_{0,n}/\partial h^2)_{h\to 0})$ etc., for different system sizes n, from their exact or approximate solutions (diagonalisation of small sizes) and retaining only the lowest levels. One then writes the recursion relations connecting the quantities of different sizes, which indicate the nature of their asymptotic variations and give the estimate of exponent values. For example

$$\Delta_{n/b}(\lambda', 0) = b^{-s/\nu} \Delta_n(\lambda), \quad \text{and} \quad \chi_{n/b}(\lambda', 0) = b^{\gamma/\nu} \chi_n(\lambda), \tag{3.6.1}$$

where ν, s and γ denote respectively the correlation length, mass gap and susceptibility exponents. Also, if we write

$$\lambda' = R_b(n, \lambda), \tag{3.6.2}$$

then the critical point can be estimated from the (extrapolated; $b \to \infty$) fixed point of the renormalisation group transformation

$$\lambda^* = R_b(n, \lambda^*), \tag{3.6.3}$$

and since the correlation length ξ diverges with $\Delta\lambda (= (\lambda - \lambda^*)/\lambda^*)$ as $(\Delta\lambda)^{-\nu}$,

$$b = \frac{\xi_n(\lambda)}{\xi'_{n/b}(\lambda')} = \left(\frac{\lambda - \lambda^*}{\lambda' - \lambda^*}\right)^{-\nu} \tag{3.6.4}$$

one gets

$$(\partial R_b/\partial b)\lambda^* \sim b^{-1/\nu}. \tag{3.6.5}$$

With various (numerical) tricks to evaluate the above quantities (Δ, χ, etc., giving in turn R) for various finite size Ising systems in dimension $d = 2$ (and 3), and utilising the above relations, various RSRG estimates of the critical point (λ_c) and exponents (ν, γ, s, etc.) have been made. Using perturbative cluster renormalisation group [150, 151]; Friedman obtained $\lambda^* = 3.43$ and $\nu = 0.92$ for triangular lattice, and $\lambda^* = 3.09$ and $\nu = 0.72$ for square lattice. Using truncated basis set method [308], Penson et al. obtained $\lambda^* = 4.73$, and $\nu = 0.95$ for triangular lattice, and $\lambda^* = 2.63$ and $\nu = 1.1$ for square lattice. Using clusters up to a maximum size $n = 4$, dos Santos et al. [111] obtained $\lambda^* = 3.74$ and $\nu = 0.93$ for triangular lattice, and $\lambda^* = 3.18$ and $\nu = 0.66$ for square lattice. One can easily see the large scatter in the estimates of critical point and the exponents obtained using various RSRG schemes and one may compare with the series estimate of Pfeuty et al. [313] and Yanase [434]: $\lambda_c = \lambda^* = 4.77$ and $\nu = 0.64$ for triangular lattice [434], and $\lambda^* = 3.04$ and $\nu = 0.63$ for square lattice [313].

3.6.2 Field-Theoretic Renormalisation Group

Using the Hubbard-Stratonovich transformations and Gaussian functional averages [286], one can write an effective (classical) Landau-Ginzburg-Wilson (LGW) free energy functional for the transverse Ising system [438]. Writing the Hamiltonian as

$$H = H_0 + V, \tag{3.6.6}$$

with

$$H_0 = -\Gamma \sum_{i=1}^{N} S_i^x \quad \text{and} \quad V = -\sum_{ij} J_{ij} S_i^z S_j^z \tag{3.6.7}$$

the partition function $Z = \mathrm{Tr}\exp(-\beta H)$ can be written as

$$Z = \frac{1}{(2\pi)^{N/2}} \int_{-\infty}^{\infty} \prod_{i'} d\psi_{i'} \exp\left[-\frac{1}{2\beta} \sum_i \int_0^{\infty} \psi_i^2(\tau)\, d\tau\right] Z(\psi) \tag{3.6.8}$$

with

$$Z(\psi) = \mathrm{Tr}\left[\exp(-\beta H_0) P \exp\left(\frac{1}{\beta} \int_0^{\beta} \sum_{ij} \psi_i(\tau) A_{ij} S_j^z(\tau)\, d\tau\right)\right], \tag{3.6.9}$$

where P is the time (τ) ordering operator, $(A^2)_{ij} = \beta J_{ij}$ and $S_z(\tau)$ is in the interaction representation. Expanding the exponential, evaluating the traces and rearranging the terms after Fourier transformations, the partition function can be written as

$$Z \sim \int_{-\infty}^{\infty} \partial S_{\boldsymbol{q},m} \exp\bigl(-\beta H_{\mathrm{eff}}(S_{\boldsymbol{q},m})\bigr), \tag{3.6.10}$$

where

$$\begin{aligned} &H_{\mathrm{eff}}(S_{\boldsymbol{q},m}) \\ &\quad= \sum_{m_1}\sum_{m_2} \int_{\boldsymbol{q}_1}\int_{\boldsymbol{q}_2} u^{(2)}(\boldsymbol{q}_1, \boldsymbol{q}_2, m_1, m_2) S_{\boldsymbol{q}_1,m_1} S_{\boldsymbol{q}_2,m_2} \delta(\boldsymbol{q}_1 + \boldsymbol{q}_2)\delta_{m_1,m_2} + O\bigl(S^4\bigr) \\ &\quad= \sum_m \int_{\boldsymbol{q}} u^{(2)}(\boldsymbol{q}, m) S_{\boldsymbol{q},m} S_{-\boldsymbol{q},-m} + O\bigl(S^4\bigr), \end{aligned} \tag{3.6.11}$$

with

$$u^{(2)}(\boldsymbol{q}, m) = r_m + q^2 + O\bigl(q^4\bigr), \tag{3.6.12a}$$

$$r_m = \frac{1}{\delta a^2}\left(\frac{1}{G(\omega_m)}\right) - J(0), \tag{3.6.12b}$$

$$J(\boldsymbol{q}) = J(0) - \delta_0 a^2 q^2 + O\bigl(q^4\bigr). \tag{3.6.12c}$$

Here, the integer m labels the Matsubara frequencies $\omega_m = 2\pi m/\beta$, and a denotes the lattice constant, δ_0 is the number of nearest neighbours and $G(\omega_m)$ the "unperturbed" propagator (for H_0) given by,

$$G(\omega_m) = \frac{4\Gamma \tanh \beta\Gamma}{(2\Gamma)^2 - (i\omega_m)^2}. \tag{3.6.13}$$

The remaining terms in (3.6.11) are related to higher order time ordered cumulant averages or semi-invariants [379]. It can be shown that the next higher order term in (3.6.11) containing $u^{(4)}$ is proportional to the temperature T as $T \to 0$, and other higher order terms containing $u^{(2n)}$, $n > 2$, are of the order T^{n-1} in the same $T \to 0$ limit. One can now identify the effect of (quantum) non-commuting operators in the effective LGW Hamiltonian (3.6.10) at finite Matsubara frequencies ω_m as well as at $\omega_m = 0$. At $T > 0$, the Matsubara frequencies being discontinuous, and the critical interval r_m being determined by the first one (r_0) becoming critical (and others becoming irrelevant due to renormalisation), the effective $u^{(2)}$ in (3.6.12a) becomes the same as that of classical Ising system ($u^{(2)} = r_0 + q^2$). The finite temperature critical behaviour is therefore the same as that of a classical Ising system in the same dimension (although the critical temperature is affected by the transverse field). At $T = 0$, however the LGW Hamiltonian (3.6.10) has a different character, and the critical behaviour is affected non-trivially. Here (for $T = 0$), the Matsubara frequencies become continuous and the sum over m in (3.6.11) is replaced by an integral over ω, giving the quadratic term in H_{eff} as

$$H_{\text{eff}} = \int_{\boldsymbol{q}} \int_{\omega} \left(r + q^2 + \alpha\omega^2\right) S_{\boldsymbol{q},\omega} S_{-\boldsymbol{q},\omega}, \tag{3.6.14}$$

where $\alpha = (4\pi\delta_0 a^2)^{-1}$ and $r = (\delta_0 a^2)^{-1}(\Gamma_0 - J(0))$. This is of the same form as that of a classical Ising model with the critical temperature interval here replaced by the critical interval $(\Gamma - J(0))$ of the transverse field, and contains an integral over an effective dimension (coming from the Matsubara frequencies) in addition to the d-dimensional integral over $\boldsymbol{q}$. The critical behaviour (or the exponents) is the same as that of a $(d+1)$-dimensional (classical) Ising model (with Γ replacing T).

Appendix 3.A

3.A.1 *Effective Classical Hamiltonian of the Transverse Ising Model*

To evaluate the effective classical Hamiltonian, let us start from the quantum Hamiltonian describing a transverse Ising model on a d-dimensional lattice

$$H = H_0 + V = -\Gamma \sum_i S_i^x - \sum_{ij} J_{ij} S_i^z S_j^z, \tag{3.A.1}$$

where S_i^α's are the Pauli spin operators, as mentioned earlier. The partition function of this quantum Hamiltonian can be written as

$$Z = \text{Tr}\exp\left[-\beta(H_0 + V)\right]. \tag{3.A.2}$$

Using the generalised Trotter formula (cf. Sect. 3.1), one can rewrite the exponential operator in Eq. (3.A.2) as

$$\exp\bigl[-\beta(H_0+V)\bigr]=\lim_{M\to\infty}\left[\exp\left(\frac{-\beta H_0}{M}\right)\exp\left(\frac{-\beta V}{M}\right)\right]^M, \quad (3.A.3)$$

which is equivalent to considering M identical replicas of the original system. The partition function of the quantum system can readily be written (inserting set of identity operators) in the form

$$Z=\lim_{M\to\infty}\operatorname{Tr}\prod_{k=1}^{M}\langle S_{1,k},S_{2,k},\ldots,S_{N,k}| \times\left[\exp\left(\frac{-\beta H_0}{M}\right)\exp\left(\frac{-\beta V}{M}\right)\right]|S_{1,k+1}S_{2,k+1},\ldots,S_{N,k+1}\rangle, \quad (3.A.4)$$

where $|S_{i,k}\rangle$'s are the eigenstates of the operators S_i^z. It is quite clear that by inserting the sets of identity operators, one has effectively introduced an additional dimension in the problem ("Trotter dimension"), which is denoted by the index k. One then uses the following relations

$$\prod_{k=1}^{M}\langle S_{1,k},S_{2,k},\ldots,S_{N,k}|\exp\left(\frac{\beta}{M}\sum_{ij}J_{ij}S_i^zS_j^z\right)|S_{1,k+1}S_{2,k+1},\ldots,S_{N,k+1}\rangle = \exp\left[\sum_{i,j=1}^{N}\sum_{k=1}^{M}\frac{\beta J_{ij}}{M}S_{i,k}S_{j,k}\right], \quad (3.A.5a)$$

$$\prod_{k=1}^{M}\langle S_{1,k},S_{2,k},\ldots,S_{N,k}|\exp\left(\frac{\beta\Gamma}{M}\sum_{i}S_i^x\right)|S_{1,k+1}S_{2,k+1},\ldots,S_{N,k+1}\rangle = \left(\frac{1}{2}\sinh\left(\frac{2\beta\Gamma}{M}\right)\right)^{NM/2}\exp\left[\frac{1}{2}\ln\coth\left(\frac{\beta\Gamma}{M}\right)\sum_{i=1}^{N}\sum_{k=1}^{M}S_{i,k}S_{i,k+1}\right]. \quad (3.A.5b)$$

In deriving the relation (3.A.5b), one has to use the relation

$$\langle S|e^{aS^x}|S'\rangle=\bigl[(1/2)\sinh(2a)\bigr]^{1/2}\exp\bigl[(SS'/2)\ln\coth(a)\bigr] \quad (3.A.6)$$

(where $|S\rangle$ and $|S'\rangle$ are the eigenstates of S^z), which can be easily derived writing the exponential operator in the form

$$e^{aS^x}=\cosh(a)+S^x\sinh(a), \quad (3.A.7)$$

which can be checked by expanding the exponential function and using $(S^x)^2=1$. One can now equate (3.A.7) with (3.A.6), putting explicitly the eigenvalues S and S'.

Using the relations (3.A.5a), (3.A.5b), one can arrive at the final form of the partition function of the quantum system, with the M-th Trotter approximation

$$Z=C^{NM/2}\operatorname{Tr}_s\exp\bigl(-\beta H_{\text{eff}}(S)\bigr);\quad C=\frac{1}{2}\sinh\left(\frac{2\beta\Gamma}{M}\right), \quad (3.A.8)$$

where

$$H_{\text{eff}}(S) = \sum_{i,j=1}^{N} \sum_{k=1}^{M} \left[-\frac{J_{ij}}{M} S_{i,k} S_{j,k} - \frac{\delta_{ij}}{2\beta} \ln \coth\left(\frac{\beta\Gamma}{M}\right) S_{i,k} S_{i,k+1} \right]. \qquad (3.A.9)$$

Since $S_{i,k}$ are classical numbers (± 1), the above Hamiltonian represents a classical Ising Hamiltonian on a $(d+1)$-dimensional lattice, with anisotropic coupling in the spatial and Trotter dimension. One must note here that, for the equivalence between the quantum and classical Hamiltonians to hold, one must consider $M \to \infty$ limit. As mentioned earlier, for infinitely large M, one must also consider the $\beta \to \infty$ limit, so that the zero-temperature phase transition in the d-dimensional quantum model is equivalent to the thermal phase transition in $(d+1)$-dimensional equivalent classical model.

3.A.2 Derivation of the Equivalent Quantum Hamiltonian of a Classical Spin System

Starting from a classical spin Hamiltonian, one can also (conversely) derive the equivalent quantum Hamiltonian in an extreme anisotropic limit [25, 231]. If one considers a classical spin Hamiltonian on a d-dimensional lattice and $\hat{T}$ denotes the transfer matrix of the classical system, the equivalent quantum Hamiltonian is defined as

$$\hat{T} = 1 - \tau H + O(\tau^2), \qquad (3.A.10)$$

where τ is a strictly infinitesimal parameter (may be considered as the lattice parameter in one particular (time) direction). The Hamiltonian H corresponds to the equivalent quantum Hamiltonian of the classical system. The free energy of the statistical mechanical system, given by the largest eigenvalue of the transfer matrix, is now related to the ground state energy of the quantum Hamiltonian and the correlation length is given by the inverse of energy gap of the quantum Hamiltonian. We shall illustrate the above mentioned equivalence using the example of a two dimensional spin-$1/2$ classical Ising system on a square lattice, with anisotropic coupling strengths J_1 and J_2. The row to row transfer matrix of the above Hamiltonian can be written as [349]

$$\hat{T} = \hat{T}_1 \hat{T}_2 \hat{T}_1, \qquad (3.A.11)$$

where

$$\hat{T}_1 = \exp\left[\sum_i \tilde{K}_1 S_i^x\right], \qquad (3.A.12)$$

$$\hat{T}_2 = \exp\left[\sum_i K_2 S_i^z S_{i+1}^z\right], \qquad (3.A.13)$$

with $K_i = \beta J_i$, $i = 1, 2$ and $\tilde{K}_1$ is given by

$$\tilde{K}_1 = -\frac{1}{2} \ln(\tanh K_1). \qquad (3.A.14)$$

Clearly the operators $\hat{T}_1$ and $\hat{T}_2$ do not commute. To write the above transfer matrix (3.A.11) in the form (3.A.10) one has to suitably define the expansion parameter T. One sets [231] $\tau = \tilde{K}_1$, and $K_2 = \lambda\tau$, where λ is finite. One can now neglect the noncommutivity of the operators $\hat{T}_1$ and $\hat{T}_2$ if one considers the extreme anisotropic limit given by

$$K_1 \to \infty\ (\tilde{K}_1 \to 0) \quad \text{and} \quad K_2 \to 0 \quad \text{with}\ \lambda = \frac{K_2}{\tilde{K}_1} \equiv O(1), \qquad (3.A.15)$$

the transfer matrix is readily written in the form (3.A.10) as $\tau \to 0$, where the equivalent quantum Hamiltonian H is given by

$$H = -\sum_i S_i^x - \lambda \sum_i S_i^z S_{i+1}^z, \qquad (3.A.16)$$

with λ playing the role of inverse temperature. One thus obtains the transverse Ising Hamiltonian from the classical Ising Hamiltonian in the extreme anisotropic limit. One should note here that this equivalence is established in the extreme anisotropic limit, τ, $K_2 \to 0$, called the "Hamiltonian" limit. The essential assumption behind this mapping is that this anisotropy does not affect the universality class of the problem.

Chapter 4
ANNNI Model in Transverse Field

4.1 Introduction

Spatially modulated periodic structures had been observed experimentally for the first time in magnetic and ferroelectric systems in the late fifties and early sixties. Subsequently, it became evident that these structures originate from competing interactions of magnetic and electric dipole moments and may be mimicked by magnetic models with regular competing interactions (frustration). The most popular model in which the effects of regular frustration on the (classical) spin model have been studied extensively is the axial next nearest neighbour Ising (ANNNI) model [122, 352, 353, 435]. The classical ANNNI model is described by a system of Ising spins with nearest neighbour interactions along all the lattice directions (x, y and z) as well as a competing next nearest neighbour interaction in one axial (say z) direction. The regular competition or frustration here gives rise to many modulated spin structures. Depending on the interaction and temperature, many commensurate and incommensurate modulated phases appear in such systems, which show very rich phase diagrams [352]. It may be mentioned that other regularly frustrated Ising models have also been constructed: e.g., the ANNNI model can be extended to include frustration along two or more axes or there may be three or more spin interaction terms. Also, one can construct frustrated Ising models with further neighbour interactions or other non-Ising frustrated models like the three state chiral clock model, etc. [352]. However, the simplest Ising model with regular frustration is the ANNNI model and it can also mirror the properties of real magnetic systems as well as many other systems with modulated structures, like in ferroelectrics, binary alloys, etc. We are interested here in the stability of such modulated phases and the phase transitions driven by quantum fluctuations at zero temperature. Again, one can have a tunable quantum fluctuation induced by presence of a transverse field. We consider therefore the ground state properties of the ANNNI model in transverse field at zero temperature. Here the presence of (regular) frustration is expected to give rise to intriguing quantum many-body phases (ground states). One can also study the stability or instability of the commensurate and incommensurate classical Ising

S. Suzuki et al., *Quantum Ising Phases and Transitions in Transverse Ising Models*, Lecture Notes in Physics 862, DOI 10.1007/978-3-642-33039-1_4,

phases occurring in the classical ANNNI models. These observations may be compared with the extensive studies, and the literature developed [20, 263], on the study of quantum magnetisation in the (frustrated) Heisenberg antiferromagnets and etc.

Before describing the detailed results on quantum ANNNI models, a brief introduction to the classical model needs to be given.

4.2 Classical ANNNI Model

The Hamiltonian for the classical ANNNI model is given by

$$H = -\frac{1}{2}\sum_{i,j,j'} J_0 S^z_{i,j} S^z_{i,j'} - \sum_{i,j} J_1 S^z_{i,j} S^z_{i+1,j} - \sum_{i,j} J_2 S^z_{i,j} S^z_{i+2,j}, \tag{4.2.1}$$

where i labels the layers perpendicular to the axial direction and j and j' denote the nearest neighbour spins within a layer. Competition is due to ferrotype nearest neighbour (positive J_1) and antiferrotype next nearest neighbour (negative J_2) interactions. The ground state (for $T = 0$) is exactly known in all dimensions: ferromagnetic type for $\kappa = |J_2|/J_1 < 0.5$ and modulated for $\kappa > 0.5$ with a period of 4 (antiphase). The $\kappa = 0.5$ point is highly degenerate, the degeneracy being equal to gN for a system of N spins where $g = (1 + \sqrt{5})/2$ is the golden ratio [352]. The system is also frustrated when both J_1 and J_2 are antiferromagnetic; the spin configurations corresponding to which can be obtained by flipping every alternate spin in those corresponding to the former case ($J_1 > 0$, $J_2 < 0$). Let us discuss in brief the features of the ANNNI model at $T \neq 0$ in different dimensions.

(a) One dimension: The one dimensional ANNNI model (here $J_0 = 0$) is exactly solvable [181, 247]. Here one uses a simple transformation: $S^z_i S^z_{i+1} \to \tau^z_i$ which transforms the Hamiltonian to that of the nearest neighbour Ising model in a longitudinal field. As the system is short ranged, there is no finite temperature phase transition, i.e., the system is paramagnetic for $T \neq 0$. However, the para phase has two distinct regions: in one the spin-spin correlations decay exponentially (cf. nearest neighbour Ising models) while in the other, oscillations are enveloped by an exponential decay. These two regions are separated by a disorder line.

(b) Two dimension: The two dimensional ANNNI model has no exact solution. However, estimates [352] of the phase diagram have been obtained using different approximations [288, 412].

The phase diagram consists of a ferromagnetic phase, a paramagnetic phase, antiphase, and in all probability a floating incommensurate phase where the spin-spin correlations decay algebraically analogous to a Kosterlitz-Thouless transition in an XY system. The paramagnetic phase is believed to exist down to $T = 0$ for $\kappa = 0.5$. A disorder line, starting from this point and touching the T axis asymptotically, divides the para phase such that on the large κ side, the exponential decay of the correlations have local periodic oscillations. The best estimate of the ferromagnetic to paramagnetic boundary is given by

$$\sinh\bigl[2\beta(J_1 + 2J_2)\bigr] \sinh 2\beta J_0 = 1. \tag{4.2.2}$$

The estimated phase diagram is shown in Fig. 4.1.

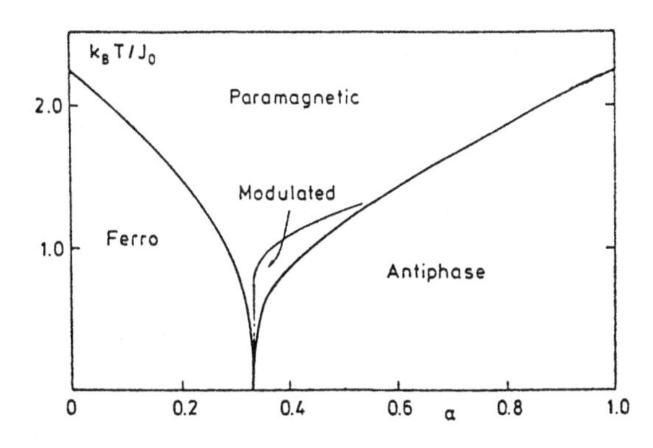

Fig. 4.1 Phase diagram of the $2d$ ANNNI model with $J_1 = (1-\alpha)J_0$ and $J_2 = -\alpha J_0$ (from [352])

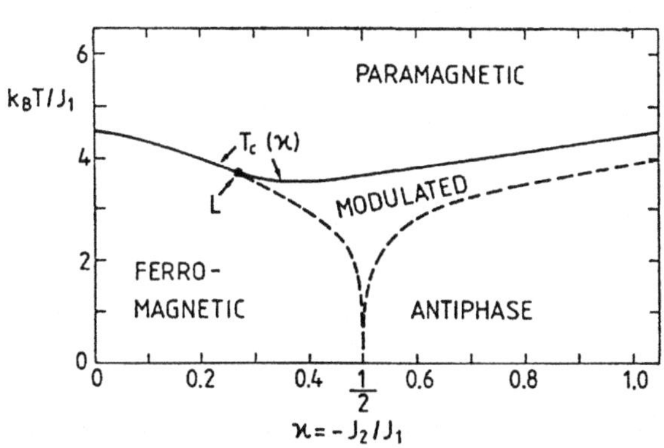

Fig. 4.2 Mean field phase diagram of three dimensional ANNNI model (from [352]). L is the Lifshitz point

(c) Three dimension: In three dimensions, the mean field phase diagram [352, 436] has been extensively studied. It consists of (commensurate or incommensurate) modulated phases in addition to the paramagnetic and ferromagnetic phases. All three phases meet at a critical point called the Lifshitz point at a finite temperature at $\kappa = 0.25$. There are neither floating phases nor disorder lines. Various phases with nontrivial periodicities are found to exist here [145, 354]. The mean field phase diagram is shown in Fig. 4.2.

4.3 ANNNI Chain in a Transverse Field

In the ANNNI model one can study the effects of quantum fluctuations by putting it in a transverse field analogous to the nearest-neighbour Ising case. The Hamiltonian is given by

$$H = -\frac{1}{2}\sum_{i,j,j'} J_0 S^z_{i,j} S^z_{i,j'} - \sum_{i,j} J_1 S^z_{i,j} S^z_{i+1,j} - \sum_{i,j} J_2 S^z_{i,j} S^z_{i+2,j} - \sum_i \Gamma S^x_i . \tag{4.3.1}$$

The interest in such quantum systems is twofold: Firstly, the zero temperature critical behaviour of a quantum spin ($S = \pm 1$) Ising system in d-dimension is usually related to the thermal critical behaviour of the corresponding classical system in $(d+1)$-dimension, and vice-versa [387]. Secondly, as mentioned before, the results for such systems will have important implications on the general role of quantum fluctuations in magnetism and can be compared to the corresponding results for the (frustrated) quantum magnets like quantum antiferromagnets etc. [20]. One can, therefore, anticipate that the quantum fluctuations in the one dimensional ANNNI model may give rise to interesting structures in the phase diagram as has been found using approximate and numerical methods in the two dimensional classical ANNNI model. In fact, the Hamiltonian for the one dimensional ANNNI model in a transverse field was first obtained as the Hamiltonian limit of the transfer matrix of the two dimensional classical ANNNI model (Rujan [336], Barber and Duxbury [24, 25, 117]). The Hamiltonian of the one dimensional ANNNI model in transverse field takes the simple form

$$H = -\sum_i J_1 S_i^z S_{i+1}^z - \sum_i J_2 S_i^z S_{i+2}^z - \sum_i \Gamma S_i^x. \tag{4.3.2}$$

The transfer matrix of the two dimensional ANNNI model can be written as

$$\hat{T} = \exp\left(\beta J_0 \sum_i S_i^x\right) \exp\beta\left(J_1 \sum_i S_i^z S_{i+1}^z + J_2 \sum_i S_i^z S_{i+2}^z\right), \tag{4.3.3}$$

where $\tanh J_0 = \exp(-2J_0)$, J_0 the interaction along one axis; J_1, J_2 are the competing interactions along the other. The mapping of the Hamiltonian of the two dimensional ANNNI model to that of the quantum one-dimensional model (i.e., taking the Hamiltonian limit of (4.3.3) where the exponential functions commute), however, is exact only in the limit $\Gamma \to \infty$, $J_1 \to 0$, $J_2 \to 0$, κ remaining finite. Therefore, it is not at all obvious that the phase diagrams of the quantum chain (4.3.2) and the classical two dimensional ANNNI model should be closely comparable. Secondly, the frustrated antiferromagnetic Heisenberg models with anisotropy in general (i.e., when the Hamiltonian includes cooperative interaction also in y and z directions (and $\Gamma = 0$) with competition between nearest and next nearest neighbour interactions) indeed exhibit [71, 129, 187, 255, 256] that zeroth order quantum fluctuation can destroy the Neel order, so that the quantum spin liquid phase is argued to be the ground state of that system. For a special one dimensional model, in which the second neighbour interaction is exactly half of that of the first neighbour ($\kappa = 0.5$), the (two fold degenerate) dimer phase has been shown to be the exact ground state (Majumdar and Ghosh [255, 256]). Although the symmetries are different, it would be interesting to compare the effect of zero point (transverse field) quantum fluctuation in the frustrated Ising system (at $\kappa = 0.5$) and check if it also destroys the classical order (phases) and leads to new quantum phases (comparable to the dimer, spin liquid or otherwise in frustrated magnetic systems) (Sen and Chakrabarti [356, 359, 360], Sen et al. [362]). The results obtained for the ANNNI model using some analytical and computational methods are described in the following sections.

4.3.1 *Some Results in the Hamiltonian Limit: The Peschel-Emery Line*

As mentioned before, this model was initially studied in order to obtain the phase diagram for the two-dimensional classical case [24, 25, 117, 336]. Barber and Duxbury [24] studied special limits of this model and obtained a phase diagram. A more complete study for the one dimensional quantum model was first attempted by Rujan [336], although the interest was in the special limits.

One can use a transformation

$$S_i^x = \sigma_{i-1}^x \sigma_i^x \tag{4.3.4}$$

and

$$S_i^z S_{i+1}^z = \sigma_i^z, \tag{4.3.5}$$

with which the Hamiltonian (4.3.2) maps onto the dual Hamiltonian H_D corresponding to a XY model with an in-plane field given by

$$H_D = -\left(J_1 \sum_i S_i^z + \Gamma \sum_i S_i^x S_{i+1}^x + J_2 \sum_i S_i^z S_{i+1}^z \right). \tag{4.3.6}$$

The expression for the mass gap Δ, which is the inverse of the correlation length, for the above model (H_D) was obtained using perturbation methods [336], and is given by

$$\Delta = J_1 \left[2(1 - 2\kappa) - 2\Gamma + \Gamma^2 \kappa / (1 - \kappa) \right], \tag{4.3.7}$$

for $\kappa < 0.5$, calculated up to the second order. For $\kappa > 0.5$, Δ is calculated in the first order only giving the result

$$\Delta = J_1 (2\kappa - 2 - \Gamma \cos 2\pi q), \tag{4.3.8}$$

from which an antiphase to "sinus" phase (modulated phase with modulation wave vector q) transition is predicted. The phase boundaries (para to ferro and antiphase to sinus) are obtained from the vanishing of Δ. Here, it is found that the paramagnetic phase indeed exists at zero Γ. No clear idea about the nature of the correlations is obtained in the paramagnetic phase. Barber and Duxbury [24] applied Rayleigh Schrödinger perturbation expansion methods about the trivial but exact limits $\Gamma \to 0$ and $\Gamma \to \infty$ and obtained the phase diagram. The Lifshitz point is obtained at a finite value of Γ here. They also used some numerical methods to find out the phase boundaries. However, the results are insufficient for $\kappa > 0.5$.

An exact solution (the only one to date for nonzero Γ) along a special line

$$\Gamma / J_1 = \kappa - 1/4\kappa \tag{4.3.9}$$

in (κ, Γ) plane has been obtained in the Hamiltonian limit (Peschel and Emery [311]). The argument runs as follows. Here, one writes the master equation for the one dimensional kinetic Ising model in such a form so that the time evolution operator can be expressed in terms of Pauli matrices and is also Hermitian. Thus

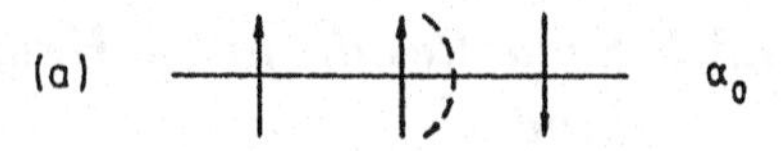

Fig. 4.3 The three basic spin flip processes along with their rates are shown

it describes, in general, a one dimensional quantum mechanical spin-1/2 problem. The master equation governing the time evolution of the probability of a certain spin configuration σ at time t for a linear chain of Ising spins with nearest neighbour interactions can be written as

$$\frac{\partial p(\sigma,t)}{\partial t} = -\sum_{\sigma'} W(\sigma \to \sigma')p(\sigma,t) + \sum_{\sigma'} W(\sigma' \to \sigma)p(\sigma',t) \tag{4.3.10}$$

where $p(\sigma,t)$ denotes the probability of finding the configuration σ at time t and $H(\sigma) = -J\sum_n \sigma_n\sigma_{n+1}$. W's are the transition probabilities. The master equation can also be written as

$$\frac{\partial \tilde{p}}{\partial t} = \sum \hat{O}\tilde{p} \tag{4.3.11}$$

where

$$\tilde{p}(\sigma,t) \sim \exp[-\beta H(\sigma)]p(\sigma,t) \tag{4.3.12}$$

and $\hat{O}$ denotes the time evolution operator. There are only three elementary spin flip processes to consider, which are shown in Fig. 4.3 along with their rates (α_0, α_+ and α_-). Assuming detailed balance, the rates for processes (b) and (c) are connected by $\alpha_- = \exp(-4K)\alpha_+$, where $K = \beta J$; β is the inverse temperature. Hence there are only two independent rate constants α_0 and α_+. The time evolution operator $\hat{O}$ can be expressed in terms of Pauli spin matrices [226]:

$$\hat{O} = \sum_n \left[A\sigma_n^x + B\sigma_{n-1}^z\sigma_n^x\sigma_{n+1}^z + C\sigma_n^z\sigma_{n+1}^z - D\sigma_n^z\sigma_{n+2}^z - E\right] \tag{4.3.13}$$

where the coefficients $A, B, \ldots, E$ are given in terms of the transition rates as given below:

$$A = \frac{1}{2}(\bar{\alpha} + \alpha_0) \tag{4.3.14a}$$

$$B = \frac{1}{2}(\bar{\alpha} - \alpha_0) \tag{4.3.14b}$$

$$C = \bar{\alpha}\sinh 2K \tag{4.3.14c}$$

$$D = \frac{1}{2}(\bar{\alpha}\cosh 2K - \alpha_0) \tag{4.3.14d}$$

$$E = \frac{1}{2}(\bar{\alpha} \cosh 2K + \alpha_0) \tag{4.3.14e}$$

where $\bar{\alpha} = \alpha + \exp(-2K)$. If B is made to vanish, so that (4.3.13) is of the same form as that of (4.3.2), one must have the condition given by (4.3.9). Thus, along this line (also known as the one dimensional line (ODL)), the Hamiltonian can be related to a soluble one-dimensional kinetic spin model, and the spin correlations in the horizontal direction here decay exponentially (from the exactly known spin correlations of $\hat{O}$, and is essentially one dimensional in character). This line touches the multiphase point $\kappa = 0.5$ at $\Gamma = 0$, proving that the disordered phase indeed extends down to $\Gamma = 0$ at $\kappa = 0.5$. The disorder line cannot be below this so called 'one dimensional line'.

4.3.2 Interacting Fermion Picture

The Hamiltonian for the ANNNI model in transverse field can be expressed in terms of interacting fermions [336, 359], following the Jordan-Wigner decoupling trick applied to (pure) transverse Ising chain in Sect. 2.2. A self-consistent Hartree-Fock method in the interacting fermion picture gives results (critical phase boundary for order-disorder transitions) for $\kappa \leq 0.5$. The Hamiltonian can be expressed in the form

$$\begin{aligned} H = -\Gamma N + 2\Gamma \sum_i c_i^\dagger c_i - J_1 \sum_i \left(c_i^\dagger - c_i\right)\left(c_{i+1}^\dagger + c_{i+1}\right) \\ - J_2 \sum_i \left(c_i^\dagger - c_i\right)\left(1 - 2c_{i+1}^\dagger c_{i+1}\right)\left(c_{i+1}^\dagger + c_{i+2}\right) \end{aligned} \tag{4.3.15}$$

where the following transformation has been made

$$S_i^x = \exp\left(-i\pi \sum_{j<i} c_j^\dagger c_j\right) c_i + c_i^\dagger \exp\left(i\pi \sum_{j<i} c_j^\dagger c_j\right) \tag{4.3.16}$$

$$S_i^z = 2c_i^\dagger c_i - 1 \tag{4.3.17}$$

and c_j's are Fermi operators satisfying

$$\left[c_i, c_j^\dagger\right] = \delta_{ij}, \qquad [c_i, c_j] = \left[c_i^\dagger, c_j^\dagger\right] = 0. \tag{4.3.18}$$

Here the model consists of a one dimensional cyclic chain of N Ising spins ($S_x = \pm 1$) so that $S_{N+1}^x = S_1^x$, $S_{N+2}^x = S_2^x$ etc. (note that here the canonical transformation $S_i^x \to S_i^z$ and $S_i^z \to -S_i^x$ has been made in (4.3.2) to use the Jordan-Wigner transformations conveniently). This Hamiltonian, in general, is not diagonalisable as it contains a four-fermion interaction coming from the $(1 - 2c_{i+1}^\dagger c_{i+1}) = -S_{i+1}^z$ term in (4.3.15). In the mean field approximation, $\langle S_i^z \rangle = 1$ in the para phase [312]. Also if J_2 is treated perturbatively, then $\langle S_i^z \rangle$ in the para phase (for $J_2 = 0$) differs slightly (less than 4 % numerically) from unity. Therefore, putting approximately

$S_i^z \simeq \langle S_i^z \rangle \simeq 1$, the Hamiltonian (4.3.15) can be expressed in a general quadratic form

$$H_d = -\Gamma N + \sum_{ij} c_i^\dagger A_{ij} c_j + \sum_{ij} c_i^\dagger B_{ij} c_j^\dagger \tag{4.3.19}$$

and can easily be put in a diagonal form

$$H_d = \sum_q \omega_q \eta_q^\dagger \eta_q - \frac{1}{2} \sum_q \omega_q, \tag{4.3.20}$$

where

$$\omega_q^2 = 4\big[\Gamma^2 + (J_1^2 + J_2^2) + \Gamma(2J_1 \cos 2\pi q + 2J_2 \cos 4\pi q) + (2J_1 J_2) \cos 2\pi q\big] \tag{4.3.21}$$

and the normal modes η_q and $\eta_q^\dagger$ are given by appropriate linear combinations of c_q and $c_q^\dagger$ [359]. Hence, from mode softening condition, the para to modulated phase boundary is given by

$$\Gamma/J_1 = |J_2|/(J_1), \tag{4.3.22}$$

and the para to ferro boundary is given by

$$\Gamma/J_1 = 1 - |J_2|/(J_1). \tag{4.3.23}$$

The phase diagram obtained from this method can be improved by using a self-consistent Hartree-Fock method in which the fermion Hamiltonian (4.3.15) can be approximately mapped [423] into the exactly solved H_d (given by (4.3.19)). When the Hamiltonian H in (4.3.15) is treated self-consistently, and the Hamiltonian is effectively written in the form of H_d, the renormalised parameters are given by [360] (see Sect. 4.A.1)

$$\Gamma' = \Gamma - 2J_2\big(\langle c_i^\dagger c_{i+1}\rangle + \langle c_i^\dagger c_{i+2}^\dagger\rangle\big) \tag{4.3.24a}$$

$$J_1' = J_1 + 4J_2\big(\langle c_i^\dagger c_{i+1}^\dagger\rangle + \langle c_i^\dagger c_{i+1}\rangle\big) \tag{4.3.24b}$$

$$J_2' = 2J_2\big(\langle c_i^\dagger c_i\rangle - 1/2\big). \tag{4.3.24c}$$

The primed variables are the unrenormalised ones. This result can also be obtained alternatively by employing a RPA-like approximation to the Hamiltonian (4.3.15)

$$\langle ABC\rangle = \langle AB\rangle\langle C\rangle + \langle AC\rangle\langle B\rangle + \langle A\rangle\langle BC\rangle$$

with proper signatures following fermion commutation rules and collecting the equivalent terms. Using relations (4.3.24a), (4.3.24b), (4.3.24c), the surfaces (4.3.22) and (4.3.23) map onto two corresponding surfaces in (Γ, κ) plane. For $\kappa'(= -J_2'/J_1') \le 0.5$, the para to ferro boundary is given by

$$\kappa = \frac{-2\pi\kappa'}{2\{\alpha(1 - \frac{1}{2}(1-\kappa')) + (-4\kappa'(1-\kappa') + 1)^{1/2}/(1-\kappa')\} + 4\kappa'\alpha}$$

$$\Gamma/J_1 = \frac{\alpha(1 - 2\kappa')}{2\{\alpha(1 - \frac{1}{2}(1-\kappa')) + (-4\kappa'(1-\kappa') + 1)^{1/2}/(1-\kappa')\} + 4\kappa'\alpha}$$

($\kappa \le 0.5$ as $\kappa' \le 0.5$ here). The essential steps and the expression for α is given in Sect. 4.A.1.

The other phase boundary (para to modulated) cannot be mapped here as all values of κ corresponding to $\kappa' > 0.5$ (where we get the para to modulated phase boundary in the earlier approximation for H) gives $\kappa = 0.5$ and $\Gamma = 0$ there.

4.3.3 Real-Space Renormalisation Group Calculations

The critical field and ground state energy are also obtained in the real-space renormalisation group (RSRG) approach [360]. The method followed here is the truncation method [183, 314] in which a number of spins are grouped in a block (see Sect. 2.4) and the Hamiltonian for a single block is solved exactly (here block size is three). Only the two lowest lying eigenstates, out of the possible states (here eight) are retained to construct an effective Hamiltonian having the same form as that of the original one. The process is iterated until a fixed point Hamiltonian is reached. It may be noted that for studying the fixed point structure for $\kappa > 0.5$, blocks with at least four spins should be constructed; otherwise even the ground state (antiphase) for at least $f = 0$ cannot be represented by the block. However, the problem then becomes difficult to tackle analytically and hence we restrict to blocks with three spins; thus restricting ourselves again to studies for $\kappa < 0.5$ only.

With the above three spin block, the effective Hamiltonian reads (see [360]; cf. [183, 314], see also Sect. 2.4.1)

$$H' = -\Gamma' \sum_i S_i^z - J_1' \sum_i S_i^x S_{i+1}^x - J_2' \sum_i S_i^x S_{i+2}^x + c \tag{4.3.25}$$

where the renormalised quantities (denoted by primes) are as follows

$$\Gamma' = -(x_o - x_1)/2 \tag{4.3.26a}$$

$$J_1' = J_1 a/b \tag{4.3.26b}$$

$$J_2' = J_2 a/b \tag{4.3.26c}$$

and

$$c = (x_o + x_1)/2 \tag{4.3.27}$$

with

$$a = \big[-2J_1(x_o + x_1)\{(x_1 - 3\Gamma)(x_o + 3\Gamma) + (x_1 + \Gamma)(x_o - \Gamma)\} + 4J_1 J_2^2 (x_o + x_1) \\ - 8J_1 J_2^3 + 4J_1 J_2\{(x_1 - 3\Gamma)(x_1 + \Gamma) + (x_o + 3\Gamma)(x_o - \Gamma)\}\big]^2 \tag{4.3.28}$$

and

$$b = \big[2\{(x_o + 3\Gamma)(x_o - \Gamma) - J_2^2\}^2 + \{-2J_1(x_o - \Gamma) + 2J_1 J_2\}^2 \\ \times \{-2J_1(x_o + 3\Gamma) + 2J_1 J_2\}^2\big]\big[2\{(x_1 - 3\Gamma)(x_1 + \Gamma) - J_2^2\}^2 \\ + \{-2J_1(x_1 + \Gamma) + 2J_1 J_2\}^2 + \{-2J_1(x_1 - 3\Gamma) + 2J_1 J_2\}^2\big] \tag{4.3.29}$$

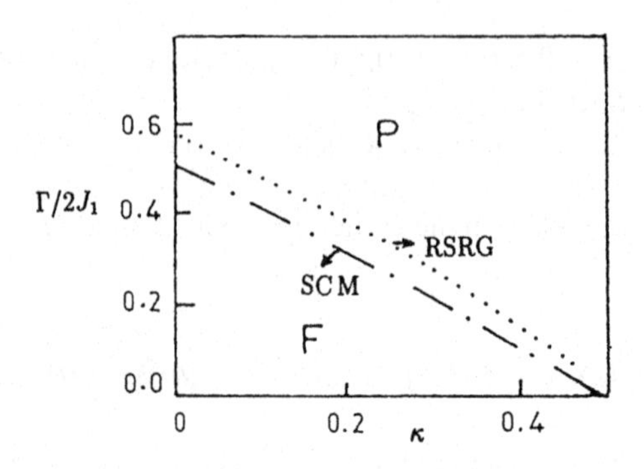

Fig. 4.4 The phase diagram for the ANNNI chain in transverse field from the RSRG and self-consistent methods for $\kappa < 0.5$ [360]. F denotes the ferromagnetic phase and P the paramagnetic phase

and x_o is the smallest root of the following equation

$$\begin{aligned} &x^3 + (\Gamma + J_2)x^2 + \left(-5\Gamma^2 - J_2^2 - 4J_1^2 + 2\Gamma J_2\right)x + 3\Gamma^3 \\ &\quad + \Gamma J_2^2 - 4(\Gamma - J_2)J_1^2 - 3J_2\Gamma^2 - J_2^3 = 0 \end{aligned} \tag{4.3.30}$$

while x_1 is the smallest root of (4.3.30) with $\Gamma \to -\Gamma$. Notice that at the zeroth iteration $J_2(0) = 2J_2$ as the contribution from the second neighbour appears twice in the inter-cell interaction when the cells consist of three spins. Also, the value of J_2/J_1 does not get renormalised at all. Thus the fixed point is determined only by the fixed point value of Γ/J_1. Above the critical value of Γ/J_1, any initial value of Γ approaches infinity, while below this value it iterates to zero; the resulting flow diagram gives the phase diagram. The phase diagram obtained from the Hartree-Fock approximation and RSRG approach are shown in Fig. 4.4.

Critical Ground State Energy The critical ground state energy values can be estimated from the self-consistent and RSRG methods. The ground state energy per site can be expressed as

$$\begin{aligned} E_g/N = &-\Gamma - J_2\left(\left\langle c_i^\dagger c_{i+2}^\dagger\right\rangle + \left\langle c_i^\dagger c_{i+2}\right\rangle\right)\left\langle c_i^\dagger c_i - 1/2\right\rangle \\ &- J_1\left(\left\langle c_i^\dagger c_{i+1}\right\rangle + \langle c_i c_{i+1}\rangle\right)/2 \\ &+ J_2\left(\left\langle c_i^\dagger c_{i+1}\right\rangle + \langle c_i c_{i+1}\rangle\right)\left(\langle c_i c_{i+1}\rangle + \left\langle c_i^\dagger c_{i+1}\right\rangle\right) \end{aligned} \tag{4.3.31}$$

so that

$$E_g/N = I_1 J_3 I_3/2\pi - \Gamma/4\pi - J_1\alpha/4\pi - J_2\alpha^2/4\pi^2 \tag{4.3.32}$$

where the expressions for the integrals I_1, I_3 and α are given in Sect. 4.A.1.

The RSRG method also gives estimate for the ground state energy along the critical line. The ground state energy per site here is given by

$$Eg/N = \sum_n \left[\frac{f(J_1(n), J_2(n))}{3^n}\right]/3 \tag{4.3.33}$$

where

$$f\left(J_1(0), J_2(0)\right) = (x_o + x_1)/2 \tag{4.3.34}$$

Notice that this method does not produce the exact results $E_g/N = -1/\pi$ and $\Gamma/J_1 = 0.5$ for $\kappa = 0$. The behaviour of E_g, however, agrees well with the results of the self-consistent method.

4.3.4 Field-Theoretic Renormalisation Group

The analytical methods discussed so far do not give the evidence for the existence of a floating phase (with algebraic decay of correlations) directly. However, the existence of a floating phase can be easily justified for this model at zero temperature; although its location cannot be found so easily. Using Gaussian functional averages [152] (over the transverse field term) for the spin correlations, the effective Landau-Ginzburg Hamiltonian may be written as (see Sect. 3.6.2).

$$H \sim \sum_m S\big[1 + JG(\omega_m)\big]S + O\big(S^4\big) \tag{4.3.35}$$

with the spin Green's function $G(\omega_m) \sim \Gamma \tanh \beta\Gamma(4\Gamma^2 + \omega_m^2)$ and Matsubara frequencies $\omega_m = 2\pi m/\beta$. Also, because of the competing interactions, we expect the fluctuations (with q and $-q$) over some modulated structure (say with wave vector q) to be dominant in the critical region, thereby effectively driving [152] an n-component competing system equivalent to a $2n$-component system without competition ($n = 1$ here for an Ising system):

$$H \sim \int d^d q \sum_{\alpha=1}^{2} \sum_m \big(r_m + q^2\big) S_\alpha(q) S_\alpha(-q) + O\big(S^4\big) \tag{4.3.36}$$

where $r_m = r + \omega_m^2$; r being the usual critical temperature interval. At $T = 0$, the Matsubara frequencies become continuous and the sum over m gives an effective additional dimension ($d \to d + 1$). The system therefore shows an effective $(d + 1)$-dimensional classical behaviour (due to integration over quantum fluctuations) for an effective 2-component (XY-like) regular magnetic system (because of integrations over the competing fluctuations in the Ising system). For our one dimensional ANNNI model in transverse field, we thus expect effectively two dimensional XY-like (power law) correlations (floating phase [288, 412]) at zero temperature. At finite temperatures, of course, the lowest value of r_m for which the field remains finite after renormalisation, is that for $m = 0$ and fields corresponding to other m values become irrelevant [152]. There is thus no dimensional increase for $T > 0$ and no floating phase is expected.

4.3.5 Numerical Methods

The phase diagram for the quantum ANNNI model has also been obtained [362] using numerical methods like exact diagonalisation for finite size and Strong Coupling

Eigenstate Method (see Sect. 2.3.2). In fact, so far, this method gives us the most detailed and accurate phase diagram.

In the exact diagonalisation method the quantum Hamiltonian (4.3.2) has been considered where the transverse field aligns spins in the x-direction for conveniently using the representation in which S^z are diagonal. The essential step is to diagonalise the $2^N \times 2^N$ Hamiltonian matrix H_N for an open chain of N spins (with interactions given by (4.3.2)). The matrix H_N is constructed from the recursion relation

$$H_N = \begin{pmatrix} H_{N-1} + D_{N-1} & -\Gamma 1_{N-1} \\ -\Gamma 1_{N-1} & H_{N-1} - D_{N-1} \end{pmatrix}. \tag{4.3.37}$$

Here H_{N-1} is the Hamiltonian matrix for $N-1$ spins, 1_{N-1} is the unit matrix of size $2^{N-1} \times 2^{N-1}$ and D_{N-1} is a diagonal matrix of the same size with diagonal elements $J_1 + J_2, \ldots, J_1 - J_2, \ldots, -J_1 + J_2, \ldots, -J_1 - J_2, \ldots$ where $\ldots$ means repetition of the preceding elements 2^{N-3} times. This form however occurs only when the arrangement of the 2^N configurations is the one that is obtained from direct product of N two-component vectors $(+,-)$. (Thus, for 3 spins the arrangement is $+++, ++-, +-+, +--, -++, -+-, --+, ---$.) Proof of (4.3.37) follows by noting that, if we add a spin at the left end to the arrangements of $N-1$ spins, then the interaction of this spin with the rest of the chain is given by

$$\begin{pmatrix} D_{N-1} & -\Gamma 1_{N-1} \\ -\Gamma 1_{N-1} & -D_{N-1} \end{pmatrix}.$$

A further simplification of (4.3.37) is also possible

$$UHU^{-1} = \begin{pmatrix} H_{N-1} - D_{N-1} + \Gamma F_{N-1} & 0 \\ 0 & -H_{N-1} - D_{N-1} + \Gamma F_{N-1} \end{pmatrix} \tag{4.3.38}$$

where

$$U = (1/\sqrt{2}) \begin{pmatrix} 1_{N-1} & -F_{N-1} \\ F_{N-1} & 1_{N-1} \end{pmatrix}$$

and F_{N-1} is a matrix (of size $2^{N-1} \times 2^{N-1}$) having 1 along the diagonal connecting top-right corner to bottom-left corner and zero elsewhere. This reduces the $2^N \times 2^N$ problem to two $2^{N-1} \times 2^{N-1}$ problems. Proof of (4.3.38) necessitates the relationships

$$H_{N-1} F_{N-1} = F_{N-1} H_{N-1},$$
$$D_{N-1} F_{N-1} = F_{N-1} D_{N-1},$$

which in turn follow from the observation that reversal of each S_i^z: (i) keeps V $(= -[\sum_i J_1 S_i^z S_{i+1}^z + \sum_i J_2 S_i^z S_{i+2}^z])$, H_0 $(= -\sum_i \Gamma S_i^x)$ and hence H_{N-1} the same but reverses the sign of D_{N-1} and (ii) is equivalent to replacing the row and column index i (of H_{N-1} and D_{N-1}) by $N-i+1$.

We would like to add that for any form of H_0 (where $H = H_0 + \Gamma \sum_i S_i^x$), a relationship of the form of (4.3.37) can be built up (with, of course, a different D_{N-1}) but the simplification by (4.3.38) will need the $S_i^z \to -S_i^z$ symmetry.

After building up the Hamiltonian matrix H_N, the eigenvector corresponding to the ground state is obtained and the r-th neighbour correlation, defined as $g(r) \equiv \langle S_1^z S_{1+r}^z \rangle$ is calculated in this state. From the nature of the correlation function, the different phases (from the data for $0 \le \kappa \le 1.0$ and with $N = 8$, for $0.2 \le \kappa \le 0.8$ with $N = 10$) have been identified. The mass gap $\Delta = \langle \psi_1 | H | \psi_1 \rangle - \langle \psi_0 | H | \psi_0 \rangle$ where ψ_0 and ψ_1 are the are the ground state and the first excited state respectively, has also been determined. From the plot of Δ vs. Γ / J_1 (for $N \le 8$), the critical fields Γ_c following the method of Hamer and Barber [166, 167] are obtained. From the spin-spin correlations five different regions in the phase diagram were clearly identified:

A: Ferromagnetic: $g(r) \sim m^2 + \exp(-r/\xi)$
B: Antiphase: $g(r) \sim m^2 + \exp(-r/\xi) \cos(\pi q r)$, $q = 1/2$
C1: Paramagnetic: $g(r) \sim \exp(-r/\xi)$
C2: Paramagnetic: $g(r) \sim \exp(-r/\xi) \cos(\pi q r)$, $q \le 1/2$
C3: Floating: $g(r) \sim r^{-\eta} \cos(\pi q r)$, $1/3 < q < 1/2$, $\eta \sim 10^{-1}$

The same plot has also been obtained using the strong coupling eigenstate method discussed next.

Strong Coupling Eigenstate Method (SCEM) As an alternative to the above method of estimating the mass gap $\Delta(\Gamma)$, an approximate diagonalisation scheme, called the Strong Coupling Eigenstate Method (see Sect. 2.3.2) has also been used. This method has been shown to lead to the exact analytical results [312] for the Ising chain in a transverse field (for continuous phase transition) and has also been used for estimating the phase boundaries for other quantum chains [129, 165, 187].

The SCEM has the advantages that the number of basis states is lowered considerably (the introduction of J_2 does not increase the number of states) and that the exact analytical results for $J_2 = 0$ can be reproduced with high accuracy. On the other hand the disadvantage is that one cannot find the eigenstates and even with the benefit of reduced system size, one cannot really go to larger chains as it quickly becomes impossible to find out the matrix elements. The essential idea of this method, as mentioned earlier in Sect. 2.3.2, is to generate a set L of strongly coupled eigenstates of H_0 $(= -\sum_i \Gamma S_i^x)$ by successive applications of the operator $V = -[\sum_i J_1 S_i^z S_{i+1}^z + \sum_i J_2 S_i^z S_{i+2}^z]$ to an unperturbed eigenstate $|0\rangle$ of H_0. The matrix (say, H_r) for the Hamiltonian operator H_0 is then constructed for the basis set L. The lowest eigenvalue of H_r gives the ground state energy provided $|0\rangle$ is chosen as the ground state eigenfunction (namely $S_i^x = 1$ for all i) of H_0. To obtain the energy of the first excited state we repeat this process choosing $|0\rangle$ as the first excited state of H_0 (namely, all $S_i^x = 1$ except any one which is -1). The efficiency of this method is demonstrated by the fact that for $N = 5$ and 7 (say) the set L consists of only 4 and 9 states. The mass gaps are measured and corresponding critical fields obtained from numerical calculations performed on system sizes $N = 4, 5, 6$ and 7 with periodic boundary conditions.

The phase diagram, obtained from the behaviour of the correlation as well as the mass gap is shown in Fig. 4.5. The SCEM seems to be reliable only for $\kappa < 0.5$ and

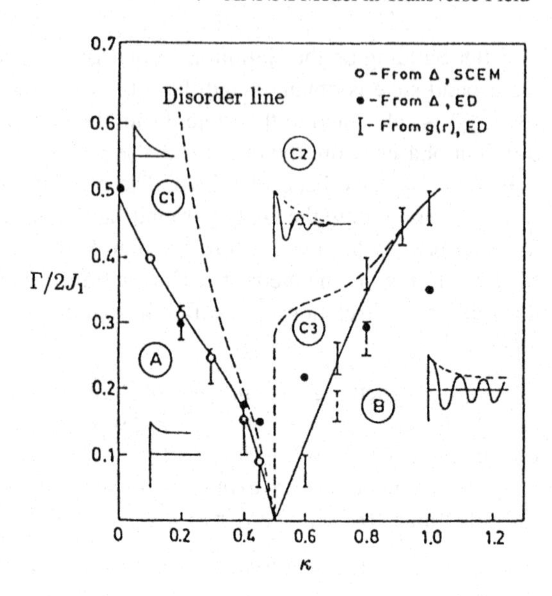

Fig. 4.5 Phase diagram of one dimensional quantum ANNNI model obtained from numerical methods indicating the presence of different phases explained in the text

in this region it agrees reasonably with the results obtained by the exact diagonalisation (ED) (see Fig. 4.5).

4.3.6 Monte Carlo Study

The quantum one dimensional ANNNI model in a transverse field can be mapped to a two dimensional classical model using the Suzuki-Trotter formula to give an equivalent Hamiltonian (see Sect. 3.1)

$$H_{\text{eff}} = \sum_{i=1}^{N}\sum_{k=1}^{M}\left[(J_1\beta/M)S_{ik}S_{i+1k} + (J_2\beta/M)S_{ik}S_{i+2k} + J_3 S_{ik}S_{ik+1}\right] \quad (4.3.39)$$

where

$$J_3 = (1/2)\left[\ln\coth(\beta\Gamma/M)\right]. \quad (4.3.40)$$

The Monte Carlo simulation has been performed [18] for the classical two dimensional ($N \times M$) model with periodic boundary conditions. Although the mapping is exact only when the Trotter dimension (M) is very large, one works with finite M values (typically M is between 30 and 70) and N (number of sites along the horizontal axis corresponding to the number of sites in the original quantum system) is varied between 60 to 200. The energy and correlation functions were evaluated to find the phase boundaries. The phase diagram again is very similar to that of the approximate two dimensional classical ANNNI model. It shows the presence of ferromagnetic, paramagnetic, antiphase and floating incommensurate phases as well as disorder line in the para phase. The position of the ferromagnetic to paramagnetic

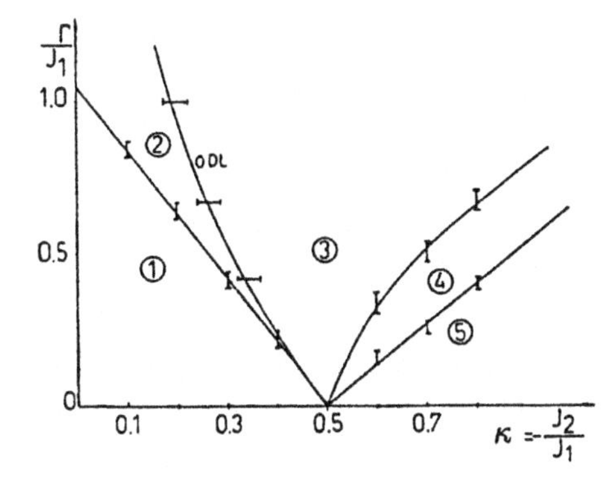

Fig. 4.6 Phase diagram of quantum one dimensional ANNNI model obtained from Monte Carlo method (from [18]). The presence of Ferromagnetic (*1*), Disordered (*2*), Disordered with oscillatory decay of correlations (*3*), Floating incommensurate (*4*) and Antiphase (*5*) are shown. *ODL* (one dimensional line) is drawn from the exact analytical result [311]

phase agrees well with the one obtained in the two-dimensional case in [352] which in the present case reads

$$\sinh\left[2\left(\frac{J_1\beta}{M}+\frac{2J_2\beta}{M}\right)\right]\sinh 2J_3 = 1. \tag{4.3.41}$$

The one dimensional line is detected through the minimum in the horizontal correlation length which also agrees with the exact result [311]. The phase diagram is shown in Fig. 4.6.

4.3.7 Recent Works

We are interested in the ground state (zero-temperature) properties of the Hamiltonian (4.3.2) with $\kappa = -J_2/J_1 \geq 1$. Exact diagonalisation studies were made for chains up to a length of 32 spins by Uimin and Rieger [330]. Quantum Monte Carlo studies were performed by Arizmendi et al. [18]. Bosonisation and a renormalisation group analysis have been used by Allen et al. [9] and later by Dutta and Sen [115]. All these studies support the presence of floating phase characterised by a spin-spin correlation in longitudinal direction that decays algebraically with distance. Careful studies using density matrix renormalisation group for $\kappa < 0.5$ [31] and $\kappa > 0.5$ [32] have also been performed, which confirm the presence of floating phase over a region of κ–Γ phase space that extends from $\kappa = 0.5$ to $\kappa = 1.5$ and beyond. Another numerical study [291] based on entanglement scaling of matrix product states however found evidence of this phase only over a narrow region.

In contrast to these approximate studies, exact results are only two in number: (i) An early study by Peschel and Emery [311] (see also [115]) showed that along the line $\Gamma/J = \kappa - 1/(4\kappa)$ the ground state has a direct product form and the correlation function decays exponentially with distance; (ii) a perturbative analysis [67] showed

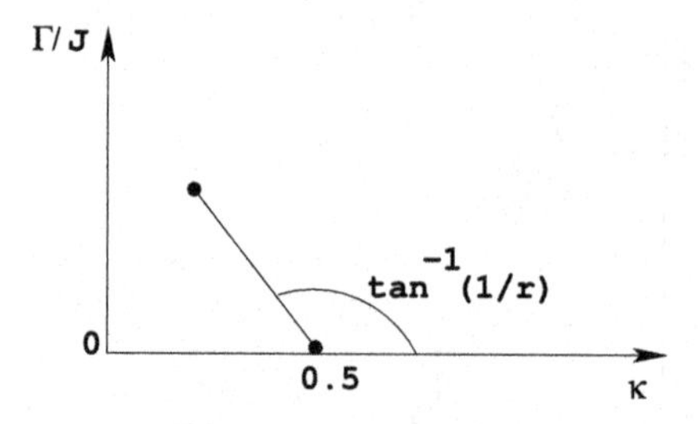

Fig. 4.7 Basic principle of the perturbation treatment. The perturbation takes the transverse field from 0 to Γ and the frustration parameter κ from 0.5 to $0.5 + r\Gamma$. Obviously, the parameter Γ should be small for perturbation expansion to be valid

that precisely two phase transition lines pass through the $(\kappa, \Gamma) = (0.5, 0)$ point, making angles $\pi/4$ and $\pi - \tan^{-1} 2$ with the κ axis. Here we shall present the basic principle of the latter work.

The basic idea is to break up the Hamiltonian as

$$H = H_{cl} + H_p \tag{4.3.42}$$

with

$$H_{cl} = -J \sum_j \left[S_j^z S_{j+1}^z - \frac{1}{2} S_j^z S_{j+2}^z \right] \quad \text{and} \quad H_p = \sum_j \Gamma \left[-S_j^x + r S_j^z S_{j+2}^z \right]$$

and then treat H_p as a perturbation over H_{cl}. (Here $r = J(\kappa - 0.5)/\Gamma$.) The perturbation takes the Hamiltonian H_A from the multiphase point $(\kappa, \Gamma) = (0.5, 0)$ to $(0.5 + r\Gamma, \Gamma)$ (Fig. 4.7). One can show [67] that the first order perturbation correction to the ground state energy eigenvalue $E^{(1)}$ can be calculated *exactly* by mapping the problem to an excited state of the nearest-neighbour transverse Ising model for which all the eigenstates are exactly known. Now, if there is a phase transition line passing through the multiphase point at an angle $\tan^{-1}(1/r_0)$ with the increasing κ axis, then the quantity $E^{(1)}$ treated as a function of r will show a non-analyticity at $r = r_0$. It is found that this occurs only at two values $r = -0.5$ and 1, where $d^2E^{(1)}/dr^2$ diverges. Hence, only two critical lines pass through the multiphase point. It is known that at small Γ, the system is in ferromagnetic state for $\kappa < 0.5$ and in antiphase state for $\kappa > 0.5$. If we assume that there does exist a floating phase adjacent to the antiphase, then the phase diagram should be either of the two types shown in Fig. 4.8. To find which of the two types represents correctly the ground state, we note that since the problem has now been mapped to a class of excited states of the nearest-neighbour model, one can calculate the mass gap correct up to first order perturbation. Also, the longitudinal spin-spin correlation for the relevant nearest-neighbour model can be calculated exactly by evaluating the related Toeplitz determinants using Szego's theorem [68]. This leads to a knowledge of the longitudinal susceptibility for the zero-th order eigenstate. It turns out that for the entire range $-0.5 < r < 1$ the mass gap vanishes and the longitudinal susceptibility diverges. This rules out the type in Fig. 4.8(a) and confirms the one in Fig. 4.8(b). Note that *three* lines meet tangentially and have the same slope (namely -0.5) at

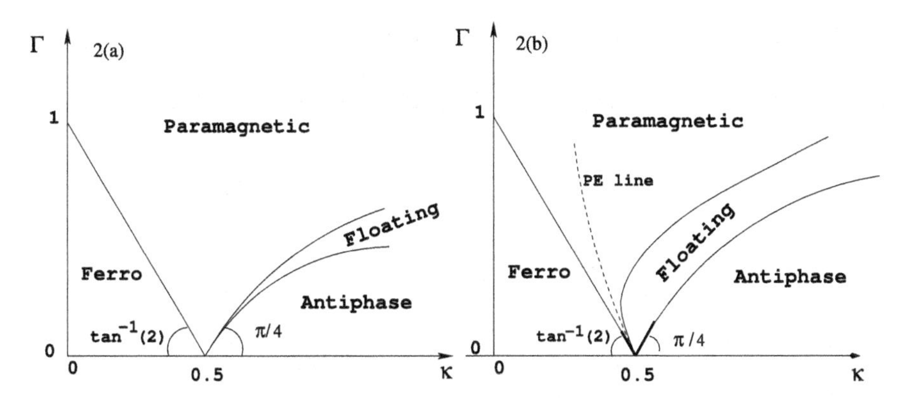

Fig. 4.8 Two types of (schematic) phase diagram consistent with the fact that there should be precisely two critical lines emanating from the multiphase point. Figure (**b**) is confirmed by further analysis. *Bold lines* in figure (**b**) indicate exact result

the multiphase point: the ferromagnetic-paramagnetic boundary, the paramagnetic-floating phase boundary and the Peschel-Emery line.

It has been mentioned before that the ground state for the transverse ANNNI chain can be mapped to 2D ANNNI model for some extreme values of the parameters. It has been shown quite convincingly that in 2D ANNNI model, the floating phase is either absent or present only over a very narrow parameter range [66, 102]. This fact can be reconciled [66] with the presence of floating phase in transverse ANNNI model by noting that the interaction parameters of the equivalent classical model is much larger than that in the other direction.

4.4 Large S Analysis

Uptill now, the results for systems with $S = \pm 1$ have been discussed. The ANNNI model in a transverse field has also been studied in the large S approximation [171, 355, 356]. In [356], the Hamiltonian is rewritten as

$$H = -\frac{1}{S}\left[\sum_i J_1 S_i^z S_{i+1}^z + \sum_i J_2 S_i^z S_{i+2}^z\right] - \sum_i \Gamma S_i^x \tag{4.4.1}$$

where for any spin S_i,

$$\sum_\alpha S_i^\alpha S_i^\alpha = S(S+1) = S_c^2. \tag{4.4.2}$$

The variable S has been introduced as the normalisation factor in (4.4.1) to ensure a proper limit for the Hamiltonian as $S \to \infty$. Under this approximation, the vector $n_\alpha = S^\alpha / S_c$ are classical (commuting) variables. An angle θ is defined, with $0 < \theta < \pi$ through the relations $\langle n_i^z \rangle = \cos\theta$ and $\langle n_i^x \rangle = \sin\theta$ denoting the expectation values of n_i in one of the classical ground state of the system (see Sect. 4.2). Three possible phases are immediately identified: (a) a paramagnetic phase in which all $\theta = \pi/2$, (b) a ferromagnetic phase in which all θ are equal but not $\pi/2$ and (c) an antiphase where the θ values look like $(\theta, \theta, \pi - \theta, \pi - \theta)$ in a single period. In terms of θ, the classical energy is easily found out:

$$E_0/S_c = -\sum_i \left[(\cos\theta_i \cos\theta_{i+1}) + \kappa(\cos\theta_i \cos\theta_{i+2})\right] - \Gamma \sum_i \sin\theta_i. \qquad (4.4.3)$$

From the minimum energy conditions in the three simple phases, the phase boundaries are obtained which give a tricritical point at $(\kappa, \Gamma) = (1/2, 1)$. The ferromagnetic phase is stable for $\Gamma < 2(1 - \kappa)$ when $\kappa < 0.5$ and the antiphase is stable for $\Gamma < 2\kappa$: when $\kappa \geq 0.5$. In the rest of the phase diagram, the paramagnetic phase wins. This simple picture is modified considerably with a spin wave analysis. In each phase (denoted generically by p), the energy of the spin waves (magnons) are obtained as a function of q quantised in units of $2\pi/N$ and $-\pi < q < \pi$. In this treatment, a diagonal form for the quantum Hamiltonian is obtained (when the ground state is any one of the three phases) describing the magnons. The total energy for any p is expressed as (see Sect. 4.A.2)

$$E(p) = E_0(p) + \frac{N}{2\pi} \int_{-\pi}^{\pi} dq\, E_1(q; p) \qquad (4.4.4)$$

where

$$E_1(q; p) = \left[A_q(p) B_q(p)\right]^{1/2} \qquad (4.4.5)$$

and will be stable if the product $A_q(p)B_q(p)$ is positive.

In each of the three simple phases, the instability to different q fluctuations gives the phase boundaries. For example, in the paramagnetic phase, the square of the energy is

$$E_1^2(q; p) = \Gamma(\Gamma - 2\kappa - 1/4\kappa). \qquad (4.4.6)$$

If $\kappa < 0.25$ this has a minimum at $q = 0$ and hence the para phase becomes unstable with respect to the $q = 0$ fluctuations (i.e., the ferromagnetic phase) at $\Gamma = 2(1 - 2\kappa)$ which is precisely the phase boundary between the paramagnetic and ferromagnetic phases for $\kappa < 0.25$. Similarly, the paramagnetic to modulated phase boundary can also be obtained as $\Gamma = 2\kappa + 1/4\kappa$ which meet at the Lifshitz point ($\kappa = 0.25$, $\Gamma = 3/2$). Some further analysis is necessary to find out the more complicated phases and the phase diagram is drawn on the basis of these results. There is a signature of infinite number of phases in the region bounded by the three major phase boundaries. The most important phases obtained from this analysis are shown in Fig. 4.9.

In another study [355], the large S analysis of the dual model given by (4.3.6) has been made. The phase diagram obtained from this study has three regions: paramagnetic, ferromagnetic and the antiphase. The Peschel-Emery curve is found to be the disorder line here. In [171], a Heisenberg Hamiltonian has been considered

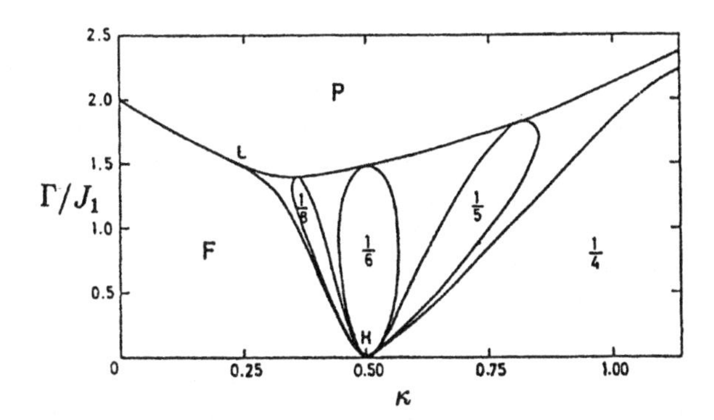

Fig. 4.9 Phase diagram from large S analysis (from [356]). The symbols P, F, $1/8$, $1/6$ and $1/4$ stand for paramagnetic, ferromagnetic and various modulated phases characterised by a rational number q. L is the Lifshitz point

which is

$$H = -\frac{1}{S^2}\left[\sum_{ij} J_1 \boldsymbol{S}_{ij}\cdot\boldsymbol{S}_{i+1,j} + \sum_{ij} J_2 \boldsymbol{S}_{ij}\cdot\boldsymbol{S}_{i+2,j} + J_0\sum_{ijj'}\boldsymbol{S}_{ij}\cdot\boldsymbol{S}_{i,j'}\right] - \frac{D}{S^2}\sum_{ij}\left(\left[S_{ij}^z\right]^2 - S^2\right) \tag{4.4.7}$$

where i labels the layers perpendicular to the axial direction and j and j' denote the nearest neighbour spins within a layer. The $D \to \infty$ limit gives the classical ANNNI model and the quantum effect is through nonzero $1/D$. The effect of quantum fluctuation was studied in the lowest order in $1/S$ at zero temperature and the phase diagram obtained in the κ–$1/D$ plane. It is observed that the degeneracy at the multiphase point $\kappa = 0.5$ is removed and transition from the state with periodicity 4 (antiphase) to the uniform ferromagnetic state occurs via a sequence of first order transitions.

4.5 Results in Higher Dimensions

In higher dimension, one can get an estimate of the phase diagram by applying mean field theory and the random phase approximations. A path integral method has also been applied which yields results for dimensions greater than one but is again restricted to $\kappa \le 0.5$. The Hamiltonian for the ANNNI model in transverse field in three dimension (with $S_z = \pm 1$) is

$$H = -(J_0/2)\sum_{ijj'} S_{ij}^z S_{ij'}^z - J_1\sum_{ij} S_{ij}^z S_{i+1,j}^z - J_2\sum_{ij} S_{ij}^z S_{i+2,j}^z - \sum_i \Gamma S_i^x. \tag{4.5.1}$$

(i) Phase Diagram in the Mean Field Approximation The phase diagram in the Γ–κ plane can be obtained from the mean field theory and random phase approximation. In the mean field theory, the effective vector field $\boldsymbol{F}_q$ on $\boldsymbol{S}_q$ and its mean field average value $\langle \boldsymbol{S}_q\rangle$ are given by

$$\boldsymbol{F}_q = \Gamma\hat{x} + J(q)\langle S_q^z\rangle\hat{z} \tag{4.5.2a}$$

and

$$\langle S_q \rangle = \left[\tanh\left(\beta |F_q|\right)\right] F_q / |F_q| \tag{4.5.2b}$$

where

$$J(q) = 4J_0 + 2J_1 \cos 2\pi q + 2J_2 \cos 4\pi q. \tag{4.5.2c}$$

Above the transition temperature $T_c(\Gamma)$

$$\langle S_q^z \rangle = 0, \qquad \langle S_q^x \rangle = \tanh(\beta\Gamma)\delta_{q,0}. \tag{4.5.3}$$

It may be noted that since there is no competition in the transverse direction, $\langle S_q^x \rangle$ exists for only the homogeneous ($q = 0$) mode. The static susceptibility is given by

$$\chi = \langle S_0^x \rangle / \left(\Gamma - J(q)\langle S_0^x \rangle\right), \tag{4.5.4}$$

which diverges at the phase boundary given by

$$\Gamma / \tanh(\beta\Gamma) = J(q), \tag{4.5.5}$$

where this occurs for each q (coming from the para phase), when $J(q)$ is maximum. Hence the phase boundary is given by (4.5.5) when $\cos 2\pi q = -J_1/(4J_2)$ i.e.,

$$\frac{\Gamma/J_1}{\tanh(\beta\Gamma)} = 4 + 1/4\kappa + 2\kappa \tag{4.5.6}$$

where $\kappa = |J_2|/J_1$ and J_0 is set to equal to J_1. The schematic phase boundary is shown in Fig. 4.10. It may be noted that the Lifshitz point occurs at the same value of $J_2/J_1 = 0.25$ as in the classical ANNNI model. Also at finite temperature, when the transverse field Γ is small, the order-disorder phase boundary equation becomes

$$T_c/J_1 = 4 + 1/4\kappa + 2\kappa, \tag{4.5.7}$$

which is identical to that for the classical ANNNI model. For finite Γ, T_c is gradually suppressed (following (4.5.5)). At $T = 0$ one can have here a non-trivial transition, when the left side of (4.5.5) becomes equal to Γ/J_1. This gives the expression $\Gamma/J_1 = 4 + 1/4\kappa + 2\kappa$ for the equation of the order-disorder phase boundary, where the transition is driven by the transverse field (see Fig. 4.10). It may be noted that the determination of the phase boundaries of the various modulated (commensurate and incommensurate) phases (denoted by M in Fig. 4.10) requires a self-consistent numerical solution for each component of the vector equation (4.5.2a), (4.5.2b), (4.5.2c) with adjustable modulation periods. Some numerical solutions for the mean field phase diagram were considered by Tentrup and Siems [399] and the results are in agreement with the above analysis.

In the para phase, the equation of motion for $\mathbf{S}$ in RPA can be written as [48]

$$\dot{\mathbf{S}}_q = 2M_q \mathbf{S}_q \tag{4.5.8}$$

with

$$M_q = \begin{pmatrix} 0 & 0 & 0 \\ 0 & 0 & J(q)\langle S_0^x \rangle - \Gamma \\ 0 & \Gamma & 0 \end{pmatrix}, \tag{4.5.9}$$

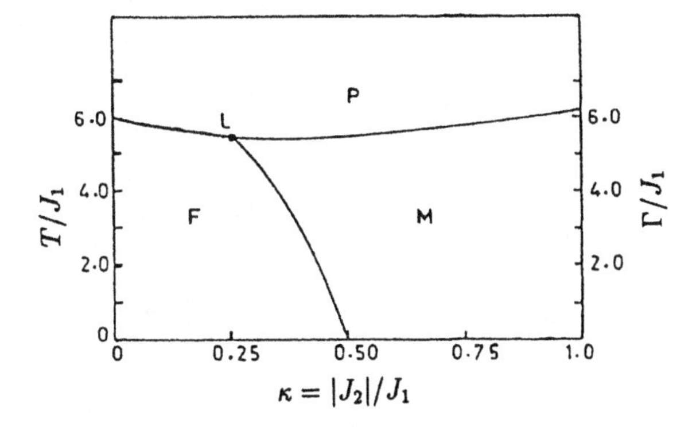

Fig. 4.10 The mean field phase diagram for quantum ANNNI model

giving the dispersion relation for the excitations

$$\omega_q^2 = 4\Gamma\left[\Gamma - J(q)\langle S_0^x\rangle\right] \tag{4.5.10}$$

where $J(q)$ is as given by (4.5.2c). These modes soften at the same phase boundary giving an identical equation (4.5.7) for the phase boundary. The mean field phase diagram is shown in Fig. 4.10.

(ii) Phase Diagram from Path Integral Approach In the path integral approach [358, 384], one can consider an Ising system in any dimension with arbitrary number of nearest and next nearest neighbours (see Sect. 3.3). The effective Hamiltonian obtained after the Suzuki-Trotter transformation is written as

$$\begin{aligned} H_{\text{eff}} = {} & -J\left(\sum_{k,\langle ij\rangle} S_{ik}S_{jk} + \kappa \sum_{k,[ij]} S_{ik}S_{jk}\right)/M + c/\beta \\ & + \ln\left[\coth(\beta\Gamma/M)\right]\sum_i\sum_k S_{ik}S_{ik+1}/2\beta. \end{aligned} \tag{4.5.11}$$

Here k is the Trotter index and M is the Trotter dimension. The constant $c = (1/2)\ln[\cosh(\beta\Gamma/M)\sinh(\beta\Gamma/M)]$ and $\langle\cdots\rangle$ and $[\cdots]$ indicate nearest and next nearest neighbours respectively. Treating the spins as M-component vector spins with components

$$S_j^k = (\pm 1, \pm 1, \ldots, \pm 1) \tag{4.5.12}$$

where $k = 1, 2, \ldots, M$, one can break the effective Hamiltonian in two parts H_0 and V, where

$$-\beta H_0 = \sum_i \mathbf{S}_i \cdot a \cdot \mathbf{S}_i + C \tag{4.5.13}$$

and

$$V = J\left(\sum_{\langle ij\rangle} \mathbf{S}_i \cdot \mathbf{S}_j + \kappa \sum_i \mathbf{S}_i \cdot \mathbf{S}_{i+2}\right)/M \tag{4.5.14}$$

where $a_{k,k'} = (1/2)\ln[\coth(\beta\Gamma/M)]\delta_{k,k'}$, and $C = NMc$. Now the full Hamiltonian can be treated perturbatively such that the free energy $F(= \ln Q;\ Q = \mathrm{Tr}\exp(-\beta H_{\mathrm{eff}}))$ is given by

$$-\beta F = -\beta F_0 + \sum_n (1/n!)(-\beta)^n C_n(V) \tag{4.5.15}$$

with F_0 the free energy corresponding to the unperturbed Hamiltonian H_0 such that

$$-\beta F_0 = \ln Q_0 \tag{4.5.16}$$

with

$$Q_0 = \sum \exp(-\beta H_0), \tag{4.5.17}$$

and the cumulants are given by

$$C_1 = \langle V\rangle_0; \qquad C_2 = \langle V^2\rangle_{0'} - \{\langle V\rangle_0\}^2 \tag{4.5.18}$$

etc.

The above expression can be regarded as an expansion in successively higher order of fluctuations. With classical systems, the first order term gives the mean field estimate and higher orders constitute fluctuation corrections. The critical field where the average magnetisation vanishes, is obtained performing calculations up to second order (following Kirkwood's prescription of classical spins) and is given by

$$\Gamma/J = \left[\delta_1 + \delta_2\kappa + \delta_1(\delta_1 - 5/2) + \kappa^2\delta_2(\delta_2 - 5/2) + 2\delta_1\delta_2\kappa\right]^{1/2}/2. \tag{4.5.19}$$

We can consider different systems (with different sets of δ_1 and δ_2 corresponding to coordination numbers for nearest and next nearest neighbours respectively) and obtain the phase diagrams. Along with the ANNNI model (in dimensions greater than one), one can also consider systems with next nearest neighbour interactions along more than one axis like the biaxial next nearest neighbour Ising (BNNNI) model or the totally isotropic next nearest neighbour model in different dimensions. Here, the values of Γ for all values of κ cannot be estimated as we are only considering $m \to 0$ critical lines. There will be certain values of κ where m is already zero at zero field (e.g. the classical modulated ground states like the antiphase with two spins up and two spins down alternately) and therefore cannot be considered here. Also, the $m \to 0$ transitions now indicate transitions from a ferromagnetic phase to either a paramagnetic phase or a modulated phase with $m = 0$. Therefore, the vanishing of the ferromagnetic order appears to be the major result of this method when applied to these systems. We take the example of the following models in one, two (square lattice) and three (cubic lattice) dimensions.

1. The ANNNI model: In all dimensions, one is restricted to $\kappa < 0.5$, as the antiphase with $m = 0$ is the classical ground state beyond this value. In one dimension, there is no solution as the right hand side of (4.5.19) becomes imaginary. In two dimensions, we find that results are obtained only up to $\kappa < 0.38$. In three dimensions, there is no problem in estimating Γ up to $\kappa < 0.5$.

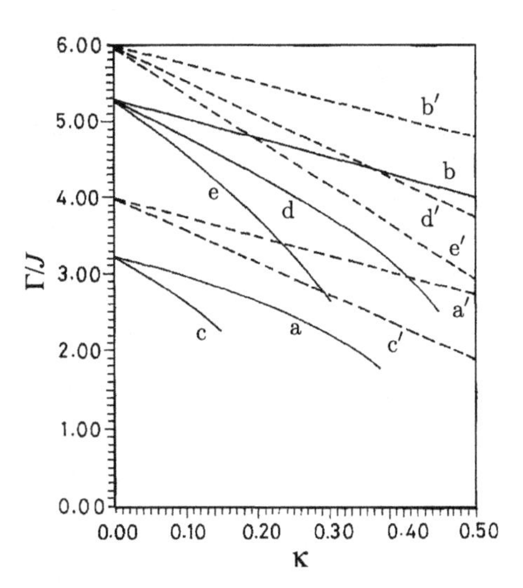

Fig. 4.11 The phase diagram for different systems from path integral method (**a**) and (**b**) for ANNNI model in two and three dimensions respectively, (**c**) and (**d**) for BNNNI model in two and three dimensions respectively, (**e**) for isotropic next nearest neighbour model in three dimensions. The *broken curves* correspond to the mean field results

2. The BNNNI model can be considered in two or three dimensions again with $\kappa < 0.5$.
3. Isotropic next nearest model in three dimensions. The results are shown in Fig. 4.11.

All these results show that their are some restricting values of κ up to which fields can be calculated. The restrictions become more important for lower dimensions and higher values of κ.

Since the approximation is known to be more accurate in higher dimensions, the increasing limitations in the lower dimensions are not surprising. Moreover, we find that this approximation also becomes weaker as the frustration parameter is made more effective.

4.6 Nearest Neighbour Correlations in the Ground State

The possibility of the existence of nearest neighbour dimers in the antiferromagnetic transverse ANNNI chain has been investigated numerically [361]. One can measure the dimer order parameter which for one dimension takes the simple form

$$G = |g_1 - g_2| \tag{4.6.1}$$

where $g_1 = \langle S^z_{2i-1} S^z_{2i} \rangle$ and $g_2 = \langle S^z_{2i} S^z_{2i+1} \rangle$ are nearest neighbour correlations for alternate pairs.

From the exact eigenvectors of system size $N = 8$, g_1 and g_2 (and hence G) was calculated for values of κ near 0.5 and small values of Γ. However, the results do not indicate the existence of nearest neighbour singlets. On the other hand, the values of g_1 and g_2 rather match with those belonging to a phase with oscillatory decay

Table 4.1 Values of nearest neighbour correlations and corresponding dimer order parameters for two frustrated models

Model	κ	Γ	$-g_1$	$-g_2$	G
ANNNI chain in transverse field	0.48	0.025	0.78692	0.71032	0.07660
		0.050	0.72266	0.60265	0.12001
		0.100	0.62595	0.42799	0.19796
		0.200	0.53507	0.37301	0.16206
	0.49	0.025	0.74796	0.66413	0.08383
		0.050	0.73996	0.65611	0.08385
		0.100	0.53899	0.28848	0.25051
		0.200	0.48542	0.30427	0.18115
	0.50	0.025	0.74143	0.64375	0.09768
		0.050	0.62955	0.40518	0.22437
		0.100	0.59737	0.36880	0.22857
		0.200	0.52765	0.35470	0.17295
	0.51	0.025	0.74782	0.66330	0.08452
		0.050	0.56542	0.28741	0.27801
		0.100	0.53547	0.28302	0.25154
		0.200	0.48216	0.30205	0.18011
	0.52	0.025	0.55518	0.27097	0.28421
		0.050	0.56542	0.28741	0.27800
		0.100	0.57566	0.32502	0.25064
		0.200	0.52091	0.33706	0.18385
Antiferromagnetic Heisenberg chain	0.50	0.0	0.98126	−0.01751	0.99701

of correlation with wave vector $\pi/6$. Similar calculations for the Majumdar-Ghosh chain [255, 256] (for which the exact result is $G = 1$) and the frustrated quantum XY models [357] yields $G \simeq 1.0$ at $\kappa = 0.5$ indicating the exactly dimerised ground state structure. Here, for these studies both J_1 and J_2 have been considered negative in all these models to explore the possibility of the existence of nearest neighbour singlets. The results for correlations in transverse ANNNI chain and the Majumdar-Ghosh (antiferromagnetic nearest and next nearest neighbour Heisenberg) chain are shown in Table 4.1.

Appendix 4.A

4.A.1 Hartree-Fock Method: Mathematical Details

The general diagonal form of the Hamiltonian in terms of free fermions can be written as

$$H_g = -\left[\Gamma' \sum_i (c_i^\dagger c_i - 1/2) + J_1' \sum_i (c_i^\dagger c_{i+1}^\dagger) + J_{11}' (c_i^\dagger c_{i+1}) + J_2' \sum_i (c_i^\dagger c_{i+2}^\dagger) \right.$$
$$\left. + J_{22}' (c_i^\dagger c_{i+2}) \right] + h.c.$$

The Hamiltonian H (4.3.15) is to be approximately written in the form of H_g such that J_1, J_2 etc. are expressed in terms of J_1', J_2' etc. The renormalised quantities J_1', J_2' etc. are obtained by minimising the free energy functional

$$F/N = F_0/N - \big(\langle H_g \rangle - \langle H \rangle\big).$$

Applying Wick's theorem, F is given by

$$\begin{aligned} F/N = F_0/N &- \big[-(\Gamma' - \Gamma)\langle c_i^\dagger c_i - 1/2 \rangle\big] \\ &- (J_1' - J_1 - 2J_2(\langle c_i^\dagger c_{i+1}^\dagger \rangle + \langle c_i^\dagger c_{i+1} \rangle)\langle c_i^\dagger c_{i+1} \rangle/2 \\ &- (J_{11}' - J_1 + 2J_2(\langle c_i c_{i+1}^\dagger \rangle + \langle c_i c_{i+1} \rangle)\langle c_i^\dagger c_{i+1}^\dagger \rangle/2 \\ &+ J_2' - 2J_2\big[\langle c_i^\dagger c_i \rangle - 1/2\big]\langle c_i^\dagger c_{i+2}^\dagger \rangle/2 \\ &+ J_{22}' - 2J_2\big[\langle c_i^\dagger c_i \rangle - 1/2\big]\langle c_i c_{i+2} \rangle/2 \end{aligned}$$

Variation of F with respect to Γ', J_1' etc. yields under stationary condition the following equation given in matrix form

$$M \times \begin{pmatrix} -(\Gamma' - \Gamma)[\langle c_i^\dagger c_{i+2} \rangle + \langle c_i c_{i+2} \rangle] \\ [J_1' - J_1 - 4J_2(\langle c_i^\dagger c_{i+1} \rangle - \langle c_i c_{i+1} \rangle)]/2 \\ [J_{11}' - J_1 + 4J_2(\langle c_i c_{i+1}^\dagger \rangle + \langle c_i c_{i+1} \rangle)]/2 \\ [J_2' + J_2(2\langle c_i^\dagger c_i \rangle - 1)]/2 \\ [J_{22}' + J_2(2\langle c_i^\dagger c_i \rangle - 1)]/2 \end{pmatrix}$$

where the elements of M are the derivatives of $\langle c_i c_k \rangle$, $\langle c_i^\dagger c_k \rangle$ etc., with $k = i, i+1$ or $i+2$, with respect to the parameters Γ', J_1' etc.

The above equation yields the self-consistent expressions (also $J_{11}' = J_1'$ and $J_{22}' = J_2'$) given by (4.3.24a), (4.3.24b), (4.3.24c). The correlations $\langle c_i^\dagger c_j \rangle$ and $\langle c_i c_j \rangle$ at zero temperature are expressed as [360, 423]

$$\begin{aligned} \langle c_i^\dagger c_j \rangle &= \delta_{ij}/2 + (1/2\pi) \int_0^\pi dk\big[\Gamma' + (J_1' \cos k)/2 - (J_2' \cos 2k)/2\big] \\ &\quad \times \big(\cos k(i-j)\big)/\lambda_k \\ \langle c_i c_j \rangle &= (1/2\pi) \int_0^\pi dk\big[(J_1' \sin k)/2 - (J_2' \sin 2k)/2\big]\big(\sin k(i-j)\big)/\lambda_k \end{aligned}$$

where

$$\lambda_k = 4\big[(\Gamma')^2 + (J_1')^2/4 + (J_2')^2/4 + \Gamma' J_1' \cos k - \Gamma' J_2' \cos 2k - J_1' J_2' \cos k/2\big].$$

In the region $\kappa' < 0.5$, we put $\Gamma'/J_1' = (1-\kappa')$ to get

$$\Gamma/J_1 = 2\big[I_1(J_1' + J_2') + J_2' I_3\big]/(J_1' I_1 - 4J_2' I_2)$$

where

$$J_1 = J_1' - 4J_2' I_2/I_1, \qquad J_2 = 2\pi J_2'/I_1$$

and the integrals I_1, I_2 and I_3 are given by

$$\begin{aligned}
I_1 &= \int_0^\pi dk\big[(1+\cos k) - \kappa'(1-\cos k)\big] \\
&= 2\big[\alpha\big\{1 - 1/\big(2(1-\kappa')\big)\big\} + \big\{-4\kappa'(1-\kappa') + 1\big\}^{1/2}/(1-\kappa')\big] \\
I_2 &= \int_0^\pi dk\big[(\cos k + 1)/(2\lambda_k)\big] = \alpha \\
I_3 &= \int_0^\pi dk\big[-(\cos 2k - 1)\kappa' + \cos k + \cos 2k\big]/(2\lambda_k) \\
&= \big[1 - 4\kappa'(1-\kappa')\big]^{1/2}/\kappa' + \alpha\big[1 - 1/(2\kappa')\big]
\end{aligned}$$

with

$$\alpha = \big[-1/\big\{\kappa'(1-\kappa')\big\}^{1/2}\big]\big[\sin^{-1} 2\big\{-4\kappa'(1-\kappa') + 1/2\big\} - \pi/2\big]/2.$$

In the other region $\kappa > 0.5$, $\Gamma'/J_1' = |\kappa'|$, and

$$J_2/J_1 = -2\pi\kappa'/\big(I_1' + 4I_2'\kappa'\big)$$

where

$$\begin{aligned}
I_1' &= \int_0^\pi dk\big[\cos k - \kappa'(1+\cos 2k)\big]/(2\lambda_k) = 0 \\
I_2' &= \int_0^\pi dk\big[\big(1/2 + \kappa'\cos k\big)\cos^2 k + \sin^2 k + \kappa'\sin^2 k\cos k\big]/\lambda_k = \pi
\end{aligned}$$

therefore $J_2/J_1 = -1/2$ for all values of κ' and $\Gamma/J_1 = 0$ identically for $\kappa = 0.5$.

4.A.2 *Large S Analysis: Diagonalisation of the Hamiltonian in Spin Wave Analysis*

In the spin wave analysis, in each phase, the angles θ_i are determined by minimising the energy given by (4.4.1). Here, new spin variables $\tilde{S}$ are introduced

$$\begin{aligned}
\tilde{S}_i^z &= S_i^z \cos\theta_i + S_i^x \sin\theta_1 \\
\tilde{S}_i^x &= S_i^x \cos\theta_i - S_i^z \sin\theta_i
\end{aligned}$$

which obey the same commutation relations of the old ones. Since

$$\big(\tilde{S}_i^x\big)^2 + \big(\tilde{S}_i^y\big)^2 + \big(\tilde{S}_i^z\big)^2 = S_c^2$$

where

$$S_c = \left[S(S+1)\right]^{1/2} = S + 1/2 + O(1/S).$$

One can expand $\tilde{S}_i^z$ in terms of $\tilde{S}_i^x$ and $\tilde{S}_i^y$:

$$\tilde{S}_i^z = S_c - \frac{1}{2S_c}\left[(\tilde{S}_i^x)^2 + (\tilde{S}_i^y)^2\right] + \mathcal{O}\left(\frac{1}{S_c^2}\right).$$

Introducing canonically conjugate variables $q_i = \tilde{S}_i^x/\sqrt{S}$ and $p_i = \tilde{S}_i^y/\sqrt{S}$ which satisfy $[q_i, p_j] = i\delta_{ij}$ to $O(1)$ in S. The Hamiltonian in phase p is then given by

$$H = E_0(p) + \frac{1}{2}\sum_i \left[f_i(p_i^2 + q_i^2) + g_i q_i q_{i+1} + h_i q_i q_{i+2}\right]$$

where

$$f_i = \cos\theta_i\left[\cos\theta_{i+1} + \cos\theta_{i-1} - \kappa(\cos\theta_{i+2} + \cos\theta_{i-2})\right] + \Gamma\sin\theta_i$$

$$g_i = -2\sin\theta_i\sin\theta_{i+1}, \qquad h_i = 2\kappa\sin\theta_i\sin\theta_{i+2}.$$

Now the Hamiltonian can be diagonalised in the momentum space. One then defines the variables

$$q_k = (1/\sqrt{N})\sum_{n=1}^{N} q_n\, e^{ikn}$$

$$p_k = (1/\sqrt{N})\sum_{n=1}^{N} p_n\, e^{ikn}.$$

Provided (f_i, g_i, h_i) are independent of i (which is true for the three elementary phases), a diagonal form of the Hamiltonian is obtained:

$$H_Q(p) = \frac{1}{2}\sum_k (A_k p_k p_{-k} + B_k q_k q_{-k})$$

where A_k and B_k are real and even functions of k.

4.A.3 Perturbative Analysis

In this Appendix we shall present a first order perturbation theory [67] where the unperturbed Hamiltonian is H_{cl} and the perturbation is H_p. We start by noting that at $\kappa = 0.5$ the ground state of H_{cl} is a state with high degeneracy and any spin configuration that has no spin-domain of length unity can be the ground state. The number of domain walls is immaterial and can be anything between 0 and $N/2$, N being the total number of spins. (Of course, for periodic boundary there can be only an even number of walls.) Let us denote the set of all such configurations as $\mathscr{S}$. Also, let the population of this set be ν which incidentally is of the order of g^N, where $g = (\sqrt{5}+1)/2$ [247]. Now, the first-order correction to the eigenvalue [347] are the eigenvalues of the $\nu \times \nu$ matrix P, whose elements are

$$P_{\alpha\beta} \equiv \langle\alpha|H_p|\beta\rangle$$

where $|\alpha\rangle$ and $|\beta\rangle$ are configurations within $\mathcal{S}$. We shall now transform P to a block diagonal structure. Note that $S_j^x|\beta\rangle \in \mathcal{S}$ if and only if the j-th spin lies at the *boundary* of a domain, and the domain too has length larger than 2. Also, in such a case, S_j^x operating on $|\beta\rangle$ will translate the wall at the left (right) of the j-th site by one lattice spacing to the right (left). This indicates that $P_{\alpha\beta} \neq 0$ if and only if $|\alpha\rangle$ and $|\beta\rangle$ have equal number of domain walls. Thus, we can break up $\mathcal{S}$ into subsets $\mathcal{S}(W)$, where $\mathcal{S}(W)$ contains all possible spin distributions with W walls ($W = 2, 4, \ldots, N/2$). Now, the $\nu \times \nu$ matrix P gets block-diagonalised into matrices $P(W)$ of size $\nu_W \times \nu_W$, where

$$P_{\alpha\beta}(W) \equiv \langle\alpha|H_p|\beta\rangle \tag{4.A.1}$$

where ν_W is the population of $\mathcal{S}(W)$ and $|\alpha\rangle, |\beta\rangle \in \mathcal{S}(W)$. Now, it can be seen that the longitudinal term in H_p only contributes a diagonal term $r\Gamma(N-4W)$ to $P_{\alpha\beta}(W)$, so that one can write

$$P(W) = M(W) + r\Gamma(N-4W)\mathbf{1} \tag{4.A.2}$$

where $\mathbf{1}$ is the $\nu_W \times \nu_W$ unit matrix, $M_{\alpha\beta}(W) \equiv \langle\alpha|H_q|\beta\rangle$ and $H_q = -\Gamma\sum_j S_j^x$ is the transverse part of H_p. Thus the non-trivial problem is to solve the eigenproblem of $M(W)$.

To solve the eigenproblem of matrices $M(W)$, let us construct from each member $|\alpha\rangle$ of $\mathcal{S}(W)$ a configuration $|\alpha'\rangle$ by removing one spin from each spin domain. The total number of spins in $|\alpha'\rangle$ will obviously be $N - W = N'$, say. Such a transformation was also used by Villain and Bak [412] for the case of two-dimensional ANNNI model. It is crucial to observe that the set $\mathcal{S}'(W)$ composed of the states $|\alpha'\rangle$ is then nothing but the set of all possible distributions of N' spins with W walls, with no restriction on the domains of length unity. Hence $\langle\alpha'|\sum_{j=1}^{N'} S_j^x|\beta'\rangle$ is non-zero only when $\langle\alpha|\sum_{j=1}^{N} S_j^x|\beta\rangle$ is non-zero and

$$M'_{\alpha\beta}(W) \equiv \langle\alpha'|H_q|\beta'\rangle = M_{\alpha\beta}(W)$$

The eigenproblem of $M'(W)$ becomes simple once we observe that $\mathcal{S}'(W)$ is nothing but the set of degenerate eigenstates of the usual classical Ising Hamiltonian

$$H_0' = -J\sum_{j=1}^{N'} S_j^z S_{j+1}^z$$

corresponding to the eigenvalue

$$E_W = -J(N' - 2W). \tag{4.A.3}$$

Thus, if we perturb H_0' by H_q, then the first-order perturbation matrix will assume a block diagonal form made up of the matrices $M'(W)$ for all possible values of W.

To solve the perturbation problem for $H_0' + H_q$, we note that this Hamiltonian is nothing but the standard transverse Ising Hamiltonian

$$H_{TI} = -\sum_{j=1}^{N'} \left[J S_j^z S_{j+1}^z + \Gamma S_j^x \right].$$

The exact solution for this Hamiltonian is known (Chap. 2). The exact expression for the energy eigenstates are

$$E = 2\Gamma \sum_k \xi_k \Lambda_k \tag{4.A.4}$$

where ξ_k may be 0, ± 1 and k runs over $N'/2$ equispaced values in the interval 0 to π. Also, Λ_k stands for $\sqrt{(\lambda^2 + 2\lambda \cos k + 1)}$, where λ is the ratio J/Γ. For $\Gamma = 0$, the energy E must be the same as E_W of (4.A.3), so that

$$2 \sum_{k=0}^{\pi} \xi_k = -\left(N' - 2W\right). \tag{4.A.5}$$

and the first order perturbation correction to this energy is

$$\left(\frac{\partial E}{\partial \Gamma} \right)_{\Gamma=0} = 2 \sum_{k=0}^{\pi} \xi_k \cos k.$$

They are therefore also the eigenvalues of the $M'(W)$ matrix. Thus the eigenvalues of the matrix $P(W)$ of (4.A.2) are

$$E_P = r\Gamma(N - 4W) + 2\Gamma \sum_{k=0}^{\pi} \xi_k \cos k. \tag{4.A.6}$$

Keeping N fixed we have to find, for which value of W and for which distribution of ξ_k, E_P is minimum subject to the constraint (4.A.5). For a given value of $\sum \xi_k$, this minimisation is achieved if -1 $(+1)$ values of ξ_k accumulate near large positive (negative) values of $\cos k$. Let the desired distribution be

$$\xi_k = \begin{cases} -1 & \text{for } k = 0 \text{ to } \theta \\ 0 & \text{for } k = \theta \text{ to } \phi \\ 1 & \text{for } k = \phi \text{ to } \pi. \end{cases} \tag{4.A.7}$$

Equation (4.A.5) now gives

$$N/N' = (4\pi - \theta - \phi)/2\pi \tag{4.A.8}$$

and one obtains,

$$E_P = -\frac{N\Gamma}{4\pi - \theta - \phi} \left[r(4\pi - 3\theta - 3\phi) + 2(\sin\theta + \sin\phi) \right].$$

Minimising E_P with respect to θ and ϕ, we find that $\theta = \phi = \theta_0$ (say) where

$$2\pi r = \sin\phi_0 + (2\pi - \phi_0)\cos\phi_0. \tag{4.A.9}$$

The minimum value of E_P is given by,

$$E^{(1)} = -N\Gamma[3r - 2\cos\phi_0]. \tag{4.A.10}$$

This is the final expression for the (exact) first order perturbation correction to ground state energy. It is easily seen that for $r < -0.5$, that is, for $\Gamma/J < (1-2\kappa)$, one has $\phi_0 = \pi$ and $W = 0$ (ferromagnetic phase), while for $r > 1$, that is $\Gamma/J < (\kappa - 0.5)$, one has $\phi_0 = 0$ and $W = N/2$ (antiphase). As r varies from -0.5 to 1, ϕ_0 gradually changes from π to 0 according to (4.A.9). It can also be seen that Also $d^2E^{(1)}/d\Gamma^2$ diverges at $r = -0.5$ and 1, indicating two critical lines there. It can be shown [67] that except for these two values of r, $E^{(1)}$ and all its higher derivatives remain finite. We can now conclude that the phase diagram is either like Fig. 4.8(a) or like Fig. 4.8(b). We shall now show that an analysis of longitudinal susceptibility points to the possibility of the former.

Let us call the eigenstate of $H_0' + H_q$ corresponding to $\theta = \phi = \phi_0$ as $|\psi'\rangle$. This state will be composed of the spin-distributions that belong to $\mathscr{S}'(W)$ and can be written as

$$|\psi'\rangle = \sum_{j'} a_{j'}|j'\rangle$$

where $|j'\rangle$ runs over all the states in $\mathscr{S}'(W)$. Let us construct from each state $|j'\rangle$ another state $|j\rangle$ by augmenting each domain by a single spin. Thus for $|j'\rangle = |{+}{+}{+}{-}{-}{+}{+}{+}{+}\rangle$, $|j\rangle$ will be $|{+}{+}{+}{+}{-}{-}{-}{+}{+}{+}{+}{+}\rangle$. Then we combine these states with the same coefficients to get a state $\sum_j a_j|j\rangle$ where $a_j = a_{j'}$. The procedure for constructing M' from M indicates that $\sum_j a_j|j\rangle$ is an eigenstate of $M(W)$ and hence of $P(W)$ (see (4.A.2)) and this eigenstate is nothing but the zero-th order eigenfunction $|\psi^{(0)}\rangle$ for the perturbed ground state of $H_{cl} + H_p$. One should observe that although the spin-spin correlation may not be equal for $|j\rangle$ and $|j'\rangle$, the longitudinal magnetisation M_z must be the same for them as equal number of positive and negative spins have been added while transforming $|j'\rangle$ to $|j\rangle$. Thus, the longitudinal susceptibility

$$\chi_z \propto \langle M_z^2\rangle - \langle M_z\rangle^2$$

of $|\psi^{(0)}\rangle$ must be the same as that of $|\psi'\rangle$. The spin-spin correlation

$$C^z(n) \equiv \langle S_i^z S_{i+n}^z\rangle$$

for $|\psi'\rangle$ may be calculated by evaluating the corresponding Toeplitz determinants [68]. In the case of $\Gamma < J$, for the entire range $0 < \phi_0 < \pi$, the correlation is

$$C^z(n) = A\frac{1}{\sqrt{n}}\cos[n(\pi - \phi_0)]$$

where A is a constant. This is clearly a floating phase since $C^z(n)$ decays algebraically with n. The susceptibility χ_z is hence infinity for both the states $|\psi'\rangle$ and $|\psi^{(0)}\rangle$. This leads us to the conclusion that the zero-th order eigenstate is in floating phase and hence, at least for small values of Γ, the ground state of transverse ANNNI chain must be a floating phase for all values of r between -0.5 and 1.

Of course, for large values of Γ the perturbation corrections may cancel the divergence of susceptibility and lead to a paramagnetic state.

One signature of floating phase or diverging correlation length is vanishing *mass-gap*. Analysis of mass gap corroborates the fact that Fig. 4.8(b) rather than Fig. 4.8(a) holds true. Let us now study the first order (in Γ) correction to the mass-gap. Clearly the first excited state is the smallest possible value of E_P other than the ground state $E^{(1)}$. To find the lowest excitation over the ground state, we note that such excitation is possible either (i) by keeping $\sum \xi_k$ fixed and rearranging the ξ_k values; or (ii) by altering θ and ϕ and thus altering $\sum \xi_k$. For (i) the lowest excitation will correspond to an interchange of $+1$ and -1 at $k = \phi_0$, which will lead to a mass gap (for the whole system)

$$\Delta^{(1)} = \frac{8\pi \Gamma \lambda \sin\phi_0}{N' \Lambda_{\phi_0}}.$$

For (ii) this quantity will be

$$\Delta^{(1)} = \frac{1}{2}\left(\frac{\partial^2 E_P}{\partial \theta^2}\right)_{\theta=\phi_0} (\delta\theta)^2$$

where $\delta\theta$ is the smallest possible deviation in θ at ϕ_0. As the smallest possible change in W, and hence in N' is 2, we get from (4.A.8)

$$\delta\theta = \frac{2(2\pi - \theta)^2}{\pi N} \sim \frac{1}{N}.$$

This shows that for both the mechanisms (i) and (ii), the mass gap $\Delta^{(1)}$ vanishes as $N \to \infty$ for all values of ϕ_0 between 0 and π. This shows that for all values of r between -0.5 and 1 there must be floating phase for small Γ.

Chapter 5
Dilute and Random Transverse Ising Systems

5.1 Introduction

The study of phase transitions in diluted magnetic systems (with random nonmagnetic impurities) and in random magnetic systems (with random interactions) has been an intriguing area of enormous theoretical as well as experimental investigations over the last half century [188, 382, 415]. Though dilution can be both "annealed" or "quenched" type, the latter is richer in the sense that it helps understanding a wide variety of novel physical systems. As is well established, the quenched dilute magnetic systems exhibit the "percolative behaviour" [375] and the lattice fluctuations induce a crossover from the thermal critical behaviour to geometrical (percolative) critical behaviour in the vicinity of the percolation threshold. We discuss here the general features of the phase diagram of the dilute Ising system in transverse field, and the transition (critical) behaviour across such phase boundary. Randomness, on the other hand, may have a crucial influence on the property of phase transitions, ever though it does not bring frustration. It is known that a weak singularity called the Griffiths singularity [161] in the free energy arises away from the critical point due to the presence of randomness. The effect of the Griffiths singularity is more marked near the quantum phase transition at zero temperature than the thermal phase transition. We see in the latter part of this chapter that unusual critical behaviours characterise the quantum phase transition in random transverse field Ising models.

5.2 Dilute Ising System in a Transverse Field

The Hamiltonian for a nearest-neighbour interacting ferromagnetic (classical) Ising Hamiltonian, with quenched dilution, is written as

$$H = -J \sum_{\langle ij \rangle} S_i^z S_j^z \eta_i \eta_j, \tag{5.2.1}$$

S. Suzuki et al., *Quantum Ising Phases and Transitions in Transverse Ising Models*,
Lecture Notes in Physics 862, DOI 10.1007/978-3-642-33039-1_5,

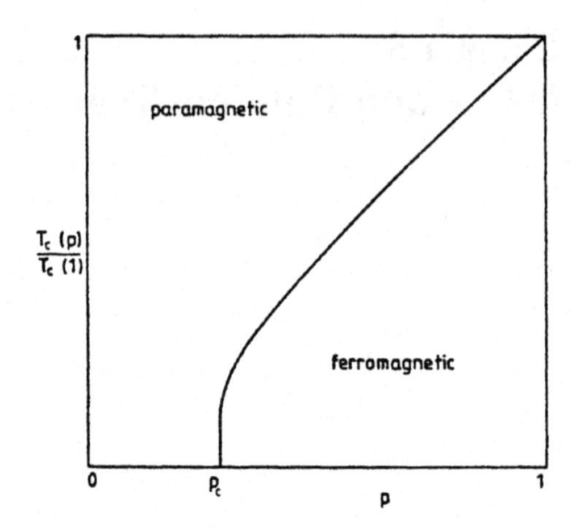

Fig. 5.1 Schematic phase diagram of quenched site-diluted Ising ferromagnet. p is the magnetic ion concentration and $T_c(p)/T_c(1)$ is the reduced transition temperature. p_c is the percolation threshold [382]

or

$$H = -J \sum_{\langle ij \rangle} S_i^z S_j^z \eta_{ij}, \tag{5.2.2}$$

where S_i^z are usual Pauli spin operators and η_i's (or η_{ij}'s) are the uncorrelated site (or bond) disorder variables taking the values 0 or 1 at each site according to the probability distribution

$$P(\eta_i) = (1 - p)\delta(\eta_i) + p\delta(\eta_i - 1), \tag{5.2.3a}$$

or

$$P(\eta_{ij}) = (1 - p)\delta(\eta_{ij}) + p\delta(\eta_{ij} - 1), \tag{5.2.3b}$$

such that $\overline{\eta}_i = p$ or $\overline{\eta}_{ij} = p$. In the site diluted case p denotes the magnetic ion concentration. Here, the over-head bar denotes the configuration averaging over the distribution of disorder given by (5.2.3a), (5.2.3b). The free energy of the quenched magnetic systems is given by

$$F = -k_B T \overline{\ln Z} \tag{5.2.4}$$

where Z is the partition function of the system for a particular realisation of disorder. The site-diluted and bond-diluted systems exhibit the same critical behaviour. The schematic phase diagram of the quenched site-diluted Ising system is shown in the Fig. 5.1, which shows that the pure system critical temperature decreases with the decrease of magnetic ion concentration (p) until it goes to zero at the percolation threshold (p_c).

Efforts have been made to study the effect of quantum fluctuations, generated by a transverse field, on the phase transitions in classical diluted magnetic Hamiltonian [170]. The Hamiltonian describing the site or bond-diluted classical Ising system in the presence of a transverse field can be written as

$$H = -J \sum_{\langle ij \rangle} S_i^z S_j^z \eta_i \eta_j - \Gamma \sum_i S_i^x \eta_i, \tag{5.2.5a}$$

$$H = -J \sum_{\langle ij \rangle} S_i^z S_j^z \eta_{ij} - \Gamma \sum_i S_i^x, \tag{5.2.5b}$$

where Γ is the strength of the transverse field. One can immediately conjecture the qualitative phase diagram of the system (see Fig. 5.2). Obviously, for dimension $d \geq 2$, the critical transverse field $\Gamma_c(T, p)$ is a function of the concentration of magnetic ions (p) and temperature (T). As p decreases the phase boundary shrinks to lower values of Γ and Γ_c vanishes at $p = p_c$. In the zero-temperature limit, the quantum phase transition due to the transverse field crosses over to a percolative phase transition at the percolation threshold $p = p_c$. The zero-temperature critical transverse field shows a discontinuous jump at the percolation threshold p_c. For the one dimensional model $T_c = 0$ and $p_c = 1$, which suggests that in this case, even in the zero-temperature limit, the long-range order will be destroyed for any Γ with infinitesimally small nonzero value of p.

5.2.1 Mapping to the Effective Classical Hamiltonian: Harris Criterion

Let us consider a nearest-neighbour site or bond-diluted transverse Ising system on a d-dimensional lattice. The zero-temperature quantum phase transition in a transverse Ising model on a d-dimensional lattice is equivalent to the finite-temperature thermal phase transition in an extremely anisotropic classical Ising Hamiltonian with one added dimension, namely the Trotter dimension. Using the Suzuki-Trotter formalism (see Sect. 3.1), we obtain the equivalent classical Hamiltonian for the quantum Hamiltonians (5.2.5a) or (5.2.5b), in the M-th Trotter approximation

$$\begin{aligned} H_{\text{eff}} = &-(J/M) \sum_{\langle ij \rangle} \sum_{k=1}^{M} S_{ik} S_{jk} \eta_i \eta_j - (1/2\beta) \ln\left[\coth(\beta\Gamma/M)\right] \\ &\times \sum_i \sum_k S_{ik} S_{i,k+1} \eta_j, \end{aligned} \tag{5.2.6}$$

where k indicates the Trotter index. In the zero temperature limit ($M \to \infty$), the quantum transition in the original quantum Hamiltonian (5.2.5a), (5.2.5b) falls in the same universality class with the thermal phase transition in the equivalent anisotropic $D = d + 1$ dimensional classical model with the disorder (distribution of magnetic atoms) correlated (striped) in the Trotter direction (i.e., we get M identical copies of the original system with the unaltered distribution of magnetic atoms, connected through ferromagnetic bonds in the Trotter direction). If one starts with a one dimensional diluted transverse Ising chain by employing Suzuki-Trotter formalism [386], one ends up with a two-dimensional classical Ising system with identical disorder in each Trotter replica, namely the McCoy-Wu model [264, 269, 270].

To see whether the dilution in the classical magnetic systems changes the universality of the magnetic phase transition in those models in the presence of dilution, one might consider the "Harris criterion" [169], which in normal cases (isotropic disorder) suggests that if the specific heat exponent (α) of the pure system is positive, the dilution changes the universality of magnetic transition and renormalisation group flow takes the system away from the nonrandom fixed point. For $\alpha < 0$, the dilution does not affect the critical behaviour and the criterion is inconclusive for $\alpha = 0$. However such quantum disordered cases require a modified Harris criterion. As mentioned earlier, the zero-temperature quantum phase transition in the dilute transverse Ising system falls in the same universality class with the thermal phase transition in $D(= d + 1)$-dimensional Ising system with striped randomness. To derive the required condition for the dilution to be a relevant parameter one must consider the "Harris criterion" [249, 382] for the systems with randomness correlated in one particular (Trotter) direction. If we consider a domain having the dimension of the order of the correlation length ξ, the fluctuation in the critical temperature due to randomness is given as

$$\Delta T_c \sim \xi^{-\frac{(D-1)}{2}} \sim (\Delta T)^{(D-1)\nu/2}, \tag{5.2.7}$$

since the randomness is correlated in a particular direction and the correlation length for the pure system diverges as $(T - T_c)^{-\nu}$. So the random field is now a relevant parameter if $(D - 1)\nu < 2$, or

$$\alpha + \nu > 0, \tag{5.2.8}$$

where both α and ν are exponents for the pure classical Ising system in $D = d + 1$ dimensions. Since for the pure classical Ising system the condition (5.2.8) is satisfied in all dimensions from $D = 2$ ($\alpha = 0$ and $\nu = 1$, for $D = d + 1 = 2$) upwards (for $D \geq 4, \alpha = 0$ and $\nu = 1$), one can readily conclude that in the dilute classical Ising system in $(d + 1)$-dimensions with correlated randomness, the fluctuations induced by the random fields are relevant and dominate the thermal fluctuations. Hence, the universality class of the magnetic transition in transverse Ising systems with dilution is expected to be different from that in the pure case.

5.2.2 Discontinuous Jump in $\Gamma_c(p, T = 0)$ at the Percolation Threshold

Using a heuristic argument, one can conclude that the zero-temperature critical field $\Gamma_c(p, T = 0)$ is discontinuous at $p = p_c$ in the case of diluted transverse Ising system with dimensionality greater than unity [170]. This can be understood in the following way: below the percolation threshold $p < p_c$, the system does not percolate and hence no long-range order exists for any finite value of Γ, giving $\Gamma_c(p, T = 0) = 0$, whereas for $p > p_c$, $\Gamma_c(p, T = 0) \geq \Gamma_c^{(1)}$, where $\Gamma_c^{(1)}$ is the zero-temperature critical field for the one-dimensional transverse Ising system (see Sect. 2.2). This finite value of critical field $\Gamma_c^{(1)}$ is the minimal requirement to destroy the order in "chain-like" structures occurring at the percolation threshold concentration. In fact, the percolation cluster is more connected than a chain [375],

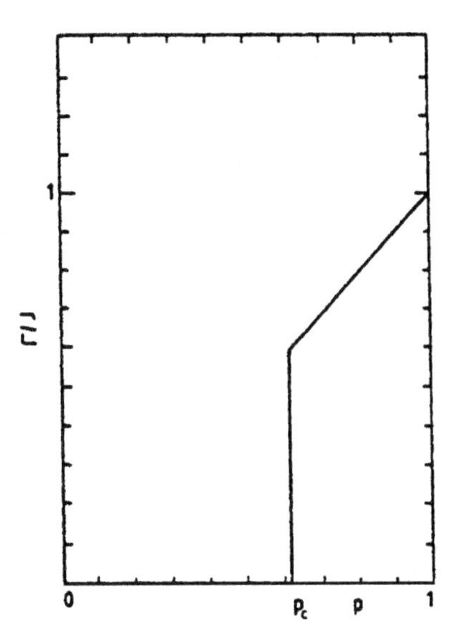

Fig. 5.2 Schematic phase boundary of the bond-diluted two-dimensional zero temperature Ising model in a transverse field Γ. The transverse field shows a discontinuous jump at the percolation threshold p_c [382]

and these additional connectivity suggests $\Gamma_c(p_c, T = 0) > \Gamma_c^{(1)}$. One thus expects the $\Gamma_c(p, T = 0)$ to jump discontinuously from zero to a value at least $\Gamma_c^{(1)}$ as p increases through p_c (see Fig. 5.2).

5.2.3 Real-Space Renormalisation Group Studies and Scaling

As discussed in Chap. 2, the zero-temperature quantum phase transition in the pure transverse Ising chain Hamiltonian

$$H = -\overline{\lambda} \sum_i S_i^z S_{i+1}^z - \sum_i S_i^x \tag{5.2.9}$$

is characterised by only one relevant parameter $\overline{\lambda} = J/\Gamma$. One therefore seeks a renormalisation-group transformation

$$\overline{\lambda} \to \overline{\lambda}' = F(\overline{\lambda}) \tag{5.2.10}$$

relating the parameter $\overline{\lambda}$ and $\overline{\lambda}'$ of the original system and one dilated by a length scale factor b. In order to capture the singular critical properties or the scaling ansatz (5.2.10) explicitly, the dependence of the diverging correlation length on the parameter $\overline{\lambda}$ (or $\overline{\lambda}'$) is to be found out. To do that [381], one has to recall the equivalence between the anisotropic classical Ising Hamiltonian to the transverse Ising Hamiltonian in the Hamiltonian limit (see Sect. 3.A.2). Let us consider an anisotropic classical Ising Hamiltonian on a square lattice

$$H = -J_1 \sum_{i,k} S_{i,k} S_{i+1,k} - J_2 \sum_{i,k} S_{i,k} S_{i,k+1}. \tag{5.2.11}$$

With $t_\alpha = \tanh \beta J_\alpha$, $\alpha = 1, 2$, the extreme anisotropic limit is given by

$$t_1 \to 1 \quad \text{and} \quad t_2 \to 0 \tag{5.2.12}$$

when one can derive the equivalent quantum Hamiltonian (see Sect. 3.A.2)

$$H = -\sum_i S_i^z S_{i+1}^z - \overline{\lambda} \sum_i S_i^x, \tag{5.2.13}$$

where the parameter $\overline{\lambda}$ is given by

$$\overline{\lambda} = \lim_{\tilde{t}_1, t_2 \to 0} \left(\frac{t_2}{\tilde{t}_1} \right), \quad \text{where } \tilde{t}_1 = \frac{1 - t_1}{1 + t_1}. \tag{5.2.14}$$

Note that $\tilde{t}_1$ tends to 0 as $t_1 \to 1$. For the anisotropic square lattice Ising model the decay of correlation function in the direction of J_2 (which becomes the chain direction of the equivalent transverse Ising model) is characterised by a correlation length ξ given by [426]

$$\exp(-a/\xi) = \left(\frac{\tilde{t}_1}{t_2} \right)^2 \quad T < T_c \tag{5.2.15}$$

$$= \left(\frac{t_2}{\tilde{t}_1} \right) \quad T > T_c \tag{5.2.16}$$

where a is the lattice constant. The correlation length of quantum transverse Ising chain is therefore, given by (since T_c in the classical transition corresponds to $\overline{\lambda} = 1$ in the quantum transition and $T < T_c$ corresponds to $t_2 > \tilde{t}_1, \overline{\lambda} > 1$ in the quantum case and vice versa)

$$\xi(\overline{\lambda}) = \frac{a}{2 \ln \overline{\lambda}} \quad \overline{\lambda} > 1 \tag{5.2.17}$$

$$= \frac{a}{\ln(1/\overline{\lambda})} \quad \overline{\lambda} < 1, \tag{5.2.18}$$

which agrees with the exact result. The renormalisation group transformation must leave the specific critical properties arising from diverging correlation length invariant. When the zero-temperature transverse Ising chain is dilated by a factor b,

$$\xi(\overline{\lambda}) = b\xi(\overline{\lambda}'), \tag{5.2.19}$$

or, equivalently,

$$\overline{\lambda}' = \overline{\lambda}^b = F(\overline{\lambda}) \tag{5.2.20}$$

This is the exact phenomenological scaling relation for the pure transverse Ising chain. Clearly it yields the exact values of the critical fixed point $\overline{\lambda}^* (= 1)$ and the critical exponent $\nu (= 1)$ [312] where the latter is obtained by linearising the renormalisation group equation (5.2.20) in the neighbourhood of the critical fixed point $\overline{\lambda}^*$. Also the scaling relations satisfy the "duality relation" [149].

The scaling relations (5.2.20), obtained for the pure transverse Ising system, can readily be generalised to treat the bond-diluted transverse Ising chain. Considering a dilation of length scale by a factor b, the scaling relation can be written as

$$\overline{\lambda}' = \prod_{i=1}^{b} \overline{\lambda}_i, \tag{5.2.21}$$

where the product is over b neighbouring bonds and the renormalised coupling is zero if any of the b bonds is absent. For the bond-diluted chain the parameters $\overline{\lambda}$'s are randomly distributed with a binary probability distribution

$$P(\overline{\lambda}_i) = (1-p)\delta(\overline{\lambda}_i) + p\delta(\overline{\lambda}_i - \overline{\lambda}). \tag{5.2.22}$$

Under renormalisation group transformation, this distribution will be transformed by virtue of the relation (5.2.21), to the new distribution

$$P'(\overline{\lambda}'_i) = \int \prod_{i=1}^{b} d\overline{\lambda}_i \prod_{i=1}^{b} P(\overline{\lambda}_i)\delta\left(\overline{\lambda}'_i - \prod_{i=1}^{b} \overline{\lambda}_i\right). \tag{5.2.23}$$

One now demands the scaled distribution of bonds to be once again binary as the original distribution so that

$$P'(\overline{\lambda}'_i) = (1-p')\delta(\overline{\lambda}'_i) + p'\delta(\overline{\lambda}'_i - \overline{\lambda}') \tag{5.2.24}$$

where

$$p' = p^b, \quad \text{and} \quad \lambda' = \lambda_b. \tag{5.2.25}$$

The essence of this approximation is that under renormalisation group transformation the dilute system scales to an equivalent dilute system with scaled parameters.

Using the recursion relations of the interaction strength and probability (5.2.25), one can readily find out the nontrivial fixed point $(\overline{\lambda}^*, p_c = p^*) = (1, 1)$, which is, as expected, the pure transverse Ising fixed point [312]. The eigenvalues of Eq. (5.2.25) linearised around the critical fixed point are given as

$$\Lambda_p \equiv \left.\frac{dp'}{dp}\right|_{p^*, \overline{\lambda}^*} = b^{\nu_p}, \qquad \Lambda_{\overline{\lambda}} \equiv \left.\frac{d\overline{\lambda}'}{d\overline{\lambda}}\right|_{p^*, \overline{\lambda}^*} = b^{\nu_{\overline{\lambda}}}. \tag{5.2.26}$$

Since the correlation length scales with the length scale, one finds the correlation length exponents given by [382]

$$\nu_p \equiv \frac{\ln b}{\ln \Lambda_p} = 1, \qquad \nu_{\overline{\lambda}} \equiv \frac{\ln b}{\ln \Lambda_{\overline{\lambda}}} = 1, \tag{5.2.27}$$

and the crossover exponent [382] is given by

$$\phi = \frac{\nu_{\overline{\lambda}}}{\nu_p}. \tag{5.2.28}$$

The exact results for the bond-diluted transverse Ising system can be put in the scaling form, in the neighbourhood of the critical fixed point, for the inverse correlation length

$$\frac{1}{\xi} = |\overline{\lambda} - 1|\Phi\left[\frac{1-p}{|1-\overline{\lambda}|}\right]. \tag{5.2.29}$$

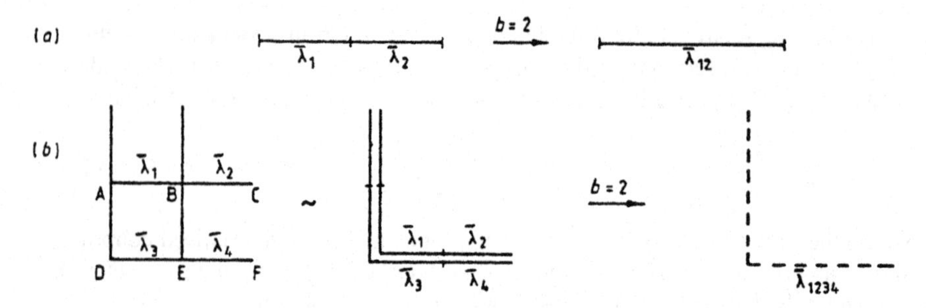

Fig. 5.3 (**a**) Scaling of two adjacent bonds of a quantum transverse Ising chain. (**b**) Approximate cluster scaling for two-dimensional case ($\bar{\lambda}_{1234} = \bar{\lambda}'$ here)

It should be noted at this point that the phenomenological scaling is exact only in the bond-diluted chain case, as the decay of correlation functions in the equivalent McCoy-Wu model [382] is exactly known. The scaling procedure is only approximate for the site-diluted (chains) or any other more general random system.

Using this concept of phenomenological scaling relations, different real-space renormalisation techniques have been employed to study the phase diagram of the bond-diluted transverse Ising systems in one and two dimensions [380]. The decimation [382] of two neighbouring bonds in the bond-diluted transverse Ising chain yields

$$\bar{\lambda}' = \bar{\lambda}_1 \bar{\lambda}_2 \tag{5.2.30}$$

Using (5.2.21)–(5.2.24) one obtains the recursion relation for p and $\bar{\lambda}$ given by

$$p' = p^2, \qquad \bar{\lambda}' = \bar{\lambda}^2, \tag{5.2.31}$$

which clearly gives the exact values of $\nu_{\bar{\lambda}}$, ν_p and ϕ for the bond-diluted transverse Ising chain. The extension to the higher dimensional systems can only be performed approximately. If one performs a bond moving [382] transformation on the two dimensional pure transverse Ising system on a square lattice with $b = 2$, one obtains the recursion relation given by (see Fig. 5.3)

$$\bar{\lambda}' = \frac{1}{2}(\bar{\lambda}_1 \bar{\lambda}_2 + \bar{\lambda}_3 \bar{\lambda}_4). \tag{5.2.32}$$

In the bond-diluted system the parameter $\bar{\lambda}_i$ is present with a probability p, the scaled parameters can readily be written as (using once again (5.2.21)–(5.2.24))

$$p' = 2p^2 - p^4; \qquad p^2 \lambda' = p' \lambda^2, \tag{5.2.33}$$

where $\lambda = 1/\bar{\lambda}$. The fixed points of the renormalisation group transformation (5.2.33) are (see Fig. 5.4)

$$(p^*, \lambda^*) = (0,0), (0,1/2), (0,\infty), (p_c, 0), (p_c, p_c), (p_c, \infty), (1,1), (1,\infty); \tag{5.2.34}$$

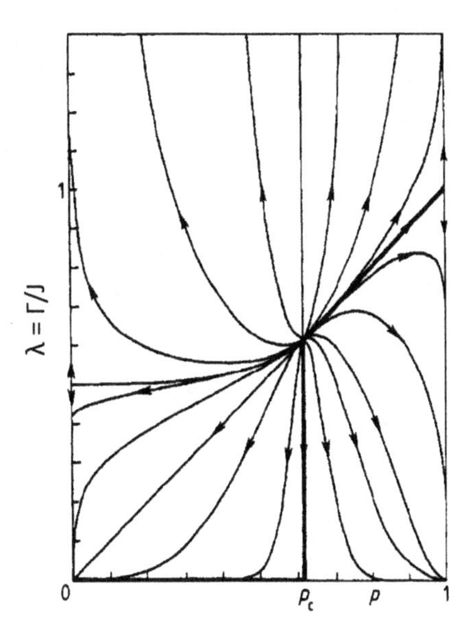

Fig. 5.4 Flow lines, fixed point and critical curve of a bond diluted two dimensional zero temperature Ising model in a transverse field Γ. The *bold line* shows the critical curve [380]

with $p_c = 0.62$. Linearising the above recursion relations (5.2.33) near these fixed points and employing the relations (5.2.26)–(5.2.28) for extracting the exponents, one gets the percolation fixed point (p_c, p_c) doubly unstable with the exponents $\nu_p = 1.62$ and $\nu_\lambda = 1$. One should also note here that the value of the exponent ν_g given for $(p^*, \lambda^*) = (p_c, p_c)$ is the same for the pure chain. This reminds us again that the real space renormalisation does not provide us with accurate values of the exponents, but it often correctly predicts the phase boundary and renormalisation group flow. Here, however, the flow diagram (Fig. 5.4) does not match the one suggested by the Harris criterion, discussed in Sect. 5.2.1. This is again a common failure of the small cell real-space renormalisation group (with parameter truncation) techniques. It is to be noted that the discontinuity in Γ_c at p_c in the phase boundary in Fig. 5.4, obtained using the real-space renormalisation group, clearly matches the expectation as discussed in Sect. 5.2.2. It should be mentioned here that the one dimensional bond-diluted transverse Ising chain has also been studied by Uzelac et al. [409] using block renormalisation group technique which preserves the duality symmetry of the original model. This method gives better result as the size of the block is increased. This method has been extended by Bhattacharya and Ray [37] to study the phase diagram of a site-diluted transverse Ising system on triangular lattice. This study also gives a qualitatively correct phase diagram but the flow diagram does not again match with the prediction of Harris criterion. In concluding this section, we would also like to mention about some recent studies on dilute transverse Ising systems with higher spin values; see e.g., the mean field studies on site dilute spin-3/2 transverse Ising system [405].

5.3 Critical Behaviour of Random Transverse Field Ising Models

The quantum phase transition in disordered systems is different in nature in many aspects from the phase transition in the classical systems, driven by thermal fluctuations. For example, the existence of Griffiths singularity in part of the disordered phase is much more prominent in the case of quantum phase transition. Griffiths [161] showed that in part of the disordered phase, rare regions, which are more strongly correlated than the average and which are locally ordered, cause the free energy to be a non-analytic function of the magnetic field. However, in a classical system, this effect is very weak and all the derivatives of free energy are finite. The random transverse field Ising model given by the Hamiltonian

$$H = -\sum_{\langle ij \rangle} J_{ij} S_i^z S_j^z - \sum_i \Gamma_i S_i^x, \tag{5.3.1}$$

where the interaction J_{ij}'s and random field Γ_i's are both independent random variables with a distribution $\pi(J)$ and $\rho(\Gamma)$, shows a prominent Griffiths phase and a quantum phase transition characterised by so-called an infinite randomness critical point. This has been established for one dimensional systems using both real space renormalisation techniques [138, 139] and numerical diagonalisation techniques [439] (see Refs. [188, 415] for reviews). More recently, several studies using quantum Monte Carlo simulations [315, 329] and numerical renormalisation group techniques [214, 235, 236, 248, 285] have shown the presence of the Griffiths phase and the infinite randomness critical point in two- or three-dimensional systems. We exclusively focus on random ferromagnetic models which do not involves frustration in the present section. Frustrated spin glass models are discussed in Chap. 6.

5.3.1 Analytical Results in One Dimension

The Hamiltonian of the random transverse Ising chain is given by

$$H = -\sum_{i=1}^{N} J_i S_i^z S_{i+1}^z - \sum_{i=1}^{N} \Gamma_i S_i^x. \tag{5.3.2}$$

Using the Suzuki-Trotter formalism (see Sect. 3.1), one can readily see that the zero-temperature transition in this model is equivalent to the thermal phase transition in a classical Ising model, where both horizontal and vertical bonds are random. This is a generalised version [368] of the original McCoy-Wu model [269]. The Hamiltonian does not incorporate frustration because the disorder is not gauge invariant. One can always make a gauge transformation to make all the interactions J_i and transverse field Γ_i at each site positive.

The analytical results obtained using real-space renormalisation group studies are given below [138, 139]. Defining

$$\Delta_\Gamma = \overline{\ln \Gamma}; \quad \text{and} \quad \Delta_J = \overline{\ln J} \tag{5.3.3}$$

where over-head bar denotes the average over disorder, the critical point is given by

$$\Delta_\Gamma = \Delta_J. \tag{5.3.4}$$

Clearly this is satisfied if the distribution of bonds and sites are identical, and the critically follows from the duality, i.e., one can perform a duality transformation from the site to bond variables (see Sect. 2.1.1) so that the roles of the transverse field and the interaction are reversed. Thus if the distribution of these are identical i.e., $\pi = \rho$, one expects to be at the critical point. For $\Delta_\Gamma > \Delta_J$, the system is paramagnetic, whereas for $\Delta_\Gamma < \Delta_J$, there is a nonzero, but unknown, spontaneous magnetisation. McCoy and Wu [264, 269] showed that there is only an essential singularity in the ground state energy density at the critical point, in contrary to the behaviour in the pure system. A convenient measure of the deviation from critically is given by Fisher [138, 139]

$$\delta = \frac{\Delta_\Gamma - \Delta_J}{\overline{(\ln \Gamma)^2} - \Delta_\Gamma^2 + \overline{(\ln J)^2} - \Delta_J^2}. \tag{5.3.5}$$

From the renormalisation group results [138, 139] one concludes that at zero temperature, the longitudinal magnetisation in the presence of a small longitudinal field h, is give by the scaling form

$$m(\delta, h) = \overline{\mu}\big[\ln(D_h/h)\big]^{2-\phi} M\left[\delta \ln\left(\frac{D_h}{h}\right)\right] \tag{5.3.6}$$

where both δ and $(1/\ln h)$ are small but their ratio tends to some constant value. Here the exponent $\phi = (1/2)(1 + \sqrt{5})$, while $\overline{\mu}$ and D_h are non-universal dimensional constants. At the critical point $\delta = 0$

$$m(h) \sim \frac{1}{|\ln h|^{2-\phi}} \tag{5.3.7}$$

for small h. In the ordered phase ($\delta < 0$), the spontaneous magnetisation scales with δ

$$m_0(\delta) \sim (-\delta)^\beta; \quad \text{with } \beta = 2 - \phi. \tag{5.3.8}$$

In the disordered phase ($\delta > 0$), the scaling function (5.3.6) yields a continuously variable power law singularity for small h

$$m(h) \sim \delta^{3-\phi} h^{2\delta} |\ln h|, \tag{5.3.9}$$

so that the linear susceptibility χ is infinite for a range of δ in the disordered phase, i.e., Griffiths singularities occur in this region. For $\delta \gg 0$ (much away from the critical point) χ is finite but still a weaker power law singularity in $m(h)$ for a wider range of δ is observed.

At the quantum critical point the scaling behaviour of the characteristic time (τ) with the characteristic length scale l is given by $\tau \sim l^z$, where z is the dynamical exponent. For the present model z diverges at the critical point

$$z = \infty \quad (\text{at } \delta = 0), \tag{5.3.10}$$

which means that the time scale varies as the exponential of the square root of the corresponding length scale. The distribution of the local relaxation time is predicted to be very broad. In the disordered phase, there is still a broad distribution of relaxation times because of Griffiths singularity and consequently, one can define a dynamical exponent z which varies with δ, and diverges as

$$z \sim \frac{1}{2\delta} \tag{5.3.11}$$

as $\delta \to 0$. Further, inside the disordered phase, when all the transverse fields are greater than all the interactions, the Griffiths singularities disappear and the distribution of the relaxation times becomes narrow. As one approaches the end of the Griffiths phase, z vanishes. Using (5.3.11) in (5.3.9) we see that the singular part of the longitudinal magnetisation behaves in the disordered phase as

$$m \sim |h|^{\frac{1}{z}}. \tag{5.3.12}$$

Real space renormalisation approach [138, 139] predicts a large fluctuation in the longitudinal spin-spin correlation function $C_{ij} = \langle S_i^z S_j^z \rangle$. The average and typical correlations behave quite differently. The average correlation function

$$C_{av}(r) = \frac{1}{L} \sum_{i=1}^{L} \overline{\langle S_i^z S_{i+r}^z \rangle}, \tag{5.3.13}$$

varies as a power law at the critically

$$C_{av}(r) \sim \frac{1}{r^{2-\phi}} \quad \text{at } \delta = 0. \tag{5.3.14}$$

ϕ being the golden mean, the power in (5.3.14) is 0.38. Away from the critically, the average correlation decays exponentially at a rate given by the true correlation length, ξ, where

$$\xi \sim \frac{1}{\delta^{\nu}} \quad \text{with } \nu = 2. \tag{5.3.15}$$

To study the "typical" behaviour it is necessary to consider the distribution of $\ln C(r)$. At the critical point

$$-\ln C(r) \sim \sqrt{r}, \tag{5.3.16}$$

with the coefficient in (5.3.16) having a distribution which is independent of r. In the disordered phase

$$-\ln C(r) \sim \frac{r}{\overline{\xi}} \tag{5.3.17}$$

for large r, the typical correlation length $\overline{\xi}$, has the behaviour

$$\overline{\xi} \sim \frac{1}{\delta^{\overline{\nu}}}; \quad \overline{\nu} = 1. \tag{5.3.18}$$

It should be noted here that the typical correlation length $\overline{\xi}$ has a different exponent from the true correlation length ξ.

5.3.2 Mapping to Free Fermions

Young and Rieger [439], mapped the Hamiltonian (5.3.2) to a free fermion Hamiltonian and diagonalised it using standard numerical diagonalisation technique. The distributions of interaction strengths $\pi(J)$ and the transverse field at each site $\rho(\Gamma)$ are given by

$$\pi(J) = \begin{cases} 1 & \text{for } 0 < J < 1 \\ 0 & \text{otherwise,} \end{cases} \tag{5.3.19a}$$

and

$$\rho(\Gamma) = \begin{cases} \Gamma_0^{-1} & \text{for } 0 < \Gamma < 1 \\ 0 & \text{otherwise.} \end{cases} \tag{5.3.19b}$$

The distribution (5.3.19a), (5.3.19b) is characterised by a single control parameter Γ_0. The system has a critical point at $\Gamma_0 = 1$, where the system is self-dual (i.e., the distributions of Γ and J are identical). The measure of deviation from critically δ (5.3.5) becomes

$$\delta = (1/2) \ln \Gamma_0. \tag{5.3.20}$$

For the distribution (5.3.19a), (5.3.19b) the Griffiths phase extends through the entire disordered region. For a finite chain of N sites one can perform a Jordan-Wigner transformation (see Sect. 2.2) to transform the spin system to a system of spinless fermions described by the Hamiltonian

$$\begin{aligned} H = & -\sum_{i=1}^{N} \Gamma_i \left(2c_i^\dagger c_i - 1\right) - \sum_{i=1}^{N-1} J_i \left(c_i^\dagger - c_i\right)\left(c_{i+1}^\dagger + c_{i+1}\right) \\ & + J_N \left(c_N^\dagger - c_N\right)\left(c_1^\dagger + c_1\right) \exp(i\pi L) \end{aligned} \tag{5.3.21}$$

where periodic boundary condition is assumed and $L = \sum_i c_i^\dagger c_i$, is the number of free fermions. The Hamiltonian (5.3.21) describes a set of fermions and can readily be put in the general quadratic form (see Sect. 2.A.2)

$$H = \sum_{ij} \left[c_i^\dagger A_{ij} c_j + \frac{1}{2} c_i^\dagger B_{ij} c_j^\dagger + h.c. \right] \tag{5.3.22}$$

where A and B are both $N \times N$ matrices with elements given as (with periodic boundary condition)

$$\begin{aligned} A_{ii} &= -\Gamma_i, \qquad A_{ii+1} = -\frac{J_i}{2}, \qquad A_{i+1i} = -\frac{J_i}{2}, \\ B_{ii+1} &= \frac{J_i}{2}, \qquad B_{i+1i} = -\frac{J_i}{2}. \end{aligned} \tag{5.3.23}$$

Clearly A is symmetric and B is antisymmetric. Young and Rieger [439] used standard diagonalisation techniques for system size $N \leq 128$ and studied the ground state energy, mass gap and the correlation functions.

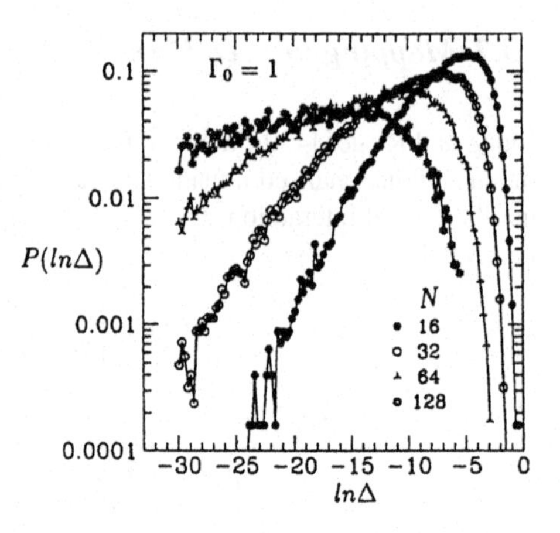

Fig. 5.5 A plot of the distribution of the log of the energy gap, Δ, at the critical point $\Gamma_0 = 1$, for $16 \leq N \leq 128$. The distribution is obtained from the value of the gap for 50000 samples for each size. One sees that the distribution gets broader as N increases [439]

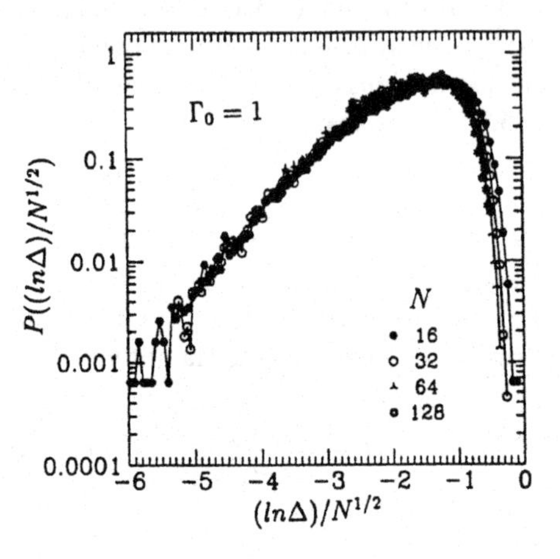

Fig. 5.6 A scaling plot of the data in Fig. 5.5, assuming that the log of the energy scale (here Δ) varies as the square root of the corresponding length [439]

For a pure transverse Ising system, the mass gap ($\Delta = E_1 - E_0$) is finite in disordered phase and tends to zero exponentially with the system size in the ordered phase. But in the random systems due to statistical fluctuations, there are finite regions, even in the disordered phase, which are locally ordered. These regions will have a very small energy gap and hence one expects large sample to sample fluctuations in the gap, especially for large system size. Young and Rieger [439], studied the distribution of $\ln \Delta$ at the critical point $\Gamma_0 = 1$ for $16 \leq N \leq 128$ (Fig. 5.5). The distribution gets broader with increasing system size which clearly indicates the divergence of z. The scaling plot for the distribution of $\ln \Delta / N^{1/2}$ (Fig. 5.6) supports the prediction that the log of the characteristic energy scale should vary as the square root of the length scale. In the disordered phase the distribution curves (Fig. 5.7) for different sizes look similar but are shifted horizontally relative to each other, implying that the data scale with a finite value of z. If one tries a scaling form

$$\ln\bigl[P(\ln \Delta)\bigr] = \frac{1}{z} \ln \Delta + \text{const}, \tag{5.3.24}$$

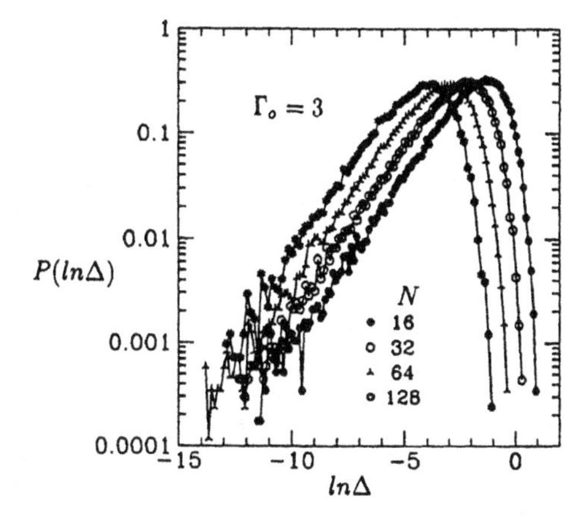

Fig. 5.7 A plot of the distribution of the log of the energy gap (Δ) in the disordered phase ($\Gamma_0 = 3$). Curves for different N are very similar and just are shifted relative to each other [439]

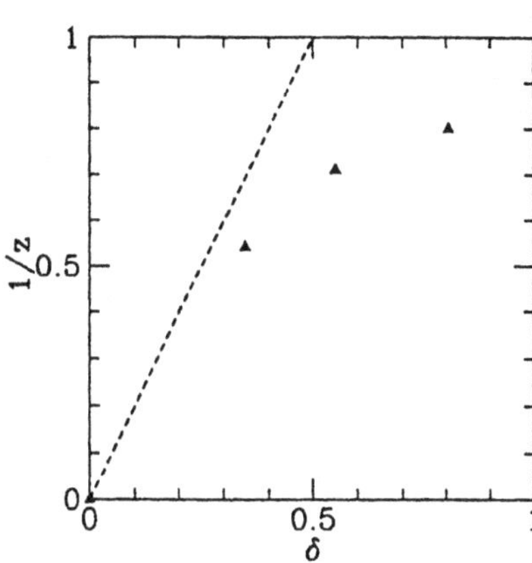

Fig. 5.8 Result for $1/z$ against δ. The *dashed line* shows the prediction, $z = 1/2\delta$, which is expected to be valid for small δ [439]

one estimates $z \sim 1.4$. The variation of $1/z$ with δ as obtained by Young and Rieger [439] is shown in the Fig. 5.8.

From the log-log plot of the average correlation function at the critical point [439] one finds that for large system size, at the critically average correlation decays with an exponent $= 0.38$, which clearly matches the prediction of Fisher [138]. The log of the typical correlation function $C(r)$ shows a linear variation with r for large system size. The scaling function for the distribution of the log of correlation function at critically (Fig. 5.9 and Fig. 5.10) is monotonic and shows an upturn as the abscissa approaches zero. This may indicate a divergence and if so, this part of the scaling function then determines the average correlation function [439].

5.3.3 Numerical Results in Two and Higher Dimensions

Random transverse Ising models in two or higher dimensions may involve frustration. We here focus on ferromagnetic models without frustration, where the random

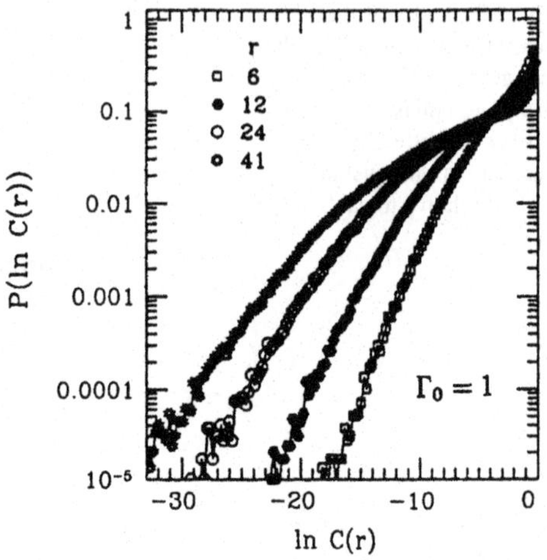

Fig. 5.9 The distribution of the log of the correlation function for different values of r at the critical point [439]

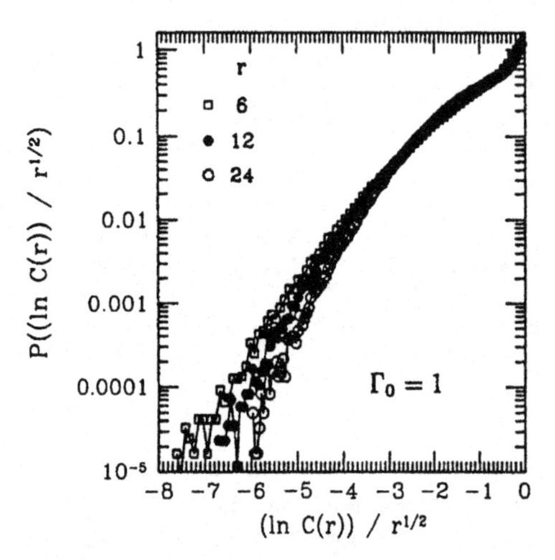

Fig. 5.10 A scaling plot of the data in Fig. 5.9. The data collapse well in the region of fairly high probability, including the upturn near the right hand edge (which may indicate a divergence as the abscissa tends to 0) [439]

coupling constant J_{ij} and transverse field Γ_i in the Hamiltonian (5.3.1) obey the distributions (5.3.19a), (5.3.19b).

In two- or higher-dimensional random transverse Ising models, neither analytic results of real-space renormalisation group nor mapping to free fermion systems is available. Pich et al. [315] studied a two-dimensional system using the quantum Monte Carlo method. As in the pure system, the present system shows a ferromagnetic long range order at low temperatures and fields. To determine the phase boundary between the ordered phase and the disordered phase, one may use the Binder ratio defined by [241]

$$g = \overline{\frac{1}{2}\left(3 - \frac{\langle M_z^4 \rangle}{\langle M_z^2 \rangle^2}\right)}, \tag{5.3.25}$$

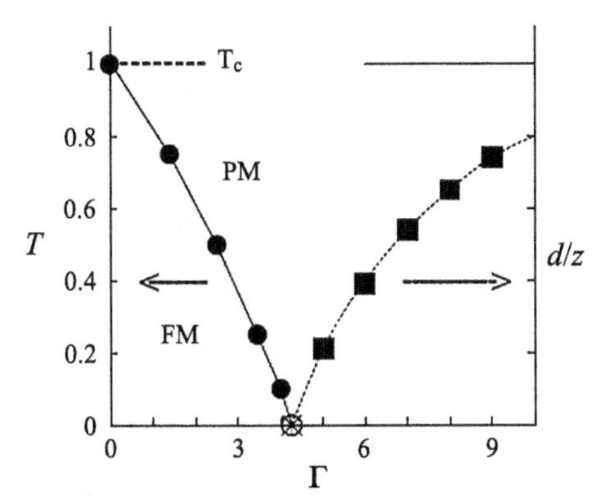

Fig. 5.11 Results of quantum Monte Carlo simulations on the random transverse field Ising ferromagnet in two dimension. The *left axis* shows the temperature, while the *right axis* shows the inverse of the dynamical exponent multiplied by the dimension ($d = 2$) in the disordered Griffiths phase. The result shows that the dynamical exponent z tends to diverge when the transverse field approaches the critical point

where the longitudinal magnetisation M_z is given by $M_z = (1/NM)\sum_i \sum_k S^z_{i,k}$ in the Suzuki-Trotter formalism with the Trotter number M and the number of spins N (see Sect. 3.1). The Binder ratio (5.3.25) has a finite value in the ordered phase, while it vanishes in the disordered phase in the thermodynamic limit. On the phase boundary, it is expected to be independent of the system size. Hence the critical point Γ_c for a given temperature T is obtained from the crossing point of g's with different sizes as functions of Γ.

Figure 5.11 (left axis) shows the phase diagram obtained by quantum Monte Carlo simulation and using the analysis of the Binder parameter g [315]. By the extrapolation to $T = 0$, the quantum critical point is estimated as $\Gamma_c = 4.2 \pm 0.2$.

The disordered Griffiths phase, if it exists, is characterised by the distribution of the energy gap Δ. The energy gap is related to the local susceptibility at the ground state $\chi_{\rm loc} = \sum_{k=1}^{M} \langle S^z_{i,k} S^z_{i,1}\rangle$ through $\ln \chi_{\rm loc} \approx -\ln \Delta + {\rm const}$. One can see this relation by observing $\chi_{\rm loc} = \sum_{n>0} |\langle \Psi_n | S^z_i | \Psi_0\rangle|^2/(E_n - E_0)$, where $|\Psi_n\rangle$ is the nth eigenstate with eigenenergy E_n, and noting that the scaling behaviour of $\chi_{\rm loc}$ is governed by $E_1 - E_0 = \Delta$. Therefore a small gap corresponds to an ordering in imaginary-time (Trotter axis) which is in turn local in real space. The probability of such a small gap is assumed to be proportional to the system size $N = L^d$. Letting λ be the exponent of Δ such that the probability distribution of $\ln \Delta$ is proportional to Δ^λ, one can write $P(\ln \Delta) \sim L^d \Delta^\lambda$. If one defines the dynamical exponent z such that $\Delta \sim L^z$, it turns out $\lambda = d/z$. Thus one reaches the scaling of the distribution of the local susceptibility [315, 333]

$$\ln\bigl[P(\ln \chi_{\rm loc})\bigr] = -\frac{d}{z} \ln \chi_{\rm loc} + {\rm const}. \tag{5.3.26}$$

Pich et al. obtained the exponent z on the basis of (5.3.26) using quantum Monte Carlo simulation with very low temperatures [315]. The field Γ dependence of d/z

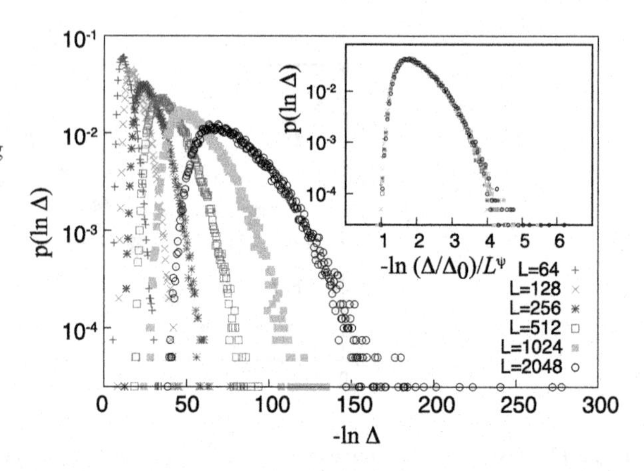

Fig. 5.12 A plot of the distribution of the log of the gap Δ at the critical point for $L = 2^6, 2^7, 2^8, 2^9, 2^{10}$, and 2^{11}. The *inset* shows a scaling plot by $-\ln(\Delta/\Delta_0)/L^{\psi}$ with $\ln \Delta_0 = -1.5$ and $\psi = 0.48$

is shown in Fig. 5.11. The exponent z seems to diverge as the quantum critical point is approached. This result clearly reveals the Griffiths singularities in the disordered phase.

At the quantum critical point, the scaling of the distribution of the gap is generalised from one-dimensional form $\ln \Delta/L^{1/2}$ [139, 141, 439] to a d-dimensional one $\ln \Delta/L^{\psi}$ [140, 188, 315, 415]. Kovács and Iglói implemented real-space renormalisation group numerically in two and higher dimensional systems and obtained the exponent ψ [235, 236]. Figure 5.12 shows the distributions of the gap for several sizes in two dimension [235]. As shown in inset, the data for different sizes collapse when one choose the scaled variable $-\ln(\Delta/\Delta_0)/L^{\psi}$ with $\ln \Delta_0 = -1.5$ and $\psi = 0.48$. Kovács and Iglói have estimated ψ as $\psi \approx 0.46$ both for three and four dimensional systems [236].

Chapter 6
Transverse Ising Spin Glass and Random Field Systems

6.1 Classical Ising Spin Glasses: A Summary

Spin glasses are magnetic systems with randomly competing (frustrated) interactions [39, 76, 136, 275]. Here, the frustration arises due to the competing (ferromagnetic and antiferromagnetic) quenched random interactions between the spins. The spins in such systems get frozen in random orientations below a certain transition temperature. Although there is no long range magnetic order, i.e., the space averages of the spin moments vanish, the spins are frozen over macroscopic scales of time and hence the time average of any spin is nonzero below the (spin glass) transition temperature. This time average is treated as a measure of the spin freezing or spin glass order parameter. Because of frustration, the ground state is (infinitely) degenerate; the degeneracy being of the order of $\exp(N)$ for a system of N spins. These ground states, as well as the local minima, are however, often separated by macroscopically large energy barriers ($O(N)$) which force the system to get trapped, depending on its history (initial configuration), in one of its degenerate (local) minima. The system thus becomes "non-ergodic" and a spin glass may be described by a nontrivial order parameter distribution [275, 304–306] in the thermodynamic limit (unlike the unfrustrated cooperative systems, where the distribution becomes trivially delta function-like in the same limit).

Several spin glass models have been studied extensively, using both analytic and computer simulation techniques. The Hamiltonian for such models can be written as

$$H = -\sum_{i<j} J_{ij} S_i^z S_j^z, \tag{6.1.1}$$

where $S_i^z = \pm 1$, $i = 1, 2, \ldots, N$, denote the Ising spins, interacting with random (quenched) exchange interactions J_{ij}, which differ in various models. We will specifically consider three extensively studied models. In the Sherrington-Kirkpatrick (SK) model [369], J_{ij}'s are long-ranged and are distributed with a Gaussian probability (centred around zero)

$$P(J_{ij}) = \left(\frac{N}{2\pi J^2}\right)^{1/2} \exp\left(-\frac{N J_{ij}^2}{2J^2}\right). \tag{6.1.2}$$

S. Suzuki et al., *Quantum Ising Phases and Transitions in Transverse Ising Models*, Lecture Notes in Physics 862, DOI 10.1007/978-3-642-33039-1_6,

In the Edwards-Anderson (EA) model [121], the J_{ij}'s are short-ranged (say between the nearest-neighbours only), but similarly distributed with Gaussian probability (6.1.2). (Here, of course, the normalisation factor N is unnecessary.) In another kind of model, the J_{ij}'s are again short-ranged, but having a binary ($\pm J$) distribution with probability p:

$$P(J_{ij}) = p\delta(J_{ij} - J) + (1 - p)\delta(J_{ij} + J). \tag{6.1.3}$$

The disorder in the spin glass system being quenched, one has to perform configurational averaging (denoted by overhead bar) over $\ln Z$, where Z ($= \mathrm{Tr}\exp(-\beta H)$; $\beta = 1/k_B T$) is the partition function of the spin glass system. To evaluate $\overline{\ln Z}$, one usually employs the replica trick, based on the representation $\ln Z = \lim_{n\to 0}[(Z^n - 1)/n]$. Now, for classical Hamiltonians (with commuting spin components), $Z^n = \prod_{\alpha=1}^{n} Z_\alpha = Z(\sum_{\alpha=1}^{n} H_\alpha)$, where H_α is the α-th replica of the Hamiltonian H in (6.1.1) and Z_α is the corresponding partition function. The spin freezing can then be measured in terms of replica overlaps, and the Edwards-Anderson order parameter takes the form $q = (1/N)\sum_{i=1}^{N} \overline{\langle S_i^z(t) S_i^z(0)\rangle}|_{t\to\infty} \simeq (1/N)\sum_{i=1}^{N} \overline{\langle S_{i\alpha}^z S_{i\beta}^z\rangle}$, where α and β correspond to different replicas.

Extensive Monte Carlo studies, together with the analytical solutions for the mean field of Sherrington and Kirkpatrick model, have revealed the nature of spin glass transition recently. It appears that the lower critical dimension for EA model d_l^c, below which the phase transition ceases to occur (with transition temperature T_c becoming zero), is between 2 and 3: $2 < d_l^c < 3$. The upper critical dimension d_u^c, at and above which mean field results (e.g., those of the Sherrington-Kirkpatrick model) apply, appears to be 6: $d_u^c = 6$. Within these dimensions d ($d_l^c < d < d_u^c$), the spin glass transition (for the Hamiltonian (6.1.1) with short-range interactions) occur, and the transition behaviour can be characterised by various exponents. Although the linear susceptibility shows a cusp at the transition point, the nonlinear susceptibility $\chi_{SG} = \sum_r g(r)$, where $g(r) = (1/N)\sum_i \overline{\langle S_i^z S_{i+r}^z\rangle^2}$, diverges at the spin glass transition point:

$$\chi_{SG} \sim (T - T_c)^{-\gamma_c}, \tag{6.1.4}$$

$$g(r) \sim r^{-(d-2+\eta_c)} f\left(\frac{r}{\xi}\right); \quad \xi \sim |T - T_c|^{-\nu_c}. \tag{6.1.5}$$

Here ξ denotes the correlation length which determines the length scaling in the spin correlation function $g(r)$ (f in $g(r)$ denotes the scaling function). Numerical simulations give $\nu_c \approx 3.45, 2.44 \pm 0.09, 1/2$ and $\gamma_c = \nu_c(2 - \eta_c) \approx 6.9, 5.8, 1$ for $d = 2$ [182, 217], 3 [218] and 6 respectively for the values of exponents. One can define the characteristic relaxation time τ through the time dependence of the spin auto-correlation

$$q(t) = \overline{\langle S_i^z(t) S_i^z(0)\rangle^2} \sim t^{-x}\tilde{q}\left(\frac{t}{\tau}\right); \quad \tau \sim \xi^{z_c} \sim |T - T_c|^{-\nu_c z_c}, \tag{6.1.6}$$

where $x = (d - 2 + \eta_c)/2z_c$ and z_c denotes the classical dynamical exponent. Numerical simulations give $z_c = 6.45 \pm 0.10$ and 5.1 ± 0.1 in $d = 3$ [258] and 4 [36] dimensions respectively. Of course, such large values of z_c (particularly in lower dimensions) also indicate the possibility of the failure of the power law variation (6.1.6) of τ with $T - T_c$ and rather suggests a Vogel-Fulcher like variation: $\tau \sim \exp[A/(T - T_c)]$.

6.2 Quantum Spin Glasses

Quantum spin glasses have the interesting feature that the transition in randomly frustrated (competing) cooperative interacting systems can be driven by both thermal fluctuations or by increasing quantum fluctuations. Quantum spin glasses can be of two types: vector spin glasses introduced by Bray and Moore [44], where of course the quantum fluctuations cannot be tuned, or a classical spin glass perturbed by some tunable quantum fluctuations, e.g., as induced by (increasing) a non-commuting transverse field [62, 195]. With the tunable amount of quantum fluctuations, this transverse Ising spin glass model is perhaps the simplest model in which the quantum effects in a random system can be and has recently been studied extensively and systematically (for a review see [361, 400] and [327]).

The interesting question in such quantum spin glass models is about the possibility of tunnelling through the (infinite) barriers of the free energy landscape in the classical spin glass model (e.g., the Sherrington-Kirkpatrick model) due to the quantum fluctuations (by the transverse field). In the classical case, the barriers separating the valleys increase in height with the macroscopic size of the system. In the thermodynamic limit, the thermal fluctuations are thus unable to let the system cross the barrier, thereby causing non-ergodicity. Quantum spin, however, should not necessarily look for the barrier height and since the barrier width (in configuration space) decreases with increasing system size, it may be able to tunnel through such barriers, provided the integrated tunnelling probability is finite. This would suggest an ergodic (replica symmetric) spin glass solution for the quantum Ising spin glass systems [323, 401]. The question whether in the spin glass state the replica symmetric solution is stable, or it is "non-ergodic" [159, 240] as in the classical case, is still unresolved. Studies on short-range transverse Ising spin glasses, using Landau-Ginzberg free energy, indicate the existence of replica symmetric ground state [325].

Much of the theoretical work on the quantum Ising spin glasses has been confined to the study of Sherrington-Kirkpatrick model in a transverse field [50, 232, 233, 316, 433], which is supposed to exhibit the limiting behaviour of the transition in short-range model in higher dimensions. Also, considerable studies have been made on the short-range Edwards-Anderson model in the transverse field [112, 163, 164, 332, 333, 417]. Exact results have been obtained in the random quantum Ising systems in one dimension [138]. This one dimensional model obviously does not have frustration (cf. Sect. 5.3). Several numerical studies have been performed in recent years on the short-range Ising spin glass model (of the Edwards-Anderson type) in a transverse field. Although, the short-range Ising spin glass models were studied analytically employing approximate renormalisation techniques [62] and also real-space renormalisation group method [112] to study the effect

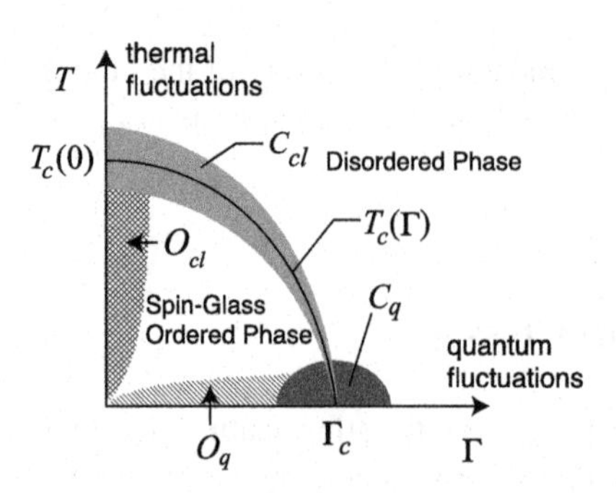

Fig. 6.1 Schematic phase diagram of transverse Ising spin glass in high dimensions. $T_c(\Gamma)$ represent the phase boundary. At low quantum fluctuations (small Γ) transition is essentially driven by thermal fluctuations. O_{cl} represents the classical spin glass ordered region. O_q is the quantum fluctuation dominated spin glass phase. C_{cl} represents the transition region where classical fluctuations dominate over thermal fluctuations due to critical slowing down. C_q represents the vicinity of zero-temperature transition where quantum fluctuations dominate (after Thill and Huse [400])

of disorder in the equivalent time direction and the phase diagram, reliable results in such models have been obtained only through the recent numerical simulations. The nature of quantum phase transition in short range transverse Ising spin glass models (in $d = 2$ and 3) and the accurate estimates for the values of the exponents associated with the transition have been obtained, using quantum Monte Carlo techniques [163, 164, 332, 333].

A schematic phase diagram for the Ising spin glass in a transverse field (in Γ–T plane) is shown in Fig. 6.1 (cf. Thill and Huse [400]). The spin glass order is found to exist in the region of low Γ and T (when both the thermal and quantum fluctuations are small) and there is a transition from the ordered state to the paramagnetic state when either type of fluctuation is increased. Critical fluctuations near enough to the transition are classical as long as they occur at a high value of critical temperature ($T_c \gg 0$). This is because, the characteristic frequency ω of critical fluctuations tends to zero due to the "critical slowing down" and $\hbar\omega \ll k_B T$, so the nature of the transition is essentially the classical and the transition belongs to the same universality class as that of classical spin glass transition. However at $T = 0$, the quantum effects are dominant (cf. Sect. 3.5) and the universality class for the spin glass to paramagnetic transition, driven by the quantum fluctuations due to the transverse field is different. As mentioned earlier, the nature of this zero-temperature transition is at the focus of recent research.

For transverse Ising spin glasses, analytical results are available only in $d = 1$ and $d = \infty$. At the zero-temperature spin glass critical point Γ_c, the spin autocorrelation functions, defined earlier, decay with a power law of $\ln t$ and as t^{-2} in $d = 1$ and infinity, respectively. For $d = 1$, both the linear and nonlinear susceptibilities diverge not only in the ordered phase, but also in some portions of the disordered phase, well away from the critical point (due to the Griffiths singularity). For the infinite dimensional model, on the other hand, the Griffiths singularity is wiped out due to the infinite-range interaction, and the linear susceptibility remains finite. However, the nonlinear susceptibility diverges as the critical point is approached from

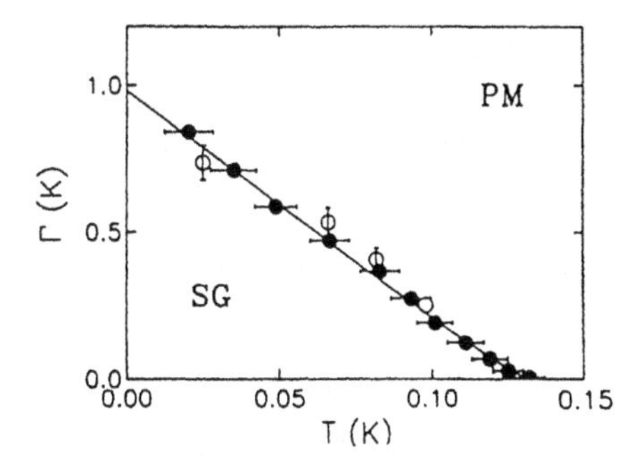

Fig. 6.2 Phase diagram of the diluted Ising spin glass from the dynamical behaviour of the linear susceptibility [427], *open circles* from measurements of nonlinear susceptibility

the paramagnetic phase with a critical exponent $\gamma = 1/2$, with multiplicative logarithmic corrections. In the quantum Monte Carlo studies of short-range transverse Ising spin glass model in $d = 2, 3$, the linear susceptibility [163, 164, 332, 333] is found to diverge for $d = 2$, but not for $d = 3$.

6.2.1 Experimental Realisations of Quantum Spin Glasses

The interest in the study of short-range Ising spin glass models in a transverse field is renewed due to the recent discovery of the zero-temperature transition in the dipolar Ising transverse-field magnet $LiHo_xY_{1-x}F_4$ [14, 321, 427, 428] (see Fig. 6.2). This, along with the proton glasses (mixture of ferroelectric and anti-ferroelectric materials such as $Rb_{1-x}(NH_4)_x(H_2P)_4$) [316] also provides very useful experimental realisations of quantum spin glasses. In mixed hydrogen bonded ferroelectrics and antiferroelectrics, the protons can tunnel between the two minima in each hydrogen bond. Generalising the transverse Ising model for such order-disorder structural transition, Pirc et al. [316] set up a pseudo-spin model where the interactions are spin glass like and the tunnelling is mimicked by the transverse field.

6.3 Sherrington-Kirkpatrick (SK) Model in a Transverse Field

We can write the generic form of the Ising spin glass Hamiltonian in a transverse field as

$$H = -\sum_{ij} J_{ij} S_i^z S_j^z - \Gamma \sum_i S_i^x. \tag{6.3.1}$$

In the Sherrington-Kirkpatrick (SK) model in a transverse field [49, 50, 108, 133, 196, 252, 406, 408, 418, 432, 437], the J_{ij}'s follow the Gaussian distribution

$$P(J_{ij}) = \left(\frac{N}{2\pi J^2}\right)^{1/2} \exp\left(-\frac{N J_{ij}^2}{2J^2}\right). \tag{6.3.2}$$

This model was first studied by Ishii and Yamamoto [195].

6.3.1 Phase Diagram

Several analytical studies have been made for obtaining the phase diagram of the transverse Ising SK model (giving in particular the zero-temperature critical field). The problem of the SK spin glass in a transverse field becomes a nontrivial one as the spin operators do not commute. This leads to a dynamical frequency dependent (spin) self-interaction.

6.3.1.1 Mean Field Estimates

One can study an effective single-spin Hamiltonian for the above quantum many-body system within the framework of mean field theory. A systematic mean field theory for the above model is carried out by Kopec [232, 233], using the thermofield dynamical approach and the short-time approximation for the dynamic spin self-interaction. Before going into the discussion of this approach, we shall briefly review the replica symmetric solution of the classical SK model ($\Gamma = 0$) in a longitudinal field (see e.g. [39, 76, 136, 275]) given by the Hamiltonian

$$H = -\sum_{\langle ij\rangle} J_{ij} S_i^z S_j^z - h \sum_i S_i^z, \tag{6.3.3}$$

where J_{ij} follows the Gaussian distribution with mean 0 and variance J, given by (6.3.2). Using the replica trick, one obtains for the configuration averaged n-replicated partition function $\overline{Z^n}$

$$\overline{Z^n} = \sum_{(S_{i\alpha}=\pm 1)} \int_{-\infty}^{\infty} dP(J_{ij})\, dJ_{ij} \exp\left[\beta \sum_{ij} J_{ij} \sum_{\alpha=1}^{n} S_{i\alpha}^z S_{j\alpha}^z + \beta h \sum_i \sum_\alpha S_{i\alpha}^z\right]. \tag{6.3.4}$$

Performing the Gaussian integral, using Hubbard-Stratonovitch transformation and finally using the method of steepest descent to evaluate integrals for a thermodynamically large system, one obtains the free energy per site

$$-\beta f = \lim_{n\to 0}\left[\frac{(\beta J)^2}{4}\left\{1 - \frac{1}{n}\sum_{\alpha,\beta} q_{\alpha\beta}^2 + \frac{1}{n}\ln \operatorname{Tr}\exp(L)\right\}\right] \tag{6.3.5}$$

where $L = (\beta J)^2 \sum_{\alpha,\beta} q_{\alpha\beta} S_\alpha^z S_\beta^z + \beta \sum_{\alpha=1}^n S_\alpha^z$ and $q_{\alpha\beta}$ is self-consistently given by the saddle-point condition $(\partial f/\partial q_{\alpha\beta}) = 0$. Considering the replica-symmetric case ($q_{\alpha\beta} = q$), one finds

$$-\beta f = \frac{(\beta J)^2}{2}\left(1 - q^2\right) + \frac{1}{\sqrt{2\pi}} \int_{-\infty}^{\infty} dr\, \mathrm{e}^{-\frac{r^2}{2}} \ln\left[2\cosh \beta h(r)\right], \tag{6.3.6}$$

where r is the excess static noise arising from the random interaction J_{ij} and the spin glass order parameter q is self-consistently given by

$$q = \frac{1}{\sqrt{2\pi}} \int_{-\infty}^{\infty} dr\, \mathrm{e}^{-\frac{r^2}{2}} \tanh^2\left(\beta h(r)\right), \tag{6.3.7}$$

and $h(r) = J\sqrt{q}\,r + h$, can be interpreted as a local molecular field acting on a site. Different sites have different fields because of disorder, and the effective distribution of $h(r)$ is Gaussian with mean value 0 and variance $J^2 q$.

At this point, one can introduce the quantum effect through the transverse term $-\Gamma \sum_i S_i^x$ (with longitudinal field $h = 0$). The effective single particle Hamiltonian in the transverse Ising quantum glass can be written as

$$H_s = -h^z(r)S^z - \Gamma S^x, \tag{6.3.8}$$

where $h^z(r)$, as mentioned earlier, is the effective field acting along the z-direction arising due to the nonzero value of the spin glass order parameter. Treating $h^z(r)$ and $\mathbf{S}$ as classical vectors (cf. Sect. 1.2) in the pseudo-spin space, one can write the net effective field acting on each spin as

$$\mathbf{h}_0(r) = -h^z(r)\hat{z} - \Gamma \hat{x}; \quad \left|h_0(r)\right| = \sqrt{h^z(r)^2 + \Gamma^2}. \tag{6.3.9}$$

One can now, readily arrive at the mean field equation for the local magnetisation

$$m(r) = p(r)\tanh\bigl(\beta h_0(r)\bigr); \quad p(r) = \frac{|h^z(r)|}{|h_0(r)|}, \tag{6.3.10}$$

and consequently, the spin glass order parameter can be written as

$$q = \frac{1}{\sqrt{2\pi}} \int_{-\infty}^{\infty} dr\, \mathrm{e}^{-\frac{r^2}{2}} \tanh^2\bigl(\beta h_0(r)\bigr) p^2(r). \tag{6.3.11}$$

The phase boundary can be obtained from the above expression by putting $q \to 0$ ($h^z(r) = J\sqrt{q}\,r$ and $h_0 = \Gamma$), when it gives

$$\frac{\Gamma}{J} = \tanh\left(\frac{\Gamma}{k_B T}\right). \tag{6.3.12}$$

This gives $\Gamma_c = J$ (see Fig. 6.3 for the phase diagram).

Ishii and Yamamoto [195] used the "reaction field" technique to construct "TAP" like equation for the free energy of the Hamiltonian (6.3.1) and perturbatively expanded the free energy in powers of Γ up to the order Γ^2 to obtain

$$k_B T_c = J\left[1 - 0.226\left(\frac{\Gamma}{J}\right)^2\right]. \tag{6.3.13}$$

The above equation also indicates that T_c decreases with increasing the transverse field, but numerical estimations of T_c including the terms higher order in Γ indicate that the quantum model is fundamentally different from the classical model: the quantum model always gives rise to a spin glass phase regardless of the strength of the transverse field Γ. Using the replica trick, introduced for the quantum spin glasses by Bray and Moore [44], Fedorov and Shender [133] showed the error in the calculation of Ishii and Yamamoto [195], and proved that quantum fluctuations do not stabilise the spin glass state in the model under consideration.

One can write the Hamiltonian (6.3.3) in the form

$$H = H_0 + V; \quad H_0 = -\Gamma \sum_i S_i^x, \quad V = -\sum_{ij} J_{ij} S_i^z S_j^z, \tag{6.3.14}$$

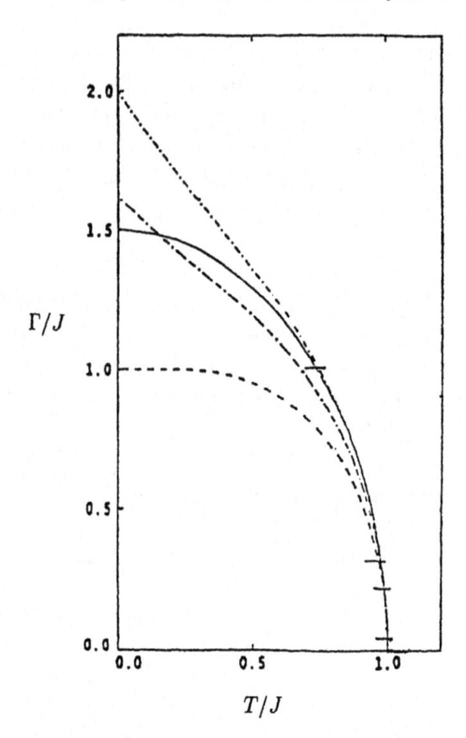

Fig. 6.3 The phase diagram for the SK model in a transverse field from different approximations: *dashed curve* from Ref. [233], *dot dash curve* from static approximation [433], *dash dash dot curve* from Ref. [408] and *solid lines* from Ref. [433]. The *horizontal dashes* denote Monte Carlo results [240]

the partition function of the system can be written as

$$Z = \operatorname{Tr} \exp(-\beta H_0) P \exp\left[\int_0^\beta d\tau \sum_{ij} J_{ij} S_i^z(\tau) S_j^z(\tau)\right], \tag{6.3.15}$$

where τ is the imaginary time, P is the time ordering operator and $S(\tau)$ are the operators in the interaction representation. Using the replica trick (see Sect. 6.A.1), the n-replicated partition function of the quantum system can be written as

$$Z^n = \operatorname{Tr} \exp\left(-\beta \sum_{\alpha=1}^n H_0(\alpha)\right) P \exp\left(\sum_{ij} J_{ij} \int_0^\beta d\tau \sum_{\alpha=1}^n S_{i\alpha}^z(\tau) S_{j\alpha}^z(\tau)\right). \tag{6.3.16}$$

Performing the Gaussian averaging (since the operators can be permutated in the P product) and rearranging terms, one gets

$$\begin{aligned}\overline{Z^n} = {} & \operatorname{Tr} \exp\left(-\beta \sum_{\alpha=1}^n H_0(\alpha)\right) P \\ & \times \exp\left[\frac{J^2}{2N} \int_0^\beta d\tau \int_0^\beta d\tau' \frac{1}{2} \sum_\alpha \left(\sum_i S_{i\alpha}^z(\tau) S_{j\alpha}^z(\tau')\right)^2 \right. \\ & \left. + \sum_{\alpha<\beta} \left(\sum_i S_{i\alpha}^z(\tau) S_{i\beta}^z(\tau')\right)^2\right].\end{aligned} \tag{6.3.17}$$

The squares appearing in the above expression can be simplified using the Hubbard-Stratonovitch transformation

$$\exp\left(\frac{\lambda a^2}{2}\right) = \left(\frac{\lambda}{2\pi}\right)^{1/2} \int_{-\infty}^{\infty} dx \exp\left[-\frac{\lambda x^2}{2} + a\lambda x\right]. \tag{6.3.18}$$

One can thus, transform the expression of free energy, found after averaging, into a functional integral in the fields $R^{\alpha\alpha}(\tau, \tau')$ and $Q^{\alpha\beta}(\tau, \tau')$

$$\beta F = -\lim_{n\to 0} \frac{1}{n}\left[\int dQ^{\alpha\beta}(\tau, \tau')\, dR^{\alpha\alpha}(\tau, \tau')\, e^{-N\phi} - 1\right], \tag{6.3.19}$$

where

$$\begin{aligned}\phi = &\int_0^\beta d\tau \int_0^\beta d\tau' \left[\sum_{\alpha,\beta} \frac{1}{2} Q^{\alpha\beta}(\tau, \tau')^2 + \sum_\alpha R^{\alpha\alpha}(\tau, \tau')^2\right] \\ &- \ln \mathrm{Tr}\left[e^{-\beta \sum_{\alpha=1}^n H_0(\alpha)} P \exp\left[J \int_0^\beta d\tau \int_0^\beta d\tau' \left(\sum_{\alpha,\beta} Q^{\alpha\beta}(\tau, \tau') S_\alpha^z(\tau) S_\beta^z(\tau')\right.\right.\right. \\ &\left.\left.\left. + \sum_\alpha R^{\alpha\alpha}(\tau, \tau') S_\alpha^z(\tau) S_\beta^z(\tau')\right)\right]\right].\end{aligned} \tag{6.3.20}$$

For a thermodynamically large system one can evaluate the above integral using the method of steepest descent. In the high temperature (paramagnetic) phase, all the functions $Q^{\alpha\beta}(\tau, \tau')$ (the spin glass order parameter), which satisfy the equations for the steepest descent, vanish. Again, the functions $R^{\alpha\alpha}(\tau, \tau')$, do not depend upon the replica index at any temperature. The saddle-point condition therefore yields

$$R^{\alpha\alpha}(\tau, \tau') = \frac{J}{2}\langle P S^z(\tau) S^z(\tau')\rangle = R(\tau, \tau'), \tag{6.3.21}$$

where the average in the above equation is taken with the Hamiltonian $H_{\mathrm{eff}} = H_0 + H_1$, where

$$-\beta H_{\mathrm{eff}} = -\beta H_0 + J \int_0^\beta d\tau \int_0^\beta d\tau'\, R(\tau, \tau') S^z(\tau) S^z(\tau'). \tag{6.3.22}$$

Hence, one obtains for the free energy in the paramagnetic phase

$$F = \frac{N}{\beta} \min\left(\int_0^\beta d\tau \int_0^\beta d\tau'\, R^2(\tau, \tau') - \ln \mathrm{Tr}\left[P \exp(-\beta H_{\mathrm{eff}})\right]\right). \tag{6.3.23}$$

To evaluate T_c, one has to expand the free energy of the system to second order in spin glass order parameter $Q^{\alpha\beta}$. The terms in F, quadratic in these fields, are

$$\begin{aligned}F = \frac{1}{2} \sum_{\alpha<\beta} &\left[\left(\int_0^\beta d\tau \int_0^\beta d\tau' \left(Q^{\alpha\beta}(\tau, \tau')\right)^2 - J^2 \int_0^\beta d\tau \int_0^\beta d\tau' \int_0^\beta d\tau'' \int_0^\beta d\tau'''\right.\right. \\ &\left.\left. \times Q^{\alpha\beta}(\tau, \tau') Q^{\alpha\beta}(\tau'', \tau''') \times \langle P S^z(\tau) S^z(\tau')\rangle\langle P S^z(\tau'') S^z(\tau''')\rangle\right)\right],\end{aligned} \tag{6.3.24}$$

where the expectation values are computed using the effective Hamiltonian (6.3.22). One can now anticipate that the "soft mode" whose amplitude becomes nonzero just below T_c is time independent (i.e., the zero frequency mode has the highest T_c). One then sets $Q^{\alpha\beta}(\tau,\tau') = Q^{\alpha\beta}$ in (6.3.22) and determines T_c from the condition that the coefficient of Q^2 vanishes such that

$$1 = \frac{J}{\beta}\int_0^\beta d\tau\, d\tau' \langle P S^z(\tau) S^z(\tau')\rangle = \frac{2}{\beta}\int_0^\beta d\tau\, d\tau' R(\tau,\tau'). \qquad (6.3.25)$$

Applying a spatially varying magnetic field, one can check that in the paramagnetic phase, the local linear susceptibility is given by

$$\chi_{\text{loc}} = \frac{1}{\beta}\int_0^\beta d\tau\, d\tau' \langle P S^z(\tau) S^z(\tau')\rangle. \qquad (6.3.26)$$

Hence, the condition for T_c becomes

$$1 = J\chi_{\text{loc}}. \qquad (6.3.27)$$

One can now expand $R(\tau,\tau')$ in a Fourier series as

$$R(\tau,\tau') = \sum_n R_n \exp\bigl(i\omega_n(\tau-\tau')\bigr) = R_0 + \delta R(\tau,\tau') \qquad (6.3.28)$$

where $\delta R(\tau,\tau')$ denotes the contribution of nonzero frequency mode and $\omega_n = (2\pi n/\beta)$. If one assumes "static approximation", $R(\tau,\tau') = R_0$, (6.3.21) reduces to (using (6.3.18)) [433]

$$0 = \int_0^\infty \bigl(x^2 - 1 - 4\beta^2 R_0^2\bigr)\exp\left(-\frac{x^2}{2}\cosh\bigl[\beta\bigl(2JR_0x^2 + \Gamma^2\bigr)^{1/2}\bigr]\right) dx, \quad (6.3.29)$$

which, at $T = T_c$, gives the phase diagram (within the static approximation) as the solution of the equation

$$\int \bigl(x^2 - 2\bigr)\exp\left(-\frac{x^2}{2}\right)\cosh\bigl[\beta J x^2 + (\beta\gamma)^2\bigr]^{1/2} dx = 0, \qquad (6.3.30)$$

which gives Γ_c $(T = 0)$ at $2J$. For $\Gamma = 0$, it can be shown using (6.3.21) that the static approximation is exact. One can mention here that, using "static" approximation and the replica method for quantum spin glasses introduced by Bray and Moore [44] (see Sect. 6.A.1), Usadel [406] mapped the above model onto a vector spin glass model and obtained the same phase boundary equation as (6.3.27). This result is the first order mean field result when one notes that this is the largest eigenvalue of the Gaussian matrix J_{ij}. This result was also obtained independently by Pirc et al. [316], using the replica method.

Yamamoto and Ishii [433], using perturbation expansion of the local linear susceptibility, found Γ_c $(T = 0)$ at about $1.51J$. Here, χ is expressed as

$$\chi = \frac{1}{N}\sum_i \int_0^\beta d\tau \overline{\left(\frac{\langle U(\beta,\tau) S_i^z U(\tau) S_i^z\rangle}{\langle U(\beta)\rangle}\right)} \qquad (6.3.31)$$

where

$$U(\beta,\tau) = P\exp\left(-\int_{\tau}^{\beta} du\, V(u)\right) \tag{6.3.32}$$

and

$$V_1(u) = \exp(uH_0)V\exp(-uH_0), \tag{6.3.33}$$

where the Hamiltonians H_0 and V are as given in (6.3.14). Equation (6.3.31) is expanded in powers of V and the terms are analysed diagrammatically. In this way writing $\chi = \sum_n a_{2n}(\beta\Gamma)(J/\Gamma)^{2n}$, the coefficients a_{2n} are evaluated for $n = 0$, 2, 4, 6, 8 (the odd ordered terms vanish). Now, the bulk nonlinear susceptibility is calculated to be

$$\chi_b = \chi_2 \frac{[1+2(J\chi)^2]}{[1-(J\chi)^2]}, \tag{6.3.34}$$

where χ_2 is the local nonlinear susceptibility [108]. Hence, the singularity in χ_b is of the form $[1-(J\chi)]^{-1}$ in (6.3.34) which is expanded as

$$\left[1-(J\chi)\right]^{-1} = \sum_n b_n(\beta\Gamma)\left(\frac{J}{\Gamma}\right)^n. \tag{6.3.35}$$

Estimating b_n's (for both $T = 0$ and $T \neq 0$) from a_{2n}'s, one obtains the series (6.3.35). The critical value Γ_c above which the system remains paramagnetic is obtained from the scaling relation

$$\left[1-(J\chi)\right]^{-1} \sim (\Gamma-\Gamma_c)^{-\gamma} \tag{6.3.36}$$

with

$$\frac{\Gamma}{J} = \left(nb_nb_{n-1} - (n-2)(b_{n-2}b_{n-3})\right)/2 \tag{6.3.37}$$

$$\gamma = 1 + \frac{b_{n-2}b_{n-3}}{[b_nb_{n-1}/(n-2) - b_{n-2}b_{n-3}/n]} \tag{6.3.38}$$

in the n-th order. The best estimate for the critical Γ and γ are obtained from the average of $n = 9$ and $n = 10$. At $T = 0$, $\Gamma_c = 1.506J$, and the critical exponent γ of the bulk susceptibility is found to be equal to 0.564 (as compared to 0.5 obtained in the static approximation in the replica method). The phase boundary (Fig. 6.3) is obtained considering the coefficients b_n as function of $\beta\Gamma$.

Transferring the d-dimensional quantum model to the equivalent $(d + 1)$-dimensional classical model using the Suzuki-Trotter formalism (cf. Sect. 3.1), Usadel and Schmitz [408] also found the phase diagram of the above model (not restricting to the static approximation). However, this method becomes cumbersome for very low temperature and the estimate for the critical field for zero temperature is made by extrapolation only. Also, they assumed zero value for the order parameter such that their result is strictly valid in the para-phase.

The phase diagram was also obtained by the cluster expansion method by Walasek et al. [418], with similar estimate of γ_c. By combining the pair approximation with the discretised path integral representation [108], the phase diagram has been obtained [252] without the replica formalism for nonzero J_0. Here also, $\Gamma_c\ (T=0)$ is found at J.

The ground state ($T=0$) properties were also investigated [432] solving numerically the mean field equations for the local magnetisation for finite system size ($N=80$). Then, fluctuations from the mean field theory were incorporated in the second order perturbation. In this approach, the Hamiltonian is split into two parts: $H=H_0+V$, where $H_0=\sum_i[\Gamma-(\sum_j J_{ij}m_j)S_i^z]+\sum_{\langle ij\rangle}J_{jj}m_im_j$ is the mean field Hamiltonian, where $m_i=\langle S_i^z\rangle$ is the mean value of the i-th spin, and $V=-\sum_{\langle ij\rangle}J_{ij}(S_i^z-m_i)(S_j^z-m_j)$ denotes the fluctuation from the mean field. The critical Γ, obtained from the vanishing of m_i, gives $\Gamma_c=2J$ from the mean field Hamiltonian. The order parameter $q=\sum_i\langle S_i\rangle^2/N$ is calculated including the fluctuations up to second order in V. But this barely improves the mean field result. Only after introducing a reaction term in energy, it was found that $\Gamma_c\ (T=0)\simeq 1.6J$ [432].

6.3.1.2 Monte Carlo Studies

Several Monte Carlo studies have also been performed [240, 323] for SK spin glass in transverse field. Applying the Suzuki-Trotter formalism (cf. Sect. 3.1, for the formulation of effective partition function of the quantum Hamiltonian), one can obtain the effective classical Hamiltonian in the M-th Trotter approximation as (cf. Sect. 6.A.2)

$$H_{\text{eff}}=-\frac{1}{M}\sum_{i,j=1}^{N}\sum_{k=1}^{M}J_{ij}S_{ik}S_{jk}-\left(\frac{1}{2\beta}\right)\ln\coth\left(\frac{\beta\Gamma}{M}\right)\sum_{i=1}^{N}\sum_{k=1}^{M}S_{ik}S_{ik+1}$$
$$-\frac{MN}{2}\ln\left[\frac{1}{2}\sinh\left(\frac{2\beta\Gamma}{M}\right)\right], \qquad (6.3.39)$$

where S_{ik} denotes the Ising spin defined on the lattice (i,k), i is the position in the original SK model and k denotes the position in the additional Trotter dimension. As mentioned in Sect. 3.1, the quantum to classical mapping is exact in the limit of infinite M.

Monte Carlo simulations of the SK model in the presence of a transverse field are performed using the Metropolis method [241] for the classical spin system given by the Hamiltonian (6.3.39) with finite M. One takes a $(N\times M)$ lattice having Ising spins on each lattice site. N spins along the spatial (x) direction in each of the Trotter rows interact with each other (within the row) with interaction strength J_{ij}, which obey a Gaussian probability distribution (6.3.2), while each of the M spins along the Trotter axis in each of the N columns interacts only with its two nearest-neighbour spins along the Trotter direction with strength $\zeta=(T/2)\ln[\coth(\Gamma/MT)]$. One usually uses periodic boundary condition in the Trotter direction. The Suzuki-Trotter approximation becomes better as one consider large values of M, otherwise a correction of the order of M^{-2} is needed for the thermodynamic quantities, e.g., $\langle Q\rangle_{\text{exact}}=\langle Q\rangle_M+O(M^{-2})$. However, with M the interaction strength ζ increases logarithmically whereas (J_{ij}/M) weakens. For $M \gg (\Gamma/T)$, the singular behaviour of ζ tends to arrange the spins parallelly along the Trotter direction whereas vanishingly small (J_{ij}/M) makes the interaction among the Trotter chains almost zero. The imbalance in the two cooperative interactions invalidates [421] the Monte Carlo process to work as an importance sampling. For certain values of T and Γ, M is to be selected such that $M > \Gamma/T$. Ray et al. [323] took $\Gamma/MT = 0.06$, which has been found to be the optimal value of the ratio in the Monte Carlo studies [421] of the (pure) transverse Ising model. In determining the phase diagram in the Γ–T plane, one finds that as Γ is increased, T_c is lowered and the condition on Γ/MT demands large M values, which virtually takes into account the enhanced quantum effects.

Ray et al. [323], took $\Gamma \ll J$ and their results indeed indicate a sharp lowering of $T_c(\Gamma)$. Such sharp fall of $T_c(\Gamma)$ with large Γ is obtained in almost all theoretical studies of the phase diagram of the model, and are also in agreement with the Monte Carlo studies of Ishii and Yamamoto [196]. One should also mention that, Ray et al. [323], obtained a non-monotonic fall of T_c with the increase of the transverse field strength Γ, a result which is not reproduced in any other Monte Carlo or analytical studies. Lai and Goldschmidt [240], studied the above system using Monte Carlo techniques on a larger system ($N \leq 100$), and studied the configuration averaged function g defined as

$$g = \frac{1}{2}\left(3 - \frac{\langle q^4 \rangle}{\langle q^2 \rangle^2}\right), \tag{6.3.40}$$

where $\langle q^n \rangle$ is the n-th moment of the order parameter distribution function $P_N(q)$. Here $P_N(q) = \langle \delta(q - (1/N)) \sum_{i=1}^{N} S_{ik}^{(1)} S_{ik'}^{(2)} \rangle$ for different replicas (1) and (2) and Trotter indices k and k' (see discussion in Sect. 6.5). The above function g is assumed to have a finite size scaling form

$$g = \overline{g}\big(N^{\alpha}(T - T_c)\big), \tag{6.3.41}$$

similar to the short-range classical Ising spin glass case [39, 76, 136, 275]; where α is some exponent and $\overline{g}$ is the scaling function. Here the Trotter size M is kept fixed. Using (6.3.41), one can find out the value T_c from the intersection of the curves of $\overline{g}$ versus T for different system sizes N. The phase diagram, thus obtained, was found to be in good agreement with the analytically obtained phase diagram (see Fig. 6.3). However, the finite size scaling form (6.3.41) is inappropriate (and incorrect) for the long-range spin glass systems. For the classical Sherrington-Kirkpatrick model, such scaling does not work (see Binder and Young [39]), and one gets the g-function intersection point continuously shifted with the system size N.

6.3.1.3 Exact Diagonalisation Results

Sen et al. [361], performed exact diagonalisation of finite sized spin glass systems ($N \leq 8$) at zero temperature and found all the eigenvalues ($E_0, E_1, \ldots, E_m$, where $m = 2^N - 1$) and the eigenvectors ($\psi_1, \psi_2, \ldots, \psi_m$). The configurational averages are performed over the physical quantities obtained after the diagonalisations. This method does not involve any mapping of the original (quantum) Hamiltonian, like the previous models, and deals precisely with the zero temperature phases of the model, as induced by the quantum fluctuations. The mass gap (difference in energy of the ground state ψ_0 and the first excited state ψ_1) and the ground state internal energy $E_g = \langle \Psi_0 | \sum_{ij} J_{ij} S_i^z S_j^z | \Psi_0 \rangle$ of this quantum spin glass system behave, on an average, essentially in the same way as in the case of the unfrustrated ferromagnetic transverse Ising system. It may be mentioned that although the eigenfunctions ψ_i and eigenenergies E_i are determined using the entire Hamiltonian (6.3.1), the ground state internal energy E_g is obtained from the average of the Ising cooperative part in the ground state. The value of the extrapolated critical field Γ_c from the disappearance of mass gap $\Delta(E_1 - E_0)$, and the peak position of the "specific heat" $C(= \partial E_g / \partial \Gamma)$ is obtained around J. The spin glass order parameter q also appears to vanish at the same value of the critical field (see Fig. 6.4). The value of $\Gamma_c(T = 0)$, obtained from the diagonalisation of finite size systems, is lower than that obtained in the perturbation treatment or in the thermofield dynamical technique. This might be interpreted as indicating Griffiths-type clustering singularity. However, these indications exist even in the configuration averaged results, which perhaps rule out the latter possibility. It may be noted that the peak position of the "linear" specific heat C need not be the right quantity to look for the glass transition and the nonlinear susceptibility should be used. The gap Δ is taken directly from the bottom-end of the excitation spectrum: see Fig. 6.4(e). It has also been directly checked for system sizes $N \leq 10$ that the scaling form (6.3.41) is inappropriate for the transverse Ising SK model, as mentioned earlier.

Apart from these, some related studies on the SK model in transverse field include the evaluation of the internal energy, susceptibility and entropy at first step Replica Symmetry Breaking [159].

6.3.2 Susceptibility and Energy Gap Distribution

Recently, Takahashi and Matsuda [394] investigated the relationship between the energy gap distribution and the susceptibilities (linear χ_l, spin glass χ_{sg} and nonlinear χ_{nl}) for the SK model put in a transverse field. They found that the energy gap itself is not self-average and the distribution of the gap between the first and the ground states determines the critical behaviour in the susceptibility. In following, we will show their argument according to the reference [394, 395] .

We first consider the quantum part of the susceptibilities as

$$\chi_l^{(q)} = \frac{2}{N} \sum_{n \neq 0} \frac{|\langle 0 | S^z | n \rangle|^2}{E_n - E_0} \tag{6.3.42}$$

and

$$\chi_{sg}^{(q)} = \frac{4}{N} \left(\sum_{n \neq 0} \frac{|\langle 0 | S^z | n \rangle|^2}{E_n - E_0} \right)^2 \tag{6.3.43}$$

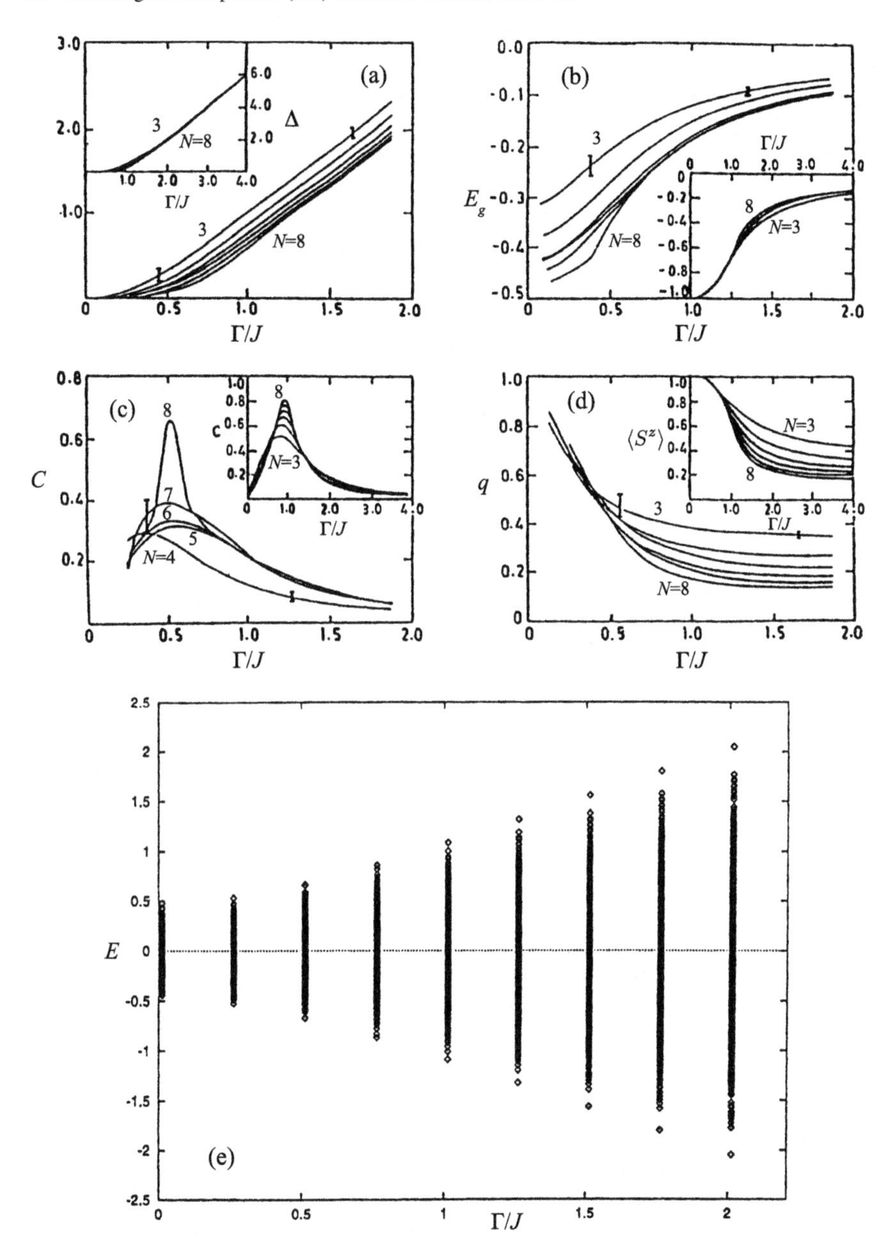

Fig. 6.4 The variation of the average mass gap Δ (**a**), ground state energy (**b**), "specific heat" ($C = \partial E_g / \partial \Gamma$) (**c**) and the order parameter (**d**) for an S-K model in transverse field, with a width J of exchange interaction distribution, obtained from finite systems of size $N \leq 8$. For comparison, the variation of these quantities (Δ, E_g, C and magnetisation m instead of the order parameter q) for a transverse Ising chain (with nearest neighbour interaction J) of comparable sizes are also shown in the respective *insets*. The *vertical bars* indicate the order of magnitude of the errors (due to the configurational fluctuations). In (**e**) the energy per site for a system of size $N = 10$ is shown for different values of Γ

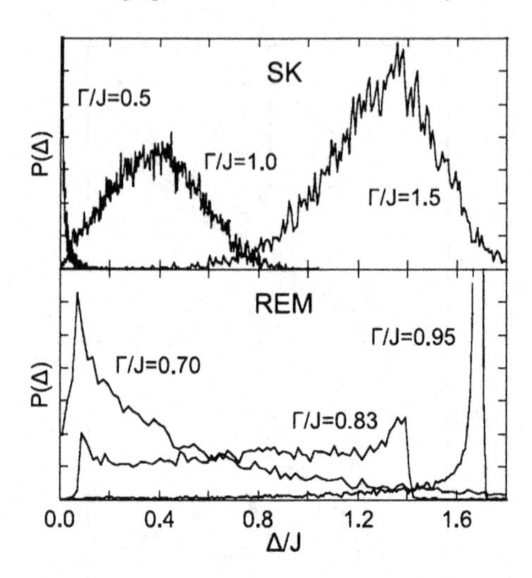

Fig. 6.5 The energy gap distributions $P(\Delta)$ for the SK model (*top panel*) and the random energy model (REM). The system size is chosen as $N=16$

$$\chi_{\mathrm{nl}}^{(\mathrm{q})}=\frac{4}{N}\sum_{n,m\neq 0}\frac{|\langle 0|S^z|n\rangle|^2}{E_n-E_0}\cdot\frac{|\langle 0|S^z|m\rangle|^2}{(E_m-E_0)^2}$$
$$-\frac{4}{N}\sum_{n,n',n''\neq 0}\frac{\langle 0|S^z|n\rangle\langle n|S^z|n'\rangle\langle n'|S^z|n''\rangle\langle n''|S^z|0\rangle}{(E_n-E_0)(E_{n'}-E_0)(E_{n''}-E_0)} \tag{6.3.44}$$

where we defined $S^z\equiv\sum_{i=1}^N S_i^z$ and $|n\rangle$ is the eigenvector of the energy E_n. $|0\rangle$ is the eigenvector of the ground state without any degeneracy. It should be noted that

$$\chi_A=\chi_A^{(\mathrm{c})}+\chi_A^{(\mathrm{q})},\quad A=\mathrm{l},\mathrm{sg},\mathrm{nl} \tag{6.3.45}$$

are satisfied and $\chi_{\mathrm{l}}^{(\mathrm{c})}=\beta(1-q)$ with spin glass order parameter q.

From the above definitions, the singularity of the susceptibility apparently comes from the quantum part. The singularity occurs when the energy gap between the first excited state and the ground state approaches zero. From Eqs. (6.3.42), (6.3.43) and (6.3.44), the susceptibilities χ_{l}, χ_{sg} and χ_{nl} behave as a function of Δ as $\chi_{\mathrm{l}}\sim 1/\Delta$, $\chi_{\mathrm{sg}}\sim 1/\Delta^2$ and $\chi_{\mathrm{nl}}\sim 1/\Delta^3$, respectively. This fact implies that the energy gap distribution might determine the critical behaviour.

In Fig. 6.5, we show the distribution of the energy gap $P(\Delta)$ for the SK model and the random energy model (REM). From this figure, we notice that the energy gap itself is not self-averaging quantity and it apparently fluctuates. Form the top panel of Fig. 6.5 for the SK model, we can assume that the gap distribution obeys the power-law $P(\Delta)\sim\Delta^k$ when Δ is small.

From this fact, for a given gap distribution $P(\Delta)\sim\Delta^k$, the average susceptibility $\chi_A^{(\mathrm{q})}$ is evaluated as

$$\chi_A^{(\mathrm{q})}\sim\int_0^\infty P(\Delta)\cdot\frac{d\Delta}{\Delta^{\kappa_A}}\sim\int_0^\infty \Delta^{k-\kappa_A}\,d\Delta \tag{6.3.46}$$

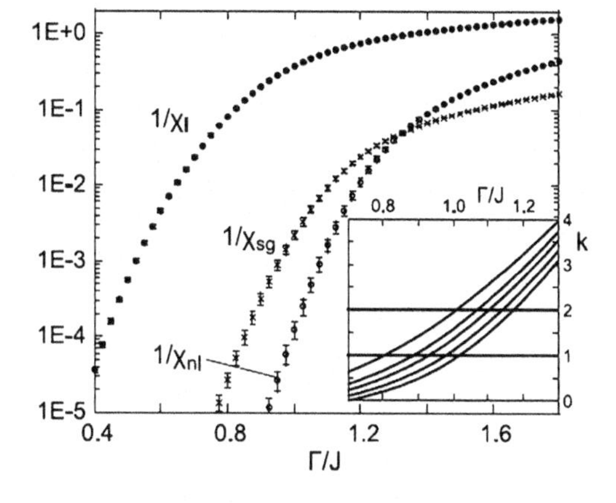

Fig. 6.6 Inverse of average susceptibility $1/\chi_A^{(q)}$ for the SK model with $N = 14$. The *inset* shows the exponent of the energy gap distribution. From the *top line* to the bottom, the system size is chosen as $N = 8, 10, 12, 14$ and 16. The gap exponent is fitted by $P(\Delta) \sim \Delta^k$ near the origin

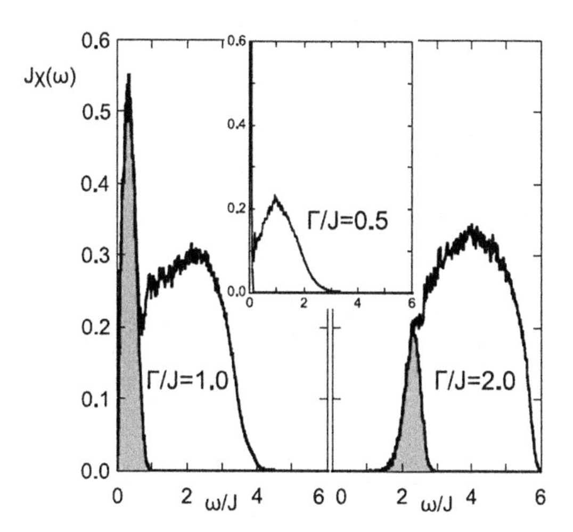

Fig. 6.7 Local correlation function $\chi(\omega)$ for the SK model with $N = 16$. The *shaded areas* are contributions of the first excited state

where we defined $A = \mathrm{l}, \mathrm{sg}, \mathrm{nl}$ and $\kappa_\mathrm{l} = 1$, $\kappa_\mathrm{sg} = 2$ and $\kappa_\mathrm{nl} = 3$. Therefore, the susceptibilities χ_l, χ_sg and χ_nl diverge for $k \leq 0$, 1 and $k \leq 2$, respectively. This implies that these susceptibilities diverge at different points. In Fig. 6.6, we find this argument is correct.

In order to confirm that the gap between the only first excited and the ground states determines the divergence of the susceptibilities, we calculate the Fourier representations of a real-time correlation function including contributions from all states:

$$\chi(\omega) = \sum_{n \neq 0} \delta(\omega - E_n + E_0) \big| \langle 0 | \sigma_i^z | n \rangle \big|^2. \tag{6.3.47}$$

We plot the result in Fig. 6.7. Thus, we conclude that the first excited state determines the critical behaviour.

6.3.3 SK Model with Antiferromagnetic Bias

In the previous subsections, we discussed SK model put in a transverse field. Here we show several interesting results on long-range Ising antiferromagnet put in a transverse field with disorder, which was recently given by Chandra et al. [70]. This model contains SK model as a special case.

The model we study here is given by the following Hamiltonian

$$H = -\frac{1}{N} \sum_{ij(j>i)} (J_0 + \tilde{J}\tau_{ij}) S_i^z S_j^z - h \sum_{i=1}^{N} S_i^z - \Gamma \sum_i S_i^x, \tag{6.3.48}$$

where J_0 is the parameter controlling the strength of the antiferromagnetic bias and $\tilde{J}$ is an amplitude of the disorder τ_{ij} in each pair interaction. Parameters h and Γ denote the longitudinal and transverse fields, respectively. Here S^x and S^z denote the x and z component of the N-Pauli spins

$$S_i^z = \begin{pmatrix} 1 & 0 \\ 0 & -1 \end{pmatrix}; \quad S_i^x = \begin{pmatrix} 0 & 1 \\ 1 & 0 \end{pmatrix}; \quad i = 1, 2, \ldots, N.$$

As such the model has a fully frustrated (infinite-range or infinite dimensional) cooperative term. When we assume that the disorder τ_{ij} obeys a Gaussian with mean zero and variance unity, the new variable $J_{ij} \equiv J_0 + \tilde{J}\tau_{ij}$ follows the Gaussian distribution, immediately we have $P(J_{ij}) = \exp[-(J_{ij} - J_0)^2/2\tilde{J}^2]/\sqrt{2\pi}\,\tilde{J}$. Therefore, we obtain the 'pure' antiferromagnetic Ising model with infinite range interactions when we consider the limit $\tilde{J} \to 0$ keeping $J_0 < 0$. Of course the model with $J_0 > 0$ and $\Gamma = 0$ is identical to the classical SK model and with $J_0 < 0$ and $\Gamma = 0$, it is the classical long-range Ising antiferromagnet model.

6.3.3.1 Mean-Field Theory

For an analytic (mean field) study of the model we define an effective magnetic field $\boldsymbol{h}_{\text{eff}}$ at each site, which is a resultant of the average cooperation enforcement in the z-direction and the applied transverse field in the x-direction, so that the above Hamiltonian can be written as

$$H = \boldsymbol{h}_{\text{eff}} \cdot \sum_{i=1}^{N} \boldsymbol{S}_i, \tag{6.3.49}$$

where

$$\boldsymbol{S}_i = S_i^z \boldsymbol{z} + S_i^x \boldsymbol{x}, \tag{6.3.50}$$

and

$$\begin{aligned} \boldsymbol{h}_{\text{eff}} &= (\boldsymbol{h}_{\text{eff}})^z \boldsymbol{z} + (\boldsymbol{h}_{\text{eff}})^x \boldsymbol{x} \\ &= \big(h + J_0 m^z + \tilde{J}\sqrt{q}\, y\big)\boldsymbol{z} + \Gamma \boldsymbol{x}, \end{aligned} \tag{6.3.51}$$

$$|\boldsymbol{h}_{\text{eff}}| = \sqrt{\big(h + J_0 m^z + \tilde{J}\sqrt{q}\, y\big)^2 + \Gamma^2}, \tag{6.3.52}$$

where $\boldsymbol{x}$, $\boldsymbol{z}$ denote unit vectors pointing to the x and z directions. We use the definitions such as $(\boldsymbol{h}_{\text{eff}})^a$ $(a = x, z)$ to indicate the a-component of the effective field $\boldsymbol{h}_{\text{eff}}$. This replacement of S_j^z by its average value $\langle S_j^z \rangle \equiv m^z$ in $(\boldsymbol{h}_{\text{eff}})^z$ should be valid for this infinite range model. The Gaussian distributed random field $\sqrt{q}\, y$ comes from the local field fluctuation given by the spin glass order parameter. The average magnetisation is then given by

$$\mathbf{m} = \frac{\text{tr}\, \mathbf{S} e^{-\beta H}}{\text{tr}\, e^{-\beta H}} = \left(\tanh \beta |\boldsymbol{h}_{\text{eff}}|\right) \cdot \frac{\boldsymbol{h}_{\text{eff}}}{|\boldsymbol{h}_{\text{eff}}|} \tag{6.3.53}$$

and hence we have

$$m^z = \int_{-\infty}^{\infty} Dy \frac{J_{\text{eff}}}{\sqrt{J_{\text{eff}}^2 + \Gamma^2}} \tanh \beta \sqrt{J_{\text{eff}}^2 + \Gamma^2} \tag{6.3.54}$$

$$m^x = \int_{-\infty}^{\infty} Dy \frac{\Gamma}{\sqrt{J_{\text{eff}}^2 + \Gamma^2}} \tanh \beta \sqrt{J_{\text{eff}}^2 + \Gamma^2} \tag{6.3.55}$$

$$q = \int_{-\infty}^{\infty} Dy \left\{ \frac{J_{\text{eff}}}{\sqrt{J_{\text{eff}}^2 + \Gamma^2}} \right\}^2 \tanh^2 \beta \sqrt{J_{\text{eff}}^2 + \Gamma^2}, \tag{6.3.56}$$

where we defined $J_{\text{eff}} \equiv h + J_0 m^z + \tilde{J} \sqrt{q}\, y$ and $m \equiv N^{-1} \sum_i \langle S_i^z \rangle$ is the magnetisation and $q \equiv N^{-1} \sum_i \langle S_i^z \rangle^2$ is the spin glass order parameter. We also used $Dy \equiv dy\, e^{-y^2/2} / \sqrt{2\pi}$. In [70], we confirm that the above mean-field equations are identical to the results obtained by the replica symmetric theory at the ground state $(\beta = \infty)$. The detail is given by Sect. 6.A.5.

For the antiferromagnetic $(J_0 < 0)$ and/or the spin glass phase (with $h = 0$), $m^z = 0$ is the only solution. We then have

$$m^x = \int_{-\infty}^{\infty} Dy \frac{\Gamma}{\sqrt{(\tilde{J} \sqrt{q}\, y)^2 + \Gamma^2}} \tanh \beta \sqrt{(\tilde{J} \sqrt{q}\, y)^2 + \Gamma^2} \tag{6.3.57}$$

$$q = \int_{-\infty}^{\infty} Dy \left\{ \frac{\tilde{J} \sqrt{q}\, y}{\sqrt{(\tilde{J} \sqrt{q}\, y)^2 + \Gamma^2}} \right\}^2 \tanh^2 \beta \sqrt{(\tilde{J} \sqrt{q}\, y)^2 + \Gamma^2}. \tag{6.3.58}$$

6.3.3.2 The Condition on Which Antiferromagnet Phase Survives

The approximate saddle point equations have already been presented in Eqs. (6.3.54), (6.3.55) and (6.3.56). The variations of m^x, q and $\tilde{q}$ are shown in Fig. 6.8. The phase boundary between the spin glass and paramagnetic phases is given by setting $m^z = 0$ and $q \simeq 0$ and we get

$$\Gamma = \tilde{J} \tanh\left(\frac{\Gamma}{T}\right). \tag{6.3.59}$$

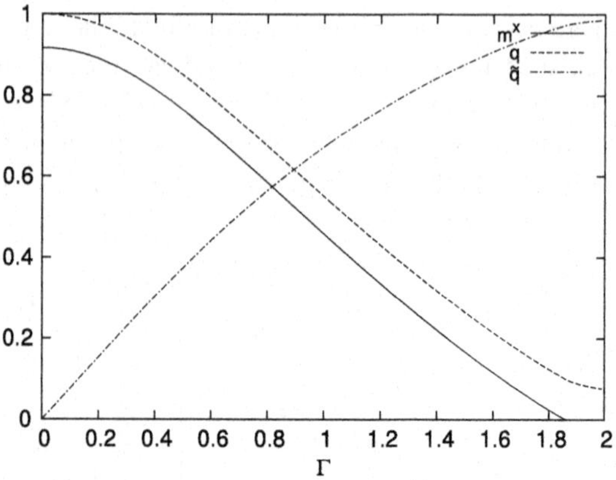

Fig. 6.8 The result of numerical calculations for the saddle point equations for m_x, q and $\tilde{q}$ as a function of Γ for $T \neq 0$

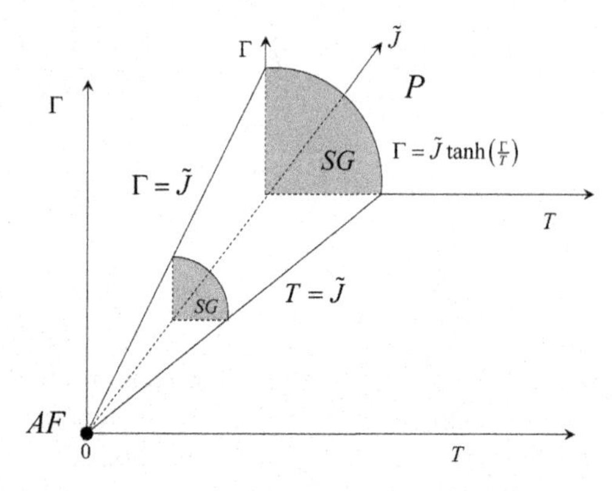

Fig. 6.9 Phase diagram for the quantum system. The antiferromagnetic order exists if and only if we set $T = \Gamma = 0$ and $\tilde{J} = 0$. As the $\tilde{J}$ decreases, the spin glass phase gradually shrinks to zero and eventually ends up at an antiferromagnetic phase at its vertex (for $\Gamma = 0 = T = \tilde{J}$)

Obviously, the boundary at $T = 0$ gives $\Gamma_{SG} = \tilde{J}$. On the other hand, when we consider the case of $\Gamma \simeq 0$, we have $T_{SG} = \tilde{J}$ (consistent with the classical result). These facts imply that there is neither an antiferromagnetic nor a spin glass phase when we consider the pure case $\tilde{J} = 0$ because the critical point leads to $T_{SG} = \Gamma_{SG} = 0$. Therefore, we conclude that the antiferromagnetic phase can exist if and only if $T = \Gamma = 0$ (Fig. 6.9).

6.4 Edwards-Anderson Model in a Transverse Field

The Edwards-Anderson Model in presence of a transverse field was first studied by Chakrabarti [62] and it was shown that the short range Ising spin glass has similar quantum effects as in the ordinary transition by approximate renormalisation group approach. Later, dos Santos et al. [112], using real space renormalisation method studied the effect of disorder in the imaginary time dimension and obtained the phase diagram in the Γ–T plane for $d = 2$. Walasek et al. [417], studied the above model using a mean field approximation along with the replica averaging technique. They introduced two order parameters, the diagonal and the off-diagonal in replica indices. The value of zero-temperature critical transverse field and phase diagram were obtained from the effective free energy. As mentioned earlier, there have been renormalisation group theoretical calculations [138] of the one dimensional model (which does not have frustration) and also detailed quantum Monte Carlo studies [163, 164, 332, 333] of the model in two and three dimensions. The analytical results are therefore available in the limiting situations with $d = 1$ and $d = \infty$ (from those for the transverse SK model). The transverse Ising chain with random bonds or in a random transverse field (or in presence of both) can be mapped onto the McCoy-Wu model [266, 267, 269, 270], for which various analytical results are known. Fisher [138] has shown with a renormalisation group calculation, that the typical and average spatial correlations behave differently: the typical correlation diverges with an exponent $\tilde{\nu} = 1$ (see also [368]) whereas the average correlation diverges with an exponent $\nu = 2$, which satisfies the inequality $2/\nu \leq d$ as an equality (see Sect. 5.2). Moreover, as mentioned earlier, both the linear and nonlinear susceptibilities diverge even in the disordered phase. This has been conjectured to be due to the presence of Griffiths singularity. In a finite size scaling analysis of the Monte Carlo simulation results for the random field chain, Crisanti and Rieger [90] found some disparities with the above mentioned results. They obtained the value of the dynamical exponent $z = 1.7$ and the correlation length exponent $\nu = 1$ and no existence of Griffiths singularity was observed. These discrepancies might arise from the finite size effect, due to which one gets a "typical" rather than the average result.

6.4.1 Quantum Monte Carlo Results

Let us now start from the Hamiltonian of the Edwards-Anderson spin glass in presence of a transverse field

$$H = -\sum_{\langle ij \rangle} J_{ij} S_i^z S_j^z - \Gamma \sum_i S_i^x, \tag{6.4.1}$$

where the random interaction is only restricted among the nearest neighbours and satisfies a Gaussian distribution with zero mean and variance J

$$P(J_{ij}) = \frac{1}{\sqrt{2\pi} J} \exp\left(-\frac{J_{ij}^2}{2J^2} \right). \tag{6.4.2}$$

Hereafter we let J be the unit of energy, making $J = 1$. With $\Gamma = 0$, the above model represents the Edwards-Anderson model with the order parameter $q = \overline{\langle S_i^z \rangle^2} = 1$ (at $T = 0$). When the transverse field term is introduced, q decreases, and at a critical value of the transverse field the order parameter vanishes. To study this quantum phase transition using quantum Monte Carlo techniques, one must remember that the ground state of a d-dimensional quantum model is equivalent to the free energy of a classical model with one added dimension which is the imaginary

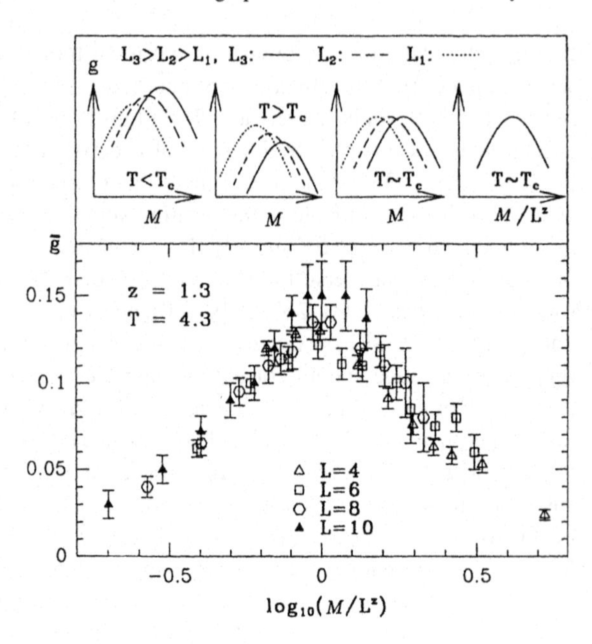

Fig. 6.10 MC results for variation of the g-function with M for different sample size L ($d = 3$) [163]. The scaled variation of $\overline{g}$ is shown in the *bottom*

time (Trotter) dimension (cf. Sect. 3.1). The effective classical Hamiltonian can be written as

$$H = -\sum_k \sum_{\langle ij \rangle} K_{ij} S_{ik} S_{jk} - \sum_k \sum_i K S_{ik} S_{ik+1} \tag{6.4.3}$$

with

$$K_{ij} = \frac{\beta J_{ij}}{M}; \qquad K = \frac{1}{2} \ln \coth\left(\frac{\beta \Gamma}{M}\right) \tag{6.4.4}$$

where S_{ik} are classical Ising spins and (i, j) denote the original d-dimensional lattice sites and $k = 1, 2, \ldots, M$ denotes a time slice. Although the equivalence between the classical and the quantum model holds strictly when $M = \infty$, one can always make an optimal choice for M. One of the reasonable choices for M is $M = \beta$. Employing this and letting $\beta_{\text{eff}} = T_{\text{eff}}^{-1} = K = \frac{1}{2} \ln \coth \Gamma$, the Boltzmann factor $\exp(-\beta_{\text{eff}} H)$ with

$$\beta_{\text{eff}} H = -\sum_k \sum_{\langle ij \rangle} J_{ij} S_{ik} S_{jk} - (1/T_{\text{eff}}) \sum_k \sum_i S_{ik} S_{i\,k+1}$$

represents the transverse Ising system at zero temperature. This equivalent classical system has been studied using standard Monte Carlo techniques. Since there exists a strong anisotropy in the spatial and Trotter (time) dimensions, one has to introduce two length scales in the problem. Near the critical point, ξ_τ, the correlation length in the Trotter direction scales as ξ^z, where ξ is the correlation length in the spatial direction and z is the dynamical exponent. Hence, any dimensionless quantity must

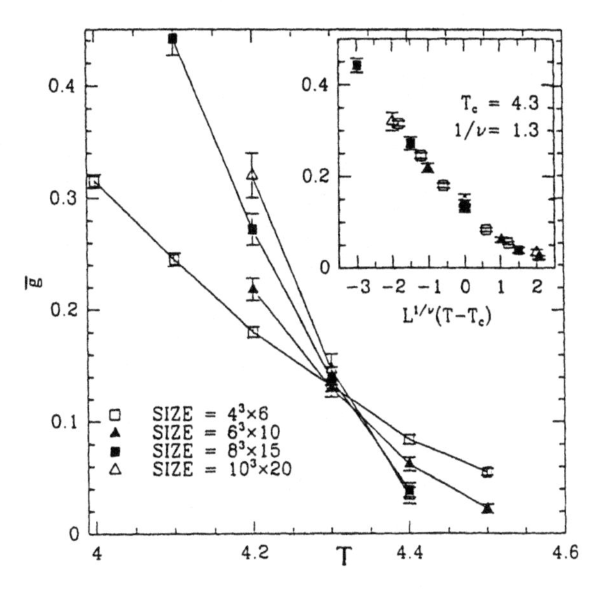

Fig. 6.11 MC results for the variation of $\overline{g}$ with T for various sample sizes ($d=3$) [163]. Crossing indicates T_c. The *inset* shows the best scaling collapse

scale as a function of two variables. The convenient choice of such a configuration averaged function is the g-function defined as [163, 332]

$$g = \overline{\frac{1}{2}\left(3 - \frac{\langle q^4 \rangle}{\langle q^2 \rangle^2}\right)} \tag{6.4.5}$$

where q is the overlap function between two replicas having the same bond distribution. It may be noted here, that this g-function is somewhat different from that defined for the classical spin glass models [39], where the numerator and denominator are averaged over disorder separately. Such averaging here appears to lead to large sample to sample fluctuations and gives worse statistics [332]. Although the exact reason for this behaviour is not known, we believe, it is related to the Trotter symmetry problem discussed later (see Sect. 6.5). The function g vanishes in the paramagnetic phase and acquires a nonzero value in the spin glass ordered phase. Near the critical point, the above function scales as

$$g = \overline{g}\left(\frac{L}{\xi}, \frac{M}{L^z}\right) = \overline{g}\left(L^{\frac{1}{\nu}}(T_{\text{eff}} - T_{\text{eff},c}), \frac{M}{L^z}\right). \tag{6.4.6}$$

For a fixed L and T_{eff}, when M is varied, the function $\overline{g}$ shows a peak at a particular value of M. At $T_{\text{eff}} = T_{\text{eff},c}$ this maximum value is not affected by the change in the system size (Fig. 6.10).

Using this, one can make an estimate of $T_{\text{eff},c}$ for the equivalent classical model, and using the scaling relation (6.4.6), one can estimate the approximate value of the dynamical exponent z, which come out to be $T_{\text{eff},c} \approx 3.3$ and 4.3 and $z \approx 1.5$ and 1.3 for two and three dimensional models respectively [163, 332]. One can also estimate the critical temperature for the classical model by studying the variation of $\overline{g}$ with temperature for different system sizes L (with M/L^z fixed). The intersection

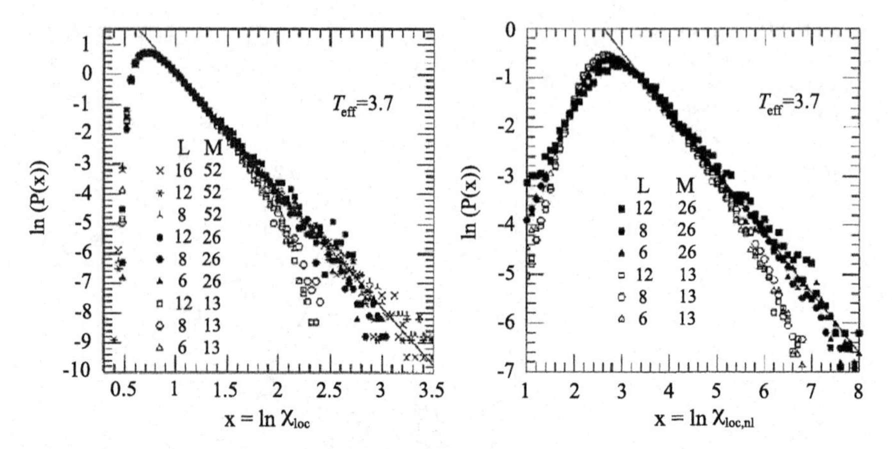

Fig. 6.12 Logarithm of the distribution of $\ln \chi_{\text{loc}}$ (*left*) and $\ln \chi_{\text{loc,nl}}$ (*right*) at effective temperature $T_{\text{eff}} = 3.7$ ($> T_{\text{eff},c}$). Results for the two-dimensional ($d = 2$) model with different L's and M's are plotted. For $M \geq 26$, results with different L's seems to collapse onto a line. The *solid line* in each panel stands for Eq. (6.4.7) with $z = 0.51$ for the *left panel* and (6.4.8) with $z = 0.54$ for the *right panel*. (Taken from [333])

of these curves (obtained for different system sizes) gives the value of critical temperature (Fig. 6.11). At this approximate T_c, the variation of $\overline{g}$ with the system size provides the value of the critical exponent ν, through the scaling relation (6.4.6) which comes out to be around 1.0 ± 0.1 and 0.8 for two and three dimensional systems respectively [163, 332].

It is intriguing to see the nature of Griffiths singularities in the zero temperature disordered phase of the present systems. As mentioned in Sect. 5.3.3, the Griffiths singularities are captured by the local susceptibility defined by $\chi_{\text{loc}} = \partial \langle S_i^z \rangle / \partial h_i$, where h_i is a local field which is coupled to S_i^z. In terms of the effective classical models, this can be written as $\chi_{\text{loc}} = \sum_{k=1}^{M} \langle S_{i1} S_{ik} \rangle$. Since the correlation in the Trotter direction decays as $\langle S_{i1} S_{ik} \rangle \sim e^{-k/\xi_\tau}$, one can relate the local susceptibility with the correlation time ξ_τ as $\chi_{\text{loc}} = e^{-1/\xi_\tau}/(1 - e^{-1/\xi_\tau}) \approx \xi_\tau$, when $M \to \infty$ and $\xi_\tau \gg 1$. In the language of the effective classical model, a phenomenological description for the Griffiths singularities is given as follows (see also an argument in Sect. 5.3.3). The Griffiths singularities come from excitations which are local in real space and ordered in the imaginary-time Trotter direction. One may naturally assume that the probability of an excitation with size $N = L^d$ is exponentially small with N and given as $e^{-aN/T_{\text{eff}}}$ in the system with temperature T_{eff}, where a is a constant. Once such an excitation with size N is created, it survives for a long imaginary-time which is exponential in N since one has to insert N domain-walls in the Trotter axis. This relaxation imaginary-time is equivalent to ξ_τ and thus one has $\xi_\tau \sim e^{bN/T_{\text{eff}}}$, where b is a constant. When the excitation is well localised, there might be order L^d number of excitations in the system. Combining this with the exponentially small probability of a size N excitation and the exponentially long relaxation imaginary-time, it turns out that the probability distribution of ξ_τ has a

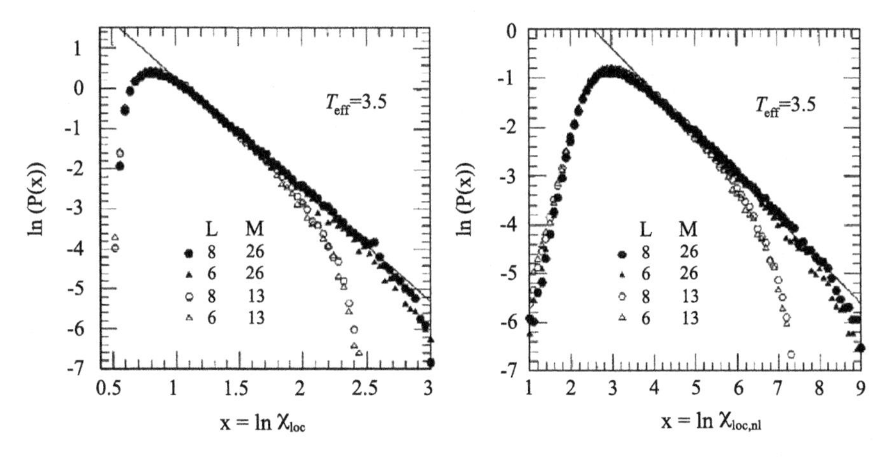

Fig. 6.13 Same plots as Fig. 6.12 but with effective temperature $T_{\mathrm{eff}} = 3.5$ which is closer to $T_{\mathrm{eff},c}$. The *solid line* in each panel stands for Eq. (6.4.7) with $z = 0.71$ for the *left panel* and (6.4.8) with $z = 0.76$ for the *right panel*. (Taken from [333])

power-law tail: $P(\ln \xi_\tau) \sim L^d e^{-(a/b)\ln \xi_\tau} = L^d \xi_\tau^{-(a/b)}$. In order to make this probability distribution dimensionless, the exponent a/b needs to be $a/b = d/z$, where z is a dynamical exponent which is introduced so as to make $L\xi_\tau^{1/z}$ a dimensionless quantity. Thus, recalling the relation $\chi_{\mathrm{loc}} \approx \xi_\tau$, one finds that the distribution χ_{loc} has the following form [333]

$$\ln\big[P(\ln \chi_{\mathrm{loc}})\big] = -\frac{d}{z} \ln \chi_{\mathrm{loc}} + \mathrm{const.} \tag{6.4.7}$$

The local nonlinear susceptibility defined by $\chi_{\mathrm{loc,nl}} = \partial^3 \langle S_i^z \rangle / \partial h_i^3$ has also a power-law tail in its distribution. A similar argument above shows that the distribution of $\chi_{\mathrm{loc,nl}}$ has the form of

$$\ln\big[P(\ln \chi_{\mathrm{loc,nl}})\big] = -\frac{d}{3z} \ln \chi_{\mathrm{loc,nl}} + \mathrm{const.} \tag{6.4.8}$$

Rieger et al. [333] and Guo et al. [164] studied the distribution of χ_{loc} and $\chi_{\mathrm{loc,nl}}$ using quantum Monte Carlo simulations for two ($d = 2$) and three ($d = 3$) dimensional Edwards-Anderson models in a transverse field respectively. The results for $d = 2$ and 3 are qualitatively the same. Here we shall see the results for $d = 2$ [333] Fig. 6.12 shows the distribution of χ_{loc} and $\chi_{\mathrm{loc,nl}}$ of the two-dimensional system with effective temperature $T_{\mathrm{eff}} = 3.7$ which is higher than the critical value 3.3. As expected, the distribution has a power-law tail. By fitting (6.4.7), (6.4.8) on the corresponding results, the dynamical exponent is estimated as $z \approx 0.51$ for χ_{loc} and 0.54 for $\chi_{\mathrm{loc,nl}}$. A small discrepancy between them might be attributed to finite size effects. Figure 6.13 shows the results for $T_{\mathrm{eff}} = 3.5$ which is closer to the critical value. Although the qualitative behaviour of the distribution is similar to that in Fig. 6.12, the exponents of the power-law tail are different. The estimated value are given as $z = 0.71$ for χ_{loc} and 0.76 for $\chi_{\mathrm{loc,nl}}$, which are larger than those obtained for $T_{\mathrm{eff}} = 3.7$. Figure 6.14 shows T_{eff} dependence of the dynamical exponent z. One

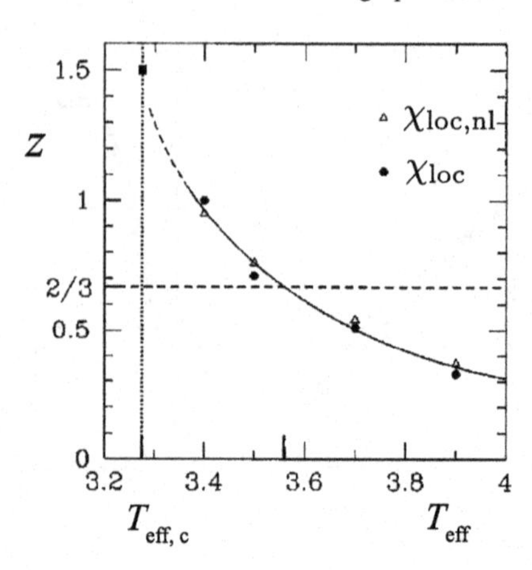

Fig. 6.14 Effective temperature dependence of the dynamical exponent z. The values estimated from the distributions of $\ln \chi_{\mathrm{loc}}$ and $\ln \chi_{\mathrm{loc,nl}}$ are plotted. These two values of z behaves almost identically. The dynamical exponent z increases with lowering T_{eff} towards $T_{\mathrm{eff},c}$. At $T_{\mathrm{eff},c}$, an estimate $z = 1.5$ has been obtained from the scaling of the g-function (6.4.6). $z = 2/3$ stands for the threshold above which the average nonlinear susceptibility diverges

can see that the dynamical exponent z increases with lowering T_{eff} towards $T_{\mathrm{eff},c}$. The value $z = 1.5$ at the critical point has been obtained from the scaling of the g-function (6.4.6). This value looks to be reached from the values for $T_{\mathrm{eff}} > T_{\mathrm{eff},c}$ by extrapolation.

We recall here that the quantum phase transitions of random transverse field Ising ferromagnets in one and two dimensions are characterised by infinite randomness fixed points, where the dynamical exponent z increases with lowering the transverse field towards the critical point in the disordered phase and diverges at the critical point (see Sect. 5.3). As shown in Fig. 6.14, the present transverse Ising spin glass models manifest a tendency of z in common with random transverse field Ising ferromagnets. One may therefore expect that the quantum phase transition of transverse Ising spin glass models in two and three dimensions are also characterised by an infinite randomness fixed point. So far only Rieger et al. and Guo et al. have studied the character of the quantum phase transitions of the present models and given $z \approx 1.5$ and 1.3 at critical points in two and three dimensions respectively. Their results, in particular those on the values of z at critical points, might be reconsidered from the viewpoint of infinite randomness fixed points.

6.5 A General Discussion on Transverse Ising Spin Glasses

For the classical Ising spin glass the lower critical dimension d_l^c is between 2 and 3; $2 < d_l^c < 3$ and the upper critical dimension d_u^c is 6 [39]. For the quantum Ising spin glass these values are not yet precisely known. Analytical studies of the short-range quantum spin glass, using the Landau-Ginzberg-Wilson (LGW) Hamiltonian [325], indicate that the upper critical dimension for such models is 8: $d_u^c = 8$.

Another important aspect of the quantum Ising spin glasses is the quantum classical correspondence. In a pure system, as mentioned earlier, the quantum phase transition in the d-dimensional quantum Hamiltonian is equivalent to the thermal phase transition in an equivalent classical Hamiltonian with one added dimension ("Trotter dimension"). But it seems, in presence of frustration the scenario is distinctively different. In fact, several studies do not indicate that the d-dimensional quantum EA model corresponds to a $(d + z)$-dimensional classical model, where z is the dynamical exponent. For the pure system the value of the dynamical exponent is 1, which readily leads to the correspondence of the d-dimensional quantum system with the $(d+1)$-dimensional classical system. Suzuki [387] conjectured that the averaging over disorder effectively introduces another additional dimension so that the d-dimensional quantum model should correspond to the $(d + 2)$-dimensional classical model. At the upper critical dimension, z being 2, this conjecture might be correct.

The Sherrington-Kirkpatrick model in presence of a transverse field has been studied extensively using both analytical and simulational techniques. The phase diagram has been obtained and from rigorous results it seems that the value of the zero-temperature critical transverse field is around $1.5J$, where J is the variance of the Gaussian distribution of the interaction between the spins. The Edwards-Anderson model has also been studied, mostly using numerical techniques for the higher dimensional model and the various exponents associated with the zero-temperature phase transition (driven by the transverse field) have been obtained using quantum Monte Carlo techniques.

6.5.1 *The Possibility of Replica Symmetric Ground States in Quantum Glasses*

The question of the existence of replica symmetric ground states in quantum spin glasses has been studied extensively. Replica symmetry restoration is a quantum phenomenon, arising due to the quantum tunnelling between the classical "trap" states separated by infinite (but narrow) barriers in the free energy surface, which is possible as the quantum tunnelling probability is proportional to the barrier area which is finite. To investigate this aspect of quantum glasses, one has to study the overlap distribution function $P(q)$

$$P(q) = \overline{\sum_{l,l'} P_l P_{l'} \delta\left(q - q^{ll'}\right)}, \qquad (6.5.1)$$

where P_l is the Boltzmann weight associated with the states l and l' and $q^{ll'}$ is the overlap between the states l and l',

$$q^{ll'} = \frac{1}{N} \sum_{i=1}^{N} \langle S_i \rangle^{(l)} \langle S_i \rangle^{(l')}. \qquad (6.5.2)$$

One can also define the overlap distribution in the following form (for a finite system size N)

$$P_N(q) = \overline{\langle \delta(q - q^{12}) \rangle}, \tag{6.5.3}$$

where q^{12} is the overlap between two sets of spins $S_i^{(1)}$ and $S_i^{(2)}$, with identical bond distribution but evolved with different dynamics,

$$q^{12} = \frac{1}{N} \sum_{i=1}^{N} S_i^{(1)} S_i^{(2)}. \tag{6.5.4}$$

$P_N(q) \to P(q)$ in the thermodynamic limit. In quantum glass problem one has to study similarly the overlap distribution function $P_N(q)$, and if the replica symmetric ground states exist, the above function must tend to a delta-function in the thermodynamic limit. In the para-phase, the distribution will approach a delta function at $q = 0$ for the infinite system.

Ray, Chakrabarti and Chakrabarti [323], performed Monte Carlo simulations, mapping the d-dimensional transverse S-K spin glass Hamiltonian to an equivalent $(d+1)$-dimensional classical Hamiltonian (6.3.39) and addressed the question of the stability of the replica symmetric solution, with the choice of the order parameter distribution function given by

$$P_N(q) = \overline{\left\langle \delta\left(q - \frac{1}{MN} \sum_{i=1}^{N} \sum_{k=1}^{M} S_{ik}^{(1)} S_{ik}^{(2)} \right) \right\rangle} \tag{6.5.5}$$

where as mentioned earlier, superscripts (1) and (2) refer to the two identical samples but evolved with different Monte Carlo dynamics. It may be noted that a similar definition for q (involving overlap in identical Trotter indices) was used by Guo et al. [163]. Lai and Goldschmidt [240] performed Monte Carlo studies with larger sample size ($N \leq 100$) and studied the order parameter distribution function

$$P_N(q) = \overline{\left\langle \delta\left(q - \frac{1}{N} \sum_{i=1}^{N} S_{ik}^{(1)} S_{ik'}^{(2)} \right) \right\rangle} \tag{6.5.6}$$

where the overlap is taken between different (arbitrarily chosen) Trotter indices k and $k'(k \neq k')$. Their studies indicate that $P_N(q)$ does not depend upon the choice of k and k' (Trotter symmetry). Rieger and Young [332] also defined q in a similar way ($q = (1/NM) \sum_{i=1}^{N} \sum_{kk'=1}^{M} S_{ik}^{(1)} S_{ik'}^{(2)}$). There are striking differences between the results Lai and Goldschmidt obtained with the results of Ray et al. In the studies of Ray et al. [323], for $\Gamma \ll \Gamma_c$, $P(q)$ is found to have an oscillatory dependence on q with a frequency linear in N (which is probably due to the formation of standing waves for identical Trotter overlaps). However, with increasing N, the amplitude of the oscillation decreases in the low-q part and the magnitude of $P(q = 0)$ decreases, indicating that $P(q)$ might go over to a delta function in the thermodynamic limit. The envelope of this distribution function appears to have a decreasing $P(q = 0)$ value as the system size is increased. Ray et al. [323] argued that the whole spin glass phase is replica symmetric due to quantum tunnelling between the classical trap states. Lai and Goldschmidt [240], on the other hand, do not find any oscillatory behaviour in $P(q)$. Contrary to the findings of Ray et al., they get a replica symmetry breaking (RSB) in the whole of the spin glass phase from the nature of $P(q)$, which, in this case, has a tail down to $q = 0$ even as N increases. According to them, the results of Ray et al. [323] are different from theirs because of the different choices of the overlap function. Goldschmidt and Lai have also obtained replica symmetry breaking solution at first step RSB and hence the phase diagram [240].

Büttner and Usadel [50], have shown that the replica symmetric solution is unstable for the effective classical Hamiltonian (6.3.39) and also estimated [49] the order parameter and the other thermodynamic quantities like susceptibility, internal energy and entropy by applying Parisi's replica symmetry breaking scheme to the above effective classical Hamiltonian. Using the static approximation, Thirumalai et al. [401], found stable replica symmetric solution in a small region close to the spin glass freezing temperature. But, as mentioned earlier, in the region close to the critical line (Fig. 6.1), quantum fluctuations are always subdued by the thermal fluctuations due to critical slowing down and hence, restoration of replica symmetry, which is essentially a quantum effect, perhaps cannot be prominent in the region close to the critical line. Yokota [437] in a numerical solution of the mean field equations for finite sizes, has obtained large number of pure states of the model. However, the results are insufficient to indicate the behaviour in the thermodynamic limit. Recent studies on short-range Ising spin glass Hamiltonian using Landau-Ginzberg-Wilson Hamiltonian [325] indicate the existence of a replica symmetric ground state in the spin glass phase (at $T = 0$).

All these (numerical) studies are for the equivalent classical Hamiltonian, obtained by applying the Suzuki-Trotter formalism to the original quantum Hamiltonian, where the interactions are anisotropic in the spatial and Trotter direction and the interaction in the Trotter direction becomes singular in the $T \to 0$ limit. Obviously, one cannot extrapolate the finite temperature results to the zero-temperature limit (and also the quantum-classical equivalence holds in the zero-temperature limit). The results of the exact diagonalisation of finite size systems ($N \leq 10$) at $T = 0$ itself [363] do not indicate any qualitative difference in the behaviour of the average (over about fifty random configurations) mass gap Δ and the internal energy E_g from that of the ferromagnetic transverse Ising case, indicating the possibility that the system might be "ergodic". On the other hand, the (zero-temperature) distribution for the order parameter does not appear to go to a delta function with increasing N as is clearly found for the corresponding ferromagnetic (random long range interaction without competition). In this case, the order parameter distribution $P(q)$ is simply the measure of normalised number of ground state configurations having the order parameter value as q. This perhaps indicates broken ergodicity for small values of Γ. The order parameter distribution also shows oscillations similar to that obtained by Ray et al. [323]. It is therefore, still an open question whether replica symmetry is broken in the whole of spin glass phase, or only at $\Gamma = 0$ region of the phase, and if not, where is the exact location of the Almeida-Thouless line [39, 76, 136, 275].

6.6 Ising Spin Glass with p-Spin Interactions in a Transverse Field

The Hamiltonian for an Ising spin glass model with p-spin interactions in a transverse field is given by

$$H = -\sum_{i_1\cdots i_p} J_{i_1\cdots i_p} S^z_{i_1}\cdots S^z_{i_p} - \Gamma \sum_{j=1}^{N} S^x_j \tag{6.6.1}$$

where the sum $(i_1 \cdots i_p)$ runs over all distinct p-plets, N is the total number of sites. The interactions $J_{i_1\cdots i_p}$ are random following a Gaussian distribution

$$P(J_{i_i\cdots i_p}) = \sqrt{\frac{N^{p-1}}{\pi p!}} \exp\left(-\frac{N^{p-1}}{p!} J^2_{i_1\cdots i_p}\right).$$

For $\Gamma = 0$, in the limit $p \to \infty$, the Hamiltonian describes the classical random energy model [103, 104, 162]. For nonzero Γ, the model is exactly solvable in the limit $p \to \infty$ using the static approximation which gives the exact results in this limit [158].

Let us first glance about the thermodynamic property of the classical Hamiltonian denoted by H_0:

$$H_0 = -\sum_{i_1\cdots i_p} J_{i_1\cdots i_p} S^z_{i_1}\cdots S^z_{i_p}.$$

We define the probability distribution $P(E)$ of an eigenenergy of H_0 as the probability distribution that a given spin configuration $(S) \equiv (S_1, S_2, \ldots)$ has an energy E:

$$P(E) = \left[\delta\big(E - H_0(S)\big)\right]_{\text{av}} = \int \delta\big(E - H_0(S)\big) \prod_{0\le i_1<\cdots<i_p\le N} P(J_{i_1,\ldots,i_p}) dJ_{i_1,\ldots,i_p}.$$

Using the representation of the Dirac's delta function, $\delta(E) = (2\pi)^{-1}\int dx\, e^{iEx}$, one can write

$$P(E) = \frac{1}{2\pi}\int_{-\infty}^{\infty} dx\, e^{iEx} \prod_{0\le i_1<\cdots<i_p\le N} \sqrt{\frac{N^{p-1}}{\pi p!}} \int_{-\infty}^{\infty} dJ_{i_1,\ldots,i_p}$$
$$\times \exp\left(-\frac{N^{p-1}}{p!} J^2_{i_1,\ldots,i_p} - iJ_{i_1,\ldots,i_p} S_{i_1}\cdots S_{i_p} x\right).$$

Performing the Gaussian integrals, a simple expression of $P(E)$ is obtained as

$$P(E) = \frac{1}{2\pi}\int_{-\infty}^{\infty} dx \exp\left(-\frac{N}{4}x^2 + iEx\right)$$
$$= \frac{1}{\sqrt{\pi N}} e^{-E^2/N}. \tag{6.6.2}$$

We next consider the pair probability of energies of H_0. It is defined by the probability that given two spin configurations $(S^{(1)}) \equiv (S_1^{(1)}, S_2^{(1)}, \dots)$ and $(S^{(2)}) \equiv (S_1^{(2)}, S_2^{(2)}, \dots)$ have energies E_1 and E_2 respectively:

$$P(E_1, E_2) = \left[\delta\left(E_1 - H_0\left(S^{(1)}\right)\right)\delta\left(E_2 - H_0\left(S^{(2)}\right)\right)\right]_{\text{av}}. \tag{6.6.3}$$

A similar algebra to $P(E)$ leads us to

$$\begin{aligned} P(E_1, E_2) &= \frac{1}{(2\pi)^2}\int dx_1\, dx_2 \exp\left[-\frac{N}{4}x_1^2 - \frac{N}{4}x_2^2 \right. \\ &\quad \left. - \frac{p!}{2N^{p-1}}\left(\sum_{1\le i_1<\dots<i_p\le N} S_{i_1}^{(1)}S_{i_1}^{(2)}\cdots S_{i_p}^{(1)}S_{i_p}^{(2)}\right)x_1x_2 - iE_1x_1 - iE_2x_2\right]. \end{aligned}$$

With $N \to \infty$, one can write

$$\frac{p!}{N^p}\sum_{1\le i_1<\dots<i_p\le N} S_{i_1}^{(1)}S_{i_1}^{(2)}\cdots S_{i_p}^{(1)}S_{i_p}^{(2)} = \frac{1}{N^p}\sum_{i_1\neq\dots\neq i_p} S_{i_1}^{(1)}S_{i_1}^{(2)}\cdots S_{i_p}^{(1)}S_{i_p}^{(2)} \to q^p,$$

where

$$q = \frac{1}{N}\sum_{i=1}^{N} S_i^{(1)}S_i^{(2)} \tag{6.6.4}$$

is the overlap parameter. Performing the Gaussian integral with respect to x_1 and x_2, one obtains the following expression of $P(E_1, E_2)$.

$$P(E_1, E_2) = \frac{1}{\pi N[(1+q^p)(1-q^p)]^{1/2}}\exp\left[-\frac{(E_1+E_2)^2}{2N(1+q^p)} - \frac{(E_1-E_2)^2}{2N(1-q^p)}\right]. \tag{6.6.5}$$

Equations (6.6.2) and (6.6.5) imply that the energies of two spin configurations becomes independent in the infinite limit of p, namely,

$$P(E_1, E_2) \xrightarrow{p\to\infty} P(E_1)P(E_2). \tag{6.6.6}$$

We focus on the case with $p \to \infty$ hereafter.

Let us consider the number of states with the energy in $[E, E + dE]$:

$$n(E)\, dE = \sum_{k=1}^{2^N} \delta\left(E - H\left(S^{(k)}\right)\right) dE. \tag{6.6.7}$$

Equations (6.6.2) and (6.6.5) lead to

$$\begin{aligned} \left[n(E)\right]_{\text{av}} dE &= 2^N P(E)\, dE \\ \left[\{n(E)\, dE\}^2\right]_{\text{av}} &= \left[\int_E^{E+dE} n(E_1)n(E_2)\, dE_1\, dE_2\right]_{\text{av}} \\ &= 2^N\left(2^N - 1\right)P(E)^2 dE^2 + 2^N P(E)\, dE. \end{aligned}$$

It follows that for $N \gg 1$

$$\left(\left[\{n(E)\,dE\}^2\right]_{\rm av} - \{[n(E)]_{\rm av}\,dE\}^2\right)^{1/2} \approx \sqrt{[n(E)]_{\rm av}\,dE}.$$

Hence the peak of the distribution of $n(E)$ is so sharp that one may approximate

$$n(E) \approx \left[n(E)\right]_{\rm av} = 2^N P(E) = \frac{1}{\sqrt{\pi N}} \exp\left\{N\left(\ln 2 - E^2/N^2\right)\right\}$$

for $|E| < N\sqrt{\ln 2}$ and $n(E) \approx 0$ for $|E| > N\sqrt{\ln 2}$. The entropy density is given by

$$s(\varepsilon) = \frac{1}{N} \ln n(E) \approx \ln 2 - \varepsilon^2$$

for $|E|/N \equiv |\varepsilon| < \sqrt{\ln 2}$. The phase transition takes place when $\varepsilon = -\sqrt{\ln 2}$ where $s(\varepsilon) = 0$. The critical temperature is determined by

$$\frac{1}{T_{\rm c}} = \left.\frac{ds(\varepsilon)}{d\varepsilon}\right|_{\varepsilon=-\sqrt{\ln 2}} = 2\sqrt{\ln 2}.$$

For $T > T_{\rm c}$, the free-energy density is given by

$$f = \varepsilon - Ts = -T \ln 2 - \frac{1}{4T}.$$

Below $T_{\rm c}$, since there is no level lower than $-\sqrt{\ln 2}$, one has a constant free-energy density:

$$f = -\sqrt{\ln 2}.$$

Now we introduce the transverse field and consider the full Hamiltonian (6.6.1). The model is cast into its equivalent classical one using the Suzuki-Trotter formula (cf. Sect. 3.1) and the free energy calculated using the replica trick [158]. The static hypothesis or ansatz is that all the order parameters are independent of the values of the Trotter indices. In the high temperature regime, the replica symmetric ansatz has been used. It is found that there exist two different regions in the paramagnetic phase (with the spin glass order parameter $q = 0$) distinguished by the transverse ordering. In one, there is no transverse ordering (classical paramagnetic region). The transverse ordering is characterised by $(\partial f/\partial \Gamma)$ where f is the free energy. In the classical paramagnetic region the free energy is given by $f = -T \ln 2 - 1/4T$ giving

$$\chi = \frac{\partial f}{\partial \Gamma} = 0. \tag{6.6.8}$$

In the other region (quantum paramagnetic), the free energy is

$$f = -T \ln 2 - T \ln\left[\cosh(\Gamma/T)\right] \tag{6.6.9}$$

giving a nonzero transverse ordering

$$\chi = (T/\Gamma) \tanh(\Gamma/T). \tag{6.6.10}$$

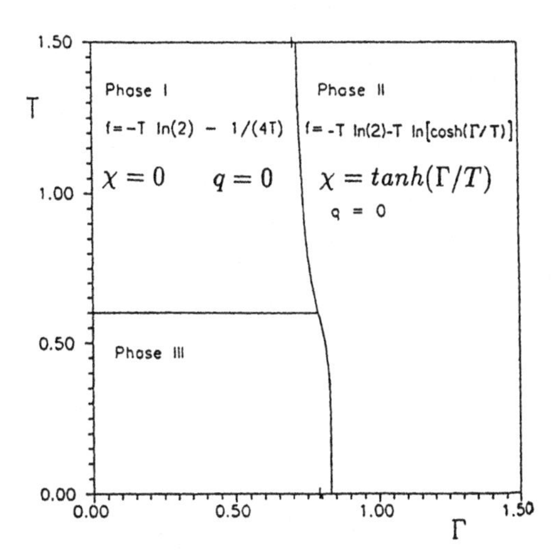

Fig. 6.15 Phase diagram of the p-spin quantum Ising spin glass in the limit of $p \to \infty$. *Phase I* and *II* are paramagnetic phases with zero and nonzero transverse ordering (χ), respectively separated by a second order transition line. All the other transitions are first order

At low temperatures, the spin glass phase exists where the replica symmetry breaking solutions have been used. The phase diagram obtained in [158] is shown in Fig. 6.15.

A systematic large-p expansion of the same model was also done [109], using again the replica trick and the Suzuki-Trotter transformation. For large but finite p, the transition line between these two phases in the high temperature regime was found to end at a critical point given by $T/(J\sqrt{p}) = 0.2593$ and $\Gamma/J = 0.7579$.

The cavity fields approach was used [61] to study the same model in the paramagnetic region. The expression of free energy agreed exactly with that of [109] and the coexistence line of the two phases is given by

$$\Gamma(T) = J^2/4T + T\ln\left[1 - \exp\left(-J^2/2T^2\right)\right]. \tag{6.6.11}$$

It was also found that while the thermodynamics at $p \to \infty$ is given exactly by the static approximation, the dynamical properties are different. Thus one can conclude that the existence of two high temperature phases is a property of large p not observed for $p = 2$.

6.6.1 p-Body Spin Glass with Ferromagnetic Bias

In this subsection, we investigate the validity of static approximation for p-body spin glass model with ferromagnetic bias in the limit of $p \to \infty$. The following argument was recently given by Obuchi et al. [300].

Let us start our argument from the following Hamiltonian:

$$H = -\sum_{I_1 < \cdots < i_p} J_{i_1 \cdots i_p} S_{i_1}^z \cdots S_{i_p}^z - \Gamma \sum_{i=1}^{N} S_i^x \tag{6.6.12}$$

where $J_{i_1 \cdots i_p}$ is the quenched randomness obeying the Gaussian distribution:

$$P(J_{i_1 \cdots i_p}) = \left(\frac{N^{p-1}}{J^2 \pi p!}\right)^{1/2} \exp\left\{-\frac{N^{p-1}}{J^2 p!}\left(J_{i_1 \cdots i_p} - \frac{j_0 p!}{N^{p-1}}\right)\right\}, \tag{6.6.13}$$

where non-zero j_0 denotes the ferromagnetic bias.

For the classical system obtained by Suzuki-Trotter decomposition, we have the following replicated partition function by means of the standard replica method

$$\begin{aligned} \ll Z_M^n \gg = \int \prod_{\mu,t} dm_t^\mu \, d\tilde{m}_t^\mu \prod_{\mu<\nu} \prod_{tt'} dq_{tt'}^{\mu\nu} \, d\tilde{q}_{tt'}^{\mu\nu} \prod_{\mu} \prod_{t \neq t'} dq_{tt'}^{\mu\mu} \, d\tilde{q}_{tt'}^{\mu\mu} \\ \times \exp N \left\{ \sum_{tt'} \sum_{\mu<\nu} \left(\frac{\beta^2 J^2}{2M^2} (q_{tt'}^{\mu\nu})^p - \frac{1}{M^2} \tilde{q}_{tt'}^{\mu\nu} q_{tt'}^{\mu\nu} \right) \right. \\ + \sum_{tt'} \sum_{\mu} \left(\frac{\beta^2 J^2}{4M^2} (q_{tt'}^{\mu\mu})^p - \frac{1}{M^2} \tilde{q}_{tt'}^{\mu\mu} q_{tt'}^{\mu\mu} \right) \\ \left. + \sum_{t,\mu} \left(\frac{\beta^2 j_0}{M} (m_t^\mu)^p - \frac{1}{M} \tilde{m}_t^\mu m_t^\mu \right) + \log \operatorname{tr} \exp(-H_{\text{eff}}) \right\} \end{aligned} \tag{6.6.14}$$

with

$$\begin{aligned} H_{\text{eff}} = -B \sum_{t,\mu} S_t^\mu S_{t+1}^\mu - \frac{1}{M} \sum_{\mu,t} \tilde{m}_t^\mu S_t^\mu - \frac{1}{M^2} \sum_{\mu<\nu} \sum_{tt'} \tilde{q}_{tt'}^{\mu\nu} S_t^\mu S_{t'}^\nu \\ - \frac{1}{M^2} \sum_{\mu} \sum_{t \neq t'} \tilde{q}_{tt'}^{\mu\mu} S_t^\mu S_{t'}^\mu. \end{aligned} \tag{6.6.15}$$

From the above expression, we immediately have the free energy density (free energy per spin) as

$$\begin{aligned} -\beta F = \sum_{tt'} \sum_{\mu<\nu} \left(\frac{\beta^2 J^2}{2M^2} (q_{tt'}^{\mu\nu})^p - \frac{1}{M^2} \tilde{q}_{tt'}^{\mu\nu} q_{tt'}^{\mu\nu} \right) \\ + \sum_{tt'} \sum_{\mu} \left(\frac{\beta^2 J^2}{4M^2} (q_{tt'}^{\mu\mu})^p - \frac{1}{M^2} \tilde{q}_{tt'}^{\mu\mu} q_{tt'}^{\mu\mu} \right) \\ + \sum_{t,\mu} \left(\frac{\beta^2 j_0}{M} (m_t^\mu)^p - \frac{1}{M} \tilde{m}_t^\mu m_t^\mu \right) + \log \operatorname{tr} \exp(-H_{\text{eff}}) \end{aligned} \tag{6.6.16}$$

with the following order parameters

$$q_{tt'}^{\mu\nu} = \langle S_t^\mu S_{t'}^\nu \rangle \tag{6.6.17}$$

$$\tilde{q}_{tt'}^{\mu\nu} = \frac{1}{2} \beta^2 J^2 p (q_{tt'}^{\mu\nu})^{p-1} \tag{6.6.18}$$

$$q_{tt'}^{\mu\mu} = \langle S_t^\mu S_{t'}^\mu \rangle \tag{6.6.19}$$

$$\tilde{q}_{tt'}^{\mu\mu} = \frac{1}{4}\beta^2 J^2 p\big(q_{tt'}^{\mu\mu}\big)^{p-1} \tag{6.6.20}$$

$$m_t^{\mu} = \langle S_t^{\mu} \rangle \tag{6.6.21}$$

$$\tilde{m}_t^{\mu} = \beta j_0 p\big(m_t^{\mu}\big)^{p-1} \tag{6.6.22}$$

where $\langle \cdots \rangle$ stands for the thermal average over the weight $e^{-H_{\text{eff}}}$. To proceed the analysis, we usually use the static approximation under the replica symmetric (RS) ansatz. Namely,

$$q_{tt'}^{\mu\mu} = R, \qquad q_{tt'}^{\mu\nu} = q, \qquad m_t^{\mu} = m. \tag{6.6.23}$$

However, here we focus on the validity of the static approximation. Hence, we do not use the static approximation and write the above order parameters as

$$q_{tt'}^{\mu\mu} = R\big(t,t'\big), \qquad q_{tt'}^{\mu\nu} = q\big(t,t'\big), \qquad m_t^{\mu} = m(t). \tag{6.6.24}$$

Then, we consider the deviation from the static approximation and rewrite the conjugate order parameters for (6.6.24) as

$$\tilde{R}\big(t,t'\big) = \tilde{R} + \Delta\tilde{R}\big(t,t'\big), \quad \tilde{q}\big(t,t'\big) = \tilde{q} + \Delta\tilde{q}\big(t,t'\big), \quad \tilde{m}(t) = \tilde{m} + \Delta\tilde{m}(t). \tag{6.6.25}$$

Substituting (6.6.24) and (6.6.25) into the effective Hamiltonian H_{eff}, we have

$$H_{\text{eff}} = H_{\text{stat}} + V\big(t,t'\big) \tag{6.6.26}$$

with

$$H_{\text{stat}} = -B\sum_{\mu,t} S_t^{\mu} S_{t+1}^{\mu} - \frac{\tilde{R}}{M^2}\sum_{\mu}\sum_{t\neq t'} S_t^{\mu} S_{t'}^{\mu} - \frac{\tilde{q}}{2M^2}\sum_{\mu\neq\nu}\sum_{tt'} S_t^{\mu} S_{t'}^{\nu} - \frac{\tilde{m}}{M}\sum_{\mu,t} S_t^{\mu} \tag{6.6.27}$$

and

$$\begin{aligned} V\big(t,t'\big) = &-\frac{1}{M^2}\sum_{\mu}\sum_{t\neq t'} \Delta\tilde{R}\big(t,t'\big) S_t^{\mu} S_{t+1}^{\mu} - \frac{1}{2M^2}\sum_{\mu\neq\nu}\sum_{tt'} \Delta\tilde{q}\big(t,t'\big) S_t^{\mu} S_{t+1}^{\nu} \\ &- \frac{1}{M}\sum_{\mu}\sum_{t} \Delta\tilde{m}(t) S_t^{\mu}. \end{aligned} \tag{6.6.28}$$

Originally, the order parameters are given as auto-correlation function between two Trotter slices t, t'. Therefore, we can naturally assume that these quantities decreases as a function of $|t-t'|$. From the definitions, the conjugate of these order parameters are given as p-th power of the order parameters. Hence, it is reasonable for us to expand the free energy density with respect to the deviation $\Delta\tilde{R}(t,t')$, $\Delta\tilde{q}(t,t')$ and $\Delta\tilde{m}(t)$, namely, we expand the f with respect to $V(t,t')$. Thus, we immediately obtain

$$\beta f = \frac{1}{2M^2}\sum_{tt'}\bigg(\frac{\beta^2 J^2}{2} q\big(t,t'\big)^p - \tilde{q}\big(t,t'\big) q\big(t,t'\big) - \Delta\tilde{q}\big(t,t'\big) q\big(t,t'\big)\bigg)$$

$$-\frac{1}{M^2}\sum_{t\neq t'}\left(\frac{\beta^2 J^2}{4}R(t,t')^p-\tilde{R}(t,t')R(t,t')-\Delta\tilde{R}(t,t')R(t,t')\right)$$
$$-\frac{1}{M}\sum_{t}\left(\beta j_0 m(t)^p-\tilde{m}(t)m(t)-\Delta\tilde{m}(t)m(t)\right)$$
$$-\lim_{n\to 0}\frac{1}{n}\left(\log \mathrm{tr}\exp(-H_{\mathrm{stat}})+\langle V\rangle_{\mathrm{stat}}\right), \tag{6.6.29}$$

where $\langle\cdots\rangle_{\mathrm{stat}}$ denotes the average by the weight $\exp(-H_{\mathrm{stat}})$. After taking the functional derivative of βf with respect to $\tilde{R}(t,t')$, we have

$$R(t,t')=\lim_{n\to 0}\frac{\mathrm{tr}(S_t^{\mu}S_{t'}^{\mu})\exp(-H_{\mathrm{stat}})}{\mathrm{tr}\exp(-H_{\mathrm{stat}})}. \tag{6.6.30}$$

To calculate the above auto-correlation function, we rewrite the $\exp(-H_{\mathrm{stat}})$ by means of the Hubbard-Stratonovitch transformation:

$$\exp(a^2)=\int Dz\,\exp(\sqrt{2}az). \tag{6.6.31}$$

When we notice

$$\frac{\tilde{R}}{M^2}\sum_{\mu}\sum_{t\neq t'}S_t^{\mu}S_{t'}^{\mu}=\exp\left\{\frac{\tilde{R}}{M^2}\left(\sum_t S_t^{\mu}\right)^2\right\} \tag{6.6.32}$$

$$\frac{\tilde{q}}{2M^2}\sum_{\mu\neq\nu}\sum_{tt'}S_t^{\mu}S_{t'}^{\nu}=\exp\left[\frac{\tilde{q}}{2M^2}\left(\sum_{\mu,t}S_t^{\mu}\right)^2-\sum_{\mu}\frac{\tilde{q}}{2M^2}\left(\sum_t S_t^{\mu}\right)^2\right] \tag{6.6.33}$$

one has

$$\exp(-H_{\mathrm{stat}})=\int Dz_1\prod_{\mu}\left\{\int Dz_2\exp\left(B\sum_t S_t^{\mu}S_{t+1}^{\mu}+\frac{A}{M}\sum_t S_t^{\mu}\right)\right\}, \tag{6.6.34}$$

where we defined

$$A=z_2\sqrt{2\tilde{R}-\tilde{q}}+z_1\sqrt{\tilde{q}}+\tilde{m}. \tag{6.6.35}$$

Here, we should notice from (6.6.34) and

$$q(t,t')=-\lim_{n\to 0}\frac{1}{n}\left\langle\sum_{\mu\neq\nu}S_t^{\mu}S_{t'}^{\mu}\right\rangle_{\mathrm{stat}}$$
$$m(t)=\lim_{n\to 0}\frac{1}{n}\left\langle\sum_{\mu}S_t^{\mu}\right\rangle_{\mathrm{stat}} \tag{6.6.36}$$

that q and m are time-independent because each replica in H_{stat} is independent and the translational invariance in the Trotter direction holds.

The denominator of (6.6.30) with (6.6.34) is rewritten in the limit of $n \to 0$ as

$$\begin{aligned}\operatorname{tr}\exp(-H_{\text{stat}}) &= \int Dz_1 \left\{ \int Dz_2 \operatorname{tr}\exp\left(B\sum_t S_t S_{t+1} + \frac{A}{M}\sum_t S_t \right) \right\}^n \\ &= 1 + \mathscr{O}(n) \end{aligned} \tag{6.6.37}$$

whereas the numerator is given as

$$\begin{aligned}\operatorname{tr}\left(S_t^{\mu} S_{t'}^{\mu}\right)\exp(-H_{\text{stat}}) &= \int Dz_1 \left\{ \int Dz_2 \operatorname{tr}\exp\left(B\sum_t S_t S_{t+1} + \frac{A}{M}\sum_t S_t \right) \right\}^{n-1} \\ &\quad \times \int Dz_2 \operatorname{tr}\left(S_t^{\mu} S_{t'}^{\mu}\right)\exp(-H_{\text{stat}}) \\ &= \int Dz_1 \frac{\int Dz_2 \operatorname{tr}(S_t^{\mu} S_{t'}^{\mu})\exp(-H_{\text{stat}})}{\int Dz_2 \operatorname{tr}\exp(B\sum_t S_t S_{t+1} + \frac{A}{M}\sum_t S_t)}. \end{aligned} \tag{6.6.38}$$

By using the inverse process of the Suzuki-Trotter decomposition, we can easily carry out the trace in the denominator:

$$\begin{aligned}\lim_{M\to\infty} \operatorname{tr}\exp\left(B\sum_t S_t S_{t+1} + \frac{A}{M}\sum_t S_t \right) &= \operatorname{tr}\exp\left(A S^z + \beta\Gamma S^x\right) \\ &= 2\cosh\sqrt{A^2 + \beta^2\Gamma^2}. \end{aligned} \tag{6.6.39}$$

For simplicity, let us define

$$Y = \int Dz_2\, 2\cosh\sqrt{A^2 + \beta\Gamma^2} = \int Dz_2\, 2\cosh\omega. \tag{6.6.40}$$

Then, the $R(t,t')$ is written by

$$R(t,t) = \int Dz_1\, Y^{-1} \int Dz_2\, G\left(t,t'\right) \tag{6.6.41}$$

with

$$G\left(t,t'\right) = \operatorname{tr}(S_t S_{t'})\exp\left(B\sum_t S_t S_{t+1} + \frac{A}{M}\sum_t S_t \right). \tag{6.6.42}$$

The $G(t,t')$ is nothing but the correlation function of one-dimensional Ising chain. It is easily computed by the transfer matrix method. The solution is given by

$$\begin{aligned} G\left(t,t'\right) &= 4x_+^2 x_-^2 \left(\lambda_+^{t-t'}\lambda_-^{M-(t-t')} + \lambda_+^{M-(t-t')}\lambda_-^{t-t'}\right) \\ &\quad + \left(2x_+^2 - 1\right)^2 \lambda_+^M + \left(2x_-^2 - 1\right)\lambda_-^M, \end{aligned} \tag{6.6.43}$$

where we defined

$$\lambda_{\pm} = \mathrm{e}^B\left(\cosh(A/M) \pm \sqrt{\cosh^2(A/M) - 1 + \mathrm{e}^{-4B}}\right) \tag{6.6.44}$$

for the eigenvalues of the transfer matrix and also defined

$$\begin{pmatrix} x_{\pm} \\ y_{\pm} \end{pmatrix} = D_{\pm} \begin{pmatrix} -\mathrm{e}^{-B} \\ \mathrm{e}^{B}(\sinh(A/M) \mp \sqrt{\sinh^2(A/M) + \mathrm{e}^{-4B}}) \end{pmatrix} \tag{6.6.45}$$

for the first component of eigenvectors ($D_{\pm}$ stands for the normalisation constant). Then, we evaluate

$$x_{\pm}^2 = \frac{1}{2} \frac{(\beta\Gamma)^2}{\omega^2 \mp A\omega}, \qquad (C\lambda_{\pm})^M = \mathrm{e}^{\pm\omega} \tag{6.6.46}$$

with $C = \{(1/2)\sinh 2\beta\Gamma/M\}^2$. Hence, we have

$$G(t,t') = G(\tau) = A^2\omega^{-2}\cosh\omega + (\beta\Gamma)^2\omega^{-2}\cosh\omega(1-2\tau) \tag{6.6.47}$$

$$\tau \equiv \lim_{M\to\infty} \frac{t-t'}{M}. \tag{6.6.48}$$

Substituting these results into (6.6.43), we finally obtain the auto-correlation function as

$$R(\tau) = \int Dz_1\, Y^{-1}\left(\int Dz_2\, A^2\omega^{-2}\cosh\omega + \beta^2\Gamma^2 \int Dz_2\, \omega^{-2}\cosh\omega(1-2\tau)\right). \tag{6.6.49}$$

Here we focus on the ferromagnetic phase. In the ferromagnetic phase, all conjugate order parameters go to infinity in the limit of $p \to \infty$. Then, we can assume that $2\tilde{R} = \tilde{q}$ from (6.6.18), (6.6.19), (6.6.23). Hence, from (6.6.35), the integral over z_2 gives 1 and we have $Y = \cosh\omega$ and

$$R_{\mathrm{F}}(\tau) = \int Dz_1 \left(A^2\omega^{-2} + \beta^2\Gamma^2\omega^{-2}\frac{\cosh\omega(1-2\tau)}{\cosh\omega}\right). \tag{6.6.50}$$

We first consider the first term appearing in the right hand side of (6.6.50). From (6.6.22), (6.6.23), we have $\tilde{m} = \beta j_0 p(m)^{p-1}$, namely, $1/\tilde{m} \propto 1/p$. Hence, in the limit of $p \to \infty$, $A^2\omega^{-2}$ is approximated by

$$\begin{aligned} A^2\omega^{-2} &= \frac{(1+\sqrt{\tilde{q}}\,z_1/\tilde{m})^2}{1 + 2\sqrt{\tilde{q}}\,z_1/\tilde{m} + (\sqrt{\tilde{q}}\,z_1/\tilde{m})^2 + (\beta\Gamma/\tilde{m})^2} \\ &\simeq 1 - \frac{\beta^2\Gamma^2}{\tilde{m}^2} = 1 - \left(\frac{\Gamma}{j_0}\right)^2 \frac{m^{-2p+2}}{p^2}. \end{aligned} \tag{6.6.51}$$

This reads

$$\int Dz_1\, A^2\omega^{-2} = 1 - \left(\frac{\Gamma}{j_0}\right)^2 \frac{m^{-2p+2}}{p^2}. \tag{6.6.52}$$

We next evaluate the second term in the right hand side of (6.6.50), namely,

$$\int Dz_1\, \omega^{-2}\frac{\cosh\omega(1-2\tau)}{\cosh\omega} \simeq \int Dz_1\, \omega^{-2}\left(\mathrm{e}^{-2\tau\omega} + \mathrm{e}^{-2\omega(1-\tau)}\right) \tag{6.6.53}$$

by the saddle point method. The result gives $(\Gamma/j_0)^2 f(\tau, p)p^{-2}$, where $f(\tau, p)$ stands for the time-dependent correlation function decreasing exponentially as p grows. As the result, we have

$$R_F(\tau) \simeq 1 - \left(\frac{\Gamma}{j_0}\right)^2 \frac{1}{p^2} + \left(\frac{\Gamma}{j_0}\right)^2 \frac{1}{p^2} f(\tau, p). \tag{6.6.54}$$

In the ferromagnetic phase, the time-dependent part of the correlation function $f(\tau, p)$ for finite p correction is exponentially small for large p as in the classical paramagnetic phase and spin glass phase. Hence, we conclude that static approximation is valid in this sense.

6.7 Random Fields

6.7.1 Classical Random Field Ising Models

The random field Ising model (RFIM), described by the Hamiltonian

$$H = -\sum_{ij} J_{ij} S_i^z S_j^z - \sum_i h_i S_i^z \tag{6.7.1}$$

where $S_i^z = \pm 1$ are the Ising spins and the h_i are independent quenched random variables with mean zero, has been subjected to rigorous theoretical and experimental investigations in recent years (for a review see [33]). The random field acts as an order-destroying field, which effectively reduces the transition temperature T_c of the classical Ising transition from the symmetry broken (ferromagnetic) phase to the symmetric phase (configuration averaged magnetisation zero) as the magnitude of the random field is increased from zero, until T_c goes to zero for a critical amplitude or width of the random field (i.e., there exists a critical line $h_r(T)$ in the h_r–T diagram). For $h_r > h_r^c$, the system is always disordered at any temperature. It has been established [46] that the RFIM does not order for $d \leq 2$, indicating the lower critical dimensionality for the system is two. The existence of long-range order in the three dimensional model, for low temperature and weak random field, has been rigorously proved [46]. It has also been established from the mean field studies of the classical model that, whenever the distribution function of the random field $P(h)$ has a minimum at zero field (e.g., the binary distribution), one obtains a tricritical point [7] on the critical line, so that the transition for the larger values of the random field is discontinuous, whereas if the distribution function $P(h)$ decreases monotonically with the increase of the magnitude of h (e.g., the Gaussian distribution), the transition is always continuous [348]. If the transition is second order, the scaling arguments [45, 137, 411] (based on the assumptions that near the critical point $T_c(h_r)$ the random field fluctuations dominate over the thermal fluctuations), suggest a modified hyperscaling relation of the form $2 - \alpha = \nu(d - \theta)$; with the exponents ν and α as the correlation length and specific heat exponents respectively. The new exponent θ is related to the exponents η and $\overline{\eta}$ (where η and $\overline{\eta}$

describe the decay of the connected and disconnected correlation functions respectively at $T_c(h_r)$) through the relation $\theta = 2 + \eta - \overline{\eta}$. Obviously there seems to exist three independent critical exponents, but recent rigorous studies [157] imply that $\theta = 2 - \eta$, and $\overline{\eta} = 2\eta$ so that the Schwartz-Soffer inequality [350] is fulfilled as an equality. One should also mention here that the recent extensive numerical studies [328, 331], using both binary and Gaussian distribution of random fields, indicate a violation of the above-mentioned simple dimensional reduction. Also, the possibility of the occurrence of a spin glass phase between the para and the ferro phases has been discussed [272, 274]. The static universal critical behaviour is found to be identical for ferromagnets in a random field and dilute antiferromagnets in a uniform field [59, 147].

6.7.2 Random Field Transverse Ising Models (RFTIM)

It has been conjectured that frustration in the RFIM gives rise to a "many valley" structure in the configuration space, similar to the situation in spin glasses [33]. Dutta et al. [116] have studied the random (longitudinal) field transverse Ising model (RFTIM) to investigate the effects of the quantum fluctuations (induced by the transverse or tunnelling field) on the transition in the RFIM. Specifically, we consider RFTIM system represented by the Hamiltonian

$$H = -\sum_{ij} J_{ij} S_i^z S_j^z - \sum_i h_i S_i^z - \Gamma \sum_i S_i^x. \tag{6.7.2}$$

6.7.2.1 Mean Field Studies

We consider a random field Ising ferromagnet (with long-range interaction), in the presence of a uniform transverse field

$$H = -\frac{J}{N} \sum_{i \neq j} S_i^z S_j^z - \sum_i h_i S_i^z - \Gamma \sum_i S_i^x, \tag{6.7.3}$$

where Γ is the strength of the tunnelling field and h_i, as mentioned earlier, is the quenched random field at each site with a probability distribution $P(h)$ having zero mean and nonzero variance. Using the replica trick and the saddle-point integration (in the $N \to \infty$ limit), one can exactly reduce the classical Hamiltonian ($\Gamma = 0$) to an effective single-site Hamiltonian of the form [348]

$$H = \sum_i H_i = -\sum_i (2mJ + h_i) S_i^z, \tag{6.7.4}$$

with the effective molecular field at each site given by $(2mJ + h_i)$, where m is the configuration averaged magnetisation.

For the quantum Hamiltonian ($\Gamma \neq 0$), in the large N limit, one can similarly construct an effective single-site Hamiltonian given by (see Sect. 6.A.3)

$$H = -\sum_i (2m^z + h_i) S_i^z - \Gamma \sum_i S_i^x, \tag{6.7.5}$$

where m_z is the configuration averaged longitudinal magnetisation. The configuration averaged magnetisation vector can be readily written [48, 379] in the self-consistent form

$$\boldsymbol{m} = \overline{\tanh \beta \Big[\sqrt{(2m^z J + h)^2 + \Gamma^2}\Big] \left(\frac{(2m^z J + h)\hat{z} + \Gamma \hat{x}}{\sqrt{(2m^z J + h)^2 + \Gamma^2}} \right)} \tag{6.7.6}$$

so that the configuration averaged longitudinal magnetisation is

$$m^z = \overline{\left[\tanh \beta \Big(\sqrt{(2m^z J + h)^2 + \Gamma^2} \Big) \frac{2m^z J + h}{\sqrt{(2m^z J + h)^2 + \Gamma^2}} \right]}, \tag{6.7.7}$$

where the over-head bar denotes a configuration average over the distribution of the random field. If one now uses a binary distribution of the random field

$$P(h) = \frac{1}{2}\delta(h - h_0) + \frac{1}{2}\delta(h + h_0), \tag{6.7.8}$$

the configuration averaged longitudinal magnetisation can be written as [48, 379]

$$\begin{aligned} m^z = {} & \frac{1}{2}\left[\tanh \beta \Big(\sqrt{(2m^z J + h_0)^2 + \Gamma^2} \Big) \frac{2m^z J + h_0}{\sqrt{(2m^z J + h_0)^2 + \Gamma^2}} \right] \\ & + \frac{1}{2}\left[\tanh \beta \Big(\sqrt{(2m^z J - h_0)^2 + \Gamma^2} \Big) \frac{2m^z J - h_0}{\sqrt{(2m^z J - h_0)^2 + \Gamma^2}} \right]. \end{aligned} \tag{6.7.9}$$

From (6.7.7), one can conclude that for any symmetric distribution $P(h)$, of the random field, $m_z = 0$ is always a solution of (6.7.7). For large enough temperature and random field, this is the only solution. At low temperature and weak random field, one finds an additional solution $m^z \neq 0$ (symmetry broken phase) with lower free energy. If the transition is continuous, one can find the transition point by expanding (6.7.7) around $m^z = 0$,

$$m^z \sim a m^z - b(m^z)^3 - x(m^z)^5 - \cdots. \tag{6.7.10}$$

A second-order transition is found when $a = 1$ as long as $b > 0$. If $b < 0$ the transition is first order and the point $a = 1$ and $b = 0$ characterises a tricritical point on the phase boundary, separating the ferromagnetic phase ($m^z \neq 0$) and the phase with $m^z = 0$ (but with nonzero value of the configuration averaged squared magnetisation). In the classical case ($\Gamma = 0$) [7], one finds

$$a = 2\beta J\big(1 - \overline{t^2}\big); \qquad b = \frac{1}{3}(2\beta J)^3 \overline{\big[(1 - t^2)(1 - 3t^2)\big]} \tag{6.7.11}$$

where $t = \tanh \beta h$. With a binary distribution of the random field one finds the tricritical point [7] at

$$\beta J = \frac{3}{4}, \qquad \tanh^2(\beta h_0) = \frac{1}{3}. \tag{6.7.12}$$

One can solve (6.7.9) (with $\Gamma = 0$) numerically, to obtain the entire phase diagram of the classical system. In the extreme quantum limit ($T = 0$), the thermal fluctuations are absent and the fluctuations induced by the random field and quantum fluctuations due to the transverse field tend to destroy the long-range order. From (6.7.9) the configuration averaged longitudinal magnetisation can be written as

$$m^z = \overline{\left[\frac{2m^z J + h}{\sqrt{(2m^z J + h)^2 + \Gamma^2}}\right]}. \tag{6.7.13}$$

Expanding the magnetisation in the form (6.7.10), we find for any symmetric distribution of the random field

$$a = \overline{\left[\frac{2J}{\sqrt{h^2 + \Gamma^2}} - \frac{2Jh^2}{(h^2 + \Gamma^2)^{3/2}}\right]}, \tag{6.7.14}$$

$$b = \overline{\left[\frac{24J^3}{(h^2 + \Gamma^2)^{3/2}} - \frac{144h^2 J^3}{(h^2 + \Gamma^2)^{5/2}} + \frac{120J^3 h^4}{(h^2 + \Gamma^2)^{7/2}}\right]}. \tag{6.7.15}$$

Specifically, if we use the binary distribution of random field

$$a = \left[\frac{2J}{\sqrt{h_0^2 + \Gamma^2}} - \frac{2Jh_0^2}{(h_0^2 + \Gamma^2)^{3/2}}\right], \tag{6.7.16}$$

$$b = \left[\frac{24J^3}{(h_0^2 + \Gamma^2)^{3/2}} - \frac{144h_0^2 J^3}{(h_0^2 + \Gamma^2)^{5/2}} + \frac{120h_0^4 J^3}{(h_0^2 + \Gamma^2)^{7/2}}\right]. \tag{6.7.17}$$

The tricritical point ($a = 1$, $b = 0$) is obtained at $\Gamma \simeq 1.4J$, $h_0 \simeq 0.74J$. The numerically obtained phase diagram is very similar to the phase diagram obtained in the classical case ($\Gamma = 0$), indicating that the transverse field behaves in the same manner as the temperature to destroy the long-range order.

When both the thermal and quantum fluctuations are present, we obtain the phase diagram (Fig. 6.16) in the Γ–h_0 plane (for various temperatures below the pure system transition temperature) by numerically solving (6.7.9) and also if the transition is second order, the transition point is given by

$$a = \overline{\left[\frac{4h^2 J\beta(1 - t^2)}{2(h^2 + \Gamma^2)} + \frac{2tJ}{(h^2 + \Gamma^2)^{1/2}} - \frac{2th^2 J}{(h^2 + \Gamma^2)^{3/2}}\right]}, \tag{6.7.18}$$

where $t = \tanh \beta h$. We find, from the numerically obtained phase diagram, that as the temperature is increased, the phase diagram shrinks to lower values of Γ and h_0 and the tricritical point on the critical line in the Γ–h_0 plane shifts to a higher value of h_0 (i.e., the second-order region on the phase boundary increases) and eventually if the temperature is higher than the value at the tricritical point of the classical phase boundary, the entire phase boundary corresponds to the continuous transition. These mean field calculations can be readily extended to obtain numerically the phase diagram when the random field distribution is Gaussian with zero mean and nonzero variance

$$P(h) = \frac{1}{\sqrt{2\pi \Delta^2}} \exp\left(-\frac{h^2}{2\Delta^2}\right). \tag{6.7.19}$$

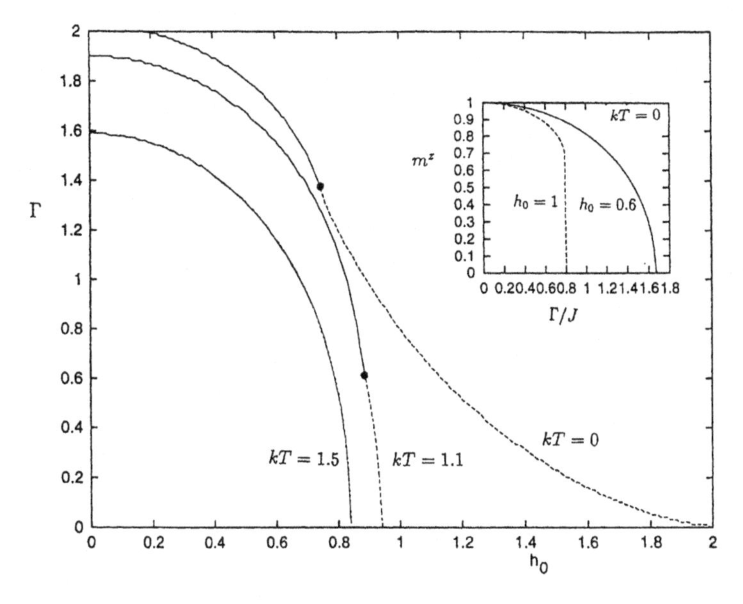

Fig. 6.16 The mean field phase diagram of RFTIM in the Γ–h_0 plane for different temperatures. The *black circle* denotes the tricritical point. The *inset* shows the nature of transition below and above the tricritical point

One can easily see in this case that the phase transition is continuous for all values of Γ and Δ (width of the Gaussian distribution) as because even in the limit of temperature and transverse field both being zero, the transition driven by the random field is continuous.

6.7.2.2 Mapping of Random Ising Antiferromagnet in Uniform Longitudinal and Transverse Fields to RFTIM

We here show that the random Ising antiferromagnet in uniform transverse and longitudinal field (RIAFTL) is in the universality class of the Ising ferromagnet with uniform transverse field and random longitudinal field (RFTIM). This equivalence is obtained, in a semiclassical approximation neglecting commutations, via a decimation of one sublattice of the RIAFTL system. We illustrate the procedure by considering first the one-dimensional model, commenting later on generalisations.

The decimation procedure is a partial trace over sites of one sublattice, e.g., that in which the site label i is odd. To rearrange the statistical weights of the remaining spins, the original (reduced) Hamiltonian

$$-\beta H = \sum_i \left(-K_{i,i+1} S_i^z S_{i+1}^z + h_i S_i^z + \Gamma S_i^x\right) = \sum_i H_i \qquad (6.7.20)$$

will be mapped into a new form

$$-\beta H' = \sum_i \left(-K'_{2i,2i+2} S^z_{2i} S^z_{2i+2} + h'_{2i} S^z_{2i} + \Gamma' S^x_{2i}\right) = \sum_i H'_{2i}. \qquad (6.7.21)$$

In (6.7.20), h and Γ are the longitudinal and transverse components of a uniform field, and the label i on h_i is there only to allow for the effects of site dilution (h_i is independent of i in the case of bond dilution). $K_{i,i+1}$ is a random antiferromagnetic exchange. The semiclassical decimation procedure which neglects commutations but is otherwise exact, is as follows

$$\begin{aligned}\prod_i \left[\mathrm{Tr}_{S_{2i+1}} \exp(H_i)\right] &= \exp\left(h_{2i} S^z_{2i} + \Gamma_{2i} S^x_{2i}\right) \mathrm{Tr}_{S_{2i+1}} \exp\left(S^z_{2i+1}\left[h_{2i+1}\right.\right.\\ &\quad \left.\left. - K_{2i,2i+1} S^z_{2i} - K_{2i+1,2i+2} S^z_{2i+2}\right] + \Gamma S^x_{2i+1}\right) \exp\left(h_{2i+2} S^z_{2i+2} + \Gamma^x_{2i+2}\right) \cdots \\ &= \text{const.} \prod_i \exp\left(H'_{2i}\right). \end{aligned} \qquad (6.7.22)$$

The trace over S_{2i+1} produces the factor

$$b\left(S^z_{2i}, S^z_{2i+2}\right) = 2\cosh\left[\left(h_{2i+1} - K_{2i,2i+1} S^z_{2i} - K_{2i+1,2i+2} S^z_{2i+2}\right)^2 + \Gamma^2\right]^{1/2}. \qquad (6.7.23)$$

This can be written as

$$\exp\left(A + B S^z_{2i} + C S^z_{2i+2} + D S^z_{2i} S^z_{2i+2}\right) \qquad (6.7.24)$$

where matching of the expression for all four possible sets of values for (S^z_{2i}, S^z_{2i+2}) gives A, B, C, D in terms of Γ, h_{2i+1}, $K_{2i,2i+1}$, $K_{2i+1,2i+2}$ (see Sect. 6.A.4). We thus arrive at the recursion relations

$$\begin{aligned} h'_{2i} &= h_{2i} + B(K_{2i,2i+1}, K_{2i+1,2i+2}) \\ &\quad + C(K_{2i-2,2i-1}, K_{2i-1,2i}) \end{aligned} \qquad (6.7.25)$$

$$\Gamma' = \Gamma \qquad (6.7.26)$$

$$K'_{2i,2i+2} = D(K_{2i,2i+1}, K_{2i+1,2i+2}). \qquad (6.7.27)$$

The particular case $h_i = 0$ of this shows that the random bond Ising antiferromagnet in a uniform transverse field maps to a random bond Ising ferromagnet in a uniform transverse field. The general case (h_i, Γ both nonzero) maps to a random longitudinal field model, along with uniform transverse field. This is most easily illustrated for the random bond case when h and Γ are both independent of site label i. For $h \ll K$, one can simplify $B(K_1, K_2)$ (where K_1 and K_2 are two neighbouring bonds) (see Sect. 6.A.4):

$$B(K_1, K_2) = -\frac{h}{2}\left[\frac{\Lambda_+}{\Omega_+} \tanh \Omega_+ + \frac{\Lambda_-}{\Omega_-} \tanh \Omega_-\right] + O\left(h^2\right), \qquad (6.7.28)$$

with

$$\Omega_\pm = \left[\Lambda^2_\pm + \Gamma^2\right]^{1/2}, \qquad \Lambda_\pm = K_1 \pm K_2. \qquad (6.7.29)$$

For the case of bond dilution, where

$$K_{i,i+1} = K \quad \text{with probability } p \tag{6.7.30}$$

$$= 0 \quad \text{with probability } (1 - p), \tag{6.7.31}$$

it is clear that Λ_+ is always positive, while Λ_- could be positive or negative with equal probability for any nonzero value of the probability p. The result is that h'_{2i} is distributed in such a way that its mean is not zero, but it divides into two parts and the part (containing Λ_-) which couples to the critical fluctuations (antiferromagnetic, in the original model) has zero mean, whereas the part with nonzero mean couples to the ferromagnetic order parameter. K', on the other hand, is a random (ferromagnetic) exchange. This confirms that the model has the universality class of the RFTIM model. The same procedure can be extended in higher dimensions using cluster approximation of the type common in decimation methods [382], with again the same conclusion. Equations (6.7.25), (6.7.26), (6.7.27) give the expected relationship between the parameters of the original system and the resulting RFTIM.

6.7.3 Concluding Remarks on the Random Field Transverse Ising Model

The phase transition behaviour of the random field Ising model in the presence of a transverse field (RFTIM) has been studied. This transverse field represents the (quantum) tunnelling fluctuations in double well systems representing the model order-disorder ferroelectric systems, Jahn-Teller systems etc [48]. The mean field phase diagram has been studied in details, in particular at zero temperature, where the transition is induced by the fluctuations induced by the random field and quantum fluctuations due to the transverse field. It has been established in a semi-classical way that the ferromagnetic transverse Ising model with random longitudinal field provides the universal critical behaviour of the random (e.g., randomly diluted) Ising antiferromagnet in a uniform field having both transverse and longitudinal components. This is shown by employing a sublattice decimation on the random antiferromagnet in a general uniform field. Although the decimation procedure is only demonstrated for a one-dimensional system, it can be generalised for the higher dimensions. This mapping also indicates the possible application of the results of the studies for RFTIM to random quantum (Ising) antiferromagnets. It may also be mentioned at this point, that the mean field studies of the RFTIM (spin-1) [35] and RFTIM (spin-1) with random bond dilution [34] have also been reported.

6.8 Mattis Model in a Transverse Field

We can write the generic form of the Ising spin glass Hamiltonian in a transverse field as

$$H = -\frac{1}{2}\sum_{ij} J_{ij} S_i^z S_j^z - \Gamma \sum_i S_i^x \tag{6.8.1}$$

where J_{ij} is the configuration dependent (random) exchange interaction. In the Mattis model [262] one chooses the configuration dependent (random) exchange interaction of the form

$$J_{ij} = J\big(|\boldsymbol{R}_i - \boldsymbol{R}_j|\big)\varepsilon_i\varepsilon_j = J_{ij}^0\varepsilon_i\varepsilon_j, \tag{6.8.2}$$

where $\varepsilon = \pm 1$ and are random. The model has (site) disorder but does not have frustration (the disorder is irrelevant and can be transformed away). For short-range interaction, the classical model (where the non-commuting tunnelling field is absent) has a finite temperature phase transition for all dimensions $d > d_l^c$ (where d_l^c is the lower critical dimension, $= 1$ in Ising case) with the longitudinal susceptibility showing a cusp like behaviour at the point of transition. For the Mattis model in a transverse field [62], one can introduce a set of pseudo-spin operators to rewrite the Hamiltonian (6.8.1) (using (6.8.2) and $\varepsilon_i^2 = 1$) in the form

$$H = -\frac{1}{2}\sum_{ij} J_{ij}^0 \tau_i^z \tau_j^z - \sum_i \tilde{\Gamma}_i \tau_i^x \tag{6.8.3}$$

where $\tau_i^z = \varepsilon_i S_i^z$, $\tau_i^x = \varepsilon_i S_i^x$ are the pseudo-spin operators and $\tilde{\Gamma}_i = \varepsilon_i \Gamma$. In recasting the Hamiltonian in terms of pseudo-spin operators, we have essentially transformed away the disorder. The operators τ_i^z have the same eigenvalue ± 1 as the operators S_i^z and satisfy the commutation relations

$$\big[\tau_i^z, \tau_i^x\big] = 2i\, S^y = \big[S^z, S^x\big]. \tag{6.8.4}$$

Since $|\tilde{\Gamma}_i| = \Gamma$, and the sign of the transverse spin magnitude does not determine the spin fluctuation probabilities, one can readily interpret, using (6.8.4), that the Hamiltonian (6.8.3) essentially represents a pure Ising system in a transverse field in terms of the pseudo-spin operators τ_i^α, and it has the same thermodynamics as that of the latter. There exists a phase transition in the pseudo-spin model from an ordered state ($\langle\tau^z\rangle \neq 0$) to a disordered state ($\langle\tau^z\rangle = 0$), although the configuration averaged longitudinal magnetisation in terms of the real spins (S) is zero ($m = \overline{\langle S^z\rangle} = \overline{\varepsilon_i}\langle\tau^z\rangle = 0$, where $\langle\cdots\rangle$ denotes the thermal average and over-head bar denotes the configuration average; $\overline{\varepsilon}_i = 0$) in either phase. This transition can be driven by both the temperature and the transverse field. For the finite temperature transition of the Mattis model in a transverse field, the exponents are the same as a pure d-dimensional classical Ising system, whereas the zero temperature transition has the critical behaviour identical to that of a pure Ising system in $(d + 1)$-dimension.

As mentioned earlier, in terms of the real spins, the magnetisation is always zero. However, the spin glass order parameter

$$Q = \overline{\big|\langle S_i^z\rangle\big|^2} = \big|\langle\tau_i^z\rangle\big|^2, \tag{6.8.5}$$

of the system shows a second-order phase transition driven by the temperature or the transverse field, with exponents related to the pure Ising exponents (in d and $d + 1$ dimensions, respectively) by (6.8.5). The zero field longitudinal susceptibility

$$\chi^{zz} = \frac{\beta}{N}\sum_{ij}\overline{\big(\langle S_i^z S_j^z\rangle - \langle S_i^z\rangle\langle S_j^z\rangle\big)} = \beta\big(1 - \langle\tau_i^z\rangle^2\big) \tag{6.8.6}$$

shows a cusp-like behaviour; the exponents for this asymmetric singularity in χ being also similarly related to the d- and $(d+1)$-dimensional Ising system exponents (for transition with respect to Γ and T) through the relation (6.8.6).

Appendix 6.A

6.A.1 The Vector Spin Glass Model

The Sherrington-Kirkpatrick model generalised to quantum spins (quantum vector spin glass) was introduced by Bray et al. [44], who applied the replica method to the Hamiltonian given by

$$H = -\sum_{ij} J_{ij} \boldsymbol{S}_i \cdot \boldsymbol{S}_j, \tag{6.A.1}$$

where the sum is over all pair of spins (the interaction is long-range), and the exchange interaction J_{ij} are independent random variables, with a symmetric Gaussian distribution

$$P(J_{ij}) = \left(\frac{N}{2\pi J^2}\right)^{1/2} \exp\left(-\frac{N J_{ij}^2}{2J^2}\right). \tag{6.A.2}$$

The spin operators satisfy the standard commutation relations,

$$\left[S_i^\alpha, S_j^\beta\right] = 2i\delta_{ij}\varepsilon_{\alpha\beta\gamma} S_k^\gamma. \tag{6.A.3}$$

Bray et al. [44] used replica method to handle the quenched disorder of the spin glass problem writing

$$\overline{\ln Z} = \lim_{n\to 0} \frac{\overline{Z^n} - 1}{n}. \tag{6.A.4}$$

The partition function of the vector spin glass model can be written as

$$Z^n = \operatorname{Tr} P \exp\left[\beta \int_0^1 d\tau \sum_{ij} J_{ij} \sum_{\alpha=1}^{n} \boldsymbol{S}_i^\alpha(\tau) \cdot \boldsymbol{S}_j^\alpha(\tau)\right]. \tag{6.A.5}$$

Performing the Gaussian average, one gets

$$\overline{Z^n} = \operatorname{Tr} P \exp\left[\frac{\beta^2 J^2}{2N^2} \int_0^1 d\tau \int_0^1 d\tau' \sum_{\alpha,\beta}\sum_{i,j} \left[\left(\boldsymbol{S}_i^\alpha(\tau) \cdot \boldsymbol{S}_j^\alpha(\tau)\right)\left(\boldsymbol{S}_i^\beta(\tau') \cdot \boldsymbol{S}_j^\beta(\tau')\right)\right]\right]. \tag{6.A.6}$$

Using the Hubbard-Stratonovitch transformation (6.3.18), one can simplify the above expression. One can then express the free energy in terms of the self-interaction $R^{\alpha\alpha}(\tau,\tau')$, spin glass order parameter $Q^{\alpha\beta}(\tau,\tau')$ and the quadrupolar order parameter $Q^{\alpha\alpha}(\tau,\tau')$. In the paramagnetic phase, the spin glass order parameter and the quadrupolar order parameter vanish, and one can derive an effective

Hamiltonian as in the case of a quantum Ising system. One then expands the total free energy in powers of $Q^{\alpha\beta}$ and obtains the value of T_c, setting the coefficient $(Q^{\alpha\beta})^2$ equal to zero, one gets the value of T_c. Bray and Moore established the existence of spin glass transition for all S and estimated the value of $T_c(S)$ given by the condition

$$J\chi_{\text{loc}} = 1; \quad \chi_{\text{loc}} = \beta \int_0^1 d\tau \int_0^1 d\tau' \langle T S^z(\tau) S^z(\tau') \rangle. \tag{6.A.7}$$

In the extreme quantum case $S = 1/2$, one can obtain, following Bray et al., the value of transition temperature from the paramagnetic phase o the spin glass ordered phase as

$$k_B T_c \sim J/4\sqrt{3}. \tag{6.A.8}$$

6.A.2 The Effective Classical Hamiltonian of a Transverse Ising Spin Glass

Let us consider the quantum transverse Ising spin glass Hamiltonian given by (6.3.1)

$$H = H_0 + V = -\Gamma \sum_i S_i^x - \sum_{ij} J_{ij} S_i^z S_j^z. \tag{6.A.9}$$

Let us consider the configuration averaged n-replicated partition function

$$\overline{Z^n} = \overline{\left[\exp(-\beta H)\right]^n}. \tag{6.A.10}$$

One can now transform the above n-replicated partition function in the following form

$$\overline{Z^n} = \overline{\text{Tr} \exp\left(-\beta \sum_{\alpha=1}^{n} H(\alpha)\right)} \tag{6.A.11}$$

where $H(\alpha)$ is the Hamiltonian of the α-th replica and it is separated in the following form

$$H(\alpha) = H_0(\alpha) + V(\alpha) = -\Gamma \sum_i S_{i\alpha}^x - \sum_{ij} J_{ij} S_{i\alpha}^z S_{j\alpha}^z. \tag{6.A.12}$$

Applying the Trotter formula (cf. Sect. 3.1), one now gets

$$\overline{Z^n} = \lim_{M\to\infty} \prod_{\alpha=1}^{n} \left[\exp\left(-\frac{\beta}{M} H_0(\alpha)\right) \exp\left(-\frac{\beta}{M} V(\alpha)\right)\right]^M. \tag{6.A.13}$$

Similarly as in the pure transverse Ising system, one introduces complete sets of eigenvectors of the operator S_α^z and using the relation

$$\langle S | \exp(\gamma S^x) | S' \rangle = \left[(1/2) \sinh 2\gamma\right]^{1/2} \exp\left(\frac{SS'}{2} \ln \coth \gamma\right), \tag{6.A.14}$$

one gets

$$\overline{Z^n} = \lim_{M\to\infty} \overline{\mathrm{Tr}\exp\left(\sum_{\alpha=1}^{n} H_{\mathrm{eff}}(\alpha)\right)} \tag{6.A.15}$$

where the effective classical Hamiltonian

$$H_{\mathrm{eff}} = \sum_{l=1}^{M}\left(K_M \sum_{i=1}^{N} S(\alpha)_{i,l} S(\alpha)_{i,l+1} + \sum_{\langle ij\rangle} K_{ij}^{(M)} S_{i,l}(\alpha) S_{j,l}(\alpha) + \ln c_M\right), \tag{6.A.16}$$

with

$$K_M = (1/2)\ln\coth(\beta\Gamma/M), \quad K_{ij}^{M} = \beta J_{ij}/M \quad \text{and}$$
$$c_M = \left\{(1/2)\sinh(2\beta\Gamma/M)\right\}^{MN/2}.$$

One can then perform the Gaussian averaging. This indicates an effective $(d+2)$-dimensional classical Hamiltonian [387].

6.A.3 *Effective Single-Site Hamiltonian for Long-Range Interacting RFTIM*

To derive the effective single-site Hamiltonian we consider the Hamiltonian of (long-range ferromagnetic) random field Ising model in a transverse field

$$H = -\frac{J}{N}\sum_{i\neq j} S_i^z S_j^z - \sum_i h_i S_i^z - \Gamma \sum_i S_i^x \tag{6.A.17}$$

where the random variable at each site satisfies a Gaussian distribution (6.7.19). The configuration averaged free energy of the system is given by

$$F = -kT\overline{\ln Z}, \tag{6.A.18}$$

where k is the Boltzmann constant and Z is the partition function for a particular realisation of the random fields. Using replica trick [39, 76, 136, 275] we can write the n-replicated free energy in the form

$$\begin{aligned} F &= -kT \lim_{n\to 0} \frac{\overline{Z^n} - 1}{n} \\ &= -kT \lim_{n\to 0}\left(\frac{1}{n}\left[\overline{\mathrm{Tr}\exp\left(-\beta\sum_{\alpha=1}^{n} H_0(\alpha)\right)}\right.\right. \\ &\quad \left.\left.\overline{\times P\exp\left(\int_0^{\beta} d\tau \sum_{\alpha=1}^{n}\sum_{ij}\frac{J}{N} S_{\alpha i}^z(\tau) S_{\alpha j}^z(\tau) + \sum_i h_i S_{\alpha i}^z(\tau)\right)}\right] - 1\right), \end{aligned} \tag{6.A.19}$$

where α denotes the α-th replica, P denotes the time ordering, $H_0(\alpha) = -\Gamma \sum_i S^x_{\alpha i}$ and $S^z(\tau)$'s are operators in the interaction representation. We can now perform the configuration averaging to obtain

$$F = -kT \lim_{n\to 0}\left(\frac{1}{n}\left[\operatorname{Tr}\exp\left(-\beta\sum_{\alpha=1}^{n} H_0(\alpha)\right)\right.\right.$$
$$\times P\exp\left(\int_0^\beta d\tau \sum_{\alpha=1}^{n}\sum_{ij}\frac{J}{N}S^z_{\alpha i}(\tau)S^z_{\alpha j}(\tau)\right.$$
$$\left.\left.\left.+\frac{\Delta^2}{2}\sum_i\left(\sum_{\alpha=1}^{n}\int_0^\beta d\tau\, S^z_{\alpha i}(\tau)\right)^2\right)\right] - 1\right). \qquad (6.A.20)$$

A Hubbard-Stratonovitch transformation simplifies the term

$$\exp\left[\int_0^\beta d\tau \sum_{\alpha=1}^{n}\sum_{ij}\frac{J}{N}S^z_{\alpha i}(\tau)S^z_{\alpha j}(\tau)\right] = \exp\left[\int_0^\beta d\tau\sum_{\alpha=1}^{n}\left|\sqrt{\frac{J}{N}}\sum_{i=1}^{N}S^z_{\alpha i}(\tau)\right|^2\right] \qquad (6.A.21)$$

(where the terms of order $(1/N)$ are neglected), so that we obtain the configuration averaged n-replicated free energy

$$F = -kT \lim_{n\to 0}\frac{1}{n}\int_{-\infty}^{\infty}\prod_{\alpha=1}^{n} dx_\alpha \operatorname{Tr}\exp\left(N\beta\sum_{\alpha=1}^{n}S^x_\alpha\right)$$
$$\times P\exp N\left(-\frac{\beta}{2}\sum_{\alpha=1}^{n}x_\alpha^2 + \sqrt{2J}\sum_{\alpha=1}^{n}x_\alpha\int_0^\beta S^z_\alpha(\tau) + \frac{\Delta^2}{2}\left(\int_0^\beta S^z_\alpha(\tau)\right)^2\right) - 1, \qquad (6.A.22)$$

where x^α s are dummy variables. In the $N\to\infty$ limit, one can readily obtain the saddle point configuration averaged free energy

$$F = -kT \lim_{n\to 0}\frac{1}{n}\left[-\frac{\beta}{2}\sum_{\alpha=1}^{n}x_\alpha^2 + \ln\operatorname{Tr}\exp(A)\right] \qquad (6.A.23)$$

where

$$\exp(A) = \exp\left(\beta\Gamma\sum_{\alpha=1}^{n}S^x_\alpha\right)$$
$$\times P\exp\left(\sqrt{2J}\sum_{\alpha=1}^{n}x_\alpha\int_0^\beta d\tau S^z_\alpha(\tau) + \frac{\Delta^2}{2}\left(\sum_{\alpha=1}^{n}\int_0^\beta d\tau S^z_\alpha(\tau)\right)^2\right). \qquad (6.A.24)$$

The square term appearing in the above expression can be simplified using once again the Hubbard-Stratonovitch transformation to obtain

$$\exp(A) = \int_{-\infty}^{\infty} \frac{ds}{\sqrt{2\pi}} \exp\left(-\frac{s^2}{2}\right) \exp\left(\beta\Gamma \sum_{\alpha=1}^{n} S_\alpha^x\right)$$
$$\times P \exp\left(\sqrt{2J} \sum_{\alpha=1}^{n} x_\alpha \int_0^\beta S_\alpha^z(\tau) d\tau + s\Delta \int_0^\beta d\tau \sum_{\alpha=1}^{n} S_\alpha^z\right), \tag{6.A.25}$$

where s is a dummy variable. Finally, one obtains the form of free energy (with $x = m^z\sqrt{2J}$ and $s\delta = h$) given by

$$F = -kT\left[-J(m^z)^2\beta + \int_{-\infty}^{\infty} \frac{dh}{\sqrt{2\pi\Delta^2}} \exp\left(-\frac{h^2}{2}\right) \ln \mathrm{Tr} \exp(\beta\gamma S^x) P\right.$$
$$\left.\times \exp\left((2m^z + h) \int_0^\beta S^z(\tau)\, d\tau\right)\right]$$
$$= -kT\left[-J(m^z)^2\beta + \int_{-\infty}^{\infty} dh\, P(h) \ln \mathrm{Tr} \exp\left(\beta\left(\Gamma S^x + (2m^z J + h)S^z\right)\right)\right]. \tag{6.A.26}$$

We have thus reduced the many-body Hamiltonian (in the $N \to \infty$ limit) to an effective single-site problem, where the molecular field at each site is given by $(2m^z J + h)$ where h is distributed with a probability distribution $P(h)$.

6.A.4 Mapping of Random Ising Antiferromagnet in Uniform Longitudinal and Transverse Fields to RFTIM

The equivalence between the transition in the random Ising antiferromagnet in uniform transverse and longitudinal fields (RIAFTL) to that in the random field transverse Ising model is obtained by employing semi-classical decimation of the one sublattice of the RIAFTL system, which neglects commutators between the spin operators. Here a partial trace is done over sites of one sub-lattice, e.g., that in which the site label i is odd. The original (reduced) Hamiltonian

$$-\beta H = \sum_i \left(-K_{i,i+1} S_i^z S_{i+1}^z + h_i S_i^z + \Gamma S_i^x\right) = \sum_i H_i \tag{6.A.27}$$

is mapped into a new form

$$-\beta H' = \sum_i \left(-K'_{2i,2i+2} S_{2i}^z S_{2i+2}^z + h'_{2i} S_{2i}^z + \Gamma' S_i^x\right) = \sum_i H'_{2i}. \tag{6.A.28}$$

The trace over S_{2i+1} produces the factors

$$b\left(S_{2i}^z, S_{2i+2}^z\right) = 2\cosh\left[\left(h_{2i+1} - K_{2i,2i+1} S_{2i}^z - K_{2i+1,2i+2} S_{2i+2}^z\right)^2 + \Gamma^2\right]^{1/2}. \tag{6.A.29}$$

This can be written as

$$\exp\bigl(A + BS^z_{2i} + CS^z_{2i+2} + DS^z_{2i}S^z_{2i+2}\bigr) \tag{6.A.30}$$

where matching of the expression for all four possible sets of values for (S^z_{2i}, S^z_{2i+2}) gives A, B, C, D in terms of Γ, h_{2i+1}, $K_{2i,2i+1}$, $K_{2i+1,2i+2}$. For example

$$\begin{aligned} B &= \frac{1}{4}\ln\left[\frac{b(1,1)b(1,-1)}{b(-1,-1)b(-1,1)}\right] \equiv B(K_{2i,i+1}, K_{2i,2i+2}) \\ &= C(K_{2i+1,2i+2}, K_{2i,2i+2}) \end{aligned} \tag{6.A.31}$$

$$D = \frac{1}{4}\ln\left[\frac{b(1,1)b(-1,-1)}{b(1,-1)b(-1,1)}\right] \equiv D(K_{2i,2i+1}, K_{2i,2i+2}), \tag{6.A.32}$$

so that we arrive at the recursion relations (6.7.25), (6.7.26) and (6.7.27). For $h \ll K$, one can evaluate $B(K_1, K_2)$ (where K_1 and K_2 are two neighbouring bonds), using the simplified relations

$$b(1,1) = 2\cosh\left[\Omega_+ - h\frac{\Lambda_+}{\Omega_+}\right] \tag{6.A.33}$$

$$b(1,-1) = 2\cosh\left[\Omega_- - h\frac{\Lambda_-}{\Omega_-}\right] \tag{6.A.34}$$

$$b(-1,-1) = 2\cosh\left[\Omega_+ + h\frac{\Lambda_+}{\Omega_+}\right] \tag{6.A.35}$$

$$b(-1,1) = 2\cosh\left[\Omega_+ + h\frac{\Lambda_-}{\Omega_-}\right] \tag{6.A.36}$$

where

$$\Omega_\pm = \left[\Lambda^2_\pm + \Gamma^2\right]^{1/2}, \qquad \Lambda_\pm = K_1 \pm K_2. \tag{6.A.37}$$

Hence

$$B(K_1, K_2) = \frac{1}{4}\ln\left[\frac{\cosh(\Omega_+ - h\frac{\Lambda_+}{\Omega_+})\cosh(\Omega_- - h\frac{\Lambda_-}{\Omega_-})}{\cosh(\Omega_+ + h\frac{\Lambda_+}{\Omega_+})\cosh(\Omega_- + h\frac{\Lambda_-}{\Omega_-})}\right]. \tag{6.A.38}$$

If we now use the relation (for small h)

$$\ln\left[\frac{\cosh(\alpha + \gamma h)}{\cosh(\alpha - \gamma h)}\right] = 2\gamma h \tanh\alpha + \cdots \tag{6.A.39}$$

we get

$$B(K_1, K_2) = -\frac{h}{2}\left[\frac{\Lambda_+}{\Omega_+}\tanh\Omega_+ + \frac{\Lambda_-}{\Omega_-}\tanh\Omega_-\right] + O\bigl(h^2\bigr), \tag{6.A.40}$$

etc.

6.A.5 Derivation of Free Energy for the SK Model with Antiferromagnetic Bias in a Transverse Field

We show the derivation of the free energy per spin for the system to be described by the Hamiltonian.

$$H = -\sum_{ij} J_{ij} S_i^z S_j^z - \Gamma \sum_i S_i^x. \tag{6.A.41}$$

Carrying out the Suzuki-Trotter decomposition, we have the replicated partition function.

$$Z_M^n = \mathrm{tr}_{\{S\}} \exp\left[\frac{\beta}{M} \sum_{ij,k} \sum_{\alpha} J_{ij} S_i^\alpha(k) S_j^\alpha(k) + B \sum_{i,k,\alpha} S_i^\alpha(k) S_i^\alpha(k+1)\right] \tag{6.A.42}$$

$$B = \frac{1}{2} \ln \coth\left(\frac{\beta\Gamma}{M}\right) \tag{6.A.43}$$

where α and k denote the replica and Trotter indices. M is the number of the Trotter slices and β is the inverse temperature. The disorder J_{ij} obeys

$$P(J_{ij}) = \frac{1}{\sqrt{2\pi J^2}} \exp\left[-\frac{(J_{ij} - j_0)^2}{2J}\right]. \tag{6.A.44}$$

In other words, the J_{ij} follows

$$J_{ij} = j_0 + Jx, \qquad P(x) = \frac{1}{\sqrt{2\pi}} \mathrm{e}^{-\frac{x^2}{2}}. \tag{6.A.45}$$

We should notice that $j_0 > 0$, $J = 0$ is pure ferromagnetic transverse Ising model, whereas $j_0 < 0$, $J = 0$ corresponds to pure antiferromagnetic transverse Ising model. Then, by using $\int_{-\infty}^{\infty} Dx\, \mathrm{e}^{ax} = \mathrm{e}^{a^2/2}$, $Dx \equiv dx\, \mathrm{e}^{-x^2/2}/\sqrt{2\pi}$, we have the average of the replicated partition function as

$$\begin{aligned}
&\langle\langle Z_M^n \rangle\rangle \\
&= \mathrm{tr}_{\{S\}} \exp\left[\frac{\beta j_0}{M} \sum_k \sum_\alpha \sum_{ij} S_i^\alpha(k) S_j^\alpha(k) + B \sum_k \sum_\alpha \sum_i S_i^\alpha(k) S_i^\alpha(k+1)\right] \\
&\quad \times \exp\left[\frac{\beta^2 \tilde{J}^2}{2M^2} \sum_{ij} \left(\sum_k \sum_\alpha S_i^\alpha(k) S_j^\alpha(k)\right)^2\right] \\
&= \mathrm{tr}_{\{S\}} \exp\left[\frac{\beta j_0}{M} \sum_k \sum_\alpha \sum_{ij} S_i^\alpha(k) S_j^\alpha(k) + B \sum_k \sum_\alpha \sum_i S_i^\alpha(k) S_i^\alpha(k+1)\right] \\
&\quad \times \exp\left[\frac{\beta^2 \tilde{J}^2}{2M^2} \sum_{kk'} \sum_{\alpha\beta} \sum_{ij} S_i^\alpha(k) S_j^\alpha(k) S_i^\beta(k') S_j^\beta(k')\right]
\end{aligned} \tag{6.A.46}$$

where the bracket was defined as $\ll \cdots \gg = \int \prod_{ij} dJ_{ij} P(J_{ij})(\cdots)$. To take a proper thermodynamic limit, we use the scaling

$$j_0 = \frac{J_0}{N}, \qquad J = \frac{\tilde{J}}{\sqrt{N}}. \tag{6.A.47}$$

For this rescaling of the parameters, the averaged replicated partition function $\ll Z_M^n \gg$ reads

$$\begin{aligned}
\ll Z_M^n \gg &= \mathrm{tr}_{\{S\}} \int_{-\infty}^{\infty} \prod_k \prod_\alpha \frac{dm_\alpha(k)}{\sqrt{2\pi M/\beta J_0 N}} \int_{-\infty}^{\infty} \prod_{kk'} \prod_{\alpha\beta} \frac{dq_{\alpha\beta}(k,k')}{\sqrt{2\pi M/\beta \tilde{J}\sqrt{N}}} \\
&\times \int_{-\infty}^{\infty} \prod_{kk'} \prod_\alpha \frac{d\tilde{q}_{\alpha\alpha}(k,k')}{\sqrt{2\pi M/\beta \tilde{J}\sqrt{N}}} \\
&\times \exp\left[-\frac{\beta J_0 N}{2M} \sum_k \sum_\alpha m_\alpha(k)^2 - \frac{(\beta\tilde{J})^2 N}{2M^2} \sum_{kk'} \sum_{\alpha\beta} q_{\alpha\beta}(k,k')^2 \right. \\
&\left. - \frac{(\beta\tilde{J})^2 N}{2M^2} \sum_{kk'} \sum_\alpha \tilde{q}_{\alpha\alpha}(k,k')^2 \right] \\
&\times \exp\left[\frac{\beta J_0}{M} \sum_k \sum_\alpha m_\alpha(k) \sum_i S_i^\alpha(k) \right. \\
&+ \left(\frac{\beta\tilde{J}}{M}\right)^2 \sum_{kk'} \sum_{\alpha\beta} q_{\alpha\beta}(k,k') \sum_i S_i^\alpha(k) S_i^\beta(k') \\
&+ \left(\frac{\beta\tilde{J}}{M}\right)^2 \sum_{kk'} \sum_\alpha \tilde{q}_{\alpha\alpha}(k,k') \sum_i S_i^\alpha(k) S_i^\alpha(k') \\
&\left. + B \sum_k \sum_\alpha \sum_i S_i^\alpha(k) S_i^\alpha(k+1) \right].
\end{aligned} \tag{6.A.48}$$

We next assume the replica symmetry and static approximations such as

$$m_\alpha(k) = \langle S^\alpha(k) \rangle = \frac{1}{N} \sum_i S_i^\alpha(k) = m \tag{6.A.49}$$

$$q_{\alpha\beta}(k,k') = \langle S^\alpha(k) S^\beta(k') \rangle = \frac{1}{N} \sum_i S_i^\alpha(k) S_i^\beta(k') = q \tag{6.A.50}$$

$$\tilde{q}_{\alpha\alpha}(k,k') = \langle S^\alpha(k) S^\alpha(k') \rangle = \frac{1}{N} \sum_i S_i^\alpha(k) S_i^\alpha(k') = \tilde{q}. \tag{6.A.51}$$

Then, we should notice the relation:

$$\begin{aligned}
&\left(\frac{\beta\tilde{J}}{M}\right)^2 q \sum_{kk'} \sum_{\alpha\beta} \sum_i S_i^\alpha(k) S_i^\beta(k') \\
&= \left(\frac{\beta\tilde{J}}{M}\right)^2 q \sum_i \left(\sum_k \sum_\alpha S_i^\alpha(k) \right)^2 - \left(\frac{\beta\tilde{J}}{M}\right)^2 q \sum_i \sum_\alpha \left(\sum_k S_i^\alpha(k) \right)^2
\end{aligned} \tag{6.A.52}$$

$$\left(\frac{\beta\tilde{J}}{M}\right)^2 \tilde{q} \sum_{kk'} \sum_{\alpha} \sum_{i} S_i^{\alpha}(k) S_i^{\alpha}(k') = \left(\frac{\beta\tilde{J}}{M}\right)^2 \tilde{q} \sum_{i} \sum_{\alpha} \left(\sum_{k} S_i^{\alpha}(k)\right)^2. \tag{6.A.53}$$

To take into account the above relations, we obtain in the limit of $N \to \infty$ as

$$\begin{aligned}
&\ll Z_M^n \gg \\
&= \int_{-\infty}^{\infty} \prod_{k} \prod_{\alpha} \frac{dm_\alpha(k)}{\sqrt{2\pi M/\beta J_0 N}} \int_{-\infty}^{\infty} \prod_{kk'} \prod_{\alpha\beta} \frac{dq_{\alpha\beta}(k,k')}{\sqrt{2\pi M/\beta \tilde{J}\sqrt{N}}} \\
&\times \int_{-\infty}^{\infty} \prod_{kk'} \prod_{\alpha} \frac{d\tilde{q}_{\alpha\alpha}(k,k')}{\sqrt{2\pi M/\beta \tilde{J}\sqrt{N}}} \\
&\times \exp\left[nN\left(-\frac{\beta J_0}{2}m^2 + \frac{(\beta\tilde{J})^2}{4}q^2 - \frac{(\beta\tilde{J})^2}{4}\tilde{q}^2\right.\right. \\
&\left.\left. + \int_{-\infty}^{\infty} Dy \ln \int_{-\infty}^{\infty} Du\, 2\cosh\beta\sqrt{(J_0 m + \tilde{J}\sqrt{q}\,y + \tilde{J}\sqrt{\tilde{q}-q}\,u)^2 + \Gamma^2}\right)\right] \\
&\simeq \exp[nNf].
\end{aligned} \tag{6.A.54}$$

Therefore, the following f is regarded as free energy per spin by the definition of replica theory

$$\begin{aligned}
f =& -\frac{\beta J_0}{2}m^2 + \frac{(\beta\tilde{J})^2}{4}q^2 - \frac{(\beta\tilde{J})^2}{4}\tilde{q}^2 \\
& + \int_{-\infty}^{\infty} Dy \ln \int_{-\infty}^{\infty} Du\, 2\cosh\beta\sqrt{(J_0 m + \tilde{J}\sqrt{q}\,y + \tilde{J}\sqrt{\tilde{q}-q}\,u)^2 + \Gamma^2}.
\end{aligned} \tag{6.A.55}$$

6.A.5.1 Saddle Point Equations

For simplicity, we define

$$b = J_0 m + \tilde{J}\sqrt{q}\,y + \tilde{J}\sqrt{\tilde{q}-q}\,u \tag{6.A.56}$$

$$\Theta = \sqrt{b^2 + \Gamma^2}. \tag{6.A.57}$$

Then, we have the following simplified free energy

$$f = -\frac{\beta J_0}{2}m^2 + \frac{(\beta\tilde{J})^2}{4}\left(q^2 - \tilde{q}^2\right) + \int_{-\infty}^{\infty} Dy \ln \int_{-\infty}^{\infty} Du\, 2\cosh\beta\Theta. \tag{6.A.58}$$

The saddle point equations are derived as follows [159, 401].

$$m = \int_{-\infty}^{\infty} Dy \left[\frac{\int_{-\infty}^{\infty} Du (\frac{b}{\Theta}) \sinh \beta\Theta}{\int_{-\infty}^{\infty} Du \cosh \beta\Theta} \right] \tag{6.A.59}$$

$$q = \int_{-\infty}^{\infty} Dy \left[\frac{\int_{-\infty}^{\infty} Du (\frac{b}{\Theta}) \sinh \beta\Theta}{\int_{-\infty}^{\infty} Du \cosh \beta\Theta} \right]^2 \tag{6.A.60}$$

$$\tilde{q} = \int_{-\infty}^{\infty} Dy \left[\frac{\int_{-\infty}^{\infty} Du \{ (\frac{b^2}{\Theta^2}) \cosh \beta\Theta + \frac{\Gamma\beta^{-1}}{\Theta^3} \sinh \beta\Theta \}}{\int_{-\infty}^{\infty} Du \cosh \beta\Theta} \right] \tag{6.A.61}$$

$$m_x = \frac{\partial f}{\partial \Gamma} = \int_{-\infty}^{\infty} Dy \left[\frac{\int_{-\infty}^{\infty} Du (\frac{\Gamma}{\Theta}) \sinh \beta\Theta}{\int_{-\infty}^{\infty} Du \cosh \beta\Theta} \right]. \tag{6.A.62}$$

6.A.5.2 At the Ground State

We first should notice that $\tilde{q}$ is always larger than q. In fact, we can easily show that

$$\begin{aligned} \tilde{q} &= \int_{-\infty}^{\infty} Dy \left[\frac{\int_{-\infty}^{\infty} Du (\frac{b^2}{\Theta^2}) \cosh \beta\Theta}{\int_{-\infty}^{\infty} Du \cosh \beta\Theta} \right] \\ &\geq \int_{-\infty}^{\infty} Dy \left[\frac{\int_{-\infty}^{\infty} Du (\frac{b^2}{\Theta^2}) \sinh \beta\Theta}{\int_{-\infty}^{\infty} Du \cosh \beta\Theta} \right] \\ &\geq \int_{-\infty}^{\infty} Dy \left[\frac{\int_{-\infty}^{\infty} Du (\frac{b^2}{\Theta^2}) \sinh \beta\Theta}{\int_{-\infty}^{\infty} Du \cosh \beta\Theta} \right]^2 = q. \end{aligned} \tag{6.A.63}$$

Then, we consider the limit of $\beta \to \infty$. If $\tilde{q} - q = \varepsilon \geq 0$ is of order 1 object, the free energy f diverges in the limit of $\beta \to \infty$ as $(\beta\tilde{J})^2(q^2 - \tilde{q}^2)/4$. Therefore, we conclude that $q = \tilde{q}$ should be satisfied in the limit of $\beta \to \infty$ and we obtain the saddle pint equation at the ground state as

$$m = \int_{-\infty}^{\infty} Dy \left(\frac{b}{\Theta} \right) = \int_{-\infty}^{\infty} Dy \frac{(J_0 m + \tilde{J}\sqrt{q}\, y)}{\sqrt{(J_0 m + \tilde{J}\sqrt{q}\, y)^2 + \Gamma^2}} \tag{6.A.64}$$

$$q = \tilde{q} = \int_{-\infty}^{\infty} Dy \left(\frac{b}{\Theta} \right)^2 = \int_{-\infty}^{\infty} Dy \left\{ \frac{(J_0 m + \tilde{J}\sqrt{q}\, y)}{\sqrt{(J_0 m + \tilde{J}\sqrt{q}\, y)^2 + \Gamma^2}} \right\}^2 \tag{6.A.65}$$

$$m_x = \int_{-\infty}^{\infty} Dy \left(\frac{\Gamma}{\Theta} \right) = \int_{-\infty}^{\infty} Dy \frac{\Gamma}{\sqrt{(J_0 m + \tilde{J}\sqrt{q}\, y)^2 + \Gamma^2}}. \tag{6.A.66}$$

Chapter 7
Dynamics of Quantum Ising Systems

7.1 Tunnelling Dynamics for Hamiltonians Without Explicit Time Dependence

A Hamiltonian with classical Ising spins has no intrinsic dynamics. It is necessary to introduce dynamics separately through equations of motion that mimic the effect of coupling the spins to other degrees of freedom connected to the heat bath. On the other hand, there is intrinsic dynamics in a quantum Ising system as it contains non-commuting terms in the Hamiltonian. Uptill now, the (equilibrium) static properties of such systems have been described and in this chapter we intend to discuss the various aspects of the "tunnelling dynamics". We first consider the cases when the transverse field is constant in time, so that there is no explicit time dependence of the Hamiltonian. We then come to the cases of quantum quench, quantum hysteresis, and etc., where the transverse field (tunnelling term) changes in time.

7.1.1 Dynamics in Ising Systems: Random Phase Approximation

Let us consider the quantum Ising model given by

$$H = -\sum_{ij} J_{ij} S_i^z S_j^z - \Gamma \sum_i S_i^x. \tag{7.1.1}$$

As discussed in Sect. 1.2, in the mean field approximation, the effective field is written as

$$\boldsymbol{h} = J(0)\langle S_i^z \rangle \hat{z} + \Gamma \hat{x}. \tag{7.1.2}$$

The equilibrium values are given by the Weiss equation

$$\langle S^z \rangle = \frac{J(0)\langle S^z \rangle}{|h|} \tanh \beta |h| \tag{7.1.3}$$

$$\langle S^x \rangle = \frac{\Gamma}{|h|} \tanh \beta |h| \tag{7.1.4}$$

S. Suzuki et al., *Quantum Ising Phases and Transitions in Transverse Ising Models*, Lecture Notes in Physics 862, DOI 10.1007/978-3-642-33039-1_7,

where $|h| = \sqrt{\Gamma^2 + (J(0)\langle S^z \rangle)^2}$ and $J(q) = \sum_{ij} J_{ij} \exp[i\boldsymbol{q} \cdot (\boldsymbol{R}_i - \boldsymbol{R}_j)]$. Hence, the transition temperature (where $\langle S^z \rangle = 0$) is given by

$$1 = \frac{J(0)}{\Gamma \tanh \beta_c \Gamma}. \tag{7.1.5}$$

Small fluctuations around the mean state are considered in random phase approximation [48]. With the fluctuating coordinates

$$\sum_i S_i^\alpha \exp(i\boldsymbol{q} \cdot \boldsymbol{R}_i) \equiv S_q^\alpha, \tag{7.1.6}$$

the equations of motion (obtained from $\dot{S}_i^\mu = i[H, S_i^\mu]$) can be written as

$$\boldsymbol{S}_q = 2\boldsymbol{M}_q \boldsymbol{S}_q \tag{7.1.7}$$

where $\boldsymbol{M}_q$ is given by

$$\boldsymbol{M}_q = \begin{pmatrix} 0 & -J(0)\langle S^z \rangle & 0 \\ J(0)\langle S^z \rangle & 0 & J(q)\langle S_0^x \rangle - \Gamma \\ 0 & \Gamma & 0 \end{pmatrix}. \tag{7.1.8}$$

The eigenfrequencies are therefore given by

$$\omega_q^2 = 4\Gamma\big(\Gamma - J(q)\langle S^x \rangle\big) + 4J^2(0)\langle S^x \rangle^2. \tag{7.1.9}$$

It is thus found that $\omega_0 \sim (T - T_c)$ as $T \to T_c$. Associated with this phenomenon (softening of the long wavelength mode), there will be a divergence in the susceptibility, like $(T - T_c)^{-1}$.

For $T < T_c$, the first term of (7.1.9) vanishes, and

$$\omega_0 = 2J(0)\langle S^z \rangle. \tag{7.1.10}$$

For finite q, ω_q has a discontinuity in slope at $T = T_c$, and $\omega_q(T = T_c) \sim q$ for small q. The modes at high temperature behave as tunnelling modes.

7.1.2 Dynamics in Dilute Ising Spin Systems

In this section we will consider the dynamics in dilute Ising systems [239]. Experimentally, it is possible to build up crystals where part or whole of the active components can be substituted by other active components with different transverse fields and Ising interactions or by inactive components which play the role of impurities diluting the active component. We can consider a crystal where each lattice site is one or other type of spin and characterised by a particular value of the transverse field and by a dipole like parameter μ_i. The direct interaction J_{ij} can then be written as $J_{ij} = \mu_i T_{ij} \mu_j$. Thus the interaction may differ from point to point. We consider then the Green function $\chi_{ij} = \langle [\mu_i S_i^z; \mu_j S_j^z(t)] \rangle$ where $[A, B]$ denotes the commutator bracket and $\langle \cdots \rangle$ the thermal average (here $S^z(t) = \exp(-iHt) S^z \exp(iHt)$

denotes the time evolved S^z). Decoupling its equations of motion in RPA (see Sect. 1.2) above the transition ($\langle S^z \rangle = 0$) one gets

$$\chi_{ij} = \phi_i \delta_{ij} - \phi_i \sum_k T_{ik} \chi_{ik} \tag{7.1.11}$$

where

$$\phi_i = 2\Gamma_i \mu_i^2 \langle S_i^x \rangle / (\omega^2 - 2\Gamma_i^2). \tag{7.1.12}$$

The above approximation is valid in the case of large transverse field and provided the system is not very near the transition point. ϕ is a local (site dependent) parameter, dependent also on the temperature and frequency. Through its coupling with χ, the physical quantities (like $\langle S^z \rangle$) of course depend on the configuration of the entire system. In the effective medium theory, the value of average χ_{ij} (written as $\overline{\chi}_{ij}$) is written as

$$\overline{\chi}_{ij} = \overline{\phi} \delta_{ij} - \overline{\phi} \sum_k T_{ik} \overline{\chi}_{ik} \tag{7.1.13}$$

and the effective medium average is determined self-consistently. For this, let us put the impurity at the origin. One can then write [239]

$$\chi_{ij}^{(0)} = \overline{\chi}_{ij} + \overline{G}_{i0} f \overline{G}_{0j} \tag{7.1.14}$$

for the Green function at the origin where $\phi_i = \overline{\phi} + \varepsilon_i$ with $\varepsilon_i = (\phi^{(0)} - \overline{\phi})\delta_{i0}$. Here

$$\overline{G} = (1 + \overline{\phi} T)^{-1}, \tag{7.1.15}$$

$$f = (1 + \varepsilon g)^{-1} \varepsilon, \tag{7.1.16}$$

with

$$g = (T\overline{G})_{00} = (1/N) \sum_q T_q / (1 + \overline{\phi} T_q). \tag{7.1.17}$$

We next complete the definition of effective medium by requiring that the average of χ_{ij} over all possible types of impurities equals the effective medium Green function; that is: $\langle f \rangle = 0$, which constitutes the desired relation between $\overline{\phi}$ and $\phi^{(0)}$. This can be written in a more explicit form

$$\overline{\phi} = \langle \phi^{(0)} / [1 + (\phi^{(0)} - \overline{\phi}) g] \rangle. \tag{7.1.18}$$

Finally, the expression of the Green function beginning or ending at the impurity site is given by

$$\chi_{0j}^{(0)} = \frac{\phi^{(0)}}{1 + (\phi^{(0)} - \overline{\phi}) g} \overline{G}_{0j}. \tag{7.1.19}$$

Using now the fluctuation-dissipation theorem

$$\frac{1}{2\pi} \int \coth(\beta\omega/2) \operatorname{Im} \chi_{00}^{(0)}(\omega)\, d\omega = \mu_0^2/4, \tag{7.1.20}$$

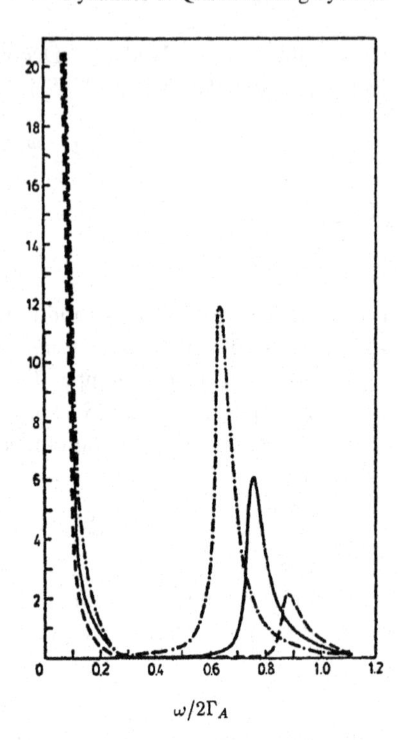

Fig. 7.1 Typical behaviour of the spectral function versus reduced frequency at $T_c(c)$ with $a = 4\Gamma_A/J(0)\mu_A^2 = 0.8$, $\alpha = \mu_B/\mu_A = 1.2$, $\gamma = \Gamma_B/\Gamma_A = 0.2$ and $c = 0.64$ (*chain curve*), $= 0.44$ (*broken curve*) (from [239])

which constitutes the sum rule, physical quantities like $\langle S_0^x \rangle$ can be calculated self-consistently. The generalised susceptibility $\overline{\chi}_q(\omega)$ can then be obtained. The spectral function $\rho(\overline{q}, \omega)$ is related to its imaginary part through

$$\rho(\overline{q}, \omega) = \left[1 - \exp(-\beta\omega)\right]^{-1} \overline{\chi}_q''(\omega). \tag{7.1.21}$$

Self-consistent calculations of (7.1.14) and (7.1.18) have been performed for a two component (A and B) system with concentrations $c_A = c$ and $c_B = 1 - c$, with $\Gamma_A > \Gamma_B$, assuming a random distribution of spins in a simple cubic lattice with nearest neighbour interaction. The spectral function and susceptibility have been obtained as functions of the concentration, ratios of the transverse fields, dipole moments of A and B and other relevant parameters at the transition temperature. Typical behaviour of the $q = 0$ spectral function is shown in Fig. 7.1.

7.1.3 Dynamics in Quantum Ising Glasses

The relaxational dynamics in glasses is of great significance as the presence of disorder is known to give rise to unusual time-dependent characteristics such as stretched exponential decay of correlation functions and non-Debye behaviour of response functions. Quantum effects become important at very low temperatures as different parts of the free energy surface can be linked by tunnelling. It has even been possible

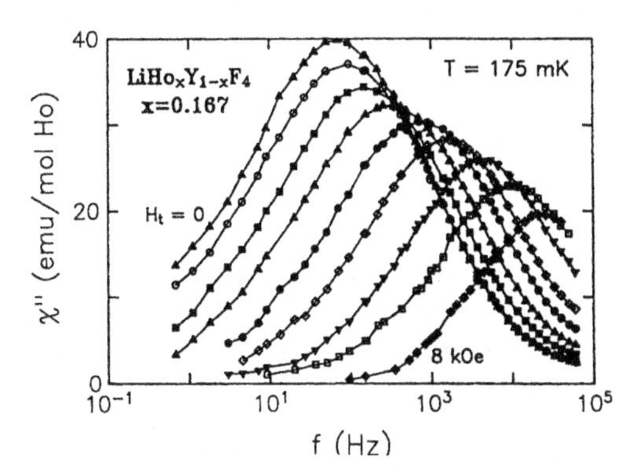

Fig. 7.2 Imaginary part of the susceptibility as a function of frequency in an experimental realisation of a quantum spin glass for a series of transverse magnetic fields in 1 kOe intervals (from [427])

to study the simplest prototype of quantum spin glasses [427] experimentally. The imaginary component of the dynamic susceptibility $\chi''(\omega)$ has been measured as a function of both ω and the strength of the transverse field. The time scale of the system's response is radically affected by the presence of the transverse field. The value of the frequency ω_p, where $\chi''(\omega)$ is peaked, increases by orders of magnitudes as the transverse field is increased (see Fig. 7.2).

The dynamics of a quantum spin glass model described by the Sherrington-Kirkpatrick model, in a transverse field has recently been analysed [21]. The relaxation dynamics has been studied by coupling the quantum subsystem to a purely dissipative heat bath. The total Hamiltonian is composed of H_S, the system Hamiltonian, H_B, the bath Hamiltonian and H_I, which describes the interaction between the spin subsystems and the heat bath. The system Hamiltonian is given by

$$H_s = -\sum_{ij} J_{ij} S_i^z S_j^z - \Gamma \sum_i S_i^x. \tag{7.1.22}$$

Here the spins i and j are connected by random exchange interactions J_{ij} which are assumed to be Gaussian distributed.

In the mean field approximation, H_s is represented as a single spin Hamiltonian:

$$H_s = -h^z S^z - \Gamma S^x \tag{7.1.23}$$

where h^z is an effective field acting along the z axis and is due to the nonzero spin glass order parameter q. Using the thermofield-dynamical approach, h^z can be represented as [234] (see also Sect. 6.3)

$$h^z(r) = \frac{1}{2} J r \sqrt{q} \tag{7.1.24}$$

where r is the excess static noise arising from the random interaction J_{ij}. The mean field equations for the local polarisation $m(r)$ and the spin glass order parameter q are

$$m(r) = p(r) \tanh\bigl[\beta \bigl|h(r)\bigr|\bigr] \tag{7.1.25}$$

and

$$q = \int_{-\infty}^{\infty} \frac{dr}{\sqrt{2\pi}} \exp\bigl[-r^2/2\bigr] m^2(r) \tag{7.1.26}$$

with

$$\left|h(r)\right| = \sqrt{\Gamma^2 + \left(h^z(r)\right)^2} \tag{7.1.27}$$

and

$$p(r) = h^z(r)/\left|h(r)\right|. \tag{7.1.28}$$

With $\Gamma = 0$, it is customary to imagine that the dynamics arises due to the additional coupling terms to the heat bath which are off-diagonal in the representation where S^z is diagonal. This is then in the spirit of the kinetic Ising model of the Glauber kind [156, 172, 219–221] (mimicking thermally activated jumps between the ground states). With $\Gamma \neq 0$, we expect the heat bath to induce not only thermal fluctuations but quantum fluctuations as well, leading to incoherence in tunnelling. The appropriate interaction Hamiltonian should, therefore, be of the form

$$H_I = g\hat{b}\left[\frac{h^z}{|h|}S^x + \frac{\Gamma}{|h|}S^z\right], \tag{7.1.29}$$

where $\hat{b}$ is an operator acting on the Hilbert space of the heat bath Hamiltonian H_B and g is a multiplicative coupling constant. The susceptibility due to an oscillatory magnetic field applied along z-axis is

$$\chi(\omega) = \frac{1}{2}\beta \lim_{\delta\to 0}\left[\frac{1}{s} - 4\tilde{c}(s)\right], \tag{7.1.30}$$

where $\tilde{c}(s)$ is the Laplace transform of the correlation function $c(t) = \langle S^z(0)S^z(t)\rangle$. The quantity s is related to the applied frequency: $s = -i\omega + \delta$ and β is the inverse temperature. Explicitly, $c(t)$ is expressed as

$$c(t) = \frac{1}{Z}\mathrm{Tr}\left[\mathrm{e}^{-\beta H_{\mathrm{tot}}} S^z(0)\mathrm{e}^{i H_{\mathrm{tot}} t} S^z(0)\mathrm{e}^{-i H_{\mathrm{tot}} t}\right] \tag{7.1.31}$$

where

$$H_{\mathrm{tot}} = H_s + H_B + H_I. \tag{7.1.32}$$

With H_{tot}, defined as in (7.1.32), one can evaluate $\tilde{c}(s)$. From $\tilde{c}(s)$, the imaginary component of the AC susceptibility is expressed as

$$\chi''(\omega) = \frac{1}{4\pi}\int_{-1}^{1} dm\, W(m)\chi''(\omega, m), \tag{7.1.33}$$

where the susceptibility for a given configuration of local polarisation m is given by

$$\begin{aligned}\chi''(\omega, m) = {} & \frac{\beta}{4}\left[\frac{\omega\lambda}{\omega^2 + \lambda^2}\right]\left[\frac{(h^z)^2}{|h|^2} - m^2\right] \\ & + \frac{2\omega\lambda|h|^2}{(|h|^2 - \omega^2)^2 + 4\omega^2\lambda^2}\frac{\Gamma^2}{|h|^2}\left[1 - \omega\frac{m}{h^z}\right].\end{aligned} \tag{7.1.34}$$

Here λ is a phenomenological relaxation rate and $W(m)$ is the local polarisation distribution given by

$$W(m) = \int dr \exp\left(-r^2/2\right)\delta\left(m - m(r)\right). \tag{7.1.35}$$

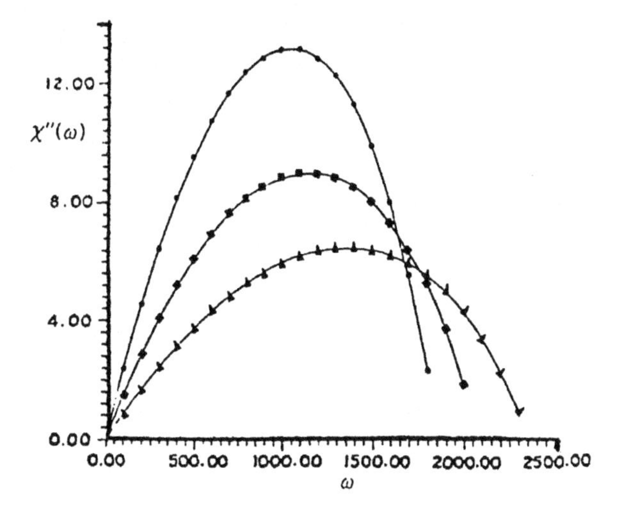

Fig. 7.3 The variation of the imaginary component of the susceptibility with the frequency ω around its peak value for $\Gamma/J = 0.3$ (*circles*), 0.38 (*squares*) and 0.45 (*triangles*) (from [21])

The second moment of $W(m)$ gives the Edwards Anderson order parameter q. It was found that for a certain range of the ratio $\alpha = \Gamma/J$, the frequency of the peak of the susceptibility shifts towards higher values of ω as α is increased. The strength of the susceptibility, on the other hand, is reduced and the peak is broadened as quantum effects are strengthened (see Fig. 7.3). This is in qualitative agreement with the experimental results of [427].

7.2 Non-equilibrium Dynamics in Presence of Time-Dependent Fields

In this section, we are interested in the response of quantum Ising systems to time dependent perturbations. These perturbations are usually in the form of time dependent transverse fields, which might be quenching, sinusoidally varying or a pulse. The dynamics in presence of a sinusoidally varying longitudinal field and time independent transverse field has also been considered.

7.2.1 Time-Dependent Bogoliubov-de Gennes Formalism

The dynamics of the transverse Ising chain in the presence of time-dependent perturbations can be formulated by means of the time-dependent Bogoliubov-de Gennes theory. We consider the transverse Ising chain given by the Hamiltonian:

$$H = -\sum_i \left(S_i^x S_{i+1}^x + g S_i^z\right), \quad g = \bar{\lambda}^{-1}. \tag{7.2.1}$$

Applying the Jordan-Wigner transformation, this Hamiltonian is written in terms of fermion operators c_q and $c_q^\dagger$ in the momentum space as

$$
\begin{aligned}
H &= -2\sum_{q>0}(g+\cos q)\left(c_q^\dagger c_q - c_{-q}c_{-q}^\dagger\right) \\
&\quad - 2i\sum_{q>0}\sin q\left(c_q^\dagger c_{-q}^\dagger - c_q c_{-q}\right) && (7.2.2) \\
&= \sum_{q>0}\mathbf{C}_q^\dagger M_q(g)\mathbf{C}_q, && (7.2.3)
\end{aligned}
$$

where we defined

$$
M_q(g) = -2\begin{pmatrix} a_q(g) & ib_q \\ -ib_q & -a_q(g)\end{pmatrix}, \quad a_q(g) = g+\cos q, \quad b_q = \sin q, \tag{7.2.4}
$$

and

$$
\mathbf{C}_q = \begin{pmatrix} c_q \\ c_{-q}^\dagger \end{pmatrix}. \tag{7.2.5}
$$

This Hamiltonian is diagonalised by the Bogoliubov transformation:

$$
\mathbf{h}_q(g) = \begin{pmatrix} \eta_q(g) \\ \eta_{-q}^\dagger(g)\end{pmatrix} = R_q^\dagger(g)\mathbf{C}_q, \tag{7.2.6}
$$

where

$$
R_q(g)^\dagger = \begin{pmatrix} u_q(g) & iv_q(g) \\ iv_q(g) & u_q(g)\end{pmatrix}, \tag{7.2.7}
$$

with

$$
\begin{aligned}
u_q(g) &= \frac{a_q(g)-\omega_q(g)}{\sqrt{2\omega_q(g)(\omega_q(g)-a_q(g))}}, \\
v_q(g) &= \frac{b_q}{\sqrt{2\omega_q(g)(\omega_q(g)-a_q(g))}}
\end{aligned} \tag{7.2.8}
$$

and

$$
\omega_q(g) = \sqrt{g^2+2g\cos q+1}. \tag{7.2.9}
$$

The inverse transformation of (7.2.6) is given by

$$
\mathbf{C}_q = R_q(g)\mathbf{h}_q(g) = \begin{pmatrix} u_q(g) & -iv_q(g) \\ -iv_q(g) & u_q(g)\end{pmatrix}\begin{pmatrix} \eta_q(g) \\ \eta_{-q}^\dagger(g)\end{pmatrix}. \tag{7.2.10}
$$

The Hamiltonian is written in a diagonalised form as

$$
\begin{aligned}
H &= \sum_{q>0}\mathbf{h}_q(g)^\dagger\begin{pmatrix} 2\omega_q(g) & 0 \\ 0 & -2\omega_q(g)\end{pmatrix}\mathbf{h}_q(g) \\
&= 2\sum_{q>0}\omega_q(g)\left(\eta_q^\dagger(g)\eta_q(g) + \eta_{-q}^\dagger(g)\eta_{-q}(g) - 1\right). && (7.2.11)
\end{aligned}
$$

Following Sect. 7.A.3, we introduce operators: $A_j = c_j^\dagger + c_j$ and $B_j = c_j^\dagger - c_j$. These operators satisfy $\{A_j, A_k\} = 2\delta_{jk}$ and $\{B_j, B_k\} = -2\delta_{jk}$. A_j and B_j can be written in terms of η_q and $\eta_q^\dagger$ using the transformation (7.2.10). We thereby define operators a_j and b_j by

$$a_j = \frac{1}{\sqrt{N}} \sum_{q>0} e^{iqj}(u_q - iv_q)\eta_q + e^{-iqj}(u_q + iv_q)\eta_{-q}, \tag{7.2.12}$$

$$b_j = \frac{1}{\sqrt{N}} \sum_{q>0} e^{iqj}(u_q + iv_q)\eta_q + e^{-iqj}(u_q - iv_q)\eta_{-q} \tag{7.2.13}$$

to have $A_j = a_j^\dagger + a_j$ and $B_j = b_j^\dagger - b_j$. One can easily show that a_j an b_j satisfy anti-commutation relations:

$$\{a_j, a_k^\dagger\} = \delta_{jk}, \qquad \{b_j, b_k^\dagger\} = \delta_{jk}, \tag{7.2.14}$$

and

$$\begin{aligned} \{a_j, b_k^\dagger\} = \{a_j^\dagger, b_k\} &= \frac{1}{N} \sum_{q>0} \{e^{iq(j-k)}(u_q - iv_q)^2 + e^{-iq(j-k)}(u_q + iv_q)^2\} \\ &\equiv -G_{j-k}. \end{aligned} \tag{7.2.15}$$

From now on we consider the time-dependent transverse field $g(t)$. We denote $g_0 = g(t_0)$ and assume that the state is in the ground state at initial time t_0. The time dependence of several quantities are studied by means of the Bogoliubov-de Gennes formalism described as follows.

We define the time-evolution operator $U(t)$ by

$$|\Psi(t)\rangle = U(t)|\Psi(0)\rangle,$$

where $|\Psi(t)\rangle$ represents the state vector. The Heisenberg operator $O^{\mathrm{H}}(t) = U^\dagger(t) O U(t)$ of an operator O obeys the Heisenberg equation of motion:

$$i\frac{d}{dt} O^{\mathrm{H}}(t) = [O^{\mathrm{H}}(t), H(t)] + U^\dagger(t)\left(i\frac{d}{dt} O\right)U(t). \tag{7.2.16}$$

Note that the second term vanishes if an operator O is independent of time. Let us now focus on $c_q^{\mathrm{H}}(t)$. The Heisenberg equation for $c_q^{\mathrm{H}}(t)$ is written as

$$\begin{aligned} i\frac{d}{dt} c_q^{\mathrm{H}}(t) &= U^\dagger(t)[c_q, H(t)]U(t) \\ &= -2a_q(g(t))c_q^{\mathrm{H}}(t) - 2ib_q(g(t))c_{-q}^{\mathrm{H}}(t)^\dagger, \end{aligned} \tag{7.2.17}$$

where the commutator between c_q and H is calculated from (7.2.3). We expend $c_q^{\mathrm{H}}(t)$ and $c_{-q}^{\mathrm{H}\dagger}$ by $\eta_q(g_0)$ and $\eta_{-q}(g_0)^\dagger$ as

$$\mathbf{C}_q^{\mathrm{H}}(t) = \begin{pmatrix} c_q^{\mathrm{H}}(t) \\ c_{-q}^{\mathrm{H}\dagger}(t) \end{pmatrix} = S_q(t)\mathbf{h}_q(g_0), \tag{7.2.18}$$

where

$$S_q(t) = \begin{pmatrix} \tilde{u}_q(t) & -\tilde{v}_q^*(t) \\ \tilde{v}_q(t) & \tilde{u}_q^*(t) \end{pmatrix}. \tag{7.2.19}$$

Note that $\tilde{u}_q(t)$ and $\tilde{v}_q(t)$ satisfies

$$\tilde{u}_{-q}(t) = \tilde{u}_q(t), \quad \tilde{v}_{-q}(t) = -\tilde{v}_q(t), \quad |\tilde{u}_q(t)|^2 + |\tilde{v}_q(t)|^2 = 1, \tag{7.2.20}$$

which are confirmed from the fermion anti-commutation relations of $c_q^{\mathrm{H}}(t)$ and those of $\eta_q(g_0)$. Applying (7.2.18) into (7.2.17) and comparing coefficients of $\eta_q(g_0)$ and $\eta_{-q}^\dagger(g_0)$ in the left and right hand sides, we obtain equations of motion for $\tilde{u}_q(t)$ and $\tilde{v}_q(t)$:

$$\begin{aligned} i\frac{d}{dt}\begin{pmatrix} \tilde{u}_q(t) \\ \tilde{v}_q(t) \end{pmatrix} &= -2\begin{pmatrix} a_q(g(t)) & ib_q(g(t)) \\ -ib_q(g(t)) & -a_q(g(t)) \end{pmatrix}\begin{pmatrix} \tilde{u}_q(t) \\ \tilde{v}_q(t) \end{pmatrix} \\ &= M_q\big(g(t)\big)\begin{pmatrix} \tilde{u}_q(t) \\ \tilde{v}_q(t) \end{pmatrix} \end{aligned} \tag{7.2.21}$$

with the initial condition, $\tilde{u}_q(t_0) = u_q(g_0)$ and $\tilde{v}_q(t_0) = -iv_q(g_0)$.

We now look into the Heisenberg representation of operators A_j and B_j. Using (7.2.18), one has

$$A_j^{\mathrm{H}}(t) = a_j(t)^\dagger + a_j(t), \qquad B_j^{\mathrm{H}}(t) = b_j^\dagger(t) - b_j(t), \tag{7.2.22}$$

where

$$a_j(t) = \frac{1}{\sqrt{N}}\sum_{q>0}\big[e^{iqj}\big(\tilde{u}_q(t) + \tilde{v}_q(t)\big)\eta_q(g_0) + e^{-iqj}\big(\tilde{u}_q(t) - \tilde{v}_q(t)\big)\eta_{-q}(g_0)\big] \tag{7.2.23a}$$

$$b_j(t) = \frac{1}{\sqrt{N}}\sum_{q>0}\big[e^{iqj}\big(\tilde{u}_q(t) - \tilde{v}_q(t)\big)\eta_q(g_0) + e^{-iqj}\big(\tilde{u}_q(t) - \tilde{v}_q(t)\big)\eta_{-q}(g_0)\big]. \tag{7.2.23b}$$

Note that $a_j(t)$ and $b_j(t)$ are not the Heisenberg operators of a_j and b_j. Obviously we have

$$\{a_j(s), a_k(t)\} = \{b_j(s), b_k(t)\} = \{a_j(s), b_k(t)\} = 0. \tag{7.2.24}$$

We define

$$G_{j-k}(t) = -\{a_j(t), b_k^\dagger(t)\} \tag{7.2.25a}$$

$$G_{j-k}(s,t) = -\{a_j(s), b_k^\dagger(t)\} \tag{7.2.25b}$$

$$G_{j-k}^{\mathrm{a}}(s,t) = -\{a_j(s), a_k^\dagger(t)\} \tag{7.2.25c}$$

$$G_{j-k}^{\mathrm{b}}(s,t) = -\{b_j(s), b_k^\dagger(t)\}. \tag{7.2.25d}$$

Due to (7.2.23a) and (7.2.23b), these are written as

$$G_{j-k}(t) = -\frac{1}{N}\sum_{q>0}\big[2\cos q(j-k)\big(|\tilde{u}_q(t)|^2 - |\tilde{v}_q(t)|^2\big) - 2i\sin q(j-k)\big(\tilde{u}_q(t)\tilde{v}_q^*(t) - \tilde{v}_q(t)\tilde{u}_q^*(t)\big)\big], \tag{7.2.26a}$$

$$G_{j-k}(s,t) = -\frac{1}{N}\sum_{q>0}\big[2\cos q(j-k)\big(\tilde{u}_q(s)\tilde{u}_q^*(t) - \tilde{v}_q(s)\tilde{v}_q^*(t)\big) - 2i\sin q(j-k)\big(\tilde{u}_q(s)\tilde{v}_q^*(t) - \tilde{v}_q(s)\tilde{u}_q^*(t)\big)\big], \tag{7.2.26b}$$

$$G^{\mathrm{a}}_{j-k}(s,t) = -\frac{1}{N}\sum_{q>0}\big[2\cos q(j-k)\big(\tilde{u}_q(s)\tilde{u}_q^*(t) + \tilde{v}_q(s)\tilde{v}_q^*(t)\big) + 2i\sin q(j-k)\big(\tilde{u}_q(s)\tilde{v}_q^*(t) + \tilde{v}_q(s)\tilde{u}_q^*(t)\big)\big], \tag{7.2.26c}$$

$$G^{\mathrm{b}}_{j-k}(s,t) = -\frac{1}{N}\sum_{q>0}\big[2\cos q(j-k)\big(\tilde{u}_q(s)\tilde{u}_q^*(t) + \tilde{v}_q(s)\tilde{v}_q^*(t)\big) - 2i\sin q(j-k)\big(\tilde{u}_q(s)\tilde{v}_q^*(t) + \tilde{v}_q(s)\tilde{u}_q^*(t)\big)\big]. \tag{7.2.26d}$$

Note that

$$G^*_{j-k}(t) = G_{j-k}(t), \qquad G^*_{j-k}(s,t) = G_{j-k}(t,s) \tag{7.2.27}$$

$$G^{\mathrm{a}*}_{j-k}(s,t) = G^{\mathrm{a}}_{k-j}(t,s), \qquad G^{\mathrm{b}*}_{j-k}(s,t) = G^{\mathrm{b}}_{k-k}(t,s). \tag{7.2.28}$$

As we shall see below, the time dependence of physical quantities can be computed through $G_l(t)$, $G_l(s,t)$, $G^{\mathrm{a}}_l(s,t)$, and $G^{\mathrm{b}}_l(s,t)$. An explicit expression of these quantities shall be given after specifying the time dependence of $g(t)$ and initial conditions for $\tilde{u}_q$ and $\tilde{v}_q$, and solving the time-dependent Bogoliubov-de Gennes equation (7.2.21).

7.2.2 Quantum Quenches

Motivated by experiments in cold atomic systems [160, 227], dynamics of isolated quantum systems subjected by a quench of a parameter has been received a growing amount of attention in the last decade, notwithstanding a theoretical study by Barouch, McCoy and Dresden in 70s [28]. To understand the fundamental nature of quench dynamics of interacting systems, the transverse Ising chain has been a subject of intense study. There are two topics in studies of quench dynamics (see for a review [69, 120, 319]). One is a sudden quench, where a parameter is changed discontinuously (Fig. 7.4(a)). The state after a quench is expected to relax towards a steady state, when the system is macroscopically large. The identification of this steady state and properties of a transient state are main targets of study. The other topic is a nearly adiabatic dynamics following a slow quench of a parameter (Fig. 7.4(b)). Scaling theory for non-adiabatic excitations induced by a quench across a quantum critical point has been developed. In this subsection, we first present studies on a sudden quench, and then move to a slow quench of the transverse field.

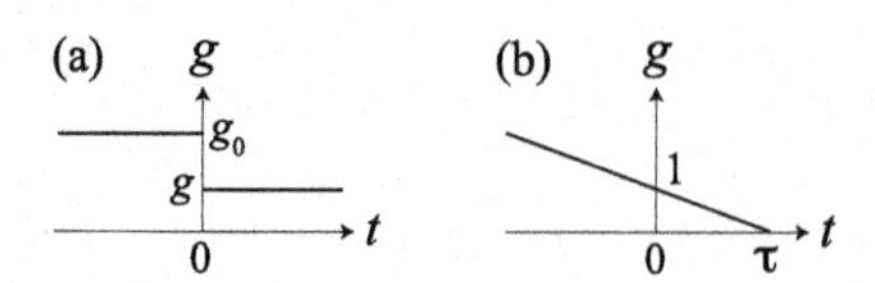

Fig. 7.4 Time dependence of the transverse field. (**a**) shows a sudden quench. The state is assume to be in the ground state of the system before the quench. (**b**) shows a linear quench. The transverse field passes a critical point at $t = 0$. The state evolves from the ground state at $t = -\infty$ where $g = \infty$

7.2.2.1 Relaxation Dynamics After Sudden Quantum Quench

Let us assume that the transverse field is changed from g_0 to g abruptly at time $t = 0$ as depicted in Fig. 7.4(a):

$$g(t) = \begin{cases} g_0 & (t < 0) \\ g & (t > 0) \end{cases}. \tag{7.2.29}$$

We assume that the system at $t = 0$ is in its ground state before the quench, namely

$$|\Psi(0)\rangle = |\Psi_{\rm gs}(g_0)\rangle = \prod_{0<q<\pi} \eta_q(g_0)\eta_{-q}(g_0)|0\rangle, \tag{7.2.30}$$

where $|\Psi_{\rm gs}(g_0)\rangle$ is the vacuum of $\eta_q(g_0)$: $\eta_q(g_0)|\Psi_{\rm gs}(g_0)\rangle = 0$ for all q, and $|0\rangle$ represents the vacuum of c_q: $c_q|0\rangle = 0$ for all q.

Under this condition, one can easily solve the time-dependent Bogoliubov-de Gennes equation (7.2.21). The solution is given by

$$\begin{pmatrix} \tilde{u}_q(t) \\ \tilde{v}_q(t) \end{pmatrix} = R_q(g) \begin{pmatrix} e^{2i\omega_q(g)t} & 0 \\ 0 & e^{-2i\omega_q(g)t} \end{pmatrix} R_q^\dagger(g) \begin{pmatrix} \tilde{u}_q(0) \\ \tilde{v}_q(0) \end{pmatrix}. \tag{7.2.31}$$

The initial condition for $\tilde{u}_q(t)$ and $\tilde{v}_q(t)$ is given by

$$\begin{pmatrix} \tilde{u}_q(0) \\ \tilde{v}_q(0) \end{pmatrix} = \begin{pmatrix} u_q(g_0) \\ -iv_q(g_0) \end{pmatrix}. \tag{7.2.32}$$

Substituting (7.2.8), the solution (7.2.31) is written explicitly as

$$\begin{pmatrix} \tilde{u}_q(t) \\ \tilde{v}_q(t) \end{pmatrix} = \begin{pmatrix} u_0 \cos 2\omega t + i\frac{au_0+bv_0}{\omega} \sin 2\omega t \\ -iv_0 \cos 2\omega t + \frac{bu_0-av_0}{\omega} \sin 2\omega t \end{pmatrix}, \tag{7.2.33}$$

where we have used shorthand notations: $a = a_q(g)$, $b = b_q(g)$, $\omega = \omega_q(g)$, $u_0 = u_q(g_0)$, and $v_0 = v_q(g_0)$.

We first focus on the transverse component of spins, S^z. The time dependence of the transverse magnetisation is expressed in the Heisenberg representation as

$$m_z(t) = \frac{1}{N}\sum_{i=1}^{N} \langle S_i^z(t)\rangle = \frac{1}{N}\sum_{i=1}^{N} \langle \Psi(0)|B_i^{\rm H}(t)A_i^{\rm H}(t)|\Psi(0)\rangle. \tag{7.2.34}$$

We recall here (7.2.22), (7.2.23a), (7.2.23b). Then one finds

$$\begin{aligned}\langle \Psi(0)|B_i^{\mathrm{H}}(t)A_i^{\mathrm{H}}(t)|\Psi(0)\rangle &= -\langle \Psi(0)|b_i(t)a_i^\dagger(t)|\Psi(0)\rangle \\ &= -\langle \Psi(0)|\{b_i(t),a_i^\dagger(t)\}|\Psi(0)\rangle \\ &= G_0(t),\end{aligned} \tag{7.2.35}$$

where we note that the initial condition leads to $a_i(t)|\Psi(0)\rangle = b_i(t)|\Psi(0)\rangle = 0$. Therefore one obtains

$$m_z(t) = G_0(t). \tag{7.2.36}$$

Substitution of the solution, (7.2.33), of the time-dependent Bogoliubov-de Gennes equation in (7.2.26a) yields

$$G_0(t) = \frac{2}{N}\sum_{0<q<\pi}\left[\frac{g_0+\cos q}{\omega_0} + \frac{(g-g_0)\sin^2 q}{\omega^2\omega_0}(1-\cos 4\omega t)\right], \tag{7.2.37}$$

where we note a shorthand notation: $\omega_0 = \omega_q(g_0)$. Taking the thermodynamic limit $\frac{1}{N}\sum_q \to \int dq/2\pi$, the time-dependent term vanishes at $t\to\infty$. Hence one obtains

$$m_z(\infty) = \int_0^\pi \frac{dq}{\pi}\left(\frac{g_0+\cos q}{\omega_0} + \frac{(g-g_0)\sin^2 q}{\omega^2\omega_0}\right). \tag{7.2.38}$$

Note that this can be expressed in terms of the complete elliptic integrals. Equation (7.2.38) implies that the transverse magnetisation converges to a value which is neither the ground state expectation value nor the thermal expectation value [28, 335]. This fact might be remarkable because the statistical mechanics teaches that the equilibrium state long after a change of a parameter obeys the Boltzmann's canonical distribution. However, since the present system involves infinite number of constants of motion (i.e., fermion occupation numbers $\eta_q^\dagger(g)\eta_q(g)$), the steady state after a quench is different from the thermal equilibrium.

The transient toward the asymptotic value is given by

$$m_z(t) - m_z(\infty) = -\frac{1}{\pi}\int_0^\pi dq\,\frac{(g-g_0)\sin^2 q}{\omega^2\omega_0}\cos 4\omega t. \tag{7.2.39}$$

This gives rise to an algebraic decay for large t. A little calculation shows that the leading power is $t^{-3/2}$ when $g_0 \neq 1$, while it is t^{-1} when $g_0 = 1$.

We next focus on the longitudinal correlation function in the presence of a sudden quench. The longitudinal correlation function is defined by

$$C^x(s,r;t) = \langle S_j^x(t+s)S_{j+r}^x(t)\rangle. \tag{7.2.40}$$

We start from investigation of the equal-time correlation function $C^x(0,r;t)$. Recalling (2.A.29), one can write

$$C^x(0,r;t) = \langle S_j^x(t)S_{j+r}^x(t)\rangle = \langle B_j^{\mathrm{H}}(t)A_{j+1}^{\mathrm{H}}(t)B_{j+1}^{\mathrm{H}}(t)\cdots A_{j+r}^{\mathrm{H}}(t)\rangle. \tag{7.2.41}$$

Due to (7.2.22) and (7.2.24), (7.2.25a), (7.2.25b), (7.2.25c), (7.2.25d), one has

$$\{a_k(t),A_l^{\mathrm{H}}(t)\} = -G_{k-l}^{\mathrm{a}}(t,t), \qquad \{b_k(t),B_l^{\mathrm{H}}(t)\} = -G_{k-l}^{\mathrm{b}}(t,t) \tag{7.2.42}$$

$$\{b_k(t),A_l^{\mathrm{H}}(t)\} = -G_{l-k}(t). \tag{7.2.43}$$

Since $G_r^{\mathrm{a,b}}(t,t) \to -\delta_{r,0}$ with $t \to \infty$, it turns out that

$$\begin{aligned}
&\{a_k(t), A_l^{\mathrm{H}}(t)\}\big|_{t\to\infty} = \{b_k(t), B_l^{\mathrm{H}}(t)\}\big|_{t\to\infty} = \delta_{k,l}, \\
&\{b_k(t), A_l^{\mathrm{H}}(t)\}\big|_{t\to\infty} = -G_{l-k}(\infty).
\end{aligned} \tag{7.2.44}$$

These relations yield

$$\begin{aligned}
C^x(0,r;\infty) &= -\langle b_j(t) A_{j+1}^{\mathrm{H}}(t) B_{j+1}^{\mathrm{H}}(t) \cdots A_{j+r}^{\mathrm{H}}(t)\rangle\big|_{t\to\infty} \\
&= G_1(\infty)\langle B_{j+1}^{\mathrm{H}}(t) A_{j+2}^{\mathrm{H}}(s) \cdots A_{j+r}^{\mathrm{H}}(t)\rangle\big|_{t\to\infty} \\
&\quad + \langle B_{j+1}^{\mathrm{H}}(t) A_{j+1}^{\mathrm{H}}(t) b_j(t) A_{j+2}^{\mathrm{H}} B_{j+2}^{\mathrm{H}}(t) \cdots A_{j+r}^{\mathrm{H}}(t)\rangle\big|_{t\to\infty} \\
&= \det \begin{pmatrix} G_1(\infty) & G_0(\infty) & \dots & G_{-r+2}(\infty) \\ G_2(\infty) & G_1(\infty) & \dots & G_{-r+3}(\infty) \\ \vdots & \vdots & \ddots & \vdots \\ G_r(\infty) & G_{r-1}(\infty) & \dots & G_1(\infty) \end{pmatrix} \\
&= \det \begin{pmatrix} D_0 & D_{-1} & \dots & D_{-r+1} \\ D_1 & D_0 & \dots & D_{-r+2} \\ \vdots & \vdots & \ddots & \vdots \\ D_{r-1} & D_{r-2} & \dots & D_0 \end{pmatrix},
\end{aligned} \tag{7.2.45}$$

where we defined

$$\begin{aligned}
D_r = G_{r+1}(\infty) &= \frac{1}{\pi}\int_0^{\pi} dq \cos(r+1)q \left(\frac{g_0 + \cos q}{\omega_0} + \frac{(g-g_0)\sin^2 q}{\omega^2 \omega_0} \right) \\
&\quad + \frac{1}{\pi}\int_0^{\pi} dq \sin(r+1)q \left(\frac{\sin q}{\omega_0} - \frac{(g-g_0)(g+\cos q)\sin q}{\omega^2\omega_0} \right).
\end{aligned} \tag{7.2.46}$$

After an algebra, one finds that (7.2.46) can be arranged into

$$D_r = \frac{1}{2\pi}\int_{-\pi}^{\pi} dq\, e^{-irq} C(e^{iq}), \tag{7.2.47}$$

where

$$C(e^{iq}) = \frac{2 + 2gg_0 + (g+g_0)(1+e^{-2iq})e^{iq}}{2(1+ge^{iq})\sqrt{(1+g_0e^{iq})(1+g_0e^{-iq})}}. \tag{7.2.48}$$

Note that $\ln e^{iq}$ is not a cyclic function of q, and thus $\ln C(e^{iq})$ is not a continuous function of q. Hence we cannot apply the Szego's theorem to the Toeplitz determinant in (7.2.45). However, this determinant is considerably simplified in the following two cases [365].

(i) Quench from $g_0 = \infty$. In this case $C(e^{iq})$ reduces to

$$C(e^{iq}) = \frac{2g + e^{iq}(1+e^{-2iq})}{2(1+ge^{iq})}. \tag{7.2.49}$$

Changing the integral (7.2.47) into an contour integral on the complex plane of $z = e^{iq}$, one can write

$$D_r = \frac{1}{2\pi i}\oint dz z^{-r-1} C(z) = \frac{1}{2\pi i}\oint dz z^{-r-1}\left(\frac{1}{z+g^{-1}} + \frac{z+z^{-1}}{2g(z+g^{-1})}\right), \tag{7.2.50}$$

where the path of the contour integral is along the unit circle and the direction is counterclockwise. Suppose here $g > 1$. Then a straightforward calculation yields

$$D_r = \begin{cases} 0 & (r \geq 1) \\ 1/(2g) & (r = 0) \\ 1 - 1/(2g^2) & (r = -1) \\ \frac{(-1)^r}{2}(1-g^2)g^{r-1} & (r \leq -2) \end{cases}. \tag{7.2.51}$$

One finds from this that the Toeplitz determinant (7.2.45) reduces to the determinant of a tridiagonal matrix. Hence one immediately gets, for $g > 1$,

$$C^x(0, r; \infty) = \left(\frac{1}{2g}\right)^r. \tag{7.2.52}$$

For $g < 1$, one has

$$D_r = \begin{cases} \frac{(-1)^r}{2} g^{r-1}(g^2-1) & (r \geq 1) \\ g/2 & (r = 0) \\ 1/2 & (r = -1) \\ 0 & (r \leq -2) \end{cases}. \tag{7.2.53}$$

In this case, Eq. (7.2.45) is written as

$$C^x(0, r; \infty) = \det\begin{pmatrix} D_0 & D_{-1} & 0 & \dots & 0 \\ D_1 & D_0 & D_{-1} & \dots & 0 \\ D_2 & D_1 & D_0 & \dots & 0 \\ \vdots & \vdots & \vdots & \ddots & \vdots \\ D_{r-1} & D_{r-2} & D_{r-3} & \dots & D_0 \end{pmatrix}. \tag{7.2.54}$$

Define now

$$\Delta_r = C^x(0, r; \infty) \tag{7.2.55}$$

$$\Delta'_r = \det\begin{pmatrix} D_1 & D_{-1} & 0 & \dots & 0 \\ D_2 & D_0 & D_{-1} & \dots & 0 \\ D_3 & D_1 & D_0 & \dots & 0 \\ \vdots & \vdots & \vdots & \ddots & \vdots \\ D_r & D_{r-2} & D_{r-3} & \dots & D_0 \end{pmatrix}. \tag{7.2.56}$$

Note that $\Delta_1 = D_0 = g/2$ and $\Delta'_1 = D_1 = \frac{1}{2}(1-g^2)$. Expanding the determinants of Δ_r and Δ'_r by cofactors of the first row, Δ_r and Δ'_r satisfy

$$\Delta_r = D_0 \Delta_{r-1} - D_{-1} \Delta'_{r-1} \tag{7.2.57}$$

$$\begin{aligned} D_r &= D_1 \Delta_{r-1} - D_{-1} \det \begin{pmatrix} D_2 & D_{-1} & 0 & \dots & 0 \\ D_3 & D_0 & D_{-1} & \dots & 0 \\ D_4 & D_1 & D_0 & \dots & 0 \\ \vdots & \vdots & \vdots & \ddots & \vdots \\ D_r & D_{r-3} & D_{r-4} & \dots & D_0 \end{pmatrix} \\ &= D_1 \Delta_{r-1} + g D_{-1} \Delta'_{r-1}, \end{aligned} \tag{7.2.58}$$

where we note $D_{r+1} = -g D_r$ for $r \geq 1$. These recurrence relations reduce to

$$\begin{aligned} \begin{pmatrix} \Delta_r \\ \Delta'_r \end{pmatrix} &= \begin{pmatrix} D_0 & -D_{-1} \\ D_1 & g D_{-1} \end{pmatrix} \begin{pmatrix} \Delta_{r-1} \\ \Delta'_{r-1} \end{pmatrix} = \begin{pmatrix} D_0 & -D_{-1} \\ D_1 & g D_{-1} \end{pmatrix}^{r-1} \begin{pmatrix} \Delta_1 \\ \Delta'_1 \end{pmatrix} \\ &= \frac{1}{2^r} R_1^{r-1} \begin{pmatrix} g \\ 1-g^2 \end{pmatrix}, \end{aligned} \tag{7.2.59}$$

where

$$R_1 = \begin{pmatrix} g & -1 \\ 1-g^2 & g \end{pmatrix}. \tag{7.2.60}$$

By diagonalisation of R_1, one obtains

$$\Delta_r = \frac{1}{2^r} \left\{ \frac{g}{2} \left(\lambda^{r-1} + \lambda^{*r-1} \right) - \frac{\sqrt{1-g^2}}{2i} \left(\lambda^{r-1} - \lambda^{*r-1} \right) \right\}, \tag{7.2.61}$$

where $\lambda = g + i\sqrt{1-g^2}$ and $\lambda^* = g - i\sqrt{1-g^2}$ are the eigenvalues of R_1. We introduce θ such that $\cos\theta = g$. Then, one can write $\lambda = e^{i\theta}$ and (7.2.61) is arranged into [365]

$$\begin{aligned} C^x(0, r; \infty) = \Delta_r &= \frac{1}{2^r} \big(\cos\theta \cos(r-1)\theta - \sin\theta \sin(r-1)\theta \big) \\ &= \frac{1}{2^r} \cos(r\theta) = \frac{1}{2^r} \cos(r \arccos g). \end{aligned} \tag{7.2.62}$$

Summarising (7.2.52) and (7.2.62), The longitudinal correlation decays exponentially with r both for $g > 1$ and $0 < g < 1$. The correlation length ξ is given by

$$\xi = \begin{cases} 1/\ln(2g) & \text{for } g > 1 \\ 1/\ln 2 & \text{for } 0 < g < 1 \end{cases}. \tag{7.2.63}$$

(ii) Quench from $g_0 = 0$. In this case, (7.2.48) reduces to

$$C(e^{iq}) = \frac{2 + g(1 + e^{-2iq})e^{iq}}{2(1 + g e^{iq})}, \tag{7.2.64}$$

and hence (7.2.47) does to

$$D_r = \frac{1}{2\pi i} \oint \frac{dz}{z^{r+1}} \left\{ \frac{z + z^{-1}}{2(z + g^{-1})} \right\}, \tag{7.2.65}$$

where we take the counterclockwise path along the unit circle on the complex z-plane. Supposing $g > 1$, the contour integral is carried out to yield

$$D_r = \begin{cases} 0 & (r \geq 1) \\ 1/2 & (r = 0) \\ 1/(2g) & (r = -1) \\ \frac{(-1)^r}{2}(g^2 - 1)g^r & (r \leq -2) \end{cases}. \tag{7.2.66}$$

Thus the Toeplitz determinant (7.2.45) reduces to the determinant of a tridiagonal matrix and is found to be

$$C^x(0, r; \infty) = \frac{1}{2^r}. \tag{7.2.67}$$

For $0 < g < 1$, the contour integral (7.2.65) is computed, so that

$$D_r = \begin{cases} \frac{(-1)^r}{2} g^r(1 - g^2) & (r \geq 1) \\ n1 - g^2/2 & (r = 0) \\ g/2 & (r = -1) \\ 0 & (r \leq -2) \end{cases}. \tag{7.2.68}$$

Same as the case (i), we consider determinants Δ_r and Δ'_r defined by (7.2.55) and (7.2.56). These determinants are written as

$$\begin{pmatrix} \Delta_r \\ \Delta'_r \end{pmatrix} = \frac{1}{2^r} R_2^{r-1} \begin{pmatrix} 2 - g^2 \\ -g(1 - g^2) \end{pmatrix}, \tag{7.2.69}$$

where

$$R_2 = \begin{pmatrix} 2 - g & -g \\ -g(1 - g^2) & g^2 \end{pmatrix}. \tag{7.2.70}$$

A straightforward calculation yields

$$\Delta_r = \frac{1}{2^r} \frac{1}{2\sqrt{1 - g^2}} \{(2 - g^2)(\lambda_+^r - \lambda_-^r) - g^2(\lambda_+^{r-1} - \lambda_-^{r-1})\}, \tag{7.2.71}$$

where $\lambda_\pm = 1 \pm \sqrt{1 - g^2}$ are the eigenvalues of R_2. Defining θ by $e^\theta = (1 + \sqrt{1 - g^2})/g$, the eigenvalues are simply written as $\lambda_\pm = g^{\pm\theta}$. With this notation and a short algebra, one finds

$$C^x(0, r; \infty) = \Delta_r = \frac{g^{r+1}}{2^r} \cosh(r + 1)\theta. \tag{7.2.72}$$

For $r \to \infty$, this reduces to [365]

$$C^x(0, r; \infty) \approx \left(\frac{1 + \sqrt{1 - g^2}}{2} \right)^{r+1}. \tag{7.2.73}$$

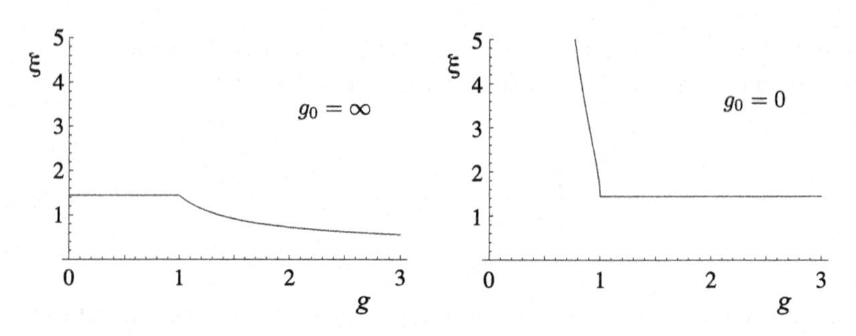

Fig. 7.5 Correlation length of the state long after time from a sudden quench as a function of the transverse field g after a quench. The *left* and *right panels* show results (7.2.63) and (7.2.74) for quenches from $g_0 = \infty$ and $g_0 = 0$ respectively

One finds again that the spatial correlation decays exponentially with separation both for $g > 1$ and $0 < g < 1$. The correlation lengths are obtained from (7.2.67) and (7.2.73) as

$$\xi = \begin{cases} 1/\ln 2 & \text{for } g > 1 \\ 1/\ln[2/(1+\sqrt{1-g^2})] & \text{for } 0 < g < 1 \end{cases}. \tag{7.2.74}$$

It is worth noting that the longitudinal correlation function (7.2.73) with $g_0 = 0$ vanish for $r \to \infty$ in spite of the completely ordered initial state. This fact implies that the quench of the transverse field is a source of fluctuations which destroy an order of the system.

Figure 7.5 shows the correlation length (7.2.63) and (7.2.74) of the state long after time after a quench ($t \to \infty$). The correlation length of the state at $t \to \infty$ is finite even if the transverse field after a quench is set at the critical point $g = 1$. The difference of the correlation length ξ from that of the initial state is larger for larger amplitude of quench. However, it is remarkable that, if the quench is across the quantum critical point, $\xi = 1/\ln 2$ is independent of the transverse field g after a quench.

We next look into the temporal and spatial correlation functions. Rossini et al. [334, 335] numerically studied the on-site autocorrelation function $C^x(s, 0; t)$ as well as the equal-time spatial correlation function $C^x(0, r; t)$ of the state long after a quench ($t \gg 1$). They found that the autocorrelation function and the spatial correlation function decay with a coherence time τ and a correlation length ξ. What is remarkable is that τ and ξ is determined by the energy gap $\Delta(g) = 2|g-1|$ of the system after quench and an effective temperature T_{eff} which controls the strength of the quench and is defined as follows. Since the state after a quench is no longer the ground state, the quench induces an energy to the system. The amount of this energy depends on how strongly the quench is done. If we assume that the system after the quench is in an equilibrium state of an effective temperature T_{eff}, the energy expectation value of the system is given by $\langle H(g)\rangle_{T_{\mathrm{eff}}} = \mathrm{Tr}(H(g)e^{-H(g)/T_{\mathrm{eff}}})/\mathrm{Tr}(e^{-H(g)/T_{\mathrm{eff}}})$, following the standard statistical mechanics. On the other hand, the system has the energy expectation value

$\langle H(g) \rangle = \langle \Psi_0(g_0)|H(g)|\Psi_0(g_0) \rangle$. We then make an equation between these two quantities:

$$\langle H(g) \rangle_{T_{\mathrm{eff}}} = \langle \Psi_0(g_0) | H(t) | \Psi_0(g_0) \rangle, \tag{7.2.75}$$

by which we define the effective temperature.

Surprisingly, the dependences of τ and ξ on T_{eff} resemble those of the relaxation time and the correlation length in the thermal equilibrium state. As we mentioned before, a quench of the transverse field induces fluctuations which destroy the long-range order. The result of Rossini et al. suggests that the fluctuation induced by a quench of the transverse field looks like the thermal fluctuation as far as the longitudinal autocorrelation function and spatial correlation function are concerned. We remark that such an effective thermalisation behaviour does not always appear in other quantities. We have mentioned that, due to the integrability (namely the presence of infinite number of constants of motion) of the present system, the transverse magnetisation is not thermalised. What kind of quantities show thermalisation behaviour and the origin of thermalisation in the integrable system are open problem to be clarified.

Before closing the present subsection we comment on the work by Calabrese and Cardy [55–57]. They applied the boundary conformal field theory to the dynamics after a quantum quench to a quantum critical point and obtained time dependence and distance dependence of several correlation functions. When the system is at a quantum critical point with a gapless linear quasiparticle dispersion, quasiparticles created by a quench of a parameter propagate with a unique speed. As a result, the system after a quench retains two-point temporal-spatial correlation of a local observable until $t \sim r/v$, where v is a speed of the quasiparticle. Beyond this, the correlation decays to a time-independent value. According to computation by Calabrese and Cardy [56, 57], two-point correlation decays exponentially with time and the characteristic scale of time includes the critical exponent, so that the ratio of characteristic time scales of different quantities yields a universal value. The consequence from the boundary conformal field theory is applicable to the transverse Ising chain. Its verification and development in the transverse Ising chain are expected.

7.2.2.2 Nearly Adiabatic Dynamics Following a Slow Quench

In the present section, we consider a dynamics following a slow quench of a parameter. Let us imagine a system in its equilibrium state. We suppose that a parameter, which specifies the equilibrium state, changes slowly with time. If the changing speed is sufficiently slow, then the system should keep track of the equilibrium state. However, if the parameter passes through a critical point of a phase transition, does the system evolve into a new phase? This fundamental question has been discussed in a wide range of area from the cosmology to the condensed matter. It should be noted that Greiner et al. [160] demonstrated an evolution of cold bosonic atoms from a superfluid to a Mott insulator. This experiment has raised a lot of interests in the dynamics across a quantum phase transition. Here we present a theory of the dynamics following a slow change of the transverse field across a quantum phase transition in the one-dimensional transverse Ising chain [78, 118].

We consider a linear quench of the transverse field:

$$g = -t/\tau + 1, \tag{7.2.76}$$

where $1/\tau$ represents a control parameter of the quench rate, and time t moves from $t = -\infty$. Note that g passes the quantum critical point at $t = 0$. We assume that the system is in the ground state initially. We show the time dependence of g in Fig. 7.4(b). With (7.2.76), the time-dependent Bogoliubov-de Gennes equation (7.2.21) is written as

$$i\frac{d}{dt}\begin{pmatrix}\tilde{u}_q(t)\\ \tilde{v}_q(t)\end{pmatrix} = 2\{(t/\tau - \alpha_q)\sigma^3 + \beta_q\sigma^2\}\begin{pmatrix}\tilde{u}_q(t)\\ \tilde{v}_q(t)\end{pmatrix}, \tag{7.2.77}$$

where we defined

$$\alpha_q = 1 + \cos q, \qquad \beta_q = \sin q. \tag{7.2.78}$$

Note $\alpha_q > 0$ and $\beta > 0$ for $0 < q < \pi$ as far as an even N is assumed (regarding the available wave number q, see Sect. 2.2). σ^1, σ^2, and σ^3 are the Pauli matrices which are written as

$$\sigma^1 = \begin{pmatrix}0 & 1\\ 1 & 0\end{pmatrix}, \qquad \sigma^2 = \begin{pmatrix}0 & -i\\ i & 0\end{pmatrix}, \quad \text{and} \quad \sigma^3 = \begin{pmatrix}1 & 0\\ 0 & -1\end{pmatrix}. \tag{7.2.79}$$

The initial conditions for $\tilde{u}_q(t)$ and $\tilde{v}_q(t)$ are given by $\tilde{u}_q(-\infty) = u_q(g = \infty)$ and $\tilde{v}_q(-\infty) = v_q(g = \infty)$, which turn out from (7.2.8) to be

$$\tilde{u}_q(-\infty) = 0 \quad \text{and} \quad \tilde{v}_q(-\infty) = -i. \tag{7.2.80}$$

Let us here introduce a unitary transformation: $U = \exp(-i\sigma^3\pi/4)$. One can easily show $U^\dagger\sigma^3 U = \sigma^3$ and $U^\dagger\sigma^2 U = \sigma^1$. Applying this transformation, (7.2.77) is written as

$$i\frac{d}{dt}\begin{pmatrix}\tilde{u}'_q(t)\\ \tilde{v}'_q(t)\end{pmatrix} = 2\{(t/\tau - \alpha_q)\sigma^3 + \beta_q\sigma^1\}\begin{pmatrix}\tilde{u}'_q(t)\\ \tilde{v}'_q(t)\end{pmatrix}, \tag{7.2.81}$$

where

$$\begin{pmatrix}\tilde{u}'_q(t)\\ \tilde{v}'_q(t)\end{pmatrix} = U^\dagger\begin{pmatrix}\tilde{u}_q(t)\\ \tilde{v}_q(t)\end{pmatrix} = \begin{pmatrix}e^{i\pi/4}\tilde{u}_q(t)\\ e^{-i\pi/4}\tilde{v}_q(t)\end{pmatrix}. \tag{7.2.82}$$

We define

$$\bar{t} \equiv 2\beta_q(t - \alpha_q\tau), \quad \bar{\tau} \equiv 2\beta_q^2\tau, \quad \begin{pmatrix}\bar{u}_q(\bar{t})\\ \bar{v}_q(\bar{t})\end{pmatrix} = \begin{pmatrix}e^{i\pi/4}\tilde{u}_q(t)\\ e^{-i\pi/4}\tilde{v}_q(t)\end{pmatrix}. \tag{7.2.83}$$

Then (7.2.81) is written as

$$i\frac{d}{d\bar{t}} = \begin{pmatrix}\bar{u}_q(\bar{t})\\ \bar{v}_q(\bar{t})\end{pmatrix} = \left(\frac{\bar{t}}{\bar{\tau}}\sigma^3 + \sigma^1\right)\begin{pmatrix}\bar{u}_q(\bar{t})\\ \bar{v}_q(\bar{t})\end{pmatrix}. \tag{7.2.84}$$

Thus one finds that our problem reduces to the Landau-Zener problem. As shown in Sect. 7.A.2, the solution of this equation with the initial condition, $\bar{u}_q(-\infty) = 0$ and $|\bar{v}_q(-\infty)| = 1$, is given up o an inessential phase factor as

$$\bar{u}_q(\bar{t}) = -i\sqrt{\frac{\bar{\tau}}{2}}e^{-\pi\bar{\tau}/8}D_{i\bar{\tau}/2-1}\big(e^{i3\pi/4}\sqrt{2/\bar{\tau}}\,\bar{t}\big) \tag{7.2.85}$$

$$\bar{v}_q(\bar{t}) = e^{i3\pi/4}e^{-pi\bar{\tau}/8}D_{i\bar{\tau}/2}\big(e^{i3\pi/4}\sqrt{2/\bar{\tau}}\,\bar{t}\big), \tag{7.2.86}$$

namely,

$$\tilde{u}_q(t) = e^{-i3\pi/4}\beta_q\sqrt{\tau}\,e^{-\pi\beta_q^2\tau/4}D_{i\beta_q^2\tau-1}\big(e^{i3\pi/4}2(t-\alpha_q\tau)/\sqrt{\tau}\big) \tag{7.2.87}$$

$$\tilde{v}_q(t) = -e^{-\pi\beta_q^2\tau/4}D_{i\beta_q^2\tau}\big(e^{i3\pi/4}2(t-\alpha_q\tau)/\sqrt{\tau}\big), \tag{7.2.88}$$

where $D_\mu(x)$ is the parabolic cylinder function.

We now focus on the density of defects as a measure for the deviation of the state after a quench from the instantaneous ground state. The density of defects is defined by

$$n = \frac{1}{N}\sum_{q>0} n_q, \tag{7.2.89}$$

$$\begin{aligned} n_q &= \big\langle\Psi(t_f)\big|\eta_q^\dagger(g_f)\eta_q(g_f) + \eta_{-q}^\dagger(g_f)\eta_{-q}(g_f)\big|\Psi(t_f)\big\rangle \\ &= \big\langle\Psi(t_f)\big|\mathbf{h}_q^\dagger(g_f)\sigma^3\mathbf{h}_q(g_f)\big|\Psi(t_f)\big\rangle + 1, \end{aligned} \tag{7.2.90}$$

where t_f denotes the final time and $g_f = g(t_f)$. Note that n_q stands for the expected number of excited quasiparticle with the wave number q in the state after a quench. Hence n corresponds to the density of excited quasiparticle in the final state. Applying the inverse Bogoliubov-de Gennes transformation and switching to the Heisenberg representation, (7.2.90) is expressed as

$$\begin{aligned} n_q &= \langle\Psi_i|\mathbf{C}_q^{\mathrm{H}\dagger}(t_f)R_q^\dagger(g_f)\sigma^3R_q(g_f)\mathbf{C}_q^{\mathrm{H}}(t_f)|\Psi_i\rangle + 1 \\ &= \langle\Psi_i|\mathbf{h}_q^{\mathrm{H}\dagger}(g_f)S_q^\dagger(t_f)R_q^\dagger(g_f)\sigma^3R_q(g_f)S_q(t_f)\mathbf{h}_q^{\mathrm{H}}(g_i)|\Psi_i\rangle + 1, \end{aligned} \tag{7.2.91}$$

where $g_i \equiv g(t = -\infty) = +\infty$ is the transverse field before the quench, and $|\Psi_i\rangle = |\Psi(t = -\infty)\rangle$ is the initial state. Due to the initial condition, $|\Psi_i\rangle$ satisfies $\eta_q(g_i)|\Psi_i\rangle = \eta{-q}(g_i)|\Psi_i\rangle = 0$. Therefore it turns out that

$$n_q = \big[S_q^\dagger(t_f)R_q^\dagger(g_f)\sigma^3R_q(g_f)S_q(t_f)\big]_{22} + 1. \tag{7.2.92}$$

We hereafter consider two special cases where the transverse field is reduced (i) until $t_f = +\infty$ (i.e., $g_f = -\infty$), and (ii) until $t_f = \tau$ (i.e., $g_f = 0$).

(i) case with $t_f = +\infty$. From (7.2.8), it turns out that $u_q(g_f = -\infty) = -1$ and $v_q(g_f = -\infty) = 0$. It follows that

$$R_q(g_f = -\infty) = \begin{pmatrix} -1 & 0 \\ 0 & -1 \end{pmatrix}. \tag{7.2.93}$$

Therefore one has

$$\begin{aligned} n_q &= \left[S_q^\dagger(t_f=+\infty)\sigma^3 S_q(t_f=+\infty)\right]_{22}+1 \\ &= \left|\tilde{v}_q(t_f=+\infty)\right|^2 - \left|\tilde{u}_q(t=+\infty)\right|^2+1 \\ &= 2\left|\tilde{v}_q(t_f=+\infty)\right|^2. \end{aligned} \tag{7.2.94}$$

Taking the limit $t \to +\infty$, (7.2.88) gives the famous Landau-Zener formula: $|\tilde{v}_q(t_f=+\infty)|^2 = e^{-2\pi\beta_q^2\tau}$ (Sect. 7.A.2). Therefore the density of defects is obtained as

$$n = \frac{1}{N}\sum_{q>0} 2\exp\left(-2\pi\beta_q^2\tau\right). \tag{7.2.95}$$

Taking the thermodynamic limit,

$$n \to 2\int_0^\pi \frac{dq}{2\pi}\exp\left(-2\pi\tau\sin^2 q\right) = 4\int_0^{\pi/2}\frac{dq}{2\pi}\exp\left(-2\pi\tau\sin^2\right), \tag{7.2.96}$$

where we have used the property that the integrand is a symmetric function with the pivot at $q=\pi/2$. For $\tau \gg 1$, (7.2.96) is dominated by the integral over the region of q where $0 \le \sin q \approx q \ll 1$. Hence one may safely approximate $\sin^2 q \approx q^2$ and expand the upper bound of q in the integral from $\pi/2$ to ∞. The result is given as

$$n \approx 4\int_0^\infty \frac{dq}{2\pi}\exp\left(-2\pi\tau q^2\right) = \frac{1}{\pi\sqrt{2\tau}}. \tag{7.2.97}$$

This result implies that the density of defects decays as $\tau^{-1/2}$ with increasing τ and that it vanishes with $\tau \to \infty$. One finds from this fact that, even if the quench is done across the quantum critical point from a disordered phase to an ordered phase in the one-dimensional transverse Ising chain, the system evolves adiabatically into an ordered state as far as the rate of the quench is infinitesimally small. The exponent $1/2$ of τ which characterises the decay rate of n is a universal number. We shall show below the universality of this exponent on the basis of a generic argument.

(ii) case with $t_f=\tau$, where the final value of the transverse field is $g_f=0$. In the present case, the density of defects can be expressed as

$$\begin{aligned} n &= \frac{1}{2N}\langle\Psi(t_f)|H(g_f)+N|\Psi(t_f)\rangle \\ &= \frac{1}{2N}\langle\Psi(t_f)|\sum_i\left(1-S_i^x S_{i+1}^x\right)|\Psi(t_f)\rangle. \end{aligned} \tag{7.2.98}$$

Note that $(1/2)\sum_i(1-S_i^x S_{i+1}^x)$ is the number of kinks between neighbouring Ising spins. Although there is no kink in the instantaneous ground state at the final time, a quench of the transverse field may produce such kinks. The density of defects n, in the present case, measures the amount of kinks present in the state after a quench. Now, substitution of $g_f=0$ in (7.2.8) leads to $u_q(g_f=0) = -\sin(q/2)$ and $v_q(g_f=0)=\cos(q/2)$. From this, one obtains

$$R_q(g_f=0) = \begin{pmatrix} -\sin\frac{q}{2} & i\cos\frac{q}{2} \\ i\cos\frac{q}{2} & -\sin\frac{q}{2} \end{pmatrix}, \tag{7.2.99}$$

and hence

$$\begin{aligned}&\left[S_q^\dagger(t_f)R_q^\dagger(g_f)\sigma^3 R_q(g_f)S_q(t_f)\right]_{22}\\&\quad=\cos q\left(\left|\tilde{u}_q(\tau)\right|^2-\left|\tilde{v}_q(\tau)\right|^2\right)\\&\qquad-i\sin q\left(\tilde{u}_q(\tau)\tilde{v}_q^*(\tau)-\tilde{u}_q^*(\tau)\tilde{v}_q(\tau)\right).\end{aligned}\tag{7.2.100}$$

The density of defects in the thermodynamic limit is then written as

$$n=\int_0^\pi\frac{dq}{2\pi}\left\{\cos q\left(1-2\left|\tilde{v}_q(\tau)\right|^2\right)+1-i\sin q\left(\tilde{u}_q(\tau)\tilde{v}_q^*(\tau)-\tilde{u}_q^*(\tau)\tilde{v}_q(\tau)\right)\right\}$$

where we have used $|\tilde{u}_q(\tau)|^2+|\tilde{v}_q(\tau)|^2=1$. We note that $\tilde{v}_q(\tau)$ is given by (7.2.88) with the substitution of $t_f=\tau$. Let us consider the contribution to the density of defects from n_q with $\pi-1/\tau^{1/4}\lesssim q\leq\pi$:

$$\begin{aligned}\int_{\pi-1/\tau^{1/4}}^{\pi}\frac{dq}{2\pi}n_q&\approx\int_{\pi-1/\tau^{1/4}}^{\pi}\frac{dq}{2\pi}\left\{\left(-1+O\left(\frac{1}{\sqrt{\tau}}\right)\right)\left(1-2\left|\tilde{v}_q(\tau)\right|^2\right)+1\right\}\\&\quad-i\int_{\pi-1/\tau^{1/4}}^{\pi}\frac{dq}{2\pi}\sin q\left(\tilde{u}_q(\tau)\tilde{v}_q^*(\tau)-\tilde{u}_q^*(\infty)\tilde{v}_q(\infty)\right).\end{aligned}\tag{7.2.101}$$

Neglecting the higher orders in $1/\tau$, the first term is arranged into

$$\int_{\pi-1/\tau^{1/4}}^{\pi}\frac{dq}{\pi}\left|\tilde{v}_q(\tau)\right|^2.\tag{7.2.102}$$

When $\tau\gg1$, $|\tilde{v}_q(\tau)|^2$ is approximated as $|\tilde{v}_q(\tau)|^2\approx\exp(-2\pi\tau\sin^2q)$ for $\pi-1/\sqrt{\tau}\lesssim q\leq\pi$. Although this approximation is not always valid for $\pi-1/\tau^{1/4}\lesssim q\lesssim\pi-1/\sqrt{\tau}$, the integrand becomes small and one can neglect its contribution to the integral. Thus Eq. (7.2.102) is estimated as

$$\begin{aligned}\int_{\pi-1/\tau^{1/4}}^{\pi}\frac{dq}{\pi}\left|\tilde{v}_q(\tau)\right|^2&\approx\int_{\pi-1/\tau^{1/4}}^{\pi}\frac{dq}{\pi}e^{-2\pi\tau\sin^2q}=\int_0^{1/\tau^{1/4}}\frac{dq}{\pi}e^{-2\pi\tau\sin^2q}\\&=\int_0^{1/\tau^{1/4}}\frac{dq}{\pi}e^{-2\pi\tau q^2}=\frac{1}{\sqrt{\tau}}\int_0^{\tau^{1/4}}\frac{dq}{\pi}e^{-2\pi q^2},\end{aligned}\tag{7.2.103}$$

where we have changed the variable as $\pi-q\to q$ and applied an approximation $\sin q\approx q$ for $0\leq q\leq1/\sqrt{\tau}$. When $\tau\gg1$, one may replace the upper bound of integral with ∞. Therefore one obtains

$$\int_{\pi-1/\tau^{1/4}}^{\pi}\frac{dq}{\pi}\left|\tilde{v}_q(\tau)\right|^2\approx\frac{1}{\sqrt{\tau}}\int_0^\infty\frac{dq}{\pi}e^{-2\pi q^2}=\frac{1}{2\pi}\frac{1}{\sqrt{2\tau}}.\tag{7.2.104}$$

A detailed calculation shows that not only the second term in (7.2.101) but also the integral of n_q over $0\leq q\lesssim\pi-1/\tau^{1/4}$ gives higher order terms. Therefore the density of defects is obtained as [118]

$$n\approx\frac{1}{2\pi}\frac{1}{\sqrt{2\tau}}.\tag{7.2.105}$$

We find from (7.2.97) and (7.2.105) that both the cases (i) and (ii) yield the same result except for the factor 2. This difference comes from the fact that the quench in the case (i) passes two quantum critical points at $g = 1$ and -1, while that in the case (ii) passes only the single critical point at $g = 1$. We recall that, when $g = 1$, the energy spectrum ω_q of the quasiparticle has a zero at $q = \pi$, whereas it does at $q = 0$ when $g = -1$. This implies that quasiparticles with the mode near $q = \pi$ and $q = 0$ are excited during the quench and they contribute to the density of defects in the case (i). However only the modes near $q = \pi$ are responsible for the density of defects in the case (ii). Thus the number of modes which contribute to the excitation is twice larger in the case (i) than the case (ii).

Kibble-Zurek Scaling We have so far investigated the quench dynamics in one-dimensional transverse Ising chain, and derived scaling relation between the density of defects n and the quench rate $1/\tau$. Hereafter we investigate the defect production due to a quench across a critical point in a generic system. Historically the defect production due to a quench was first discussed for a thermodynamic phase transition in the context of the evolution of the universe by Kibble [223]. After that, Zurek [444] proposed an experiment for defect production in the universe using a ^{4}He superfluid. A theory on a quench across a quantum critical point was developed by Zurek et al. [445] and Polkovnikov [318].

Let us consider a system with a parameter λ. We assume that the ground state is disordered for $\lambda > 0$ and ordered for $\lambda < 0$. The system undergoes a quantum phase transition of the second order at $\lambda = \lambda_c = 0$. We suppose that the energy gap Δ from the ground state to the first excited state is scaled by λ as $\Delta \sim \lambda^{z\nu}$, and the quasiparticle has a dispersion relation $\omega_{\mathbf{q}} \sim |\mathbf{q}|^z$ at the critical point, where $\mathbf{q}$ is the wave-number vector of the quasiparticle.

Now we assume a linear quench of the parameter λ:

$$\lambda = -t/\tau, \tag{7.2.106}$$

where $1/\tau$ is the quench rate. The time t is supposed to move from $t = -\infty$ to $t > 0$.

The coherence time of the system, τ_{coh}, is determined by the inverse of the energy gap: $\tau_{\text{coh}} = \Delta^{-1} \sim \lambda^{-z\nu}$. Roughly speaking, if the normalised quench rate $|\dot{\Delta}|/\Delta$ is smaller than τ_{coh}^{-1}, then the system maintains the coherence and evolves adiabatically. However, one the other hand, if $|\dot{\Delta}|/\Delta > \tau_{\text{coh}}^{-1}$, the adiabatic evolution breaks down. On the basis of this consideration, we make an equation so that it determines the time $\hat{t}$ (namely, the parameter $\hat{\lambda} = \lambda(\hat{t})$) at which the adiabaticity breaks:

$$\frac{|\dot{\Delta}|}{\Delta} = \frac{1}{\tau_{\text{coh}}}, \tag{7.2.107}$$

which is followed by

$$\frac{1}{\hat{\lambda}\tau} = \hat{\lambda}^{z\nu}, \tag{7.2.108}$$

where we have neglected inessential factors. The solution of this equation is

$$\hat{\lambda} = \tau^{-1/(z\nu+1)}. \tag{7.2.109}$$

$\hat{\lambda}$ gives an energy scale $\hat{\Delta} \sim \hat{\lambda}^{z\nu}$. Equation (7.2.109) implies that quasiparticles with energy $\lesssim \hat{\Delta} \sim \tau^{-z\nu/(z\nu+1)}$ are excited during a quench. Since the quasiparticle has the dispersion $\omega_{\mathbf{q}} \sim |\mathbf{q}|^z$, the largest wave number $\hat{q} = |\hat{\mathbf{q}}|$, that is likely to contribute to the excitation, satisfies

$$\hat{q}^z \sim \hat{\Delta} \sim \tau^{-z\nu/(z\nu+1)}. \tag{7.2.110}$$

Once quasiparticle are excited, they never relax. The density of defects after a quench is evaluated by the volume in the $\mathbf{q}$-space which contributes to the excitation. Therefore we obtain

$$n \sim \int_{|\mathbf{q}|\leq\hat{q}} d\mathbf{q} \sim \hat{q}^d \sim \tau^{-d\nu/(z\nu+1)}, \tag{7.2.111}$$

where d is the dimension of the system. Equation (7.2.111) provides a generic scaling relation between the density of defects and the quench rate with critical exponents ν and z and the dimension d. In the case of the one-dimensional transverse Ising chain, one has $d = z = \nu = 1$. Thus one has $n \sim \tau^{-1/2}$ which is in consistence with (7.2.97) and (7.2.105).

7.2.3 Oscillating Fields: Quantum Hysteresis

When the external perturbation, on a cooperatively interacting system, is sinusoidal in time, the system cannot respond instantaneously and the (cooperative) response gets delayed leading to hysteresis, which is a typical non-equilibrium phenomenon. At a rudimentary level, hysteresis manifests itself as the competition between experimental time scales, determined by the frequencies of the applied perturbation, and internal time scales which are governed by relaxation phenomena, activated rate processes, decay of metastable states and etc. Since metastable states occur quite naturally in connection with first order phase transitions, hysteresis is particularly noticeable near such transitions, and is often used to mark their onset.

The interest in studying hysteresis in quantum systems is because of the fact that quantum systems are characterised by new routes to relaxation. Thus a quantum system has additional time scales associated with tunnelling, which can link different minima of the free energy surface.

As far as hysteresis in classical systems is concerned, there have been several recent attempts to provide a satisfactory statistical mechanical treatment. In most of the model studies, the scaling of the hysteresis loop area with the probe frequency and temperature as well as the evidence of a dynamic phase transition have been the issues of interest [2–4, 22].

The simplest quantum Ising model which is expected to show hysteresis phenomena is the Ising model in a sinusoidally varying transverse field. The study of hysteresis in Ising models in the presence of a oscillating transverse field is also motivated by the fact that it can actually be experimentally realised in the laboratory by pressure modulation in the KDP crystal or by simply applying an oscillatory magnetic field to the rare earth system of $LiRF_4$.

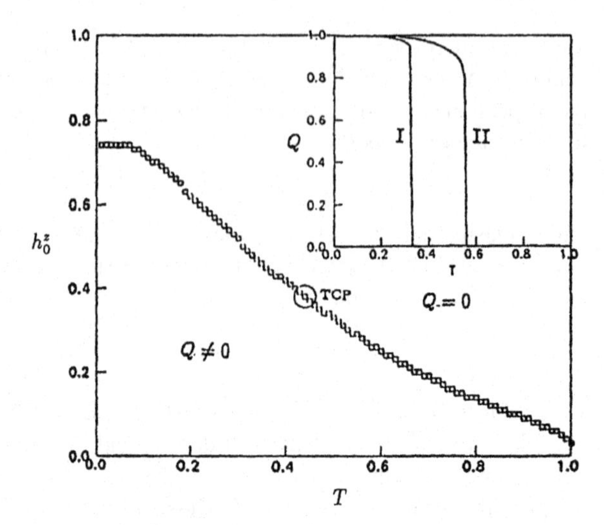

Fig. 7.6 Projection of the dynamic phase boundary (and TCP) in the h_0^z–T plane. The *inset* shows the change of the nature of the phase transition just above (*I*: $h_0^z = 0.48$) and below (*II*: $h_0^z = 0.28$) the tricritical point (TCP). $\Gamma = 0.2$ and $\omega = 0.0314$ here

The interesting points are mainly twofold: (a) The scaling of the loop area $A_x = \oint m^x\, d\Gamma$ with the transverse field and temperature. This loop area, being related to the energy loss, is directly measurable in the laboratory. (b) The dynamic phase transition. This is obtained from the vanishing of the order parameter defined by $Q = \oint m^z\, dt$, which acquires nonzero values (below the phase boundary) for asymmetric variations at low T, Γ etc. Two different cases, leading to hysteresis, have been considered here in quantum Ising systems: (i) The longitudinal field is oscillatory, the transverse field is time independent and (ii) the longitudinal field is zero, the transverse one is oscillatory. In all the studies, the Hamiltonian is treated in the mean field approximation and the equations of motion for the average magnetisation obtained.

In general, in presence of both time dependent longitudinal and transverse fields, the Ising Hamiltonian can be written generally as

$$H = -\sum_{\langle ij\rangle} J_{ij} S_i^z S_j^z - \Gamma(t)\sum_i S_i^x - h^z(t)\sum_i S_i^z. \tag{7.2.112}$$

Here $\Gamma(t)$ and $h^z(t)$ represent the time dependent transverse or longitudinal field respectively. Acharyya, Chakrabarti and Stinchcombe [4] proposed the phenomenological extension and generalisation of the mean field equation of motion for average magnetisation $\boldsymbol{m}$, following that of the classical Ising system in contact with a heat bath:

$$\tau\frac{d\boldsymbol{m}}{dt} = -\boldsymbol{m} + \tanh\left(\frac{|h|}{T}\right)\frac{\boldsymbol{h}}{|h|}, \tag{7.2.113}$$

where

$$|h| = \sqrt{\left(m^z + h^z(t)\right)^2 + \left(\Gamma(t)\right)^2}. \tag{7.2.114}$$

Banerjee, Dattagupta and Sen [22] have recently given a more microscopic derivation of the equations of motion (see Sect. 7.2.3; case III).

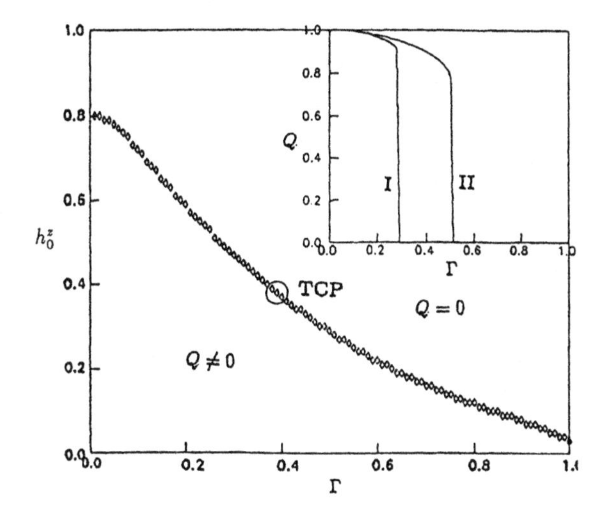

Fig. 7.7 Projection of the dynamic phase boundary (and TCP) in the h_0^z–Γ plane. The *inset* shows the change of the nature of the phase transition just above (*I*: $h_0^z = 0.48$) and below (*II*: $h_0^z = 0.28$) the tricritical point (TCP). $T = 0.2$ and $\omega = 0.0314$ here

Case I Ising system in transverse field and oscillating longitudinal field.

In presence of an oscillating longitudinal field and transverse field, we put $\Gamma(t) = \Gamma$ and $h^z(t) = \cos(\omega t)$ in (7.2.112) so that $|h| = \sqrt{(m^z + h^z(t))^2 + \Gamma^2}$ in (7.2.113). Dynamic phase transitions and AC susceptibility have been studied numerically by solving the equation of motion with these expressions.

Dynamic Phase Transition Using a fourth-order Runge-Kutta method (in single precision; the value of the time differential (dt) was taken to be 10^{-6}), the above dynamical equations (7.2.113) were solved. The dynamic order parameter Q ($= \oint m^z \, dt$) was also evaluated.

For the model considered here (represented by Hamiltonian (7.2.112)), where Γ is constant in time and $h^z(t) = \cos(\omega t)$, there is a dynamic phase boundary $T_d(h_0^z, \omega, \Gamma)$ (separating $Q \neq 0$ phase from $Q = 0$ phase). The projection of this boundary in the h_0^z–T plane for a fixed Γ and ω (Fig. 7.6) is found. A crossover of this transition across $T_d(h_0^z, \omega, \Gamma)$, from a discontinuous to a continuous one, at a tricritical point (TCP) $T_d^{\text{TCP}}(h_0^z, \omega, \Gamma)$ (see inset of Fig. 7.6), showing the nature of the transition just below and above the TCP at a particular ω and Γ has been observed. Similar behaviour of Q has been observed for other projections of the phase boundary. Figure 7.7 shows the phase boundary line $\Gamma_d(h_0^z, \omega, T)$ in the h_0^z–Γ plane (separating $Q \neq 0$ phase from $Q = 0$ phase). The position of the tricritical point $\Gamma_c^{\text{TCP}}(h_0^z, \omega, T)$ is also indicated on the phase boundary line. Inset of Fig. 7.7 shows the nature of the transition just below and above the TCP at a particular ω and T.

The dynamic phase transition, in fact, arises due to the coercivity property. In the $Q \neq 0$ phase, because of the failure of the external field to provide for the coercive field, the m–h loop is not symmetric about the field axis and lies in the upper half (or lower half) of the m–h plane depending upon the initial magnetisation. So, the

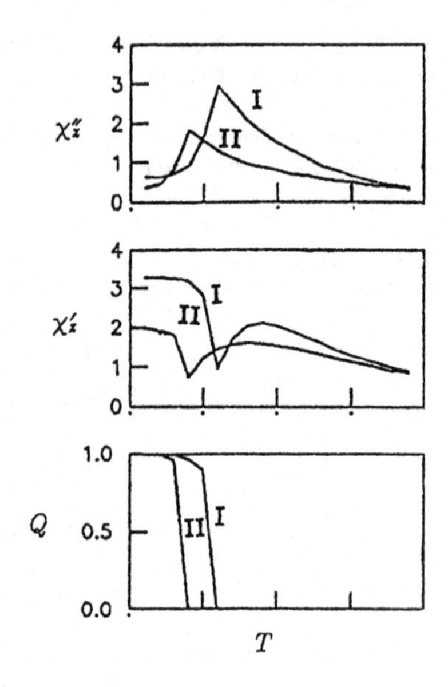

Fig. 7.8 Temperature variation of χ_z', χ_z'', and Q for two different values of h_0^z and a fixed value of $\omega = 0.0314$ and $\Gamma = 0.1$. (*I*) For $h_0^z = 0.3$ and (*II*) $h_0^z = 0.5$

phase boundary equation $T_d(h_0^z, \omega, \Gamma)$ for the dynamic phase transition, when expressed as $h_0^z(T, \omega, \Gamma)$, gives in effect the coercive field variation with respect to T, Γ and ω. The tricritical point $T_d^{\mathrm{TCP}}(h_0^z, \omega, \Gamma)$ appears because of the failure of the system to relax within the time period $(2\pi/\omega)$ of the external field. The intrinsic relaxation time τ_{eff} in the ferro phase decreases with lowering of temperature and below $T_d^{\mathrm{TCP}}(h_0^z, \omega, \Gamma)$: $\tau_{\mathrm{eff}}(h_0^z, T) \geq 2\pi/\omega$ (equality at $T = T_d^{\mathrm{TCP}}$), so that the magnetisation changes sign (from m^z to $-m^z$) abruptly and continuously and Q changes from a value very near to unity to 0 discontinuously. This indicates that $T_d^{\mathrm{TCP}}(h_0^z, \omega, \Gamma)$ should decrease with increasing frequency as is indeed observed (see the inset of Fig. 7.6). The same is also true for $\Gamma_d^{\mathrm{TCP}}(h_0^z, \omega, T)$ (see the inset of Fig. 7.7).

The AC Susceptibility Following the successful introduction of AC susceptibility [2] for (classical) Ising system, the properties of similarly defined (linear) AC susceptibility in transverse Ising system have been studied.

In this case, by solving the mean-field equation of motion the time variation of m^x and m^z is obtained. Both the transverse and longitudinal magnetisation showed that the responses are delayed but having the same frequency of the perturbing oscillating field. The amount of delay is different for transverse and longitudinal magnetisation. The response magnetisation $m^\alpha(t)$ can be expressed as $P^\alpha(\omega(t - \tau_{\mathrm{eff}}))$ where P^α denotes the periodic function with the same frequency ω of the perturbing field and $\tau_{\mathrm{eff}}^\alpha$ denotes the effective delay for the α-th component $(\alpha = x, z)$ of the response. The susceptibilities are defined in a "linear" way: assuming a linear response $m^\alpha(t) \sim m_0^\alpha \exp(i\omega t - \phi^\alpha)$; $\phi^\alpha = \omega_{\mathrm{eff}}^\alpha$, for a perturbation $h^z(t) \sim h_0^z \exp(-i\omega t)$, the AC susceptibility χ_α is defined as $(m_0^\alpha/h_0^z)\exp(-i\phi^\alpha)$. This defines then the in-phase and the out-of-phase components of the AC suscepti-

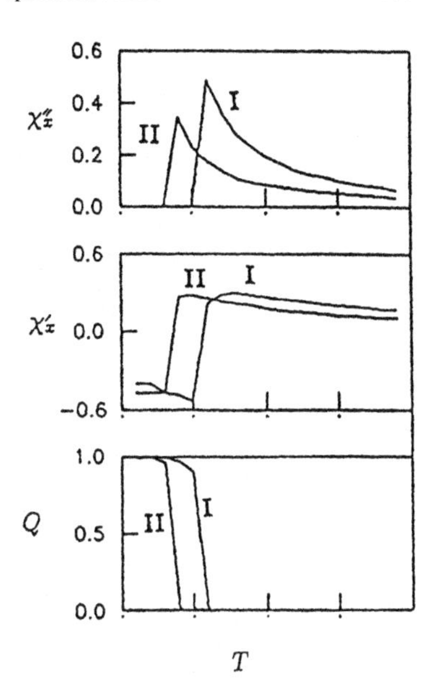

Fig. 7.9 Temperature variation of χ_x', χ_x'', and Q for two different values of h_0^z and a fixed value of $\omega = 0.0314$ and $\Gamma = 0.1$. (*I*) For $h_0^z = 0.3$ and (*II*) $h_0^z = 0.5$

bilities: $\chi_z' = (m_0^z/h_0^z)\cos(\phi_z)$, $\chi_z'' = (m_0^z/h_0^z)\sin(\phi_z)$ and $\chi_x' = (m_0^x/h_0^z)\cos(\phi_x)$, $\chi_x'' = (m_0^x/h_0^z)\sin(\phi_x)$.

The components of the transverse and longitudinal susceptibilities are plotted against the temperature (T) (see Fig. 7.8 and Fig. 7.9). At the dynamic transition point, where the dynamic order parameter Q vanishes, the χ_x' and χ_z' give sharp dips and χ_x'' and χ_z'' give sharp peaks. Both χ_x' and χ_z' have another smeared peak at some higher temperature ($T > T_c(\Gamma)$) indicating the high temperature decay of magnetisation.

In the limits $T^{-1} \to 0$ and $\Gamma \to 0$, the equation of motion for z-component gets completely decoupled suggesting that the dynamic transition etc. for the z-component remains qualitatively the same as in the classical mean field case [2].

Case II Hysteresis due to time dependent transverse field and zero longitudinal field.

Here the longitudinal field $h^z(t) = 0$ and the transverse field $\Gamma(t)$ is sinusoidally varying with time. As mentioned before, there have been two different approaches to study the case when the transverse field is explicitly time dependent and the longitudinal field is zero. In the first study [4], one again considers (7.2.113) putting $\Gamma(t) = \Gamma_0\cos(\omega t)$, so that $|h| = \sqrt{(m^z)^2 + (\Gamma(t))^2}$. The relaxation rates here have been assumed to be the same for both the components m^x and m^z.

The variation of the longitudinal and transverse magnetisation loop areas $A_x = \oint m^x\, d\Gamma$ and $A_z = \oint m^z\, d\Gamma$ respectively, and the dynamic order parameter $Q = \oint m^z\, dt$, as functions of the frequency (ω) and amplitude (Γ_0) of the periodically varying transverse field and the temperature (T) of the system have been studied [4].

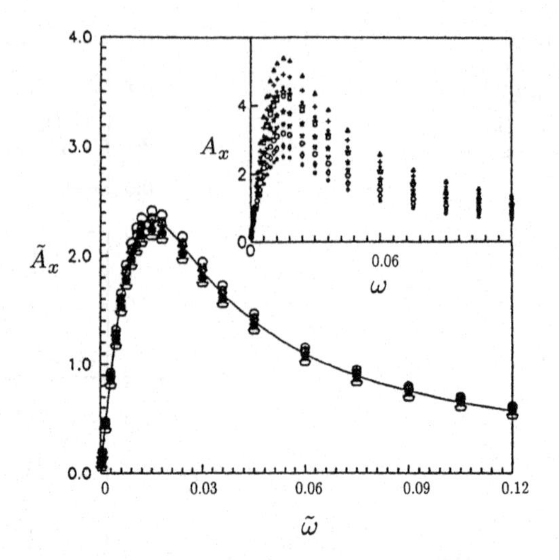

Fig. 7.10 The variation of the scaled loop area $\tilde{A}_x = A_x \Gamma_0^{-\alpha} T^{\beta}$ with the scaled frequency $\tilde{\omega} = \omega/(\Gamma_0^{\gamma} T^{\delta})$ using $\alpha = 7/4$, $\beta = 1/2$ and $\gamma = 0 = \delta$. Different symbols correspond to different Γ_0 and T: (○) $\Gamma_0 = 1.25$ and $T = 1.25$; (□) $\Gamma_0 = 1.50$ and $T = 1.25$; (△) $\Gamma_0 = 1.75$ and $T = 1.25$; (◇) $\Gamma_0 = 1.25$ and $T = 1.50$; (★) $\Gamma_0 = 1.50$ and $T = 1.50$; (+) $\Gamma_0 = 1.75$ and $T = 1.50$; (∗) $\Gamma_0 = 1.25$ and $T = 1.75$; (×) $\Gamma_0 = 1.50$ and $T = 1.75$; (⋆) $\Gamma_0 = 1.75$ and $T = 1.75$. The *solid line* indicates the proposed Lorentzian scaling function. The *inset* shows the variation of A_x with ω at different Γ_0 and T

This has been done by solving the mean field equations (7.2.113) of motion numerically for the average magnetisation $\boldsymbol{m} = \langle \boldsymbol{S}_i \rangle = m^x \hat{x} + m^z \hat{z}$.

It was found that the variation of the loop area A_x with frequency ω for different parameters (Γ_0 and T) can be expressed in a scaling form (see Fig. 7.10)

$$A_x \sim \Gamma_0^{\alpha} T^{-\beta} g\left(\frac{\omega}{\Gamma_0^{\gamma} T^{\delta}}\right) \tag{7.2.115}$$

with a Lorentzian scaling function

$$g(x) \sim \frac{x}{1 + cx^2}. \tag{7.2.116}$$

The best fit values for the exponents α, β, γ and δ were found to be around 1.75 ± 0.05, 0.50 ± 0.02, 0 ± 0.02 and 0 ± 0.02 respectively. The mean field equation of motion was solved analytically in three different (linearised) limits: (i) High temperature limit where $\alpha = 2$, $\beta = 1$, $\gamma = 0$ and $\delta = 0$; (ii) $\alpha = 2$, $\beta = 0 = \gamma = \delta$ in the low tunnelling field amplitude limit; and (iii) $\alpha = 1$, $\beta = 0 = \gamma = \delta$ in the (adiabatic) limit of very slowly varying transverse field; with the Lorentzian scaling function (7.2.115). These limiting results (of effectively linear analysis) for the exponent values give the useful bounds for the observed exponent values. (For a detailed discussion of these limits, see Sect. 7.A.1.1.)

The dynamic phase transition from $Q = 0$ (for high Γ_0 and T) to $Q \neq 0$ (beyond critical values of Γ_0 and T) occurs across the Γ_0–T line and the phase diagram

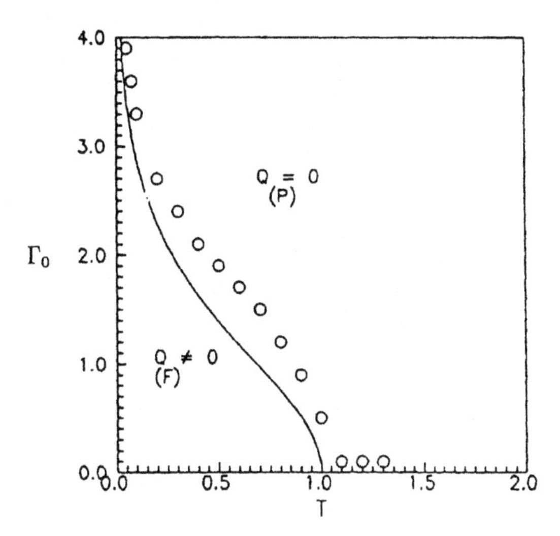

Fig. 7.11 Phase diagram for the dynamic phase transition: below the critical (T) line indicated by the symbols, Q acquires a nonzero value in the 'F' phase and $Q = 0$ in the 'P' phase. The *points* (symbols) correspond to $Q = 0$ for $\omega = 0.003$. The analytical estimate (7.2.117) of the phase boundary has been indicated by the *solid line*

(in the Γ_0–T plane) for this transition had been obtained (shown in Fig. 7.11). An analytic estimate for the phase boundary line (see Sect. 7.A.1.2)

$$T = \tilde{\Gamma}_0 / \sinh(\tilde{\Gamma}_0); \qquad \tilde{\Gamma}_0 = (\pi/2)\Gamma_0 \tag{7.2.117}$$

was also obtained [4], which gave a fair agreement with the numerical results.

Case III Quantum hysteresis with microscopic equations of motion.

In this third (general) case [22] the transverse field is considered to be rotating in the transverse plane. The equations of motion for the different components of magnetisation is derived here in a system plus heat bath approach, which reduce to the usual Glauber equations in the absence of the field. The system Hamiltonian for this case is

$$H_s = -\sum_{\langle i,j \rangle} J_{ij} S_i^z S_j^z - \Gamma_0 \cos 2\omega t \sum_i S_i^x - \Gamma_0 \sin 2\omega t \sum_i S_i^y, \tag{7.2.118}$$

where Γ_0 is the strength of the transverse field rotating with frequency 2ω (note the change in the frequency value, which is introduced for some later convenience). The rotating field selected as above introduces an x–y symmetry in the Hamiltonian in addition to making it quantum mechanical. In the mean field approximation, the system Hamiltonian can be expressed as

$$H_s \simeq -h^z S^z - \Gamma^x(t) S^x - \Gamma^y(t) S^y, \tag{7.2.119}$$

where the site-independent mean field $h^z = \sum_j J_{ij} \langle S_j^z \rangle$, $\Gamma^x(t) = \Gamma_0 \cos 2\omega t$ and $\Gamma^y(t) = \Gamma_0 \sin 2\omega t$.

Here the system and the heat bath coupling is considered in such a way, that in absence of any field the dynamic equations reduce to that given by Glauber kinetics. The starting point here is therefore the total Hamiltonian

$$H_{\text{tot}} = -h^z S^z - \Gamma_0 \big(S^x \cos 2\omega t + S^y \sin 2\omega t \big) + H_I + H_B. \tag{7.2.120}$$

The system Hamiltonian (containing the oscillatory terms) can be diagonalised by first a rotation ωt about the z-axis

$$U_z = \exp(-i\omega t S^z), \tag{7.2.121}$$

followed by the transformation R_y, which reads

$$R_y = \exp\left[-\frac{1}{2} S^y \tan^{-1}\left(\frac{\Gamma_0}{h^z + \omega}\right)\right]. \tag{7.2.122}$$

The transformed system Hamiltonian is:

$$\tilde{H}_s = -|h| S^z, \tag{7.2.123}$$

where $|h| = \sqrt{(h^z)^2 + \Gamma_0^2}$. As the transverse coupling is to be treated exactly, it is evident that the Hilbert space of the subsystem has to be enlarged now in order to incorporate both the terms proportional to h^z and Γ_0 within the Hamiltonian H_s. It is expected then that the interaction H_I with the heat bath ought to be such as to induce relaxation in both these terms. The minimal form of H_I to bring out the requisite physics is

$$\tilde{H}_I \sim \hat{b}(S^x + S^z), \tag{7.2.124}$$

where $\hat{b}$ represents the operator form of the bath variables. In the new representation of the rotated quantisation axis $\tilde{H}_I$ is entirely off-diagonal and therefore, responsible for inducing Glauber kinetics. It is also easy to see that in the original laboratory frame, the corresponding form of H_I is then (employing the inverse rotations U_z and R_y)

$$H_I = g\hat{b}\left[\left(\frac{h^z + \omega}{h}\right)(h^z S^x - \Gamma_0 S^z) + S^y\right]. \tag{7.2.125}$$

Under the two rotations given by (7.2.121) and (7.2.122), $\tilde{H}_I$ contains time dependent terms. However, within the Markovian limit of heat bath induced relaxation, the $\omega = 0$ limit of the coupling with the heat bath is the only important term, and in this limit

$$\tilde{H}_I = g\hat{b}(S^x + S^y), \tag{7.2.126}$$

which matches with the earlier form (cf. (7.2.124)). Collecting all the above mentioned facts together, the full time-dependent problem reduces to a time-independent one in the rotated frame, governed by the Hamiltonian

$$\tilde{H}_{\text{tot}} = \tilde{H}_s + \tilde{H}_I + H_B. \tag{7.2.127}$$

The rate equations for the components of the magnetisation are now derived to study the changes in the components of magnetisation in the formalism described in Sect. 7.A.3. After lengthy algebra involving the properties of Pauli matrices, regrouping of terms and using a short time approximation for the bath correlations characterised by λ the following rate equations for $\boldsymbol{m}$ are obtained [22] (with the Planck's constant $\hbar$ set equal to unity)

$$\frac{dm^x}{dt} = m^x\left[-\lambda - \lambda\frac{\Gamma_0^2}{|h|^2}\cos^2(2\omega t)\right] + m^y\left[2h^z - \lambda\frac{\Gamma_0^2}{|h|^2}\cos(2\omega t)\sin(2\omega t)\right]$$
$$+ m^z\left[-\lambda\frac{\Gamma_0}{|h|^2}(h^z+\omega)\cos(2\omega t) - 2\Gamma_0\sin(2\omega t)\right]$$
$$+ 2\lambda\frac{\Gamma_0}{|h|}\cos(2\omega t)\tanh\beta|h|, \tag{7.2.128}$$

$$\frac{dm^y}{dt} = m^x\left[-\lambda\frac{\Gamma_0^2}{|h|^2}\cos(2\omega t)\sin(2\omega t) - 2h^z\right] + m^y\left[-\lambda\frac{\Gamma_0^2}{|h|^2}\sin^2(2\omega t) - \lambda\right]$$
$$+ m^z\left[-\lambda\frac{\Gamma_0}{|h|^2}(h^z+\omega)\sin(2\omega t) + 2\Gamma_0\cos(2\omega t)\right]$$
$$+ 2\lambda\frac{\Gamma_0}{|h|}\sin(2\omega t)\tanh\beta|h|, \tag{7.2.129}$$

$$\frac{dm^z}{dt} = m^x\left[-\lambda\frac{\Gamma_0}{|h|^2}(h^z+\omega)\cos(2\omega t) + 2\Gamma_0\sin(2\omega t)\right]$$
$$+ m^y\left[-\lambda\frac{\Gamma_0}{|h|^2}(h^z+\omega)\sin(2\omega t) - 2\Gamma_0\cos(2\omega t)\right]$$
$$+ m^z\left[-2\lambda + \lambda\frac{\Gamma_0^2}{|h|^2}\right] + 2\lambda\frac{(h^z+\omega)}{|h|}\tanh\beta|h|. \tag{7.2.130}$$

It may be noted that the relaxation rates come out to be different for m^z and $m^x(m^y)$. Although (7.2.128)–(7.2.130) reduce to (7.2.113) in the limit $\omega \to 0$, the nonlinear equations (7.2.128)–(7.2.130) admit other unphysical solutions as well. Solving these equations numerically, the hysteresis loop area A ($= \int \boldsymbol{m}\cdot d\boldsymbol{\Gamma} = \int m^x\, d\Gamma^x + \int m^y\, d\Gamma^y$) is obtained for the physical solution. In the low frequency limit ($\omega \to 0$), the scaling of the area as functions of temperature T and transverse field Γ_0 can be fitted, in the high temperature limit, in the form

$$A \sim \Gamma_0^a T^{-b} \tag{7.2.131}$$

with $a \simeq 2.03 + 0.03$ and $b \simeq 0.12 + 0.02$. It may be noted that the scaling form (7.2.115) can be reduced to the above form in the $\omega \to 0$ limit with $a = \alpha - \gamma$ and $b = \beta + \delta$ in the high temperature limit. In this limit, the analytic solution of (7.2.113) gave the values of $a = 2.0$ and $b = 1.0$, compared to the above values obtained here.

The data indicate that the area initially increases with ω, goes through a maximum and decreases to zero as ω approaches infinity, when the system can no longer respond to the fast varying field. However, the scaled area $\tilde{A}$ is skewed as a function of ω (shown in Fig. 7.12) and is not a Lorentzian as seen in the previous study [4] (using the equation of motion (7.2.113); see case II above).

7.2.3.1 Exact Results for a Transverse Ising Chain

In the preceding sections, we have studied hysteresis in classical and quantum Ising systems. The case of quantum hysteresis was however studied using mean field

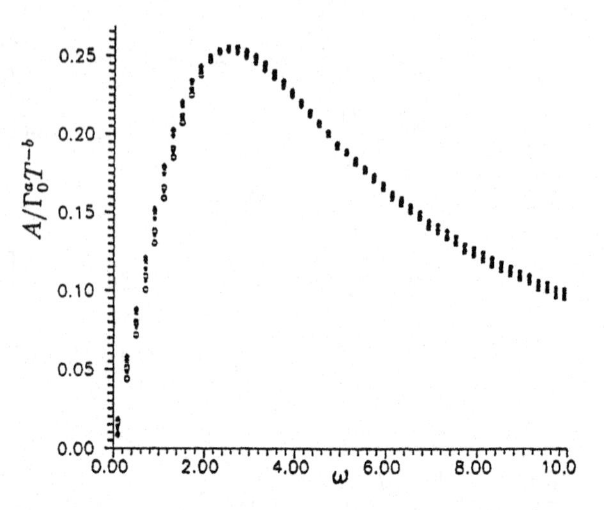

Fig. 7.12 Scaled area versus the frequency of the rotating field is shown for six different sets of values of temperature and strength of transverse field: $\Gamma_0 = 0.1$ (×), $\Gamma_0 = 0.3$ (∗), $\Gamma_0 = 0.5$ (◊) (all for $T = 4.0$) $\gamma_0 = 0.1$ (∘), $\Gamma_0 = 0.3$ (□), $\Gamma_0 = 0.5$ (□) (all for $T = 6.0$)

approximation only. Recent studies have shown that this approximation is inadequate to study quantum hysteresis and exact solutions reveal spectacular aspects not observed in classical models. Till now such studies have been made only for the case of an Ising chain with nearest-neighbour ferromagnetic interaction (along the longitudinal direction) subjected to a transverse field that oscillates with time [38, 92].

Classical dynamical hysteresis are characterised by the fact that they can be inferred from comparison of two time-scales: one is the intrinsic relaxation time τ of the system, which is independent of the driving field, and the other is the time period $\mathcal{T}$ of the driving field. The system falls out of equilibrium when $\mathcal{T} \ll \tau$, i.e. when the driving is too fast for the state of system to adjust with the time-dependent field. Thus, the dynamics will always be a *monotonic* function of $\mathcal{T}$. This robust picture has been observed ubiquitously in many studies in classical thermal systems, and is summarised in [63]. For example, in case of sinusoidally driven classical Ising magnet in three-dimension, the ordered phase shrinks monotonically as $\mathcal{T}$ is increased, and finally becomes a line coincidental with the temperature-axis ending at the critical temperature.

But in a periodically driven quantum many-body system there can be another kind of process which surprisingly violates this classical intuition of monotonic competition between two kinds of time-scales and give rise to sharp *non-monotonic freezing behaviour* as a function of the driving period $\mathcal{T}$ [38, 92]. It has been shown for transverse Ising chain driven at zero temperature by a periodic field that there can be surprising non-monotonic peak-valley structures in the profile of the order parameter Q (infinite-time average of the transverse magnetisation) when plotted against time period or amplitude of the external field (Fig. 7.13). The position of the peaks can be derived analytically to be

$$\frac{\Gamma_0 \mathcal{T}}{\pi \hbar} = m \tag{7.2.132}$$

where Γ_0 is the amplitude of the field and $m = 1, 2, 3, \ldots$ for square-wave field and $\frac{1}{2}z_1, \frac{1}{2}z_2, \frac{1}{2}z_3, \ldots$ for sinusoidal field. Here z_i are the zero-s of Bessel function of the

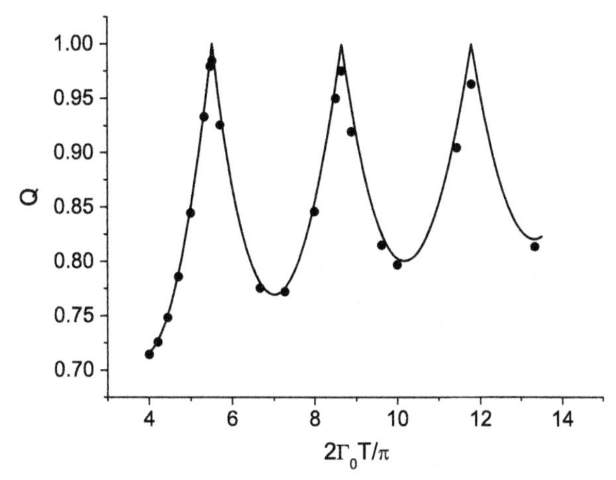

Fig. 7.13 Plot of order parameter Q for $\Gamma_0 = 10$, in the case of sinusoidal field. The position of the peaks are z_i (see (7.2.132)). The *points* are obtained by solving numerically the time-dependent Schrödinger equation and the *line* is from an analytic expression derived in [92]

first kind and zero-th order. When this condition is satisfied, one gets $Q \approx 1$ and the system remains nearly "frozen" to the initial state so that the dynamics is negligible. Note that such freezing is *independent* of the initial state—whether ordered or not. This excludes the possibility of some accidental symmetry suppressing certain transitions, resulting in the freezing. Another remarkable thing is that, this confinement does not arise from any kind of energy barrier—unlike the classical cases, where a freezing always has a long (macroscopic) but finite life-time after which transition happens, but here the freezing sustains actually in the true $t \to \infty$ limit. One must, however, note that this effect is observable only when $\frac{1}{\mathcal{T}}, \Gamma_0 \gg J$, where J is the order of the interaction strength between the spins.

7.2.4 Response due to a Pulsed Transverse Field in Absence of a Longitudinal Field

In order to study another interesting dynamic response of quantum systems where the perturbation is time dependent, the response of pulsed transverse field on an Ising system has been studied. In the Hamiltonian (7.2.112) the time variation of the transverse field is taken as

$$\begin{aligned} \Gamma(t) &= \Gamma + \delta\Gamma, \quad \text{for } t_0 \le t \le t_0 + \delta t \\ &= \Gamma \quad \text{elsewhere.} \end{aligned} \tag{7.2.133}$$

Accordingly, the mean field dynamic equation of motion (7.2.113) for magnetisation m is solved numerically [3]. As mentioned before, such a (tunnelling term) pulse can be applied to order-disorder ferroelectrics by applying a pressure pulse. Here, the pulse has been applied at equilibrium.

First, the system was allowed to relax to its equilibrium state at any temperature (T) and then the pulse of small amplitude and short duration (δt) (compared to the

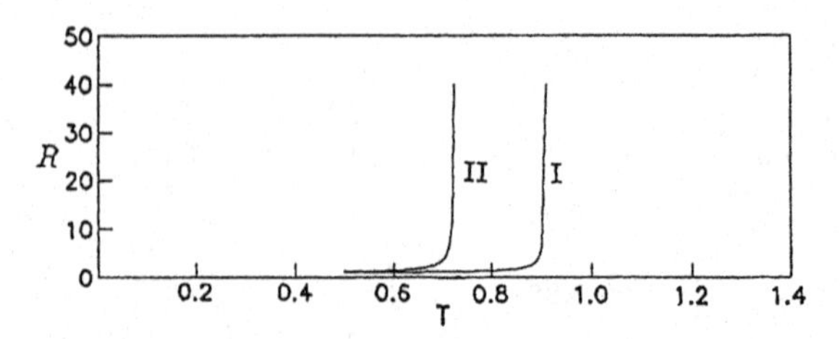

Fig. 7.14 Temperature variation of the width ratio R for different Γ for pulsed variation in Γ. *(I)* $\Gamma = 0.5$, $\delta\Gamma = 0.025$ and $\delta t = 50 \times dt$ and *(II)* $\Gamma = 0.8$, $\delta\Gamma = 0.025$ and $\delta t = 50 \times dt$ (dt is the time interval chosen for the Runge-Kutta solution of (7.2.113))

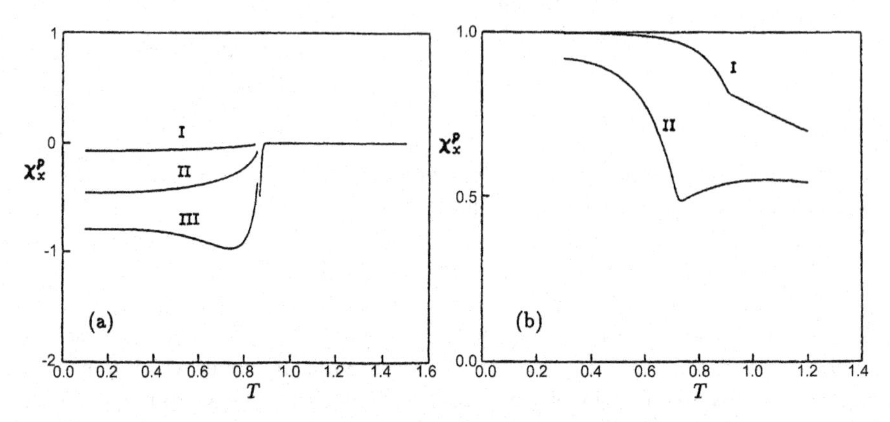

Fig. 7.15 (**a**) Temperature variation of χ_z^p for different values of the pulse width δt of the transverse field. *(I)* $\delta t = 2 \times dt$, *(II)* $\delta t = 10 \times dt$ and *(III)* $\delta t = 50 \times dt$. In all these cases $\Gamma = 0.6$ and $\delta\Gamma = 0.05$ (dt is the same as in the previous figure). (**b**) Temperature variation of χ_x^p for different Γ: *(I)* $\Gamma = 0.5$, *(II)* $\Gamma = 0.8$. The *kinks* indicate the transition points

relaxation time of the system) was applied. The response in transverse magnetisation was observed. Here, the transverse magnetisation m^z shows a dip of width Δt (starting at the time t_0 when the pulse has been applied). The ratio $R = \Delta t/\delta t$ has been measured and the temperature variation of R was studied. The observation (see Fig. 7.14) shows that R has a very sharp variation, almost diverging, at the order-disorder transition temperature ($T_c(\Gamma)$).

The pulse susceptibility has been defined as $\chi_z^p \equiv m_p^z/\delta\Gamma$, where m_p^z is the amplitude of the response magnetisation m^z. This is found to increase (possibly diverge) at the same point. It may be noted that χ_z^p becomes the static susceptibility as the pulse width becomes infinitely large ($\delta t \to \infty$). The transverse susceptibility defined as χ_x^p so shows significant change in the temperature variation at the same transition point (see Fig. 7.15).

The variation of χ_z^p with temperature for different values of the pulse width δt has also been noted to observe the effect of the pulse width. For small values of δt, the response is not prominent as the relaxation time is large compared to δt, for higher values there is some response which vanishes as one approaches the transition point where relaxation time also increases. For very large values of δt, the response is similar to that of static perturbations (static susceptibility has a sharp dip at $T_c(\Gamma)$) (see Fig. 7.15).

These studies on the responses to pulses (of the tunnelling field) on Ising system use the mean field equation of motion (7.2.113), where dynamics arise solely from the contact with heat bath. However, at extremely low temperatures (and for very high frequency measurements), the quantum dynamics should become very prominent. For example, in case of a pulse in transverse field, the change in the energy ($\delta\Gamma$) of the system will introduce a "quantum relaxation" time $\Delta t \sim \hbar/\delta\Gamma$, as dictated by the uncertainty relation. In such cases, this quantum time scale $\Delta t \sim \hbar/\delta\Gamma$, will compete with δt, the pulse width of the tunnelling field.

Appendix 7.A

7.A.1 Mean Field Equation of Motion

7.A.1.1 Some Analytic Solutions in the Linearised Limit

Linearisation of (7.2.113) can be carried out in three limits: namely, in the high temperature ($T^{-1} \to 0$), the low tunnelling field amplitude ($\Gamma_0 \to 0$) and in the adiabatic ($\omega \to 0$) limits.

In the first two cases, the linearised equations of motion take the forms:

$$\tau \frac{dm^x}{dt} = -m^x + \frac{\Gamma(t)}{T} \tag{7.A.1}$$

and

$$\tau \frac{dm^z}{dt} = -\left(1 - \frac{1}{T}\right)m^z, \tag{7.A.2}$$

in the $T^{-1} \to 0$ limit, and

$$\tau \frac{dm^x}{dt} = -m^x + \tanh\left(\frac{m^z}{T}\right)\frac{\Gamma(t)}{m^z} \tag{7.A.3}$$

and

$$\tau \frac{dm^z}{dt} = -m^z + \tanh\left(\frac{m^z}{T}\right), \tag{7.A.4}$$

in the $\Gamma_0 \to 0$ limit. For $T > 1$, m^z decays to zero in the first case (7.A.1). The long time solution of the dynamical equation for m^z in [2] can be obtained easily as the attractive fixed point of the map describing the discretised form of the differential equation: $m^z(t+\tau) = \tanh[m^z(t)/T]$. This gives $m^z = \tanh(m^z/T) \neq 0$ for $T < 1$ and $m^z = 0$ for $T > 1$. Putting the solution for m^z for $T < 1$ into the dynamical equation for m^x in (7.A.2), reduces it to the same form as (7.A.1) but with $\Gamma(t)/T$

now replaced by $\Gamma(t)$ in the second term. Defining now $s \equiv \omega t$ and $\lambda = \omega t$, one can express the general solution of $m^x(t)$ as

$$m^x(t) = J\cos(s) + K\sin(s) \tag{7.A.5}$$

with $K = \lambda J = (\Gamma_0/\Gamma)[\lambda/(1+\lambda^2)]$ for $T > 1$ from (7.A.1) as well as from (7.A.2), and Γ_0/T replaced by Γ_0 for $T < 1$ (from (7.A.2)). The loop area $A_x = \oint m^x d\Gamma = 2K\Gamma_0$ can then be expressed in the form given in (7.2.116) (containing the full Lorentzian scaling function $g(\lambda)$). The resulting exponents are $\alpha = 2, \beta = 1$ and $\gamma = 0 = \delta$ in the $T^{-1} \to 0$ limit (from (7.A.1) and (7.A.2) for $T > 1$) and $\alpha = 2$, $\beta = 0 = \gamma = \delta$ in the $\Gamma_0 \to 0$ limit (from (7.A.2) for $T < 1$).

In the adiabatic ($\omega^{-1} \gg \tau$, or small λ) limit, one can write $\boldsymbol{m} = \boldsymbol{m}_0 + \delta\boldsymbol{m} + \mathcal{O}(\lambda^2)$, where $\boldsymbol{m}_0 \sim \mathcal{O}(\lambda^0)$ and $\delta\boldsymbol{m} \sim \mathcal{O}(\lambda^1)$: $\boldsymbol{m}_0 = [\tanh(|h|/T)](h/|h|)$. Collecting now the linear terms in λ, one gets

$$\delta m^x = -\lambda\frac{dm_0^x}{ds} \tag{7.A.6}$$

and

$$\delta m^z = \frac{\delta m^z}{T}\tanh\left(\frac{\Gamma}{T}\right). \tag{7.A.7}$$

For $T > 1$, $\delta m^z = 0$ (of course $m_0^z = 0$). Using this to obtain m_0^x for the right hand side of the other equation, one gets $\delta m^x = -\lambda\frac{d}{ds}\tanh[\Gamma(s)/T]$. Hence the loop area $A_x = \oint m^x d\Gamma = \oint(m_0^x + \delta m_x)d\Gamma + \mathcal{O}(\lambda^2)$, where $\Gamma = \Gamma_0\cos(s)$, can be expressed as

$$A_x \sim \omega\frac{\Gamma_0^2}{T}I\left(\frac{\Gamma_0}{T}\right) + \mathcal{O}(\omega^2), \quad \text{where } I(y) = \int_0^{2\pi}\sin^2(s)\,\mathrm{sech}^2\big[y\cos(s)\big]\,ds \tag{7.A.8}$$

where $y = \Gamma_0/T$. As $I(y)$ goes to π and $4/y$ in the small and large y limits respectively, the loop area A_x can again be expressed in the form (7.2.116) with $g(x) \sim x$ in the $x \to 0$ limit, and with $\alpha = 2$, $\beta = 1$, $\gamma = 0 = \delta$ in the $\Gamma_0 \ll T$ limit and with $\alpha = 1$, $\beta = 0 = \gamma = \delta$ in the $\Gamma_0 \gg T$ limit. Although these limiting values for the exponents are not observed (because of the inaccuracy of the Linearisation approximations in the range of the study), they provide useful bounds for the observed values, and the Lorentzian scaling function appears quite naturally here.

7.A.1.2 Approximate Analytic Form of Dynamic Phase Boundary

Using the mean field equation for the z-component of magnetisation (7.2.113)

$$\tau\frac{dm^z}{dt} = -m^z + \tanh\left(\frac{|h|}{T}\right)\frac{m^z}{|h|} \tag{7.A.9}$$

one can estimate approximately the dynamic phase boundary. For small m^z, $|h| \sim \Gamma(s)$, where $s = \omega t$. Defining, $\lambda = \omega\tau$, the above equation can be approximated as

$$\lambda\frac{dm^z}{dt} = -m^z + \tanh\left(\frac{\Gamma(s)}{T}\right)\frac{m^z}{\Gamma(s)} = \left[\frac{1}{\Gamma(s)}\tanh\left(\frac{\Gamma(s)}{T}\right) - 1\right]m^z \tag{7.A.10}$$

or

$$\frac{\lambda}{2\pi}\oint \frac{dm^z}{m^z} = \frac{1}{2\pi}\int_0^{2\pi}\left[\frac{1}{\Gamma(s)}\tanh\left(\frac{\Gamma(s)}{T}\right) - 1\right]ds. \tag{7.A.11}$$

The right hand side of the above equation gives the logarithm of the factor by which m^z grows over a cycle. So, the equation of the phase boundary can be written as

$$\frac{1}{\Gamma_0}F\left(\frac{\Gamma_0}{T}\right) - 1 = 0, \tag{7.A.12}$$

where

$$F(y) = \frac{1}{2\pi}\int_0^{2\pi}\frac{\tanh(y\cos(s))}{\cos(s)}ds; \quad y = \Gamma_0 T. \tag{7.A.13}$$

For large y, $y\cos(s) \sim y(s-\pi/2)$, and since the integrand in $F(y)$ contributes significantly near $s=\pi/2$, we expand the tanh term as $\tanh x = x/(1+x^2)^{1/2}$, and get

$$\begin{aligned} F(y) &= \frac{1}{2\pi}\int_0^{2\pi y}\frac{\tanh(z)}{z}dz = \frac{1}{\pi}\int_0^{\pi y}\frac{dz}{(1+z^2)^{1/2}} \\ &= \frac{2}{\pi}\sinh^{-1}(z)\Big|_{-\pi y/2}^{\pi y/2}. \end{aligned} \tag{7.A.14}$$

So the approximate analytic form of phase boundary is

$$T = \frac{\pi\Gamma_0/2}{\sinh(\pi\Gamma_0/2)}. \tag{7.A.15}$$

7.A.2 Landau-Zener Problem and Parabolic Cylinder Functions

Let us consider a time-dependent Hamiltonian of a two-level system,

$$H_{\mathrm{LZ}}(t) = -\alpha t\sigma^3 + \sigma^1 = \begin{pmatrix} -\alpha t & 1 \\ 1 & \alpha t\end{pmatrix}, \tag{7.A.16}$$

where σ^1 and σ^3 are Pauli matrices:

$$\sigma^1 = \begin{pmatrix} 0 & 1 \\ 1 & 0\end{pmatrix}, \quad \sigma^3 = \begin{pmatrix} 1 & 0 \\ 0 & -1\end{pmatrix}. \tag{7.A.17}$$

Assume that the time t moves from $t=-\infty$ to ∞. We express the basis vectors of the system as $|0\rangle = (1\ 0)^{\mathrm{T}}$ and $|1\rangle = (0\ 1)^{\mathrm{T}}$. The eigenstates of this system are $|0\rangle$ and $|1\rangle$ when $t=\pm\infty$, while those at any finite times are superpositions of $|0\rangle$ and $|1\rangle$. We assume that the initial state of the system is $|\Psi(-\infty)\rangle = |1\rangle$. The time evolution is determined by the Schrödinger equation:

$$i\frac{d}{dt}\big|\Psi(t)\big\rangle = H_{\mathrm{LZ}}(t)\big|\Psi(t)\big\rangle. \tag{7.A.18}$$

We expand $|\Psi(t)\rangle$ as

$$|\Psi(t)\rangle = \psi_0(t)|0\rangle + \psi_1(t)|1\rangle. \tag{7.A.19}$$

The Schrödinger equation yields

$$i\frac{d}{dt}\psi_0(t) = -\alpha t\psi_0(t) + \psi_1(t), \tag{7.A.20}$$

$$i\frac{d}{dt}\psi_1(t) = \psi_0(t) + \alpha t\psi_1(t). \tag{7.A.21}$$

Our problem is to solve this set of equations under the initial condition, $\psi_0(-\infty) = 0$ and $|\psi_1(-\infty)| = 1$, up to an unessential phase factor. This problem is called the Landau-Zener problem and the solution is given by the parabolic cylinder functions [91, 242, 243, 388, 413, 443].

Eliminating ψ_1 in (7.A.21) by using (7.A.20), one obtains an equation of $\psi_0(t)$ as

$$\frac{d^2}{dt^2}\psi_0(t) + \left(1 - i\alpha + \alpha^2 t^2\right)\psi_0(t) = 0. \tag{7.A.22}$$

Define a new variable z by

$$t = \frac{e^{i\pi/4}}{(2\alpha)^{1/2}}z. \tag{7.A.23}$$

Then Eq. (7.A.22) is arranged as

$$\frac{d^2}{dz^2}\psi_0\left(\frac{e^{i\pi/4}}{(2\alpha)^{1/2}}z\right) + \left(\frac{i}{2\alpha} + \frac{1}{2} - \frac{z^2}{4}\right)\psi_0\left(\frac{e^{i\pi/4}}{\sqrt{2\alpha}}z\right) = 0. \tag{7.A.24}$$

Writing $U(z) = \psi_0(e^{i\pi/4}z/(2\alpha)^{1/2})$, one obtains

$$\frac{d^2}{dz^2}U(z) + \left(p + \frac{1}{2} - \frac{z^2}{4}\right)U(z) = 0, \quad p = i/2\alpha. \tag{7.A.25}$$

The solutions of (7.A.25) are parabolic cylinder functions denoted by $D_p(z)$. If $D_p(z)$ is a solution, $D_p(-z)$, $D_{-p-1}(iz)$, and $D_{-p-1}(-iz)$ are also solutions of the same equation. These four functions are linearly dependent. For instance, $D_p(-z)$ and $D_{-p-1}(-iz)$ are expressed by $D_p(z)$ and $D_{-p-1}(iz)$ as

$$D_p(-z) = e^{ip\pi}D_p(z) + \frac{\sqrt{2\pi}}{\Gamma(-p)}e^{i(p+1)\pi/2}D_{-p-1}(iz), \tag{7.A.26}$$

$$D_{-p-1}(-iz) = \frac{\sqrt{2\pi}}{\Gamma(p+1)}e^{ip\pi/2}D_p(z) + e^{i(p+1)\pi}D_{-p-1}(iz). \tag{7.A.27}$$

$D_p(z)$ has following asymptotic expansions for $|z| \gg 1$ and $|z| \gg |p|$:

$$D_p(z) \approx e^{-z^2/4}z^p\left(1 + O\left(z^{-2}\right)\right) \quad \left[\text{for } |\arg z| < \frac{3}{4}\pi\right], \tag{7.A.28}$$

$$D_p(z) \approx e^{-z^2/4} z^p \left(1 + O\left(z^{-2}\right)\right) - \frac{\sqrt{2\pi}}{\Gamma(-p)} e^{ip\pi} e^{z^2/4} z^{-p-1} \left(1 + O\left(z^{-2}\right)\right)$$
$$\left[\text{for } \frac{1}{4}\pi < \arg z < \frac{5}{4}\pi\right], \tag{7.A.29}$$

$$D_p(z) \approx e^{-z^2/4} z^p \left(1 + O\left(z^{-2}\right)\right) - \frac{\sqrt{2\pi}}{\Gamma(-p)} e^{-ip\pi} e^{-z^2/4} z^{-p-1} \left(1 + O\left(z^{-2}\right)\right)$$
$$\left[\text{for } -\frac{5}{4}\pi < \arg z < -\frac{1}{4}\pi\right]. \tag{7.A.30}$$

$D_p(z)$ satisfies following recursion relations:

$$D_{p+1}(z) - zD_p(z) + pD_{p-1}(z) = 0, \tag{7.A.31}$$
$$\frac{d}{dz} D_p(z) + \frac{1}{2} z D_p(z) - pD_{p-1}(z) = 0, \tag{7.A.32}$$
$$\frac{d}{dz} D_p(z) - \frac{1}{2} z D_p(z) + D_{p+1}(z) = 0. \tag{7.A.33}$$

Now, the initial condition for $U(z)$ is written as $U(z \to e^{\pm i3\pi/4} \times \infty) \to 0$, where upper and lower signs correspond to $\alpha > 0$ and $\alpha < 0$ respectively. To find the solution which meets this condition, one needs to look into the asymptotic values $D_p(z)$, $D_p(-z)$, $D_{-p-1}(iz)$, and $D_{-p-1}(-iz)$. Equations (7.A.28)–(7.A.30) are followed by

$$\left|D_p(z)\right| \to e^{\mp 3\pi/8\alpha} \neq 0 \quad \left(z \to e^{\pm i3\pi/4} \times \infty\right), \tag{7.A.34}$$
$$\left|D_p(-z)\right| \to e^{\pm \pi/8\alpha} \neq 0 \quad \left(-z \to e^{\mp i\pi/4} \times \infty\right), \tag{7.A.35}$$

$$\left|D_{-p-1}(iz)\right| \to \begin{cases} \left|\frac{\sqrt{2\pi}}{\Gamma(i/2\alpha+1)} e^{-\pi/8\alpha}\right| \neq 0 & (iz \to e^{-i3\pi/4} \times \infty,\ \alpha > 0) \\ 0 & (iz \to e^{-i\pi/4} \times \infty,\ \alpha < 0) \end{cases}, \tag{7.A.36}$$

$$\left|D_{-p-1}(-iz)\right| \to \begin{cases} 0 & (-iz \to e^{i\pi/4} \times \infty,\ \alpha > 0) \\ \left|\frac{\sqrt{2\pi}}{\Gamma(i/2\alpha+1)} e^{\pi/8\alpha}\right| \neq 0 & (-iz \to e^{i3\pi/4} \times \infty,\ \alpha < 0) \end{cases} \tag{7.A.37}$$

where note that $p = i/2\alpha$ is a pure-imaginary number. With this observation, one finds that $D_{-p-1}(-iz)$ and $D_{-p-1}(iz)$ are appropriate solutions for $\alpha > 0$ and $\alpha < 0$ respectively. Therefore

$$\psi_0(t) = AD_{-p-1}(\mp iz) = AD_{-p-1}\left(\mp ie^{-i\pi/4}(2\alpha)^{1/2} t\right), \tag{7.A.38}$$

and

$$\begin{aligned} \psi_1(t) &= i\dot{\psi}_0(t) + \alpha t \psi_0(t) \\ &= \pm A(2\alpha)^{1/2} e^{-i\pi/4} \left\{ \frac{d}{d(\mp iz)} D_{-p-1}(\mp iz) - \frac{1}{2}(\mp iz) D_{-p-1}(\mp iz) \right\} \\ &= \pm A(2\alpha)^{1/2} e^{i3\pi/4} D_{-p}(\mp iz), \end{aligned} \tag{7.A.39}$$

where a recursion relation, (7.A.33), is used in the last equality. The factor A is determined by the initial condition: $|\psi_1(-\infty)| = 1$. An asymptotic expansion of $D_{-p}(\mp iz)$ for $z \to e^{\pm i3\pi/4} \times \infty$ yields

$$D_{-p}(-iz) \approx e^{z^2/4}(\mp iz)^{-p} + O\big(z^{-1}\big), \tag{7.A.40}$$

which is followed by

$$\big|D_{-p}\big(-iz \to -ie^{i3\pi/4} \times \infty\big)\big| = \big|e^{\mp ip\pi/4}\big| = e^{\pm\pi/8\alpha} = e^{\pi/8|\alpha|}. \tag{7.A.41}$$

Therefore A is fixed as $A = e^{-\pi/8|\alpha|}/(2\alpha)^{1/2}$ and the solution is obtained as

$$\psi_0(t) = \frac{e^{-\pi/8\alpha}}{\sqrt{2\alpha}} D_{-i/2\alpha-1}\big(e^{-i3\pi/4}\sqrt{2\alpha}\,t\big) \tag{7.A.42}$$

$$\psi_1(t) = e^{-\pi/8\alpha} e^{i3\pi/4} D_{-i/2\alpha}\big(e^{-i3\pi/4}\sqrt{2\alpha}\,t\big) \tag{7.A.43}$$

for $\alpha > 0$, and

$$\psi_0(t) = \frac{e^{-\pi/8|\alpha|}}{i\sqrt{2|\alpha|}} D_{i/2|\alpha|-1}\big(e^{i3\pi/4}\sqrt{2|\alpha|}\,t\big) \tag{7.A.44}$$

$$\psi_1(t) = e^{-\pi/8\alpha} e^{i3\pi/4} D_{i/2|\alpha|}\big(e^{i3\pi/4}\sqrt{2|\alpha|}\,t\big) \tag{7.A.45}$$

for $\alpha < 0$.

Finally we investigate the limit of $z \to e^{\mp i\pi/4} \times \infty$ corresponding to $t \to \infty$. From asymptotic expansions (7.A.29) and (7.A.30), one has

$$\begin{aligned} &\big|D_{-p-1}\big(-iz \to e^{\mp i3\pi/4} \times \infty\big)\big|^2 \\ &= \left|\frac{\sqrt{2\pi}}{\Gamma(p+1)} e^{\pm ip\pi} e^{\mp ip3\pi/4}\right|^2 = \left|\frac{\sqrt{2\pi}}{\Gamma(p+1)} e^{\pm ip\pi/4}\right|^2 \\ &= 2\alpha\big(e^{\pi/2\alpha} - e^{-\pi/2\alpha}\big)e^{\mp\pi/4\alpha}, \end{aligned} \tag{7.A.46}$$

$$\big|D_{-p}\big(-iz \to e^{\pm ip3\pi/4} \times \infty\big)\big|^2 = \big|e^{\pm ip3\pi/4}\big|^2 = e^{\mp 3\pi/4\alpha}, \tag{7.A.47}$$

where we have used an identity regarding the Gamma function: $|\Gamma(1 + iy)|^2 = 2\pi y/(e^{\pi y} - e^{-\pi y})$. Thus the Landau-Zener formula on the non-adiabatic transition probability is obtained as

$$\big|\psi_1(t \to \infty)\big|^2 = e^{-\pi/|\alpha|}. \tag{7.A.48}$$

7.A.3 Microscopic Equation of Motion for Oscillatory Transverse Field

We start from the Liouville equation of motion (with Planck's constant $\hbar = 1$)

$$\frac{d\rho}{dt} = -i\big[H_{\text{tot}}, \rho(t)\big] \tag{7.A.49}$$

where H_{tot} is the total Hamiltonian defined in (7.2.120). In the interaction picture the evolution is governed by

$$i\frac{d\rho_I(t)}{dt} = \left[V_I, \rho_I(t)\right] = V_I^{\times}(t)\rho_I(0) \tag{7.A.50}$$

where

$$\rho_I(t) = \exp\left[i(H_s + H_B)t\right]\rho(t)\exp\left[-i(H_s + H_B)t\right] \tag{7.A.51}$$

and

$$V_I(t) = \exp\left[i(H_s + H_B)\right]H_I \exp\left[-(H_s + H_B)\right]. \tag{7.A.52}$$

$V_I^{\times}(t)$ is the Liouville operator associated with $V_I(t)$. The solution of (7.A.50) can be formally written as

$$\rho_I(t) = \exp_T\left[-i\int_0^t V_I^{\times}(t')\,dt'\right]\rho_I(0), \tag{7.A.53}$$

with $\exp_T$ denoting a time-ordered series with the operators for the latest time at the left. Note that at $t = 0$, $\rho_I(0) = \rho(0)$. Combining (7.A.51) and (7.A.53), we have

$$\rho(t) = \exp\left[-i(H_s + H_B)t\right]\exp_T\left[-i\int_0^t V_I^{\times}(t')\,dt'\right]\rho(0)\exp\left[i(H_s + H_B)t\right]. \tag{7.A.54}$$

The rate equations for magnetisation m^{μ} ($\mu = x, y, z$) are obtained from

$$\frac{dm^{\mu}}{dt} = \text{Tr}\left[\frac{d\rho(t)}{dt}S^{\mu}\right] \tag{7.A.55}$$

where

$$m^{\mu} = \text{Tr}\big(\rho(t)S^{\mu}\big) \tag{7.A.56}$$

and $\rho(t)$ is given in (7.A.54). However, it is easier to work in terms of a reduced density matrix for the spin system alone:

$$\rho_s(t) = \text{Tr}_b\,\rho(t) \tag{7.A.57}$$

where Tr_b denotes a trace operation over the degrees of freedom of the heat bath. Thus, from (7.A.54)

$$\begin{aligned}\rho_s(t) &= {}^{-iH_st}\text{Tr}_T\left[-i\int_0^t dt'\,V_I^{\times}(t')\right]\rho(0)\mathrm{e}^{iH_st} \\ &\simeq \mathrm{e}^{-iH_st}\exp_T\left[-i\int_0^t dt'\langle V_I^{\times}(t')\rangle - i\int_0^t dt'\int_0^{t'} dt''\langle V_I^{\times}(t')V_I^{\times}(t'')\rangle_T\right] \\ &\quad\times\rho(0)\mathrm{e}^{iH_st}.\end{aligned} \tag{7.A.58}$$

The angular brackets $\langle\cdots\rangle$ refer to an averaging over the bath degrees of freedom. It has been assumed that the density matrix can be factorised as

$$\rho(0) \simeq \rho_b \otimes \rho_s, \tag{7.A.59}$$

and the cumulant expansion theorem has been used. The physical ground for writing (7.A.59) is that at $t = 0$, the spin-system is assumed to be decoupled from the heat bath; it is at that instant that the perturbation H_I which couples the spin system to the bath is switched on. The subsequent time evolution of $\rho_s(0)$ is what we are interested in. Assuming invariance under time translation, we can write

$$\rho_s(t) = \mathrm{e}^{-iH_s t} \exp_T\left(-\int_0^t (t-\tau)\langle V_I^\times(\tau) V_I^\times(\tau)\rangle d\tau\right)\rho(0)\mathrm{e}^{iH_s t}. \qquad (7.A.60)$$

There has been an additional assumption in writing the above equation, viz.,

$$\langle V_I^\times(t')\rangle = \langle V_I^\times(0)\rangle = 0. \qquad (7.A.61)$$

This can always be ensured by an appropriate choice of the coupling term H_I. This assumption is necessitated by the physical requirement of the model that at a large enough time, the system should equilibrate to a situation governed by the Hamiltonian H_s alone. Using a short-time approximation for the bath correlation functions of the kind $\langle\hat{b}(\tau)\hat{b}(0)\rangle$, viz., that correlation functions die out after a time short compared to any other "times" of physical interest in H_s, the upper limit in the integrals in (7.A.60) can be extended to ∞. This enables us to further write the master equation as

$$\frac{d\rho_s(t)}{dt} = -i[H_s, \rho_s(t)] - \mathrm{e}^{-iH_s t}\left[\int_0^\infty d\tau\langle V_I^\times(\tau) V_I^\times(0)\rangle\right]\mathrm{e}^{iH_s t}\rho_s(t). \quad (7.A.62)$$

While using the above equation in the rotated frame, one must replace V by $\tilde{V}$, H_s by $\tilde{H}_s$, $\rho(t)$ and $\tilde{\rho}(t)$ by etc. where, for example,

$$\tilde{\rho}(t) = R_y^{-1} U_z^{-1} \rho(t) U_z R_y. \qquad (7.A.63)$$

The rate equations for the components of magnetisation are obtained using the above formalism where we have

$$\begin{aligned}\tilde{V}_I(\tau) &= \mathrm{e}^{i(\tilde{H}_s + H_B)\tau}\tilde{H}_I \mathrm{e}^{-i(\tilde{H}_s + H_B)\tau} \\ &= \frac{1}{2} g\hat{b}(\tau)\left[S^+ \mathrm{e}^{-2ih\tau} + S^{-1}\mathrm{e}^{2ih\tau}\right] \\ &\quad + \frac{1}{2} g\hat{b}(\tau)\left[S^+ \mathrm{e}^{-2ih\tau} - S^- \mathrm{e}^{2ih\tau}\right].\end{aligned} \qquad (7.A.64)$$

We then use $m^\mu = \mathrm{Tr}_s[\rho(t)S^\mu] = \mathrm{Tr}_s[U_z R_y(t)\tilde{\rho}(t) R_y^{-1} U_z^{-1} S^\mu]$ and (7.A.62) to get the rate equations (7.2.128)–(7.2.130) (see [22] for further details). The bath correlations which will appear in the calculations are not calculated but parametrised in terms of a phenomenological relaxation rate by making use of Kubo relations. In order to parametrise the bath correlations, we use the following Kubo relation

$$\int_{-\infty}^{\infty} d\tau\, \mathrm{e}^{ih\tau}\langle\hat{b}(\tau)\hat{b}(0)\rangle = \mathrm{e}^{\beta h}\int_{-\infty}^{\infty} d\tau\, \mathrm{e}^{-ih\tau}\langle\hat{b}(\tau)\hat{b}(0)\rangle \qquad (7.A.65)$$

so that we can write

$$g^2 \int_{-\infty}^{\infty} d\tau\, \mathrm{e}^{\pm ih\tau}\langle\hat{b}(\tau)\hat{b}(0)\rangle = \lambda\frac{\mathrm{e}^{\pm\beta h/2}}{\mathrm{e}^{\beta h/2} + \mathrm{e}^{-\beta h/2}}, \qquad (7.A.66)$$

where

$$\lambda \equiv g^2 \int_{-\infty}^{\infty} d\tau \left(e^{+ih\tau} + e^{-ih\tau}\right)\langle \hat{b}(\tau)\hat{b}(0)\rangle$$

$$= g^2 \int_{-\infty}^{\infty} d\tau \, e^{ih\tau} \left[\langle \hat{b}(\tau)\hat{b}(0)\rangle + \langle \hat{b}(0)\hat{b}(\tau)\rangle\right] \tag{7.A.67}$$

is the phenomenological relaxation rate.

Chapter 8
Quantum Annealing

8.1 Introduction

Let us consider the energy minimisation of a random Ising model like an Ising spin glass. We suppose that its energy landscape has a complex multimodal structure as shown in Fig. 8.1(a). In such a system we sometimes encounter a difficulty of finding out the ground state. The easiest way to conquer it is to examine the energy of all the states. However such a brute force method is useless if the number of spins N exceeds 10^2, because of the exponential increase in the number of states. Combinatorial optimisation like the present problem is an important issue in computer science. A close relation between basic optimisation problems and spin glasses has already been discovered [275] and fruitful collaborations have been done in the interdisciplinary field between computer science, information theory, and statistical physics [254, 273, 297]. An important optimisation method, which originates from statistical physics, is simulated annealing [60, 228]. One utilises the thermal fluctuation in simulated annealing. The thermal fluctuation helps the state of a system hop from one energy minimum to another, overcoming an energy barrier. It is not difficult to simulate the canonical ensemble by means of the Monte Carlo or the molecular dynamics method as far as the temperature is sufficiently high. One may expect that the ground state could be obtained by decreasing the temperature slowly. In fact, simulated annealing has been proved mathematically to yield the ground state with the probability one if the temperature is decreased according to an appropriate schedule [155].

A potential power of quantum mechanics in computation was pointed out by Feynman [134]. After the discovery of an efficient algorithm using quantum mechanics for factorisation by Shor [370], quantum computation has attracted a lot of interests in computer science as well as physics and has become an absorbing topic in today's science [295]. In combinatorial optimisation problems, quantum mechanics may offers a different route to the target from classical optimisation algorithms, in particular, simulated annealing. Intuitively, the quantum tunnelling effect helps the system hop from a state to another. Let us think of an energy barrier with height h and width w. While the classical escape rate over the barrier is given by $\exp(-h/T)$,

S. Suzuki et al., *Quantum Ising Phases and Transitions in Transverse Ising Models*, Lecture Notes in Physics 862, DOI 10.1007/978-3-642-33039-1_8,

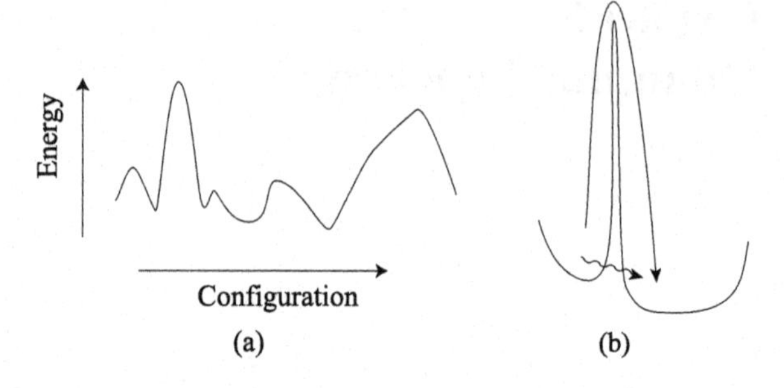

Fig. 8.1 (**a**) A complex multimodal energy landscape of a random Ising model. (**b**) A tall but thin energy barrier. A quantum fluctuation might be more useful than a thermal fluctuation in order to overcome this barrier

quantum tunneling probability is approximately given by $\exp(-\sqrt{h}w/\Gamma)$, where T and Γ are the temperature and the magnitude of quantum fluctuations respectively. This ensures the clear advantage of quantum tunneling when $w \ll \sqrt{h}$ (Fig. 8.1(b)) [95, 323, 372]. We note that such a thorn-shaped structure in the energy landscape may underlie a spin glass. It has been established that the Sherrington-Kirkpatrick model gives rise to the replica symmetry breaking in the spin-glass phase. The possibility of the recovery of the replica symmetry in the presence of the transverse field has been discussed [49, 159, 224, 240, 323, 401], notwithstanding the absence of conclusive results. If this happens, quantum fluctuation should heal the pathological energy landscape and help us attain the energy minimisation.

The method of quantum annealing is analogous to that of simulated annealing [11, 17, 69, 93, 94, 130, 135, 211, 341]. A quantum fluctuation can be introduced to an Ising model simply by the transverse field [130, 211]. The transverse field is made strong at first, and the system is initialised at the ground state. Then, the transverse field is weakened with time slowly. When the transverse field vanishes, one extracts the ground state of a random Ising model. Note that the time evolution of the system is governed by the Schrödinger equation of motion. In spite of the similarity with simulated annealing, the different nature of the quantum fluctuation from the thermal fluctuation and the different rule of dynamics make us expect that quantum annealing outperforms simulated annealing.

In the present chapter, we present several properties of quantum annealing. First we mention basic properties of combinatorial optimisation problems in next section. We then introduce quantum annealing in Sect. 8.3. The essence of quantum annealing is an adiabatic time evolution with a time-dependent Hamiltonian. We review the non-crossing rule of energy levels and the quantum adiabatic theorem which are basic theories of an adiabatic time evolution. We then give an explanation on the mechanism of quantum annealing in this section. In Sect. 8.4, we give a brief review on experiments and numerical tests of quantum annealing. Quantum annealing is one of the algorithms which are implemented in a quantum computer. Several attempts to realise a quantum computation by quantum annealing have been made. On the other hand, classical computers have also significant roles for quantum annealing. They serve as not only a simulator of a quantum computer but also a solver of optimisation problems by an application of quantum annealing. Section 8.5 is devoted to the study on the scaling of energy gaps. As we show in Sect. 8.3, the

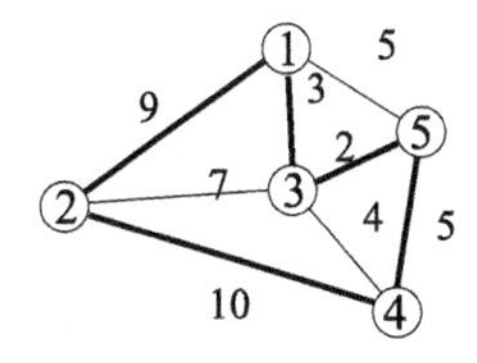

Fig. 8.2 A map for a travelling salesman. A salesman needs to depart from the city *1*, visit all the cities once and return with the shortest time. A length is given to each path between cities. What paths should he choose? The answer is depicted by *thick lines* in the picture

success of quantum annealing is governed by the magnitude of the minimum energy gap between the instantaneous ground state and the first excited state of the time-dependent Hamiltonian. The exponential scaling of the minimum energy gap with respect to the size (the number of spins) N implies the exponential scaling of the time to solve the problem. We discuss this issue using some specific models. Following the energy gap, we focus our attention to the scaling of errors with respect to the runtime in Sect. 8.6. Quantum annealing inevitably yields errors, as far as it is performed with a finite runtime. The scaling of errors with runtime is one of the measure for the validity of the algorithm. We investigate the scaling of errors for a specific model, and contrast quantum annealing with simulated annealing in this section. Finally, we mention convergence theorems in Sect. 8.7. A mathematical basis of quantum annealing is given here. The present chapter concludes in Sect. 8.8.

8.2 Combinatorial Optimisation Problems

Let us start from the travelling salesman problem.

> Suppose that there are five cities named as 1, 2, 3, 4, and 5. Depicted in Fig. 8.2, the city 1 connects with cities 2, 3, and 5, and the path lengths are given as 9, 3, and 5 respectively. Similarly, other cities connects with others and path lengths are given as shown in Fig. 8.2. What is the shortest path for a salesman to visit all cities once and return to the original place?

One can formulate the travelling salesman problem in terms of an Ising model as follows [179]. We refer to a path between cities i and j by a link $\langle ij\rangle$. In the example of Fig. 8.2, there are 16 links as follows.

$$\langle 12\rangle,\quad \langle 13\rangle,\quad \langle 15\rangle,\quad \langle 21\rangle,\quad \langle 23\rangle,\quad \langle 24\rangle,\quad \langle 31\rangle,\quad \langle 32\rangle,$$
$$\langle 34\rangle,\quad \langle 35\rangle,\quad \langle 42\rangle,\quad \langle 43\rangle,\quad \langle 45\rangle,\quad \langle 51\rangle,\quad \langle 53\rangle,\quad \langle 54\rangle.$$

We write the path length of the link $\langle ij\rangle$ as $d_{\langle ij\rangle}$. In Fig. 8.2, for example, $d_{\langle 12\rangle} = d_{\langle 21\rangle} = 9$, $d_{\langle 13\rangle} = d_{\langle 31\rangle} = 3$, and so on. We assign the Ising spin $S_{\langle ij\rangle}$ to the link $\langle ij\rangle$ and suppose that $S_{\langle ij\rangle} = +1$ ($S_{\langle ij\rangle} = -1$) indicates that the path $\langle ij\rangle$ is (not) passed. Since all the cities must be visited once, one has to be exposed by constraints: $\sum_j (S_{\langle ji\rangle} + 1)/2 = 1$ and $\sum_j (S_{\langle ij\rangle} + 1)/2 = 1$ for all i. The former means

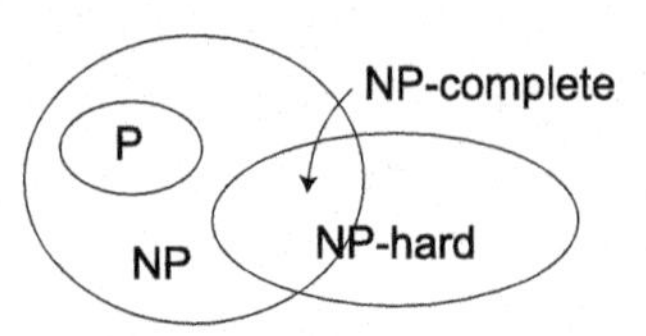

Fig. 8.3 Conceptual picture showing the relation between the classes P, NP, NP-hard, and NP-complete. Note that $P \neq NP$ is the conjecture

that a salesman goes to city i from another city, while the latter means that he goes out from the city i to another. The travelling salesman problem reduces to finding a spin configuration which minimises the Hamiltonian [179]

$$H_{\mathrm{TSP}} = \sum_{\text{all links}} d_{\langle ij\rangle} \frac{S_{\langle ij\rangle}+1}{2}, \tag{8.2.1}$$

under the constraints

$$\sum_j \frac{S_{\langle ji\rangle}+1}{2} = 1 \quad \text{and} \quad \sum_j \frac{S_{\langle ij\rangle}+1}{2} = 1. \tag{8.2.2}$$

The problem with five cities as depicted in Fig. 8.2 is easy to solve. However if thousands of cities are involved, it is hard to obtain an exact solution.

According to the computational complexity theory, combinatorial optimisation problems are classified into several classes [153]. First we consider a decision problem. A decision problem is a problem which asks one whether a given statement is true or not. The class P consists of the decision problems which can be solved in a polynomial time by a Turing machine (which is a usual computer). The class NP consists of the decision problems which can be solved using a non-deterministic Turing machine[1] in a polynomial time. The class NP includes easy problems and hard ones. Here easy problems mean P problems, while hard problems are those for which no polynomial algorithm using a Turing machine is known. It is clear that the class P is included in the class NP. However, whether the classes P and NP are the same or not is an open problem, though many of us actually believe in the conjecture $P \neq NP$. This problem is known as the Millennium Prize Problem of the Clay Mathematics Institute [79]. In addition to P and NP, there are important classes named as NP-hard and NP-complete. Let us consider a problem p. Suppose that p has a property that any problem in the class NP can be mapped to p by a polynomial time of operation. The class NP-hard consists of the problems which have this property. It includes non-decision problems as well as decision problems. The travelling salesman problem mentioned above belongs to the non-decision NP-hard class. The class NP-complete is the subclass of the classes NP and NP-hard. If an NP-complete problem is proved to be a P problem, it is proved that $P = NP$. On the other hand, if an NP-complete problem is proved not to be a P problem, it is proved that $P \neq NP$. Therefore an NP-complete problem has a significant role to discuss the $P \neq NP$ conjecture. Figure 8.3 shows a conceptual picture regarding important classes of optimisation problems.

[1]A computer which enables one to carry out a non-deterministic computation. See Ref. [153] for details.

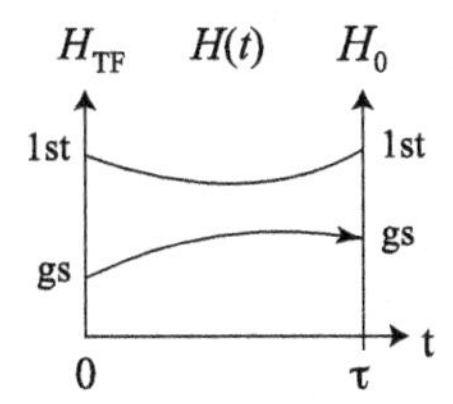

Fig. 8.4 Schematic picture showing an adiabatic time evolution. The state starting from the ground state of H_{TF} keeps track of the instantaneous ground level and goes to the ground state of H_0

8.3 Optimisation by a Quantum Adiabatic Evolution

We suppose that a Hamiltonian $H_0(\{S_i\})$ with N Ising spins $(S_1, S_2, \ldots, S_N)$ represents a combinatorial optimisation problem. Our problem is to find the ground state of $H_0(\{S_i\})$. To this end, we first replace Ising spins $\{S_i\}$ to z-components of the Pauli spin $\{S_i^z\}$. We next introduce a quantum fluctuation by the transverse field: $H_{TF}(\{S_i^x\}) = -\sum_i S_i^x$. Then we make a time-dependent Hamiltonian as

$$f(t)H_0\bigl(\{S_i^z\}\bigr) + g(t)H_{TF}\bigl(\{S_i^x\}\bigr), \tag{8.3.1}$$

where $f(t)$ and $g(t)$ are time-dependent factors which control weights of H_0 and H_{TF}. We impose $f(0)/g(0) \gg 1$ at $t = 0$ and $f(\tau)/g(\tau) \ll 0$ at $t = \tau$, where τ stands for the final time or the runtime. A simple choice for $f(t)$ and $g(t)$ introduced by Farhi et al. [130] is $f(t) = 1 - t/\tau$ and $g(t) = t/\tau$, namely

$$H(t) = \left(1 - \frac{t}{\tau}\right)H_0 + \frac{t}{\tau}H_{TF}. \tag{8.3.2}$$

Employing (8.3.2), $H(t)$ interpolates H_{TF} and H_0 linearly in time. The state vector $|\Psi(t)\rangle$ obeys the Schrödinger equation,

$$i\frac{d}{dt}\big|\Psi(t)\big\rangle = H(t)\big|\Psi(t)\big\rangle. \tag{8.3.3}$$

As an initial condition, we prepare the ground state of H_{TF} for $|\Psi(0)\rangle$. Note that it is easy to make this state, since all the spins are simply aligned along S^x axis.

The spin state starting from the ground state of H_{TF} evolves adiabatically to the ground state of H_0, if a finite energy gap lies between the ground state and excited state and the changing speed of $H(t)$ is infinitely slow. Figure 8.4 illustrates an adiabatic time evolution from the ground state of H_{TF} to that of H_0.

The success of the optimisation by an adiabatic time evolution depends on the presence of a finite energy gap and how the time-dependent Hamiltonian changes with time. In order to discuss this issue, let us look into the two important properties of the quantum mechanics.

8.3.1 Non-crossing Rule

Let us consider a Hamiltonian H defined on a finite-dimensional Hilbert space. We assume that no observable commutes with the Hamiltonian. We suppose that the

Hamiltonian contains independent real parameters $p_1, p_2, \ldots$. The non-crossing rule by von Neumann and Wigner [416] states that, if the eigenvalues of H are initially different from each other, one must adjust three parameters to have a doubly degenerated eigenvalue. This prohibits crossing of the energy eigenlevels by changing only one parameter. This non-crossing rule results from the following theorem.

Theorem 8.1 *Let H be an Hermite matrix of dimension d. If H has f different eigenvalues and each of them has m_i $(i = 1, 2, \ldots, f)$ multiplicity, then the number of the free real parameters to fix H is given by*

$$d^2 + f - \sum_{i=1}^{f} m_i^2. \tag{8.3.4}$$

Proof We count the number of free real parameters to fix an Hermite matrix H. Because of the Hermiticity, one can write H as

$$H = UDU^\dagger, \tag{8.3.5}$$

where U is a unitary matrix and D is a real diagonal matrix. We remark here that the unitary matrix which satisfies (8.3.5) is not unique. Indeed, supposing V to be a unitary matrix such that $[D, V] = DV - VD = 0$, one has $H = UVDV^\dagger U^\dagger$. Now let n_H, n_U, and n_D be the number of free real parameters to fix H, U, and D respectively, and n_V be that to fix a unitary matrix V which commutes with D. One has $n_H + n_V = n_U + n_D$, namely,

$$n_H = n_U + n_D - n_V. \tag{8.3.6}$$

Obviously D has d free parameters, so that

$$n_D = d. \tag{8.3.7}$$

A unitary matrix is generally composed of $2d^2$ real numbers. However, due to the unitarity, they are conditioned by $2 \times \frac{1}{2}d(d-1) + d = d^2$ relations among themselves. ($2 \times \frac{1}{2}d(d-1)$ relations from $\mathrm{Re}\{[U^\dagger U]_{ij}\} = \mathrm{Im}\{[U^\dagger U]_{ij}\} = 0$ with $i > j$, and d relations from $[U^\dagger U]_{ii} = 1$. Note that $[U^\dagger U]_{ii}$ has no imaginary part.) Therefore a unitary matrix has d^2 free real parameters:

$$n_U = d^2. \tag{8.3.8}$$

The number n_V depends on the matrix D. The commutability of D and V yields $D_i V_{ij} = D_j V_{ij}$, where D_i represents a diagonal elements of D. From this, one can see that $V_{ij} = 0$ $(i \neq j)$ when all diagonal elements of D are different from each other. Therefore V is a diagonal matrix in such a case. In addition, if V is unitary, it is shown that the number of free real parameters in V is $n_V = d$. When two elements of D are identical to have

$$D = \begin{bmatrix} D_1 & & & & \\ & D_1 & & 0 & \\ & & D_3 & & \\ & 0 & & \ddots & \\ & & & & D_d \end{bmatrix},$$

one can easily see that V must be

$$V = \left[\begin{array}{cc|cccc} V_{11} & V_{12} & & & 0 & \\ V_{21} & V_{22} & & & & \\ \hline & & V_{33} & & & 0 \\ & 0 & & \ddots & & \\ & & & 0 & & V_{dd} \end{array}\right].$$

Therefore, if V is unitary, one finds that $n_V = 2^2 + d - 2 = d + 2$. In general, when the elements of D are divided into f values, $D_1, D_2, \ldots, D_f$, and each of them has the multiplicity m_i, the number of free real parameters n_V is given by

$$n_V = m_1^2 + m_2^2 + \cdots + m_f^2. \tag{8.3.9}$$

Equations (8.3.6), (8.3.7), (8.3.8), and (8.3.9) prove the theorem. □

We remark that, in the case of a real symmetric matrix, (8.3.4) is replaced by

$$\frac{1}{2}d(d-1) + f - \sum_{i=1}^{f} \frac{1}{2}m_i(m_i - 1). \tag{8.3.10}$$

This is understood by replacing the unitary matrix by the orthogonal matrix in the above proof and noting that there are $\frac{1}{2}d(d-1)$ free parameters in the orthogonal matrix with dimension d.

Now we consider the consequence of Theorem 8.1. At first we suppose that all energy eigenvalues of a Hamiltonian H on the Hilbert space of dimension d are different from each other. Then, making $f = d$ and $m_i = 1$ $(i = 1, \ldots, d)$, one has d^2 free real parameters to fix H, according to Theorem 8.1. Next, when a doubly degenerated eigenvalue is present, (8.3.4) with $f = d - 1$, $m_1 = 2$, and $m_i = 1$ $(i = 2, \ldots, d - 1)$ leads to $d^2 - 3$ free real parameters. Therefore one needs to adjust three parameters to make a Hamiltonian with a doubly degenerated eigenenergy from a Hamiltonian without degenerated eigenenergies. Hence it is concluded that the level crossing of any two eigenlevels does not take place by the continuous change of one parameter.

The simplest example is a two-dimensional Hermite matrix. In general, a two-dimensional Hermite matrix H is written using four real parameters a, b, g, and h as

$$H = \begin{bmatrix} a & g + ih \\ g - ih & b \end{bmatrix}.$$

The eigenvalues of this matrix are given by

$$\frac{1}{2}\left\{a + b + \sqrt{(a-b)^2 + |g + ih|^2}\right\}, \quad \frac{1}{2}\left\{a + b - \sqrt{(a-b)^2 + |g + ih|^2}\right\},$$

and the difference between them is

$$\sqrt{(a-b)^2 + |g + ih|^2}.$$

It is clear that three of four parameters are adjusted to be $a = b$ and $g = h = 0$, so that two eigenvalues collapse into a single value. Also one can see that, in general, a change of one parameter does not lead to a degenerated eigenvalue. Note that, although a degenerated eigenvalue is obtained by an adjustment of one parameter in the case with $g = h = 0$, such a case is excluded from the application of the non-crossing rule since a conserving observable (which is the Pauli matrix σ^3 for instance) is involved in H.

Now we return to quantum annealing. We consider the time-dependent Hamiltonian (8.3.2) for simplicity. This Hamiltonian contains the parameter $s = t/\tau$. When $s = 1$, the Hamiltonian $H(\tau) = H_0$ could have degenerated energy levels. However, when s is infinitesimally smaller than 1, the perturbation by a transverse field lifts the degeneracy in H_0, so that all the eigenenergies are separated from each other. According to the non-crossing rule, these eigenlevels never cross with moving s except for special points at which the Hamiltonian shares the eigenstates with other observables. Such special points are located only at $s = 0$ and 1 usually in our Hamiltonian (8.3.2). Therefore, as far as no commutable observable is present, all the eigenenergies are separated by finite energy gaps for $0 < s < 1$.

8.3.2 Quantum Adiabatic Theorem

We consider a time-dependent Hamiltonian. A system, which is initially in an eigenstate of a Hamiltonian, stays in an instantaneous eigenstate if the change in a Hamiltonian is sufficiently slow. This property is described by the quantum adiabatic theorem [43, 215, 271].

Hereafter we assume that the Hamiltonian $\bar{H}(s)$ depends on $s = t/\tau$. Note that τ controls the changing speed of $\bar{H}(s)$. The larger τ is, the slower $\bar{H}(s)$ changes with t. Let $\bar{\varepsilon}_n(s)$ and $|\bar{\phi}_n(s)\rangle$ $(n = 0, 1, 2, \ldots)$ be an eigenvalue and eigenstate of $\bar{H}(s)$ respectively:

$$\bar{H}(s)\big|\bar{\phi}_n(s)\big\rangle = \bar{\varepsilon}_n(s)\big|\bar{\phi}_n(s)\big\rangle. \tag{8.3.11}$$

We impose without loss of generality

$$\big\langle\bar{\phi}_n(s)\big|\frac{d}{ds}\big|\bar{\phi}_n(s)\big\rangle = 0 \quad (n = 0, 1, 2, \ldots). \tag{8.3.12}$$

Note that one can make $|\bar{\phi}_n(s)\rangle$ satisfy (8.3.12) as follows. Consider an eigenstate $|\bar{\phi}_n(s)\rangle\rangle$ with $\langle\langle\bar{\phi}_n(s)|\frac{d}{ds}|\bar{\phi}_n(s)\rangle\rangle \neq 0$. We remark that $\langle\langle\bar{\phi}_n(s)|\frac{d}{ds}|\bar{\phi}_n(s)\rangle\rangle$ is pure imaginary since $\langle\langle\bar{\phi}_n(s)|\frac{d}{ds}|\bar{\phi}_n(s)\rangle\rangle + \text{c.c.} = \frac{d}{ds}\langle\langle\bar{\phi}_n(s)|\bar{\phi}_n(s)\rangle\rangle = 0$. Define $|\bar{\phi}_n(s)\rangle$ by $|\bar{\phi}_n(s)\rangle\rangle$ multiplied by the phase factor $\exp(-\int_0^s \langle\langle\bar{\phi}_n(s')|\frac{d}{ds'}|\bar{\phi}_n(s')\rangle\rangle ds')$. Then one can easily show that (8.3.12) is satisfied.

Now let $|\Psi_\tau(s)\rangle$ be the state vector. The Schrödinger equation is written as

$$i\frac{d}{ds}\big|\Psi_\tau(s)\big\rangle = \tau\bar{H}(s)\big|\Psi_\tau(s)\big\rangle. \tag{8.3.13}$$

We assume the initial condition:

$$|\Psi_\tau(0)\rangle = |\bar{\phi}_l(0)\rangle. \tag{8.3.14}$$

In addition, we introduce three assumptions as follows.

(i) Let

$$\bar{\Delta}_{mn}(s) = \bar{\varepsilon}_m(s) - \bar{\varepsilon}_n(s). \tag{8.3.15}$$

There exists a positive value $\Delta_0 > 0$ and $|\bar{\Delta}_{mn}(s)|$ with $m \neq n$ is bounded from below by Δ_0:

$$|\bar{\Delta}_{mn}(s)| \geq \Delta_0 > 0. \tag{8.3.16}$$

Note that the condition (8.3.16) is consistent with the non-crossing rule.

(ii) Define

$$f_{mn}(s) = \begin{cases} \frac{1}{\bar{\Delta}_{mn}(s)} \langle\bar{\phi}_m(s)|\frac{d\bar{H}(s)}{ds}|\bar{\phi}_n(s)\rangle & (m \neq n) \\ 0 & (m = n) \end{cases}. \tag{8.3.17}$$

Let

$$f_{mn} = \max_s |f_{mn}(s)|. \tag{8.3.18}$$

There exists a finite positive value $f_1 < \infty$ such that for all m and n

$$f_{mn} \leq f_1. \tag{8.3.19}$$

(iii) Define

$$g_{mn}(s) = \begin{cases} \frac{f_{mn}(s)}{\bar{\Delta}_{mn}(s)} & (m \neq n) \\ 0 & (m = n) \end{cases}. \tag{8.3.20}$$

$\mathrm{Re}\{g_{mn}(s)\}$ and $\mathrm{Im}\{g_{mn}(s)\}$ are fragmentarily monotonic functions of s and the number of monotonic regions between $s = 0$ and 1 is at most $N_1 < \infty$.

The quantum adiabatic theorem is stated as follows

Theorem 8.2 *Let $\omega_m(s) = \int_0^s \bar{\varepsilon}_m(s')\,ds'$. When $\tau \to \infty$, the probability amplitude of finding $\bar{\phi}_m(s)$ in the state at time $t = s\tau$ $(0 \leq s \leq 1)$ obeys*

$$\left|e^{i\tau\omega_m(s)}\langle\bar{\phi}_m(s)|\Psi_\tau(s)\rangle - \delta_{ml}\right| \leq O\left(\frac{1}{\tau}\right). \tag{8.3.21}$$

Proof We expand the state vector by the eigenbasis $\{e^{-i\tau\omega_n(s)}|\bar{\phi}_n(s)\rangle\}$ as

$$|\Psi_\tau(s)\rangle = \sum_n c_n(s)e^{-i\tau\omega_n(s)}|\bar{\phi}_n(s)\rangle, \tag{8.3.22}$$

where $c_n(s) = e^{i\tau\omega_n(s)}\langle\bar{\phi}_n(s)|\Psi_\tau(s)\rangle$. Substituting (8.3.22) in the Schrödinger equation (8.3.13), one obtains an equation for $c_n(s)$:

$$i\frac{d}{ds}c_m(s) = -i\sum_{n\neq m}\langle\bar{\phi}_m(s)|\frac{d}{ds}|\bar{\phi}_n(s)\rangle \exp\left[i\tau\int_0^s \bar{\Delta}_{mn}(s')\,ds'\right]c_n(s). \tag{8.3.23}$$

We here recall (8.3.11) and differentiate it by s

$$\frac{d\bar{H}(s)}{ds}|\bar{\phi}_n(s)\rangle + \bar{H}(s)\frac{d}{ds}|\bar{\phi}_n(s)\rangle = \frac{d\bar{\varepsilon}_n(s)}{ds}|\bar{\phi}_n(s)\rangle + \bar{\varepsilon}_n(s)\frac{d}{ds}|\bar{\phi}_n(s)\rangle. \quad (8.3.24)$$

This yields for $m \neq n$

$$\langle\bar{\phi}_m(s)|\frac{d}{ds}|\bar{\phi}_n(s)\rangle = -\frac{\langle\bar{\phi}_m(s)|\frac{d\bar{H}(s)}{ds}|\bar{\phi}_n(s)\rangle}{\bar{\Delta}_{mn}(s)} = -f_{mn}(s). \quad (8.3.25)$$

Applying (8.3.25) and (8.3.20) to (8.3.23), one obtains

$$\frac{d}{ds}c_m(s) = \sum_{n\neq m} g_{mn}(s)\bar{\Delta}_{mn}(s)\exp\left[i\tau\int_0^s \bar{\Delta}_{mn}(s')\,ds'\right]c_n(s). \quad (8.3.26)$$

Taking the initial condition $c_m(0) = \delta_{ml}$ into account, (8.3.26) is integrated to yield

$$\begin{aligned}
c_m(s) &= \delta_{ml} + \sum_{n\neq m}\int_0^s ds_1 g_{mn}(s_1)\bar{\Delta}_{mn}(s_1)e^{i\tau\varphi_{mn}(s_1)}c_n(s_1) \\
&= \delta_{ml} + \int_0^s ds_1\, K_{ml}(s_1,\tau) \\
&\quad + \sum_{n_1}\int_0^s ds_2\int_0^{s_2} ds_1\, K_{mn_1}(s_2,\tau)K_{n_1 l}(s_1,\tau) + \cdots \\
&= \delta_{ml} + \sum_{k=1}^{\infty}\int_0^s ds_k\cdots\int_0^{s_3} ds_2\int_0^{s_2} ds_1\left[K(s_k,\tau)\ldots K(s_2,\tau)K(s_1,\tau)\right]_{ml},
\end{aligned} \quad (8.3.27)$$

where we have defined

$$\varphi_{mn}(s) = \int_0^s \bar{\Delta}_{mn}(s')\,ds', \quad (8.3.28)$$

$$K_{mn}(s,\tau) = g_{mn}(s)\bar{\Delta}_{mn}(s)e^{i\tau\varphi_{mn}(s)}, \quad (8.3.29)$$

and $[\cdot]_{ml}$ to represent the ml-component of a matrix. We focus on the integral

$$\int_0^s ds'\, K_{mn}(s',\tau) = \int_0^s ds'\, g_{mn}(s')\bar{\Delta}_{mn}(s')e^{i\tau\varphi_{mn}(s')}. \quad (8.3.30)$$

Suppose that $\mathrm{Re}\{g_{mn}(s)\}$ is a monotonic function of s in between $a \leq s \leq b$. Due to the second mean value theorem for integration, there exists a number c such that $a < c < b$ and

$$\begin{aligned}
&\int_a^b ds\,\mathrm{Re}\{g_{nm}(s)\}\bar{\Delta}_{mn}(s)\cos\{\tau\varphi_{mn}(s)\} \\
&\quad = \mathrm{Re}\{g_{nm}(a)\}\int_a^c ds\,\bar{\Delta}_{mn}(s)\cos\{\tau\varphi_{mn}(s)\} \\
&\quad\quad + \mathrm{Re}\{g_{nm}(b)\}\int_c^b ds\,\bar{\Delta}_{mn}(s)\cos\{\tau\varphi_{mn}(s)\}
\end{aligned}$$

$$= \mathrm{Re}\{g_{mn}(a)\}\left[\frac{\sin\{\tau\varphi_{mn}(s)\}}{\tau}\right]_a^c + \mathrm{Re}\{g_{mn}(b)\}\left[\frac{\sin\{\tau\varphi_{mn}(s)\}}{\tau}\right]_c^b, \tag{8.3.31}$$

where we note $\int ds\,\bar{\Delta}_{mn}(s)\cos\{\tau\varphi_{mn}(s)\} = \sin\{\tau\varphi_{mn}(s)\}/\tau$. By assumption (i) and (8.3.18), one has

$$\left|\mathrm{Re}\{g_{mn}(s)\}\right| = \frac{|\mathrm{Re}\{f_{mn}(s)\}|}{|\bar{\Delta}_{mn}(s)|} \le \frac{f_{mn}}{\Delta_0}. \tag{8.3.32}$$

Therefore Eq. (8.3.31) leads to

$$\begin{aligned}
&\left|\int_a^b ds\,\mathrm{Re}\{g_{mn}(s)\}\bar{\Delta}_{mn}(s)\cos\{\tau\varphi_{mn}(s)\}\right| \\
&\quad \le \frac{1}{\tau}\left[\left|\mathrm{Re}\{g_{mn}(a)\}\left[\sin\{\tau\varphi_{mn}(s)\}\right]_a^c\right| + \left|\mathrm{Re}\{g_{mn}(b)\}\left[\sin\{\tau\varphi_{mn}(s)\}\right]_c^b\right|\right] \\
&\quad \le \frac{f_{mn}}{\tau\Delta_0}\left(\left|\left[\sin\{\tau\varphi_{mn}(s)\}\right]_a^c\right| + \left|\left[\sin\{\tau\varphi_{mn}(s)\}\right]_c^b\right|\right) \le \frac{4f_{mn}}{\tau\Delta_0}.
\end{aligned} \tag{8.3.33}$$

Similarly it is shown that

$$\left|\int_a^b ds\,\mathrm{Re}\{g_{mn}(s)\}\bar{\Delta}_{mn}(s)\sin\{\tau\varphi_{mn}(s)\}\right| \le \frac{4f_{mn}}{\tau\Delta_0}, \tag{8.3.34}$$

which with (8.3.33) yields

$$\left|\int_a^b ds\,\mathrm{Re}\{g_{mn}(s)\}\bar{\Delta}_{mn}(s)e^{i\tau\varphi_{mn}(s)}\right| \le \frac{8f_{mn}}{\tau\Delta_0}. \tag{8.3.35}$$

By assumption (iii), there are at most N_1 regions in $0 \le s' \le s$ where $\mathrm{Re}\{g_{mn}(s')\}$ is monotonic. Hence one reaches

$$\left|\int_0^s ds'\,\mathrm{Re}\{g_{mn}(s')\}\bar{\Delta}_{mn}(s')e^{i\tau\varphi_{mn}(s')}\right| \le \frac{8N_1 f_{mn}}{\tau\Delta_0}. \tag{8.3.36}$$

The same result is obtained for $\mathrm{Im}\{g_{mn}(s)\}$. Therefore, it turns out from (8.3.30) that

$$\begin{aligned}
\left|\int_0^s ds'\,K_{mn}(s',\tau)\right| &\le \left|\int_0^s ds'\,\mathrm{Re}\{g_{mn}(s')\}\bar{\Delta}_{mn}(s')e^{i\tau\varphi_{mn}(s')}\right| \\
&\quad + \left|\int_0^s ds'\,\mathrm{Im}\{g_{mn}(s')\}\bar{\Delta}_{mn}(s')e^{i\tau\varphi_{mn}(s')}\right| \\
&\le \frac{16N_1 f_{mn}}{\tau\Delta_0}.
\end{aligned} \tag{8.3.37}$$

Applying this inequality to the integral by s_1 and $|K_{mn}(s,\tau)| = |f_{mn}(s)| \le f_{mn}$ to the integrals by s_k ($k \ge 2$) in (8.3.27), one finally obtains

$$\left|c_{ml}(s) - \delta_{ml}\right| \le \sum_{k=1}^{\infty}\frac{16N_1}{\tau\Delta_0}\frac{s^{k-1}}{(k-1)!}\left[F^k\right]_{ml} = \frac{16N_1}{\tau\Delta_0}\left[Fe^F\right]_{ml}, \tag{8.3.38}$$

where F denotes the matrix composed of $\{f_{mn}\}$. By assumption (iii), $[Fe^F]_{ml}$ is bounded from above by $f_1 e^{f_1}$. Therefore it is shown that

$$\left|c_{ml}(s) - \delta_{ml}\right| \leq \frac{16N_1 f_1 e^{f_1}}{\tau \Delta_0}, \tag{8.3.39}$$

so that (8.3.21) is proved. □

The quantum adiabatic theorem guarantees the adiabatic time evolution when $\tau \to \infty$. We shall next derive a criterion with respect to τ for the adiabatic approximation. From now on we assume $|\Psi_\tau(0)\rangle = |\bar{\phi}_0(0)\rangle$ as we do in quantum annealing. Equation (8.3.27) leads to the excitation probability amplitude at s:

$$c_m(s) = \sum_{k=1}^{\infty} \int_0^s ds_k \cdots \int_0^{s_3} ds_2 \int_0^{s_2} ds_1 \left[K(s_k, \tau) \cdots K(s_2, \tau) K(s_1, \tau)\right]_{m0}$$
$$(m \geq 1). \tag{8.3.40}$$

According to inequality (8.3.37), this quantity vanishes in the adiabatic limit, $\tau \to \infty$. The correction to the adiabatic limit at finite τ within the leading order is given by

$$c_m(s) \approx \int_0^s ds_1\, K_{m0}(s_1, \tau) = \int_0^s ds_1\, g_{m0}(s_1) \bar{\Delta}_{m0}(s_1) e^{i\tau\varphi_{m0}(s_1)}. \tag{8.3.41}$$

Note that $e^{i\tau\varphi_{m0}(s)}$ oscillates intensely with s for large τ, while $g_{m0}(s)$ is independent of τ and behaves moderately. Therefore one may replace $g_{m0}(s)$ with $g_{m0}(s')$ at a point s' in between 0 and s, and draw it out from the integral. Thus $|c_m(s)|$ follows

$$\begin{aligned} \left|c_m(s)\right| &\lesssim \max_{s'}\left\{\left|g_{m0}(s')\right|\right\} \left| \int_0^s ds_1\, \bar{\Delta}_{m0}(s_1) e^{i\tau\varphi_{m0}(s_1)} \right| \\ &= \max_{s'}\left\{\left|g_{m0}(s')\right|\right\} \left| \frac{1}{i\tau}\left(e^{i\tau\varphi_{m0}(s)} - 1\right) \right| \\ &\leq \max_{s'}\left\{\left|g_{m0}(s')\right|\right\} \frac{2}{\tau}. \end{aligned} \tag{8.3.42}$$

When the right hand side of (8.3.42) is small, the present perturbation analysis is valid. It follows that the criterion for the adiabatic approximation is given by

$$\tau \gg \max_{s,m}\left[\frac{|\langle\bar{\phi}_m(s)|\frac{d\bar{H}(s)}{ds}|\bar{\phi}_0(s)\rangle|}{\{\bar{\Delta}_{m0}(s)\}^2}\right] \geq \frac{\max_{s,m}\{|\langle\bar{\phi}_m(s)|\frac{d\bar{H}(s)}{ds}|\bar{\phi}_m(s)\rangle|\}}{[\min_s\{\bar{\Delta}_{10}(s)\}]^2}, \tag{8.3.43}$$

where the variable s ranges from 0 to a given time (normalised by τ), and we note that the energy gap $\bar{\Delta}_{10}(s)$ is the smallest of all $\bar{\Delta}_{m0}(s)$. In quantum annealing, τ is given as runtime. If τ meets the condition (8.3.43) for $0 \leq s \leq 1$, quantum annealing provides the solution with high probability. We emphasise that, when the middle or the right hand side of (8.3.43) is large, one needs to give a still larger τ. In this sense, the magnitude of

$$\max_{s,m}\left[\frac{|\langle\bar{\phi}_m(s)|\frac{d\bar{H}(s)}{ds}|\bar{\phi}_0(s)\rangle|}{\{\bar{\Delta}_{m0}(s)\}^2}\right] \quad \text{or} \quad \frac{\max_{s,m}\{|\langle\bar{\phi}_m(s)|\frac{d\bar{H}(s)}{ds}|\bar{\phi}_m(s)\rangle|\}}{[\min_s\{\bar{\Delta}_{10}(s)\}]^2} \tag{8.3.44}$$

have a crucial role in the success of quantum annealing.

8.4 Implementation of Quantum Annealing

Quantum annealing has been implemented in several way in practice and its validity has been investigated. In this section we review some experimental works on quantum annealing.

8.4.1 Numerical Experiments

The validity of quantum annealing has been tested numerically for several optimisation problems.

Kadowaki and Nishimori in the seminal paper [211] applied both quantum annealing and simulated annealing to an optimisation problem given by the spin glass Hamiltonian:

$$H_0 = -\sum_{i,j} J_{ij} S_i^z S_j^z - h \sum_i S_i^z. \tag{8.4.1}$$

To apply quantum annealing, they added the transverse-field term $-\Gamma(t)\sum_i \sigma_i^x$ to this Hamiltonian and solved the time-dependent Schrödinger equation. As for simulated annealing, they considered the master equation:

$$\frac{dP_\mu(t)}{dt} = \sum_\nu L_{\mu\nu}(t)P_\nu(t) - \sum_\nu L_{\nu\mu}(t)P_\mu(t), \tag{8.4.2}$$

where μ and ν denote spin states which diagonalise the Ising Hamiltonian H_0, and $P_\mu(t)$ is the probability of the system in the state μ. The transition matrix is given by $L_{\mu\nu}(t) = e^{-E_\mu/T(t)}/(e^{-E_\mu/T(t)} + e^{-E_\nu/T(t)})$ for μ and ν which are different only in a single spin and $L_{\mu\nu}(t) = 0$ otherwise, where E_μ is the eigenenergy H_0 with respect to the state μ and $T(t)$ is the temperature in unit of k_B. Figure 8.5 shows the results for the Sherrington-Kirkpatrick type model with 8 spins, in which each spin connects with all other spins and J_{ij} is randomly drawn from the Gaussian distribution with mean 0 and variance 1/8. The probability $P_{\mathrm{QA}}(t)$ that the ground state of H_0 is found in the time-dependent state during quantum annealing is defined by $P_{\mathrm{QA}}(t) = |\langle\Psi_{\mathrm{gs}}|\Psi(t)\rangle|^2$ where $|\Psi_{\mathrm{gs}}\rangle$ is the ground state of H_0 and $|\Psi(t)\rangle$ is the time-dependent state obeying the Schrödinger equation. The probability $P_{SA}(t)$ stands for $P_\mu(t)$ in (8.4.2) with μ of the ground state. The transverse field and the temperature

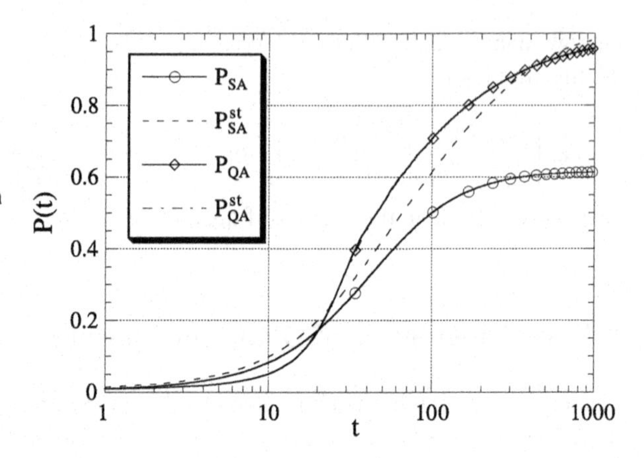

Fig. 8.5 Time evolution of the probability that the ground state of H_0 is found during quantum annealing and simulated annealing for the Sherrington-Kirkpatrick type model with 8 spins. Both the transverse field and the temperature are decreased as $\Gamma(t) = T(t) = 3/\sqrt{t}$. (Taken from Ref. [211])

are assumed to decrease with time as $\Gamma(t) = T(t) = 3/\sqrt{t}$. The initial conditions for $|\Psi(t)\rangle$ and $P_\mu(t)$ are given by $|\Psi(0)\rangle = (1/\sqrt{2^8})\sum_\mu |\mu\rangle$ and $P_\mu(0) = 2^{-8}$ respectively. It is found in the figure that $P_{QA}(t)$ is larger than $P_{SA}(t)$ for large t. In Ref. [211], non-random configurations of J_{ij} with different time schedules of $\Gamma(t)$ and $T(t)$ are also studied and similar results are reported.

Farhi et al. [131] simulated quantum annealing using Exact Cover with spins up to $N = 20$. Exact Cover is one of the NP-complete problems. An instance of Exact Cover is composed of M clauses. Each clause involves three bits chosen randomly from N bits. We say that the clause is satisfied if one of these three bits is 1 and the other two are 0. The problem is to decide whether all of given M clauses are satisfied at least by one configuration of N bits or not. The parameter $\alpha = M/N$ characterises the property of this problem. When $\alpha \ll 1$, the problem is easy and the answer should be "satisfied" (SAT) in almost all cases. On the other hand, when $\alpha \gg 1$, the problem is also easy but the answer should be "unsatisfied" (UNSAT). The SAT solutions decrease with increasing α. In the limit $N \to \infty$, it is conjectured that the SAT solutions abruptly vanish at a certain critical value α^* [5, 279, 324] and one encounters a phase transition between SAT and UNSAT phases at this point. The problem is considered to be most difficult at around α_c.

In order to formulate this problem in the language of spins, we assign a cost h_C to a clause C. h_C is defined as $h_C = 0$ if the constraint is satisfied, and $h_C = 1$ if it is not satisfied. For instance, supposing that three bits of a clause C is labelled by $C1$, $C2$, and $C3$, the cost is written as

$$h_C(S_{C1}, S_{C2}, S_{C3}) = \frac{(S_{C1} + S_{C2} + S_{C3})^2 - 1}{8} + \frac{\{(S_{C1} + S_{C2} + S_{C3})^2 - 9\}(S_{C1} + S_{C2} + S_{C3} - 1)}{16},$$

where S_i is an Ising-spin variable. The problem Hamiltonian is thus written as $H_0(\{S_i\}) = \sum_C h_C(S_{C1}, S_{C2}, S_{C3})$. To apply quantum annealing, the Ising spin S_i is replaced with S_i^z and a quantum fluctuation is introduced by the transverse field.

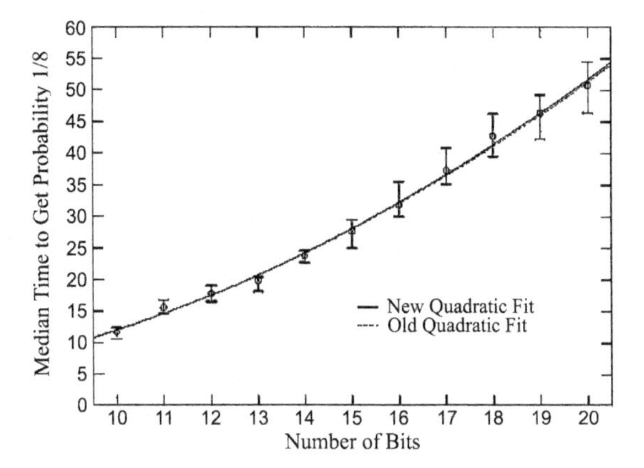

Fig. 8.6 Runtime to achieve the success probability $1/8$ by quantum annealing in solving Exact Cover. 75 instances are studied and median values are plotted. The *error bars* show 95 % confidence limits for each median. (Taken from Ref. [131])

The time-dependent Hamiltonian employed in Ref. [131] is given by (8.3.2). In numerical simulation, Farhi et al. generated tens of instances of Exact Cover which has the unique satisfied solution, solved the Schrödinger equation, and measured the probability that the solution is found in the final state at $t = \tau$. Figure 8.6 shows the median runtime to achieve the probability $1/8$. To obtain the runtime to achieve the probability $1/8$ for a fixed Exact Cover instance, simulation is repeated changing the runtime till the probability becomes between 0.12 and 0.13. The solid line indicates a quadratic fit to the data. The agreement between a quadratic curve and the results of simulation suggests that the mean time to get the solution of Exact Cover grows quadratically with the number of bits. We remark here that the results in Fig. 8.6 does not imply that Exact Cover can be solved by quantum annealing in a time scaled polynomially by the number of bits. In order to reveal whether the cost of quantum annealing is polynomial or not, one needs to argue the worst case, namely, the scaling of the longest runtime to achieve the probability $1/8$ must be studied. It is worth noting that the largest size studied in Ref. [131] is not necessarily sufficient to convince us the size scaling of the runtime. Since it would be impractical to solve the time-dependent Schrödinger equation for sizes beyond 20 spins using a classical computer, one has to rely on other numerical or analytical methods. We argue this issue in Sect. 8.5 from a different point of view.

From a practical point of view, one may not necessarily restrict oneself to the Schrödinger equation of motion. If one employs another rule of dynamics, one can apply quantum annealing to much larger sizes using classical computers. Martoňák et al. [260] demonstrated quantum annealing by a quantum Monte-Carlo method for the travelling salesman problem. The Hamiltonian to be minimised of the travelling salesman problem is given by (8.2.1). To apply quantum annealing to the travelling salesman problem, one needs to take into account the constraints given by (8.2.2). A standard way to incorporate them is to use the 2-opt move of the path. A 2-opt move consists of a removal and creation of two links. Suppose that N cities are labelled by $i = 1, 2, \ldots, N$, and a path 1-2-$\cdots$-N-1 is given initially. Let us choose two links $\langle i, i+1\rangle$ and $\langle j, j+1\rangle$ from this path. We then remove these links and create new links as $\langle i, j\rangle$ and $\langle i+1, j+1\rangle$ or $\langle i, j+1\rangle$ and $\langle j, i+1\rangle$. In the language of spins, this motion can be realised by the application of spin operators: $S^+_{\langle i+1\, j+1\rangle} S^+_{\langle ij\rangle} S^-_{\langle j\, j+1\rangle} S^-_{\langle i\, i+1\rangle}$ or $S^+_{\langle j\, i+1\rangle} S^+_{\langle i\, j+1\rangle} S^-_{\langle j\, j+1\rangle} S^-_{\langle i\, i+1\rangle}$, where $S^\pm_{\langle l\,m\rangle} = \frac{1}{2}(S^x_{\langle l\,m\rangle} \pm i\, S^y_{\langle l\,m\rangle})$ stand for the spin-flip operators. Hence the natural choice for the quantum Hamiltonian should be

$$H(t) = \sum_{\langle ij\rangle} d_{\langle ij\rangle} \frac{S^z_{\langle ij\rangle} + 1}{2} - \frac{1}{2} \sum_{\langle ij\rangle} \sum_{\langle i'j'\rangle} \Gamma_1(i, j, i', j') \big(S^+_{\langle ii'\rangle} S^+_{\langle jj'\rangle} S^-_{\langle i'j'\rangle} S^-_{\langle ij\rangle} + H.c.\big)$$
$$- \frac{1}{2} \sum_{\langle ij\rangle} \sum_{\langle i'j'\rangle} \Gamma_2(i, j, i', j') \big(S^+_{\langle ij'\rangle} S^+_{\langle i'j\rangle} S^-_{\langle i'j'\rangle} S^-_{\langle ij\rangle} + H.c.\big). \tag{8.4.3}$$

The amplitudes of quantum coupling Γ_1 and Γ_2 may depend on the distance between cities involved in a 2-opt move. Note that under this Hamiltonian the magnetisation $\sum_{\langle ij\rangle} S^z_{\langle ij\rangle}$ is conserved. Hence with this Hamiltonian the state moves in a subspace with a fixed magnetisation. The quantum Monte-Carlo simulation requires mapping of this quantum Hamiltonian to a classical one. However the presence of four-spin interactions in (8.4.3) causes a difficulty in such a mapping. To avoid this difficulty, Martoňák et al. replaced (8.4.3) to a transverse field Ising model:

$$H(t) = \sum_{\langle ij\rangle} d_{\langle ij\rangle} \frac{S^z_{\langle ij\rangle} + 1}{2} - \Gamma \sum_{\langle ij\rangle} S^x_{\langle ij\rangle}. \tag{8.4.4}$$

This Hamiltonian can be immediately mapped to a coupled classical systems by the procedure described in Sect. 3.1. The usual Monte-Carlo dynamics of the classical Hamiltonian from (8.4.4) could violate the 2-opt move and break the constraint of (8.2.2). However, the constraint can be fulfilled if one implements the 2-opt move in each Trotter slice. Thus quantum annealing is carried out by decreasing the transverse field Γ with the Monte-Carlo step in a fixed sufficiently low temperature. The travelling salesman problem studied in Ref. [260] is the instance pr1002 of the TSPLIB [326], which is a benchmark test problem composed of 1002 cities and the exact solution is known by another algorithm. Figure 8.7 shows results of quantum annealing and simulated annealing. In quantum annealing, the number of Trotter slices P is set at $P = 30$ and the temperature is kept at $T = 10/3$ so as to make the effective temperature PT of each Trotter slice $PT = 100$. The transverse field Γ is weakened linearly in a Monte-Carlo step τ from an initial value $\Gamma_i = 300$ to zero. Before annealing, equilibration is done under Γ_i. In simulated annealing, the temperature is lowered linearly from $T_i = 100$ to zero in a Monte-Carlo step τ. In both cases, 2-opt moves are implemented with a neighbourhood pruning, which restricts 2-opt moves within 20 nearest neighbours, namely one Monte-Carlo step consists of 20×1002 ($\times 30$ in the case of quantum annealing) 2-opt moves. The average is taken by the best path in Trotter slices over 96 independent runs. The results of simulation clearly show that the excess length after quantum annealing decays faster than simulated annealing. Since the quantum Monte-Carlo simulation consumes more computational cost than the classical one, it might be fair to replace the Monte-Carlo step in quantum annealing by the machine time $P\tau$. The red curve in Fig. 8.7 is the result of quantum annealing with the machine time. It is obvious

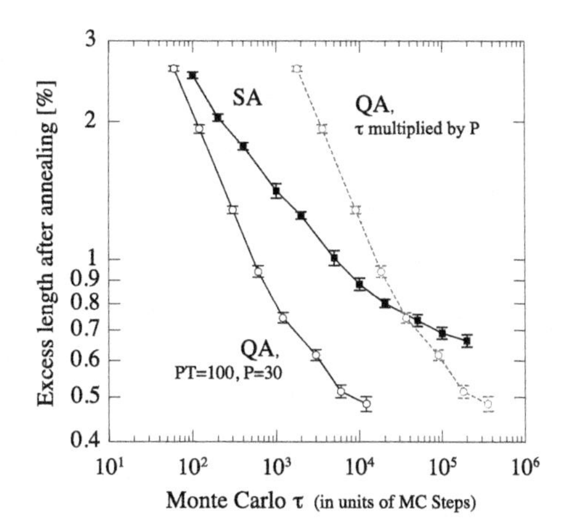

Fig. 8.7 Excess lengths after quantum annealing and simulated annealing of a travelling salesman problem. Both quantum annealing and simulated annealing are implemented by means of the Monte-Carlo method. The *dashed lines* are obtained just by multiplying τ by P in the result of quantum annealing, where $P = 30$ is the number of Trotter slices. The excess length decays faster in quantum annealing than in simulated annealing

that quantum annealing still gets advantage for large machine time. These results give a numerical evidence that quantum annealing brings a faster convergence to the target than simulated annealing.

Quantum annealing by means of the quantum Monte-Carlo method has been applied other models. Santoro et al. [342] showed better performance of quantum annealing over simulated annealing in the two-dimensional spin glass model. Sarjala et al. [343] studied the decay rate of an error after quantum annealing in random field Ising models in one, two and three dimensions. Battaglia et al. [30] focused on the 3-satisfiability problem. The 3-satisfiability problem is one of the NP-complete problems, and explained as follows. We consider N bits denoted by $(\sigma_1, \sigma_2, \ldots, \sigma_N)$. An instance of the 3-satisfiability problem (3-SAT) is composed of M clauses. A clause C involves three bits randomly chosen from N, which we label by $C1$, $C2$, and $C3$. A value z_i taken from 0 and 1 with probability $1/2$ is given to each bit. Thus a clause C is represented as (z_{C1}, z_{C2}, z_{C3}). We here consider the exclusive distinction between z_{Ci} and the corresponding bit σ_{Ci} $(i = 1, 2, 3)$. When at lease one of the three values of exclusive distinction is 0, we say that a clause C is satisfied. The problem of 3-satisfiability is to decide whether given M clauses can be satisfied by a bit configuration. In terms Ising spins, this problem can be translated into a problem to find a zero energy ground state of the Hamiltonian:

$$H = \frac{1}{8} \sum_C (1 - \zeta_{C1} S_{C1})(1 - \zeta_{C2} S_{C2})(1 - \zeta_{C3} S_{C3}), \tag{8.4.5}$$

where ζ_i and S_i have been defined as $\zeta_i = 2z_i - 1$ and $S_i = 2\sigma_i - 1$ respectively. It is shown in Ref. [30] that quantum annealing performs worse than simulated annealing, if it is applied to the 3-satisfiability problem. Matsuda et al. [261] studied quantum annealing in the Villain model defined on the square lattice. The vertical bonds are ferromagnetic whereas horizontal ones alternate between ferromagnetic and antiferromagnetic. What is special in this model is that one has an exponentially large number of degenerated ground states. Matsuda et al. observed that quantum annealing does not work well in comparison with simulated annealing in this model.

All these numerical experiments using quantum Monte-Carlo simulations show that quantum annealing basically works, but it does not always outperforms simulated annealing.

8.4.2 Experiments

In order to implement quantum annealing, one needs to make a system realising a transverse Ising model, initialise the spin state at the ground state of the transverse field term, move the state without decoherence by weakening the strength of the transverse field, and read the final state. A quantum computer naturally performs this procedure. Although a quantum computer is still beyond our reach, some elementary experiments of quantum computation by means of quantum annealing have been carried out.

Steffen et al. [376] first demonstrated quantum annealing for a three-bit MAX-CUT problem using an NMR quantum computer. The MAXCUT problem is one of the graph partitioning problem and belongs to the NP-hard class. Although the problem challenged in Ref. [376] is quite small, it has been observed that quantum annealing brings the solution.

A system of coupled superconducting flux qubits is another candidate for a quantum computer. D-Wave Systems, a Canadian company, is developing a commercial quantum computer made from superconducting flux qubits. It has been reported in Ref. [201] that the ground state of an Ising spin chain with eight spins is obtained by quantum annealing. Although a large gap must be filled before the realisation of a commercial quantum computer, the challenge of D-Wave Systems [201, 309] is noteworthy.

Trapped atomic ions also provide a route toward the realisation of a quantum computer. Kim et al. [225] reported that the ground state of a frustrated Ising-spin system with three spins made from trapped ions is attained by the adiabatic evolution following a change of the transverse field. What was done here is essentially the same as quantum annealing. The development of a quantum computer using a large number of trapped ions is expected.

Experiments mentioned above are concerned with microscopic systems. Brooke et al. [47] in turn studied quantum annealing and simulated annealing using a bulk material $\mathrm{LiHo}_x\mathrm{Y}_{1-x}\mathrm{F}_4$. When one applies a magnetic field perpendicular to the c axis, this material realises an disordered Ising system in a transverse field. Brooke et al. conducted two protocols of annealing and compared resulting states. The one protocol is simulated annealing, where the temperature is reduced at first at the weak transverse field and then the field is lifted toward the target. In the other protocol, one lifts the field at a high temperature, reduces the temperature with keeping the field, and then reduces the field at the low temperature. The latter corresponds to quantum annealing. In both cases, the initial and final conditions for the temperature and field are the same. The measurements of the ac susceptibility reveals that the state after quantum annealing involves shorter time scales. This means that quantum annealing brings states in narrower energy minima than simulated annealing. Although it is not clear whether the true ground state is in such narrow energy minima, experiments suggest that quantum annealing should have a virtue in extracting the global minimum from the sharp energy landscape.

8.5 Size Scaling of Energy Gaps

As we mentioned in Sect. 8.3.2, the success of quantum annealing rests on the magnitude of

$$\frac{\max_{s,m}\{|\langle\bar{\phi}_m(s)|\frac{d\bar{H}(s)}{ds}|\bar{\phi}_0(s)\rangle|\}}{[\min_s\{\bar{\Delta}_{10}(s)\}]^2}. \tag{8.5.1}$$

The runtime τ must be greater than (8.5.1) in order to obtain the solution with high probability. One may assume that the numerator of (8.5.1) is $O(N)$ since the Hamiltonian is an extensive quantity. Therefore, if $\Delta_{\min} = \min_s\{\Delta_{10}(s)\}$ is scaled by a polynomial of N^{-1} then the runtime grows polynomially with N as well, while the exponential scaling of $\Delta_{\min}$ gives rise to the exponential increase of the runtime. The knowledge on the scaling of $\Delta_{\min}$ is crucially important to reveal the ability of quantum annealing. In this section, we shall argue the scaling of $\Delta_{\min}$ in several models.

8.5.1 Simple Case

As a simple case, let us focus on the pure transverse Ising chain. Because of the parity conservation of the fermion number, the lowest excited state accessible during the time evolution from the ground state is not actually the first excited state. The energy gap from the ground state to the lowest accessible excited state is given by

$$\Delta(\Gamma) = 4\sqrt{\left(\cos\frac{\pi}{N} - \Gamma\right)^2 + \sin^2\frac{\pi}{N}},$$

where we have assumed a system with size N and with the periodic boundary condition. It is easy to show that this energy gap takes its minimum $\Delta_{\min} = 4\sin\frac{\pi}{N}$ when $\Gamma = \cos\frac{\pi}{N}$. For $N \gg 1$, it follows that

$$\Delta_{\min} \approx \frac{4\pi}{N}. \tag{8.5.2}$$

Hence the minimum energy gap closes as $1/N$ with $N \to \infty$. In the thermodynamic limit, this occurs at the quantum critical point $\Gamma_c = 1$. The scaling relation (8.5.2) is reproduced from the dynamical scaling relation in the vicinity of the quantum critical point:

$$\tau(\Gamma) \sim |\Gamma - \Gamma_c|^{-z\nu} \sim \xi^z$$

where $\tau(\Gamma)$ is the coherence time defined by $\tau(\Gamma) = \hbar/\Delta(\Gamma)$ and the ξ is the correlation length of the ground state. In finite d-dimensional systems, one may replace the correlation length with the linear size $N^{1/d}$ and $\Delta(\Gamma)$ with Γ_{min}. Using the critical exponents $z = \nu = 1$ and $d = 1$, one can obtains $\Delta_{\text{min}} \sim 1/N$. In general, d-dimensional systems with a quantum phase transition characterised by the dynamical exponent z have

$$\Delta_{\text{min}} \sim N^{-z/d}. \tag{8.5.3}$$

Therefore, as far as a standard second order quantum phase transition is concerned, the minimum energy gap Δ_{min} is scaled polynomially with the system size.

8.5.2 Annealing over an Infinite Randomness Fixed Point

We move on to disordered systems. Since the non-trivial combinatorial optimisation problem involves disorder, it is significant to investigate quantum annealing of disordered systems. In this subsection, we focus on the random transverse field Ising chain, though it is still trivial from the viewpoint of the optimisation problem.

Let us consider the random transverse field Ising chain:

$$H = -\sum_i J_i S_i^z S_{i+1}^z - \Gamma \sum_i h_i S_i^x, \tag{8.5.4}$$

where the coupling constant J_i and the random parameter h_i are assumed to follow uniform distributions,

$$P(J_i) = \theta(J_i)\theta(1 - J_i),$$
$$P(h_i) = \theta(h_i)\theta(1 - h_i),$$

respectively. Same as the pure system, the model (8.5.4) is self-dual at $\Gamma = 1$, where a quantum phase transition takes place. The critical point $\Gamma_{\text{c}} = 1$ is known as an infinite randomness fixed point. The critical properties are revealed by the strong disorder renormalisation group [138]. According to detailed calculations [139, 188, 415], the distribution of the energy gap Δ from the ground state to the first excited state at the critical point of the system with size N is given as a function of $x = -\ln(\Delta/\Delta_0)/\sqrt{N}$, where Δ_0 is the largest energy scale of the system. In fact, the distribution function is obtained as [139]

$$P(\Delta, N)\,d\Delta = P(x)\,dx \tag{8.5.5}$$

$$P(x) \approx \begin{cases} \frac{2}{\sqrt{\pi}} e^{-x^2/4} & \text{for } x \gg 1, \\ \frac{2\pi}{x^3} e^{-\pi^2/(4x^2)} & \text{for } x \ll 1. \end{cases} \tag{8.5.6}$$

Let x_{max} be the value of x that maximise $P(x)$. The typical energy gap Δ_{typ} of the system at the critical point is estimated by $-\ln(\Delta_{\text{typ}}/\Delta_0)/\sqrt{N} \approx x_{\text{max}}$. This implies that the scaling of the typical energy gap is given by

$$\Delta_{\text{typ}} \sim e^{-\text{const.} \times \sqrt{N}}. \tag{8.5.7}$$

The average of the energy gap is given by

$$\Delta_{\text{av}} = \int_0^\infty \Delta P(\Delta, N)\, d\Delta = \int_0^\infty \Delta_0 e^{-x\sqrt{N}} P(x)\, dx.$$

When $\sqrt{N} \gg 1$, the integrand vanishes for $x \gtrsim 1/\sqrt{N} \ll 1$, so that one may safely approximate $P(x)$ as $P(x) \approx (2\pi/x^3)e^{-\pi^2/(4x^2)}$ to yield

$$\Delta_{\text{av}} \approx \int_0^\infty \frac{2\pi}{x^3} e^{-x\sqrt{N}-\pi^2/(4x^2)}\, dx.$$

One can see that the integrand has a sharp peak at $x^* = (\pi^2/2N^{1/2})^{1/3}$. The width of the peak can be estimated as $2x^{*2}/\pi$ by making $x = x^* + \delta$ and expanding $-x\sqrt{N} - \pi^2/(4x^2)$ up to the order of δ^2. Therefore one obtains

$$\Delta_{\text{av}} \approx \frac{1}{x^*} e^{-x^*\sqrt{N}-\pi^2/(4x^{*2})} \approx N^{1/6} \exp\left\{-\frac{3}{2}\left(\frac{\pi^2}{2}N\right)^{1/3}\right\}. \tag{8.5.8}$$

We recall here that the lowest excited state accessible during the time evolution from the ground state is not the first excited state. We then call for an ansatz that the energy gap $\Delta_{\text{acc}}(\Gamma)$ from the ground state to the lowest accessible excited state also obeys a distribution with the variable $x = -\ln(\Delta_{\text{acc}}(\Gamma_{\text{c}})/\Delta_0)/\sqrt{N}$ at the critical point. The numerical study by Caneva et al. [58] has confirmed that this ansatz seems actually true. In the present system, the energy gap of a fixed sample is not always minimised at the critical point. One can bound the minimum energy gap Δ_{min} by

$$\Delta_{\text{min}} \le \Delta_{\text{acc}}(\Gamma_{\text{c}}).$$

From this inequality and the scaling analysis on the energy at the critical point, we reach that the typical runtime to achieve optimisation by quantum annealing is estimated as

$$\tau_{\text{typ}} \gtrsim e^{\text{const.}\times N^{1/2}}, \tag{8.5.9}$$

whereas the averaged one is estimated as

$$\tau_{\text{ave}} \gtrsim e^{\text{const.}\times N^{1/3}}. \tag{8.5.10}$$

Now let us generalise these results to a system with a generic infinite randomness fixed point. We suppose that the distribution of the energy gap to the first excited state is given with an exponent ψ by

$$P(\Delta, N)\, d\Delta = P(x)\, dx, \quad x = -\ln(\Delta/\Delta_0)/N^\psi \tag{8.5.11}$$

at the critical point. Note $\psi = 1/2$ for the random transverse field Ising chain. Most recent numerical study has shown $\psi \approx 0.48$ for the two-dimensional random transverse Ising ferromagnet [235]. With generic ψ, the argument presented above shows that the typical runtime is estimated as

$$\tau_{\text{typ}} \gtrsim e^{\text{const.}\times N^\psi}. \tag{8.5.12}$$

An estimate for the averaged runtime τ_{ave} is not available since the detailed form of the distribution function $P(x)$ with generic ψ is not known. If one assumes the same form as the random transverse Ising chain, one has

$$\tau_{\text{ave}} \gtrsim \exp\bigl(\text{const.} \times N^{2\psi/3}\bigr).$$

In any case, it turns out that the disordered Ising chain under consideration gives rise to subexponential complexity for quantum annealing, though it serves as a trivial optimisation problem.

It is significant to comment on the role of the frustration in the infinite randomness fixed point. One might naturally think that Ising spin glass models in a transverse-field involve an infinite randomness fixed point. In fact, quantum Monte Carlo simulations have shown its manifestation in a two-dimensional transverse field Ising spin glass [333]. However the character of a quantum phase transition of this model is still uncertain. The strong disorder renormalisation group is concerned with the strongest interaction energy and transverse field and its procedure is not influenced by the presence of frustration. This suggests that the frustration might be an irrelevant factor to the infinite randomness fixed point. If this is true, not only trivial optimisation problems but also nontrivial ones share the same scaling of runtime in quantum annealing. Notwithstanding this speculation, the infinite randomness fixed point in frustrated models and quantum annealing over such a critical point remain to be clarified.

8.5.3 *Annealing over a First Order Quantum Phase Transition*

We have so far analysed size scaling of the energy gap at continuous quantum phase transitions in the previous subsections. We next consider first order quantum phase transitions in the present subsection.

8.5.3.1 Fully Connected p-Body Ising Ferromagnet in a Transverse Field

Let us consider the fully connected p-body ferromagnetic Ising model in a transverse field. The Hamiltonian is written as

$$H = -\frac{1}{N^{p-1}} \sum_{i_1,\dots,i_p=1}^{N} S_{i_1}^z \cdots S_{i_p}^z - \Gamma \sum_i S_i^z, \tag{8.5.13}$$

where N is the number of spins in the whole system and p denotes the number of spins which are interacting. We assume that p is odd and consider $p \to \infty$ limit. As we showed in Sect. 3.4.2, this model exhibits a first order quantum phase transition at $\Gamma_c = 1$.

We begin with the perturbation theory of the ground state. Let us first choose the ground state of the classical Hamiltonian H_0 as an unperturbed state. The Rayleigh-Schrödinger perturbation theory gives the perturbed ground-state energy as

$$E_{\mathrm{g}}(\Gamma) = E_{\mathrm{g}}^{(0)} + \langle \mathrm{g}| \sum_{n=0}^{\infty} (-\Gamma V) \left\{ \frac{Q}{E_{\mathrm{g}}^{(0)} - H_0} \left(E_{\mathrm{g}}^{(0)} - E_{\mathrm{g}}(\Gamma) - \Gamma V \right) \right\}^n |\mathrm{g}\rangle, \tag{8.5.14}$$

where $|\mathrm{g}\rangle$ and $E_{\mathrm{g}}^{(0)}$ denotes the unperturbed state and the corresponding energy, $|k\rangle$ $(k = 1, 2, \ldots)$ stands for an eigenstate of the classical Hamiltonian H_0, and $Q = \sum_{k\geq 1} |k\rangle\langle k|$ is the projection operator onto the subspace orthogonal to $|\mathrm{g}\rangle$. We define $V = \sum_i S_i^x$ and $V_{kl} = \langle k|V|l\rangle$. One can easily see that $V_{kk} = 0$ and $V_{kl} \neq 0$ for $|k\rangle$ and $|l\rangle$ which differ by a single spin flip. Hence (8.5.14) reduces to

$$\begin{aligned} E_{\mathrm{g}}(\Gamma) &= -N - \frac{\Gamma^2}{N} \sum_{k\geq 1} |V_{\mathrm{g}k}|^2 + O(\Gamma^4) \\ &= -N - \Gamma^2 + O(\Gamma^4), \end{aligned} \tag{8.5.15}$$

where we note $E_{\mathrm{g}}^{(0)} = -N$ and $\sum_{k\geq 1} |V_{\mathrm{g}k}|^2 = N$. Next, we consider the ground state of $-\Gamma V$ as the unperturbed state and deal with H_0 as the perturbation. Suppose that $|\mathrm{g}\rangle\rangle$ and $|k\rangle\rangle$ $(k = 1, 2, \ldots)$ represent the ground state and an eigenstate of $-\Gamma V$ respectively. The perturbed energy is written as

$$\tilde{E}_{\mathrm{g}}(\Gamma) = -N\Gamma + \langle\langle \mathrm{g}|H_0|\mathrm{g}\rangle\rangle + \sum_{k\geq 1} \frac{|\langle\langle k|H_0|\mathrm{g}\rangle\rangle|^2}{-N\Gamma - E_k^{\Gamma}} + O(\Gamma^{-2}), \tag{8.5.16}$$

where E_k^{Γ} is an eigenenergy of $-\Gamma V$. Since the ground state of $-\Gamma V$ is the superposition of all the eigenstates of H_0 with the same weight, the second term is evaluated as

$$\langle\langle \mathrm{g}|H_0|\mathrm{g}\rangle\rangle = \sum_k \left|\langle k|\mathrm{g}\rangle\rangle\right|^2 E_k^{(0)} = \frac{1}{2^N} \sum_k E_k^{(0)} \approx O\left(\frac{N}{2^N}\right),$$

where $E_k^{(0)}$ $(k = 1, 2, \ldots)$ denotes the eigenenergy of H_0 with respect to $|k\rangle$. Similarly, the matrix element in the third term is computed as $|\langle\langle k|H_0|\mathrm{g}\rangle\rangle| \approx N2^{-N}$. Noting that the summation is dominated by the terms with $E_k^{\Gamma} \approx O(1)$, one finds that

$$\sum_{k\geq 1} \frac{|\langle\langle k|H_0|g\rangle\rangle|^2}{-N\Gamma - E_k^{\Gamma}} \approx -\frac{1}{N\Gamma} \left(\frac{N}{2^N}\right)^2 \frac{N!}{\{(N/2)!\}^2} \approx -\frac{1}{N\Gamma} \frac{N^2}{2^N} = \frac{1}{\Gamma} O\left(\frac{N}{2^N}\right).$$

Therefore (8.5.16) is estimated as

$$\tilde{E}_{\mathrm{g}}(\Gamma) = -N\Gamma + O\left(\frac{N}{2^N}\right). \tag{8.5.17}$$

Equations (8.5.15) and (8.5.17) imply that the ground-state energies with $\Gamma \ll 1$ and $\Gamma \gg 1$ are not modified from the unperturbed energies within the leading order (which is proportional to N).

In order to reveal the size scaling of the energy gap at the quantum critical point, one needs an elaborate analysis. We consider the following Hamiltonian which approximates (8.5.13) at the critical point Γ_c:

$$H_c = -N|g\rangle\langle g| - N\Gamma_c|g\rangle\rangle\langle\langle g|. \tag{8.5.18}$$

Note that the first and the second terms represent the classical term H_0 and the transverse-field term $-\Gamma_c V$ respectively. In the first term, we have neglected the highest energy level of H_0. As for the second term, we adopt an approximation that all the eigenenergies except the ground energy vanish, because the zero-energy states dominate all the eigenstates of $-\Gamma V$ in their number. One can show that corrections to this approximation do not change the result [205]. We solve the eigenvalue problem of (8.5.18):

$$H_c|\psi\rangle = \lambda|\psi\rangle, \tag{8.5.19}$$

where λ and $|\psi\rangle$ represent the eigenvalue and the eigenstate respectively. Substituting (8.5.18) and applying $\langle\langle g|$ and $\langle\langle k|$ to the left, one obtains

$$\begin{aligned} -N\langle\langle g|g\rangle\langle g|\psi\rangle - N\Gamma_c\langle\langle g|\psi\rangle &= \lambda\langle\langle g|\psi\rangle, \\ -N\langle\langle k|g\rangle\langle g|\psi\rangle &= \lambda\langle\langle k|\psi\rangle \quad (k = 1, 2, \ldots). \end{aligned}$$

These equations are arranged into

$$\langle\langle g|\psi\rangle + \frac{N}{N\Gamma_c + \lambda}\langle\langle g|g\rangle\langle g|\psi\rangle = 0, \tag{8.5.20}$$

$$\langle\langle k|\psi\rangle + \frac{N}{\lambda}\langle\langle k|g\rangle\langle g|\psi\rangle = 0 \quad (k = 1, 2, \ldots). \tag{8.5.21}$$

Multiplying the first and the second equations by $\langle g|g\rangle\rangle$ and $\langle g|k\rangle\rangle$ respectively, (8.5.20) and (8.5.21) are summed into

$$\langle g|\psi\rangle + \frac{N}{N\Gamma_c + \lambda}\left|\langle\langle g|g\rangle\right|^2\langle g|\psi\rangle + \frac{N}{\lambda}\sum_{k\geq 1}\left|\langle\langle k|g\rangle\right|^2\langle g|\psi\rangle = 0,$$

where we have used $|g\rangle\rangle\langle\langle g| + \sum_{k\geq 1}|k\rangle\rangle\langle\langle k| = 1$. Since $|\langle\langle g|g\rangle|^2 = |\langle\langle k|g\rangle|^2 = 2^{-N}$ and $\sum_{k\geq 1} = 2^N - 1$, the above equation reduces to

$$1 + \frac{N}{2^N}\left(\frac{1}{N\Gamma_c + \lambda} + \frac{2^N - 1}{\lambda}\right) = 0.$$

Noting that $\Gamma_c = 1$, this equation is easily solved to yield

$$\lambda = -N \pm \frac{N}{2^{N/2}}.$$

Thus one obtains the size scaling of the energy gap at the quantum critical point:

$$\Delta_{\min} = \frac{2N}{2^{N/2}}. \tag{8.5.22}$$

This result implies that the present system suffers from the exponential scaling of the minimum energy gap, although the classical Hamiltonian H_0 has a trivial ground state.

8.5.3.2 Quantum Random Energy Model

The second model we investigate is the p-body random Ising model in the transverse field:

$$H(\Gamma) = H_0 - \Gamma V$$
$$H_0 = \sum_{1 \le i_1 < \cdots < i_p \le N} J_{i_1,\ldots,i_p} S_{i_1}^z S_{i_2}^z \cdots S_{i_p}^z, \quad V = \sum_i S_i^x, \tag{8.5.23}$$

where we assume that the $J_{i_1,\ldots,i_p}$ obeys the Gaussian distribution,

$$P(J_{i_i,\ldots,i_p}) = \sqrt{\frac{N^{p-1}}{\pi p!}} \exp\left(-\frac{N^{p-1}}{p!} J_{i_1,\ldots,i_p}^2\right).$$

As mentioned in Sect. 6.6, the classical Hamiltonian H_0 is reduced to the random energy model by taking $p \to \infty$ [103, 104], where the eigenenergy of H_0 obeys the distribution

$$P(E) = \frac{1}{\sqrt{\pi N}} e^{-E^2/N}. \tag{8.5.24}$$

In the present subsection, we shall mainly discuss this particular case by means of the perturbation theory [204].

First, we choose $\Gamma = 0$ as a starting point of the perturbation theory. Let $|k\rangle$ and $E_k^{(0)}$ be an eigenstate and an eigenenergy of H_0 respectively. We suppose that $E_k^{(0)}$ is of order N. According to the Rayleigh-Schrödinger perturbation theory, the energy perturbed by $-\Gamma V$ is written as

$$E_k(\Gamma) = E_k^{(0)} + \langle k| \sum_{n=0}^{\infty} (-\Gamma V) \left\{ \frac{Q}{E_k^{(0)} - H_0} \left(E_k^{(0)} - E_k(\Gamma) - \Gamma V\right) \right\}^n |k\rangle, \tag{8.5.25}$$

where $Q = \sum_{l \neq k} |l\rangle\langle l|$ is the projection operator. Let $V_{ij} = \langle i|V|j\rangle$ be the matrix element of V. Equation (8.5.25) is then written up to the second order of Γ as

$$E_k(\Gamma) = E_k^{(0)} + \sum_{l \neq k} \frac{\Gamma^2 |V_{kl}|^2}{E_k^{(0)} - E_l^{(0)}} + \cdots, \tag{8.5.26}$$

where we remark that the first order term of Γ vanishes since $V_{kk} = \sum_i \langle k|S_i^x|k\rangle = 0$. Notice that $|V_{kl}|^2 = 1$ if $|l\rangle$ differs from $|k\rangle$ by a single-spin flip and $|V_{kl}|^2 = 0$ otherwise. It follows that

$$E_k(\Gamma) = E_k^{(0)} + \frac{\Gamma^2}{E_k^{(0)}} \left(N + \frac{1}{E_k^{(0)}} \sum_{l \neq k} |V_{kl}|^2 E_l^{(0)} + \cdots \right).$$

Since each $E_l^{(0)}$ obeys the distribution (8.5.24), one can estimate $\sum_{l \neq k} |V_{kl}|^2 E_l^{(0)} \approx O(\sqrt{N})$. For higher orders, one can see $\sum_{l \neq k} |V_{kl}|^2 (E_l^{(0)})^n = O(N^{n/2})$. Therefore one obtains

$$E_k(\Gamma) = E_k^{(0)} + \Gamma^2 \frac{N}{E_k^{(0)}} + \Gamma^2 O\left(N^{-3/2}\right) + \cdots, \tag{8.5.27}$$

where we have used $E_k^{(0)} = O(N)$. The ground energy in particular is written as

$$E_g(\Gamma) \approx -N\sqrt{\ln 2} - \frac{\Gamma^2}{\sqrt{\ln 2}}. \tag{8.5.28}$$

We remark that the second term in (8.5.27) is of order $\Gamma^2 \times 1$. Higher orders in Γ are shown to give $O(1/N)$ contributions. This implies that the energy $E_k^{(0)}$ of order N is not perturbed by the transverse field $-\Gamma V$ within the leading order in N. Consequently the free-energy density and the entropy density are independent of Γ.

We next look upon the perturbation expansion around $\Gamma = \infty$. We here focus on the ground state $|\text{g}\rangle\rangle$ of $-\Gamma V$ as the unperturbed state. The eigenenergy $-\Gamma V$ with respect to $|\text{g}\rangle\rangle$ is $-N\Gamma$. We suppose that $|k\rangle\rangle$ and E_k^Γ denote an eigenstate and an eigenenergy of $-\Gamma V$. The perturbation expansion of the ground energy around $\Gamma = \infty$ is given by

$$\tilde{E}_\text{g}(\Gamma) = -N\Gamma + \langle\langle \text{g}|H_0|\text{g}\rangle\rangle + \sum_{k\geq 1} \frac{|\langle\langle k|H_0|\text{g}\rangle\rangle|^2}{-N\Gamma - E_k^\Gamma} + \cdots. \tag{8.5.29}$$

The second term is arranged as

$$\langle\langle \text{g}|H_0|\text{g}\rangle\rangle = \sum_{k=1}^{2^N} \left|\langle\langle \text{g}|k\rangle\right|^2 E_k^{(0)} = \frac{1}{2^N}\sum_{k=1}^{2^N} E_k^{(0)},$$

where note that $|k\rangle$ and $E_k^{(0)}$ are an eigenstate and an eigenenergy of H_0. Since the eigenenergies of H_0 obey the distribution (8.5.24), the second term is estimated as

$$\langle\langle \text{g}|H_0|\text{g}\rangle\rangle \approx \frac{1}{2^N}\sqrt{N}2^{N/2} = \sqrt{N}2^{-N/2},$$

which turns out to be exponentially small. Regarding the third term, $|k\rangle\rangle$ must differ from $|\text{g}\rangle\rangle$ by p spins in order to produce a finite contribution to the summation. This is because H_0 flips p spins in the σ^x basis if it is applied to $|\text{g}\rangle\rangle$. The energy E_k^Γ of such states is given by $(-N + 2p)\Gamma$. Hence one has

$$\sum_{k\geq 1} \frac{|\langle\langle k|H_0|\text{g}\rangle\rangle|^2}{-N\Gamma - E_k^\Gamma} = -\frac{1}{2p\Gamma}\sum_{k\geq 1}\left|\langle\langle k|H_0|\text{g}\rangle\rangle\right|^2 = -\frac{1}{2p\Gamma}\langle\langle \text{g}|H_0^2|\text{g}\rangle\rangle.$$

Due to the distribution (8.5.24) of the eigenenergies of H_0, the expectation value of H_0^2 is estimated as

$$\langle\langle \text{g}|H_0^2|\text{g}\rangle\rangle = \frac{1}{2^N}\sum_{k=1}^{2^N}(E_k^{(0)})^2 \approx \int_{-\infty}^{\infty} E^2 \frac{1}{\sqrt{\pi N}} e^{-E^2/N}\,dE = \frac{N}{2}.$$

Thus one obtains

$$\sum_{k\geq 1} \frac{|\langle\langle k|H_0|\text{g}\rangle\rangle|^2}{-N\Gamma - E_k^\Gamma} \approx -\frac{N}{4p\Gamma}.$$

One can show that higher order corrections are negligible. Hence the perturbed ground state energy is obtained as

$$\tilde{E}_{\rm g}(\Gamma) \approx -N\Gamma - \frac{N}{4p\Gamma}. \tag{8.5.30}$$

To obtain the expression for generic eigenenergy, one needs to go around with the perturbation theory of degenerated states of $-\Gamma V$. However the matrix elements of H_0 between degenerated states are tiny and lift the degeneracy by a small amount. Hence the argument for the ground state can be basically applicable and yields

$$\tilde{E}_k(\Gamma) \approx -E_k^\Gamma - \frac{N}{4p\Gamma} \tag{8.5.31}$$

for a generic unperturbed eigenenergy E_k^Γ of $-\Gamma V$. The second terms in (8.5.30) and (8.5.31) vanish with $p \to \infty$. Therefore the energies are not affected by the perturbation within the leading order in N. It turns out that the free-energy density and the entropy density are not modified by the perturbation as well.

The quantum critical point can be identified from the crossing point of $E_{\rm g}(\Gamma)$ and $\tilde{E}_{\rm g}(\Gamma)$. Equations (8.5.28) and (8.5.30) yield

$$\Gamma_{\rm c} = \sqrt{\ln 2}. \tag{8.5.32}$$

The quantum phase transition between the ground states for $\Gamma < \Gamma_{\rm c}$ and $\Gamma > \Gamma_{\rm c}$ is the first order transition. The same result can be obtained using the static approximation [158], which has been shown to be exact when $p \to \infty$ [109].

The energy gap between the lowest two states at the critical point is estimated as follows. We consider two ground energies $E_{\rm g}(\Gamma)$ and $\tilde{E}_{\rm g}^\Gamma$. These two energy levels come close to each other with Γ approaching $\Gamma_{\rm c}$. However, level crossing is avoided at $\Gamma_{\rm c}$ because of quantum tunnelling between two states. This is what the non-crossing rule states. To obtain the energy gap, we introduce the following Hamiltonian:

$$\begin{aligned} H &= E_{\rm g}(\Gamma_{\rm c})|{\rm g}\rangle\langle{\rm g}| + \tilde{E}_{\rm g}(\Gamma_{\rm c})|{\rm g}\rangle\rangle\langle\langle{\rm g}| \\ &= -N\sqrt{\ln 2}\big(|{\rm g}\rangle\langle{\rm g}| + |{\rm g}\rangle\rangle\langle\langle{\rm g}|\big). \end{aligned} \tag{8.5.33}$$

Here we have restricted ourselves to these two levels and neglected the effect of higher levels. In addition, we have made an approximation that the ground states for $\Gamma < \Gamma_{\rm c}$ and $\Gamma > \Gamma_{\rm c}$ do not change from the unperturbed ground states $|{\rm g}\rangle$ and $|{\rm g}\rangle\rangle$ respectively. This approximation is reasonable because of the above perturbative analysis and the discontinuous character of the transition. Now, let $|{\rm e}\rangle$ be the normalised state which is orthogonal to $|{\rm g}\rangle$, such that $\langle{\rm g}|{\rm e}\rangle = 0$ and $\langle{\rm e}|{\rm e}\rangle = 1$. We expand $|{\rm g}\rangle\rangle$ using $|{\rm g}\rangle$ and $|{\rm e}\rangle$ as

$$|{\rm g}\rangle\rangle = \alpha|{\rm g}\rangle + \beta|{\rm e}\rangle. \tag{8.5.34}$$

The Hamiltonian (8.5.33) can be written as

$$\begin{aligned} H = -N\sqrt{\ln 2}\big\{&\big(1+|\alpha|^2\big)|{\rm g}\rangle\langle{\rm g}| + |\beta|^2|{\rm e}\rangle\langle{\rm e}| \\ &+ \alpha\beta^*|{\rm g}\rangle\langle{\rm e}| + \alpha^*\beta|{\rm e}\rangle\langle{\rm g}|\big\}. \end{aligned}$$

By diagonalising this Hamiltonian, one finds that the energy gap is given by [204]

$$\Delta = 2N\sqrt{\ln 2}|\alpha|^2 = 2N\sqrt{\ln 2}\langle g|g\rangle\rangle = 2\sqrt{\ln 2}N2^{-N/2}, \tag{8.5.35}$$

where note that $\langle g|g\rangle\rangle = 2^{-N/2}$. This shows that the energy gap at the quantum critical point closes exponentially with system size N.

8.5.3.3 Numerical Studies

In order to clarify the scaling of the minimum energy gap in systems representing practical optimisation problems, we need to rely on numerical studies. Young et al. investigated a quantum Hamiltonian of Exact Cover by means of the quantum Monte Carlo method [440, 441]. The quantum Monte Carlo might be the most suitable method for our purpose, considering its applicability to various systems and large sizes. Let us consider the imaginary-time autocorrelation function of S_i^z, $\langle S_i^z(0)S_i^z(u)\rangle$, where $S_i^z(u) = e^{-uH}S_i^z e^{uH}$ is the Heisenberg operator with the imaginary time u. Let Δ_{10} be the energy gap between the ground and first excited states. When $\beta = T^{-1} \to \infty$ and $u \gg \Delta_{10}$, this autocorrelation function reduces to

$$\langle S_i^z(0)S_i^z(u)\rangle \to q + \sum_{n\geq 1} \left|\langle n|S_i^z|g\rangle\right|^2 e^{-u(E_n - E_g)}, \tag{8.5.36}$$

where q is the spin glass order parameter defined by

$$q = \left|\langle g|S_i^z|g\rangle\right|^2. \tag{8.5.37}$$

We note that $|g\rangle$ and $|n\rangle$ ($n = 1, 2, \ldots$) stand for the ground and excited states of H with eigenenergies E_g and E_n respectively. Equation (8.5.36) turns out to be $\langle S_i^z(0)S_i^z(u)\rangle - q \approx e^{-u\Delta_{10}}$ for $u \gg \Delta_{10}$. Therefore one finds Δ_{10} by the decreasing rate of $\ln[\langle S_i^z(0)S_i^z(u)\rangle - q]$ with respect to u. Now we consider the time-dependent Hamiltonian $H(s) = (1-s)H_0 - s\sum_i S_i^x$, where $s = t/\tau$ is the time normalised by the runtime and H_0 is the problem Hamiltonian. What Young et al. showed by a quantum Monte Carlo simulation in Ref. [441] are as follows. (1) For some instances of Exact Cover, the spin glass order parameter q as a function of s exhibits a discontinuity between different values. When this discontinuous transition occurs, q drops abruptly and then rises up to a different value. (2) The energy gap takes a minimum simultaneously with q. Its value are much smaller compared to those for instances in the absence a discontinuous transition of q. (3) The ratio of the instances with a discontinuous transition increases up to unity with increasing the system size.

We recall that Farhi et al. have reported a polynomial scaling of the median runtime to get a solution in the same problem on the basis of simulation for system sizes up to $N = 20$ [131] (see Sect. 8.4.1). In this small size regime, the probability that an instance with a discontinuous transition is drawn is so low that almost of all instances should be easy problems with large energy gaps. Assuming that the energy gap at a discontinuous transition point closes exponentially with the system size, the result of quantum Monte Carlo simulation implies that the complexity of Exact Cover by quantum annealing turns into an exponential one with the system size.

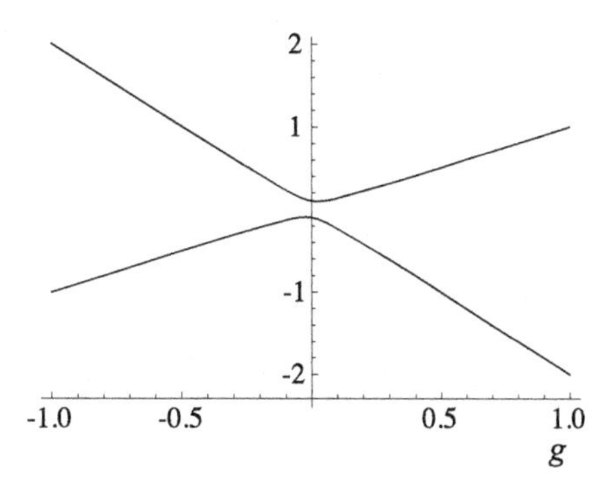

Fig. 8.8 The eigenenergies of the Hamiltonian (8.5.38) with $a = 1$, $b = 2$, and $\delta = 0.1$, which bring about anti-crossing. The energy gap is minimised at $g = 0$ and given by $|\delta|^2$

8.5.4 Anderson Localisation

In the previous subsection, we have discussed a first order quantum phase transition. Here in this section we discuss scaling of the gap at a level anti-crossing point with the aid of the notion of the Anderson localisation.

Let us consider a correspondence between a transverse-field Ising model and a tight-binding model. In this correspondence, an eigenstate of the Ising Hamiltonian with N spins is regarded as a vertex site of the N-dimensional hypercube. Then it is easy to see that an Ising Hamiltonian and a transverse field correspond to a potential energy on site and hopping between nearest neighbour sites. Now we consider a transverse-field Ising model representing quantum annealing of Exact Cover. By the analogy to the Anderson localisation [15], it is expected that the randomness in the potential energy makes the wavefunction of an eigenstate localised in the N-dimensional space and decay exponentially with distance from the centre of localisation. As one changes the transverse field, the energy of such localised states changes. For simplicity we consider two levels which are close in energy. According to the non-crossing rule described in Sect. 8.3.1, level crossing of two levels does not occur as far as no special symmetry or conserving quantity is present. The Hamiltonian expressing this level anti-crossing can be written effectively as

$$H = \begin{bmatrix} ag & \delta \\ \delta & -bg \end{bmatrix}, \tag{8.5.38}$$

where we chose localised states as the basis. ag and $-bg$ are the energies of localised states without tunnelling energy. g is a parameter representing the transverse field so as to vanish at the crossing point of two levels. δ stands for the tunnelling energy. Figure 8.8 shows two eigenvalues of (8.5.38). The minimum energy gap is obtained, when anti-crossing takes place, as $\Delta_{\min} = 2|\delta|^2$. The tunnelling energy δ decays exponentially as the distance between the positions of two localised states increases. Thus the minimum energy gap closes exponentially with the size N if the distance of two lowest levels is $O(N)$.

Let us elaborate on above argument a little further. We consider an instance of Exact Cover which has a unique satisfied solution. We assume that this instance consists of M clauses. The ratio $\alpha = M/N$ is very close to the critical value α^*

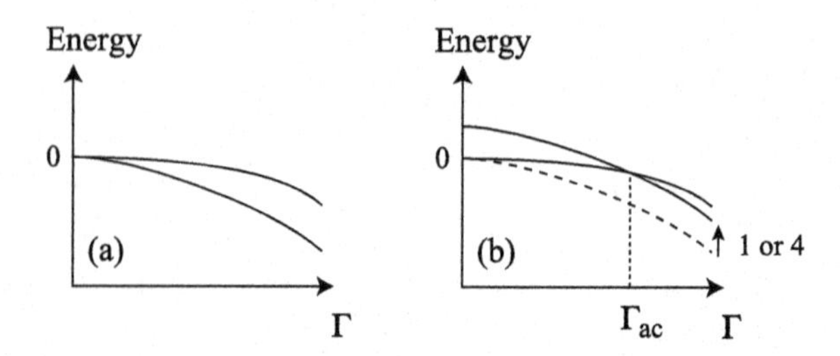

Fig. 8.9 Schematic of energy levels of transverse field Ising models representing Exact Cover. (**a**) Lowest two energy levels of a Hamiltonian with $M-1$ clauses which give rise to two satisfied solutions. (**b**) Lowest two energy levels of a Hamiltonian with M clauses. The addition of one clause lifts one of the degenerated ground states at $\Gamma=0$ by 1 or 4. For finite Γ, the lowest energy level of the Hamiltonian of $M-1$ clauses shifts upward by about 1 or 4 with a finite probability. As a result, a level anti-crossing takes place at $\Gamma=\Gamma_{\mathrm{ac}}$

which divides the SAT and UNSAT phases. It is known that problems with α close to α^* are particularly hard and that satisfied solutions, if they exist, are different from each other in $O(N)$ bits when $\alpha\approx\alpha^*$. Now we focus on $M-1$ clauses for which there are two satisfied solutions. Adding the transverse field to the Ising Hamiltonian of $M-1$ clauses, the doubly degenerated ground level separates into two levels. When the transverse field Γ is small, these two eigenenergies decreases with the transverse field (Fig. 8.9(a)). Let $E_{\mathrm{g}}(\Gamma)$ and $E_1(\Gamma)$ denote the ground and first excited eigenenergies. We define Γ_* such that at this point the difference of two eigenenergies becomes 4, namely, $E_{\mathrm{g}}(\Gamma_*)+4=E_1(\Gamma_*)$. We now add the remaining clause. Then the degeneracy of the ground state at $\Gamma=0$ is resolved and one of the degenerated states acquires the energy 1 or 4. When the transverse field is switched on, one may think that the lowest level of the Hamiltonian with $M-1$ clauses is lifted with a non-zero probability by about 1 or 4 so as to connect with the excited state at $\Gamma=0$ (Fig. 8.9(b)). As a result, it is likely that two levels cross at a certain $\Gamma=\Gamma_{\mathrm{ac}}$. Due to the non-crossing rule, this level crossing is avoided in practice and a level anti-crossing takes place. If the shift in energy by the addition of one clause is 4, then the anti-crossing takes place at Γ_*, namely, $\Gamma_{\mathrm{ac}}\approx\Gamma_*$. On the other hand, if the energy shift is 1, one has $\Gamma_{\mathrm{ac}}<\Gamma_*$. Thus one finds that $\Gamma_{\mathrm{ac}}\lesssim\Gamma_*$. Altshuler et al. [10] applied the perturbation theory to this energy-level analysis, and obtained size scaling of Γ_* and Δ_{min} at the anti-crossing point. The application of perturbation theory is allowed if the energy levels of the Hamiltonian are discrete and no continuous spectrum is involved in the thermodynamic limit. On the basis of the theory of the Anderson localisation [1], we guess that this condition is given by $\Gamma<\Gamma_{\mathrm{cr}}=O(N/d\ln d)=O(1/\ln N)$, where d is the spatial dimension and note that $d=N$ in the present system. The results of the perturbation theory are written as $\Gamma_*\sim N^{-1/8}$ and $\Delta_{\mathrm{min}}\sim\exp(-cN\ln N)$, where c is a constant. We refer to Ref. [10] for the details of the perturbation theory. We note that $\Gamma_*\ll\Gamma_{\mathrm{cr}}$ for $N\gg 1$ implies the consistency of the perturbation theory. The results show that a level anti-crossing takes place at Γ_{ac} close to 0, and that the energy gap at anti-crossing decays as an exponential of $N\ln N$. We comment that Farhi et al. [132]

have discussed an avoidance of the anti-crossing by randomising the transverse field, and that Choi [75] has proposed a different choice for the problem Hamiltonian which might change the exponential scaling of the minimum energy gap.

The anti-crossing at an infinitesimal $\Gamma_{\rm ac}$ is contrasted with a first order quantum phase transition which takes place at $\Gamma_{\rm c} = O(1)$. In practice, a quantum Monte Carlo simulation of Exact Cover has shown a transition point which is rather close to 1 [441] (see Sect. 8.5.3.3). In addition to this, Jörg et al. [206] have shown for an random optimisation problem related to 3-SAT the presence of a first order quantum phase transition with a transition point around $\Gamma_{\rm c} = 1$ by means of a quantum cavity method and a quantum Monte Carlo simulation. The consistency of the anti-crossing at an infinitesimal $\Gamma_{\rm ac}$ with the first order quantum phase transition has not been clarified. One needs further investigation about this.

8.6 Scaling of Errors

As far as quantum annealing is performed with a finite runtime, it is inevitable to suffer from a finite amount of errors. The decay rate of errors with runtime is an important measure to characterise the validity of quantum annealing. In the present section, we discuss the decay rate of errors by quantum annealing.

In the presence of a non-zero minimum energy gap, the error by quantum annealing vanishes in the infinite limit of the runtime $\tau \to \infty$. However, the runtime must be taken to be finite in the practical situation. When the runtime is finite, according to the quantum adiabatic theorem, the probability that an excited state is detected in the final state is given by

$$\left|\langle \phi_k | \Psi(\tau)\rangle\right|^2 \sim \tau^{-2} \quad (k = 1, 2, \ldots) \tag{8.6.1}$$

for $\tau \gg \Delta_{\rm min}^{-2}$, where $|\phi_k\rangle$ is the k-th excited state of the problem Hamiltonian H_0, $|\Psi(\tau)\rangle$ is the final state after annealing, and $\Delta_{\rm min}$ is the minimum energy gap above the instantaneous ground state. The residual energy is estimated as [390]

$$\begin{aligned} E_{\rm res} &= \langle \Psi(\tau) | H_0 | \Psi(\tau)\rangle - E_{\rm g} \\ &= E_{\rm g}\left(\left|\langle \phi_{\rm g} | \Psi(\tau)\rangle\right|^2 - 1\right) + \sum_{k\geq 1} E_k \left|\langle \phi_k | \Psi(\tau)\rangle\right|^2 \\ &\sim \tau^{-2}, \end{aligned} \tag{8.6.2}$$

where E_g and E_k are the ground and excited energies of H_0, and $|\phi_{\rm g}\rangle$ is the ground state of H_0. Hence one finds that the excitation probability and residual energy decay as τ^{-2} with runtime τ, as far as $\tau \gg \Delta_{\rm min}^{-2}$.

As we have seen in the previous section, quantum phase transitions have a significant role in understanding of quantum annealing. Let us consider a simple case that a standard second order quantum phase transition occurs on the way of quantum annealing. We assume a critical point characterised by an exponent ν for the correlation length and a dynamical exponent z for the coherence time. Suppose that quantum annealing is performed by changing the transverse field as $\Gamma(t) = -t/\tau$

with rate $1/\tau$ and time t moving from $t=-\infty$ to 0. Then the density of defects after annealing is scaled as $\tau^{-d\nu/(z\nu+1)}$ [318], as we mentioned in Sect. 7.2.2.2. The residual energy density obeys the same power-law scaling. if the ground state of the problem Hamiltonian is separated from excited states by finite energy gaps.

Apart from the standard second order quantum phase transition, the problem is not so simple. One of the tractable systems is a random transverse field Ising chain given by (8.5.4). Caneva et al. studied quantum annealing of this model using the Landau-Zener transition formula [58]. Let us assume that the transverse field is changed as $\Gamma=-t/\tau$ with time t moving from $t=-\infty$ to 0. The state starting from the ground state of the transverse-field term is likely to excite when the energy gap to the first accessible excited state Δ_{acc} takes the minimum. The excitation probability during the time evolution is well approximated by the Landau-Zener formula: $e^{-\pi\Delta_{\min}^2\tau/4\alpha}$, where $\Delta_{\min}$ is the minimum of Δ_{acc} and α is the slope of the approaching lowest two eigenenergies as functions of $s=t/\tau$. Supposing that Δ_{acc} is minimised at the quantum critical point and the slope is independent of the randomness, then the average of the excitation probability at the critical point is estimated as

$$P_{\text{ex,cr}}(N)=\int_0^\infty e^{-\pi\Delta_{\text{acc}}^2\tau/4\alpha}P(\Delta_{\text{acc}},N)\,d\Delta_{\text{acc}}, \tag{8.6.3}$$

where $P(\Delta_{\text{acc}},N)$ is the distribution function of Δ_{acc} with system size N. As we have mentioned in Sect. 8.5.2, the distribution of the energy gap Δ between the ground and the first excited states is a function of scaled parameter $x=-\ln(\Delta/\Delta_0)/\sqrt{N}$ at the quantum critical point, where Δ_0 is the largest energy scale of the system. We here make an ansatz that Δ_{acc} has the same property as Δ, namely, $P(\Delta_{\text{acc}},N)\,d\Delta_{\text{acc}}=P(x)\,dx$ with $x=-\ln(\Delta_{\text{acc}}/\Delta_0)/\sqrt{N}$. It follows that

$$P_{\text{ex,cr}}(N)=\int\exp\{-\kappa\tau e^{-2x\sqrt{N}}\}P(x)\,dx, \tag{8.6.4}$$

where $\kappa=\pi\Delta_0^2/4\alpha$. Defining $x_{\text{c}}=\ln(\kappa\tau)/2\sqrt{N}$, this is written as

$$P_{\text{ex,cr}}(N)=\int\exp\{-e^{-2\sqrt{N}(x-x_{\text{c}})}\}P(x)\,dx. \tag{8.6.5}$$

For $N\gg 1$, the double-exponential part almost vanishes for $x<x_{\text{c}}$ and becomes unity for $x>x_{\text{c}}$. Hence following approximation is allowed:

$$P_{\text{ex,cr}}(N)\approx\int\theta(x-x_{\text{c}})P(x)\,dx=\int_{x_{\text{c}}}^\infty P(x)\,dx\equiv\Pi(x_{\text{c}}). \tag{8.6.6}$$

We remark here that the location where Δ_{acc} takes the minimum distributes around the critical point. Therefore the average of the excitation probability $P_{\text{ex}}(N)$ during quantum annealing satisfies

$$P_{\text{ex}}(N)\gtrsim P_{\text{ex,cr}}(N)\approx\Pi(x_{\text{c}}). \tag{8.6.7}$$

Let ε be a positive small number. We define a size N_ε such that $P_{\text{ex}}(N_\varepsilon)=\varepsilon$. Then Eq. (8.6.7) leads to

$$N_\varepsilon=P_{\text{ex}}^{-1}(\varepsilon)\lesssim P_{\text{ex,cr}}^{-1}(\varepsilon)=\left(\frac{\ln\kappa\tau}{2x_{\text{c},\varepsilon}}\right)^2=\left(\frac{\ln\kappa\tau}{2\Pi^{-1}(\varepsilon)}\right)^2, \tag{8.6.8}$$

where $x_{c,\varepsilon} = \ln \kappa\tau / 2\sqrt{N_\varepsilon}$. By the definition of N_ε, the system of size N_ε after quantum annealing with a given τ should remain in the ground state. Therefore one may regard N_ε as the averaged size of ferromagnetic domains after quantum annealing. The inverse of this size corresponds to the density of kinks created during quantum annealing. As a result the density of kinks follows [58]

$$\rho \sim \frac{1}{N_\varepsilon} \gtrsim \left(\frac{2\Pi^{-1}(\varepsilon)}{\ln \kappa\tau} \right)^2 \sim (\ln \tau)^{-2}. \tag{8.6.9}$$

To evaluate the residual energy, consider a classical chain ($\Gamma = 0$) with length N. Since the distribution of the coupling constant J_i is uniform between 0 and 1, the lowest value of J_i in the chain should be of order $1/N$. Therefore the lowest excitation energy from the ground state should be $O(1/N)$. Now in the infinite system, if there are kinks with density $1/N_\varepsilon$, then each kink should cost an energy of order $1/N_\varepsilon$. It turns out that the excitation energy per spin to such a state amounts to $1/N_\varepsilon^2$. Thus we find that the residual energy density per spin after quantum annealing obeys

$$\varepsilon_{\text{res}} \sim N_\varepsilon^{-2} \gtrsim (\ln \tau)^{-4}. \tag{8.6.10}$$

The same result in essence for the density of kinks can be obtained by Dziarmaga [119] using the Kibble-Zurek argument (see Sect. 7.2.2.2). The energy gap above the ground state at the critical point typically is scaled as $-\ln \Delta_{\text{typ}} \sim N^{1/2}$ [139] with the system size N, namely

$$\Delta_{\text{typ}} \sim e^{-\text{const.} \times N^{1/2}}. \tag{8.6.11}$$

The correlation length ξ is scaled as [139]

$$\xi \sim \gamma^{-2}, \tag{8.6.12}$$

where $\gamma = (\Gamma - \Gamma_c)/\Gamma_c$ is the dimensionless parameter. Supposing the schedule of annealing as $\gamma = -t/\tau$, the equality between the coherence time Δ_{typ}^{-1} and the remaining time to the critical point $|t|$ yields

$$\tau\gamma = e^{\text{const.} \times \gamma^{-1}}, \tag{8.6.13}$$

where we remark $N \approx \xi$. This equation of γ cannot be solved analytically, but one has

$$\hat{\gamma} = \frac{A}{\ln \tau\hat{\gamma}}, \tag{8.6.14}$$

where $\hat{\gamma}$ represents the solution of (8.6.13) and A is a constant. Equations (8.6.12) and (8.6.14) yield

$$\hat{\xi} \sim \hat{\gamma}^{-2} \sim (\ln \tau)^2. \tag{8.6.15}$$

Supposing that the growth of the correlation length stops at $\hat{\xi}$, the density of kinks after annealing is estimated as

$$\rho \sim \hat{\xi}^{-1} \sim (\ln \tau)^2. \tag{8.6.16}$$

It is significant to compare the results for scaling of kink density and residual energy after quantum annealing to those after simulated annealing. Assuming the Glauber model as the spin dynamics at finite temperature [105, 156] and applying the Kibble-Zurek argument, one obtains the scaling of the density of kinks as [389]

$$\rho^{\mathrm{SA}} \sim (\ln \tau)^{-1}, \tag{8.6.17}$$

and the scaling of residual energy per spin as

$$\varepsilon_{\mathrm{res}}^{\mathrm{SA}} \sim (\ln \tau)^{-2}. \tag{8.6.18}$$

Comparing (8.6.9) and (8.6.10) with (8.6.17) and (8.6.18) respectively, it turns out that quantum annealing brings faster decays of density of kinks and residual energy than simulated annealing. In this sense, quantum annealing performs better than simulated annealing in the present system.

8.7 Convergence Theorems

We mentioned in Sect. 8.3.2 on the basis of the quantum adiabatic theorem that quantum annealing succeeds in optimisation if the change in the time-dependent Hamiltonian is sufficiently slow. In this section, we focus on the degree of freedom on the time-dependence of the transverse field and provide conditions which ensures convergence of quantum annealing in a generic optimisation problem.

8.7.1 Sufficient Condition of the Schedule

We consider an Ising model in a transverse field

$$H(t) = H_0 - \Gamma(t) \sum_i^N S_i^x, \tag{8.7.1}$$

where H_0 denotes an Ising model. We do not specify the detail of H_0 but assume that the range of the eigenvalues of H_0 is of order of the system size N. We assume that $\Gamma(t)$ is a monotonic decreasing function of time t. Initially $\Gamma(t)$ is so large that the transverse-field term dominates the total Hamiltonian, and it vanishes for $t \to \infty$. Let the system be in the ground state initially. Morita and Nishimori [281] provided the following theorem on a condition of $\Gamma(t)$ that ensures the adiabatic time evolution.

Theorem 8.3 *Let ε be a nonzero positive small number such that $0 < \varepsilon \ll 1$. For a given positive number t_0, a sufficient condition of $\Gamma(t)$ for the adiabatic time evolution is given by*

$$\Gamma(t) = A(\varepsilon t + B)^{-1/(2N-1)} \quad \textit{for } t > t_0, \tag{8.7.2}$$

where A and B are constants which are of $O(N^0)$ and exponentially large in N respectively.

We note that under this condition the excitation probability is bounded by ε^2 at any time $t > t_0$.

Proof We recall the condition (8.3.43) for the adiabatic approximation. The condition for the adiabatic evolution with a time-dependent Hamiltonian $H(t)$ is written as

$$\frac{1}{\{\Delta_{m0}(t)\}^2}\left|\langle\phi_m(t)|\frac{dH(t)}{dt}|\phi_0(t)\rangle\right| \le \varepsilon \ll 1 \quad \text{for all } t \ge 0, \tag{8.7.3}$$

where $\Delta_{m0}(t)$ denotes the energy gap between the instantaneous ground state $|\phi_0(t)\rangle$ and the mth excited state $|\phi_m(t)\rangle$ of $H(t)$, and ε is a nonzero positive small number.

We first focus on the matrix element of $dH(t)/dt$. It is easy to see

$$\left|\langle\phi_j(t)|\frac{dH(t)}{dt}|\phi_0(t)\rangle\right| \le -N\frac{d\Gamma(t)}{dt}, \tag{8.7.4}$$

since only the transverse-field $\Gamma(t)$ in (8.7.1) depends on time and the matrix element of S_i^x is at most unity in the absolute value. Note that $d\Gamma(t)/dt$ is negative.

We next look into the instantaneous energy gap $\Delta_{m0}(t)$. In order to give a bound on $\Delta_{m0}(t)$, we make use of the Hopf's theorem stated as follows.

For a strictly positive matrix M (i.e., $M_{mn} > 0$ for any m and n), there exists a positive eigenvalue λ_0 which satisfies

$$|\lambda| \le \frac{\kappa-1}{\kappa+1}\lambda_0 \quad \textit{for any other eigenvalue } \lambda \textit{ of } M, \tag{8.7.5}$$

where κ is defined by

$$\kappa = \max_{l,m,n}\left\{\frac{M_{ln}}{M_{mn}}\right\}. \tag{8.7.6}$$

This theorem is a sharpened version of the Perron-Frobenius theorem for a non-negative matrix M (i.e., $M \ge 0$ for any i and j). We refer to Sect. 8.A.1 for the proof.

Now let us choose $M = (E_{\max}^{(0)} + \Gamma_0 - H(t))^N$, where $E_{\max}^{(0)}$ is the largest eigenvalue of H_0 and $\Gamma_0 = \Gamma(t_0)$. Note that $\Gamma(t) < \Gamma_0$ for $t > t_0$. With the basis which makes H_0 diagonal, M is strictly positive for $N > 1$ and $\Gamma(t) > 0$ since all elements of $M^{1/N}$ are non-negative and irreducible (which means that any eigenstate can move to any other by applying $M^{1/N}$ at most N times). The Hopf's theorem leads to the following inequality:

$$\left(E_{\max}^{(0)} + \Gamma_0 - \varepsilon_m(t)\right)^N \le \frac{\kappa-1}{\kappa+1}\left(E_{\max}^{(0)} + \Gamma_0 - \varepsilon_0(t)\right)^N \quad \text{for } m \ge 1, \tag{8.7.7}$$

where $\varepsilon_0(t)$ and $\varepsilon_m(t)$ are the instantaneous ground and mth eigenenergies of $H(t)$ respectively.

κ in (8.7.7) is evaluated as follows. At first, notice that the minimum element of M is attained by a pair of states which have mutually opposite spins. It is given by

$$\min_{m,n}\{M_{mn}\} = N!\Gamma^N(t), \tag{8.7.8}$$

where note that $N!$ comes from the number of paths which connects two states with N times of single-spin flips. Regarding the maximum element of M, one has the following inequality

$$\begin{aligned}\left[E_{\max}^{(0)}+\Gamma_0-H(t)\right]_{mn} &\leq \left[E_{\max}^{(0)}+\Gamma_0-E_{\min}^{(0)}+\Gamma(t)\sum_i^N\sigma_i^x\right]_{mn}\\ &\leq E_{\max}^{(0)}+\Gamma_0-E_{\min}^{(0)}+\Gamma(t)N<E_{\max}^{(0)}+\Gamma_0-E_{\min}^{(0)}+N\delta,\end{aligned}$$

which yields an upper bound for M:

$$\max_{i,j}\{M_{ij}\}<\left(E_{\max}^{(0)}-E_{\min}^{(0)}+(N+1)\Gamma_0\right)^N, \tag{8.7.9}$$

where $E_{\min}^{(0)}$ is the smallest eigenvalue of H_0. From (8.7.8) and (8.7.9), we obtain an upper bound for κ:

$$\kappa\leq\frac{(E_{\max}^{(0)}-E_{\min}^{(0)}+(N+1)\Gamma_0)^N}{N!\Gamma^N(t)}. \tag{8.7.10}$$

On the other hand, since $\max_{m,n}\{M_{mn}\}\geq E_{\max}^{(0)}+\Gamma_0-E_{\min}^{(0)}\sim O(N)$, κ is bounded from below as

$$\kappa\geq\frac{(E_{\max}^{(0)}+\Gamma_0-E_{\min}^{(0)})^N}{N!\Gamma_0^N}>1.$$

With this bound and inequality $\{(1-x)/(1+x)\}^{1/N}\leq(1-x)^{1/N}\leq 1-x/N$ for $0\leq x\leq 1$, (8.7.7) yields

$$\begin{aligned}E_{\max}^{(0)}+\Gamma_0-\varepsilon_m(t) &\leq\left(\frac{1-1/\kappa}{1+1/\kappa}\right)^{1/N}\left(E_{\max}^{(0)}+\Gamma_0-\varepsilon_0(t)\right)\\ &\leq\left(1-\frac{1}{N\kappa}\right)\left(E_{\max}^{(0)}+\Gamma_0-\varepsilon_0(t)\right).\end{aligned} \tag{8.7.11}$$

This inequality is arranged into

$$\Delta_{m0}(t)\geq\frac{2}{N\kappa}\left(E_{\max}^{(0)}+\Gamma_0-\varepsilon_0(t)\right)>\frac{2}{N\kappa}\left(E_{\max}^{(0)}-\varepsilon_0(t)\right), \tag{8.7.12}$$

where we note $\Delta_{m0}(t)=\varepsilon_m(t)-\varepsilon_0(t)$. Combining (8.7.10) and (8.7.12), we obtain

$$\Delta_{m0}(t)>A_N\Gamma^N(t) \tag{8.7.13}$$

with

$$\begin{aligned}A_N &=\frac{2N!(E_{\max}^0-\varepsilon_0(t))}{N(E_{\max}^{(0)}-E_{\min}^{(0)}+(N+1)\Gamma_0)^N}\\ &\approx 2\sqrt{2\pi/N}e^{-N}\left(E_{\max}^{(0)}-E_{\min}^{(0)}\right)\left(\frac{N}{E_{\max}^{(0)}-E_{\min}^{(0)}+(N+1)\Gamma_0}\right)^N,\end{aligned} \tag{8.7.14}$$

where we used the Stirling's approximation, $N! \approx \sqrt{2\pi N}e^{-N}N^N$, and made $E^{(0)}_{\max} - \varepsilon_0(t) \approx E^{(0)}_{\max} - E^{(0)}_{\min}$ since $E^{(0)}_{\min} - \varepsilon_0(t) \approx O(\Gamma^2) \ll 1$ while $E^{(0)}_{\max} - E^{(0)}_{\min} \gg 1$.

Now we return to the adiabatic condition (8.7.3). Inequalities (8.7.4) and (8.7.13) yields

$$\frac{1}{\Delta_j^2(t)}\left|\langle\phi_j(t)|\frac{dH(t)}{dt}|\phi_0(t)\rangle\right| < -\frac{N}{A_N^2\Gamma^{2N}(t)}\frac{d\Gamma(t)}{dt}. \tag{8.7.15}$$

Therefore it turns out that a sufficient condition for the adiabaticity is

$$-\frac{N}{A_N^2\Gamma^{2N}(t)}\frac{d\Gamma(t)}{dt} = \varepsilon \ll 1. \tag{8.7.16}$$

Integrating this equation, one obtains

$$\Gamma(t) = A\big(\varepsilon(t-t_0)+B\big)^{-1/(2N-1)}, \tag{8.7.17}$$

where $A = A_N^{-2/(2N-1)}\{(2N-1)/N\}^{-1/(2N-1)}$ and $B = N/\{(2N-1)A_N^2\Gamma_0^{2N-1}\}$. It turns out from (8.7.14) that $A \sim O(N^0)$ and that B is exponentially large in N for $N \gg 1$. □

8.7.1.1 Complexity

We showed above that a sufficient condition of $\Gamma(t)$ for the adiabatic evolution is its schedule which has the time dependence $t^{-1/(2N-1)}$ for $t \to \infty$. However this does not mean that the optimisation problem can be solved in a polynomial time. Here we show that the computational complexity is still exponential with respect to the size of problem.

Assuming the schedule (8.7.17), the system evolves adiabatically and the solution is obtained with high probability when the transverse field becomes $\Gamma(t) \ll 1$. Let δ_1 be a positive small number, i.e., $\delta_1 \ll 1$. The computational complexity is defined by the time t^* such that $\Gamma(t^*) = \delta_1$, namely,

$$A\big(\varepsilon(t^*-t_0)+B\big)^{-1/(2N-1)} = \delta_1. \tag{8.7.18}$$

Recalling the definitions of A and B, one obtains

$$t^* - t_0 = \frac{\Gamma_0}{\varepsilon A_N^2\Gamma_0^{2N}}\frac{N}{2N-1}\left\{\left(\frac{\Gamma_0}{\delta_1}\right)^{2N-1} - 1\right\}. \tag{8.7.19}$$

Since, $A_N\Gamma_0^N$ is exponentially small in N for large N, it is shown that t^* grows exponentially with N.

8.7.2 Convergence Condition of Quantum Annealing with Quantum Monte Carlo Dynamics

One of the possible ways to implement quantum annealing using a classical computer is to make use of a quantum Monte Carlo method. Of course, the stochastic dynamics generated by the quantum Monte Carlo method is different from the Schrödinger dynamics, and the quantum adiabatic theorem does not serve as a base of quantum annealing. In this subsection we focus on quantum annealing using the quantum Monte Carlo dynamics. The convergence of quantum annealing with quantum Monte Carlo dynamics was studied by Morita and Nishimori [280]. In a quantum Monte Carlo method, a transverse Ising Hamiltonian is mapped to an effective classical Hamiltonian [386]. Quantum annealing is then implemented by changing a parameter, which is related to the transverse field with the Monte Carlo step [259]. Morita and Nishimori extended the convergence theorem of simulated annealing by Geman and Geman [155] to dynamics of such effective classical systems.

We consider a time-dependent transverse Ising model of the form of (8.7.1). As is mentioned in Sect. 3.1, the transverse Ising model in d-dimension can be mapped to an Ising model in $(d+1)$-dimension. The effective Hamiltonian after mapping can be written as (see Eq. (3.1.8))

$$H_{\mathrm{eff}} = \frac{\beta}{M}\sum_{k=1}^{M} H_0^{(k)} - K_M(t)\sum_i S_{i,k}S_{i,k+1},$$

where $H_0^{(k)}$ stands for the Ising model part in the original Hamiltonian, k is an index representing the Trotter slice, M is the number of Trotter slices, and $K_M(t) = \frac{1}{2}\ln\coth\frac{\beta}{M}\Gamma(t)$. The effective partition function of the system is given by

$$Z(t) = \mathrm{Tr}\exp(-H_{\mathrm{eff}}) = \mathrm{Tr}\exp\left[-\frac{\beta}{M}\sum_{k=1}^{M} H_0^{(k)} + K_M(t)\sum_{k=1}^{M}\sum_i S_{i,k}S_{i,k+1}\right], \tag{8.7.20}$$

where this partition function coincides with the exact partition function of the original system when $M \to \infty$. We now define effective temperatures $T_0 = M/\beta$ and $T_1(t) = 1/K_M(t)$. Using these notations, the partition function (8.7.20) can be written as

$$Z(t) = \mathrm{Tr}\exp\left[-\frac{F_0}{T_0} - \frac{F_1}{T_1(t)}\right],$$

where

$$F_0 = \sum_{k=1}^{M} H_0^{(k)}, \qquad F_1 = -\sum_{k=1}^{M}\sum_i S_{i,k}S_{i,k+1}.$$

From now on, we use an abstract notation x for the Ising-spin state $\{S_{i,k}\}$. We substitute $F_{0,x}$ and $F_{1,x}$ for F_0 and F_1 to show x-dependence of these quantities.

We next define sets of the Ising-spin states. Let $\mathscr{S}_x$ be a set of the states which are accessible from a state x in a single Monte Carlo step, and $\mathscr{S}_\Lambda$ be a set of the states x's which give $F_{1,x}$ such that $F_{1,x} \geq F_{1,x'}$ for all $x' \in \mathscr{S}_x$. $\bar{\mathscr{S}}_\Lambda$ is the complementary set to $\mathscr{S}_\Lambda$.

We define the maxima of energy differences:

$$L_0 = \max_x \left\{ \max_{x' \in \mathscr{S}_x} |F_{0,x'} - F_{0,x}| \right\}, \tag{8.7.21}$$

and

$$L_1 = \max_x \left\{ \max_{x' \in \mathscr{S}_x} |F_{1,x'} - F_{1,x}| \right\}. \tag{8.7.22}$$

We furthermore define

$$R = \min_{x \in \bar{\mathscr{S}}_\Lambda} \left\{ \max_{x'} d_{x'x} \right\}, \tag{8.7.23}$$

where $d_{x'x}$ is the number of the smallest Monte Carlo steps needed to bring the state from x to x'. Note that $d_{x'x}$ can be seen as a distance between x and x'. Roughly speaking, R is the smallest number of steps needed to move from a state to an arbitrary state.

Hereafter we follow the convention that the time t stands for the Monte Carlo step. Let $G_{x'x}(t)$ be the time-dependent transition probability from a state x to x' given by

$$G_{x'x}(t) = \begin{cases} p_{x'x} A(Q_{x'}(t)/Q_x(t)) & \text{for } x' \neq x \\ -\sum_{x''} p_{x''x} A(Q_{x''}(t)/Q_x(t)) & \text{for } x' = x \end{cases}, \tag{8.7.24}$$

where $p_{x'x}$ is the probability that x' is chosen in the single Monte Carlo step from x. $p_{x'x}$ must satisfy $p_{x'x} \geq 0$, $p_{xx} = 0$, $\sum_{x'} p_{x'x} = 1$. In addition, we assume $p_{xx'} = p_{x',x}$ and that there exists a positive number s such that

$$\sum_{x_1,\ldots,x_{s-1}} p_{x'x_{s-1}} \cdots p_{x_1 x} > 0$$

for any x and x'. $A(u)$ is a monotonically increasing function, for which we assume

$$0 < A(u) \leq 1 \quad \text{and} \quad A(1/u) = A(u)/u \quad \text{for } u > 0. \tag{8.7.25}$$

$Q_x(t)$ is defined by

$$Q_x(t) = \frac{1}{Z(t)} \exp\left[-\frac{F_{0,x}}{T_0} - \frac{F_{1,x}}{T_1(t)} \right], \tag{8.7.26}$$

The convergence theorem of quantum annealing with the quantum Monte Carlo dynamics is stated as follows.

Theorem 8.4 *If the time dependence of the effective temperature $T_1(t)$ follows*

$$T_1(t) \geq \frac{R L_1}{\ln(t+2)}, \tag{8.7.27}$$

or equivalently, if the transverse field follows

$$\Gamma(t) \geq \frac{M}{\beta} \tanh^{-1} \frac{1}{(t+2)^{2/RL_1}}, \tag{8.7.28}$$

then the distribution generated by $G_{x'x}(t)$ converges to the distribution

$$Q_x^\star = \frac{1}{Z^*} \exp\left(-\frac{F_{0,x}}{T_0}\right), \qquad Z^\star = \sum_x \exp\left(-\frac{F_{0,x}}{T_0}\right). \tag{8.7.29}$$

We note that (8.7.28) reduces to

$$\Gamma(t) \geq \frac{M}{\beta} \frac{1}{(t+2)^{2/RL_1}} \tag{8.7.30}$$

for $t \gg 1$. This is contrasted with the logarithmic schedule of the temperature in simulated annealing [155]: $T(t) \geq c/\ln t$, where c is a constant.

The proof of this theorem is attained by showing that this Markov chain possesses the strong ergodicity described below.

We first show following lemmas. Let w be the minimum of $p_{x'x}$ with respect to $x' \in \mathscr{S}_x$ and x, namely,

$$p_0 = \min_x \left\{ \min_{x' \in \mathscr{S}_x} p_{x'x} \right\}. \tag{8.7.31}$$

Lemma 8.1 *For any $x, x' \in \mathscr{S}_x$, and t, one has*

$$G_{x'x}(t) \geq p_0 A(1) \exp\left(-\frac{L_0}{T_0} - \frac{L_1}{T_1(t)}\right). \tag{8.7.32}$$

Lemma 8.2 *For any $x \in \bar{\mathscr{S}}_\Lambda$, there exists t_1 such that*

$$G_{xx}(t) \geq p_0 A(1) \exp\left(-\frac{L_0}{T_0} - \frac{L_1}{T_1(t)}\right) \quad \textit{for any } t \geq t_1. \tag{8.7.33}$$

Proof of Lemma 8.1 We define

$$a = \frac{F_{0,x'} - F_{0,x}}{T_0} + \frac{F_{1,x'} - F_{1,x}}{T_1(t)} = \ln \frac{Q_x(t)}{Q_{x'}(t)},$$

namely $e^{-a} = Q_{x'}(t)/Q_x(t)$. From Eqs. (8.7.21) and (8.7.22), one can see that

$$a \leq \frac{L_0}{T_0} + \frac{L_1}{T_1(t)},$$

and hence

$$e^{-a} \geq \exp\left(-\frac{L_0}{T_0} - \frac{L_1}{T_1(t)}\right). \tag{8.7.34}$$

Now, the definition (8.7.31) of p_0 leads to

$$p_{x'x} \geq p_0 \tag{8.7.35}$$

for any x and $x' \in \mathscr{S}_x$. Using (8.7.25), (8.7.34), (8.7.35) and the monotonicity of $A(u)$, if $a > 0$, one has

$$G_{x'x}(t) = p_{x'x}A\left(e^{-a}\right) \geq p_0 e^{-a} A\left(e^{a}\right) > p_0 e^{-a} A(1) \geq p_0 A(1) \exp\left(-\frac{L_0}{T_0} - \frac{L_1}{T_1(t)}\right).$$

If $a \leq 0$ on the other hand, since $A(e^{-a}) \geq A(1)$, one has

$$G_{x'x}(t) \geq p_0 A\left(e^{-a}\right) \geq p_0 A(1) \geq p_0 A(1) \exp\left(-\frac{L_0}{T_0} - \frac{L_1}{T_1(t)}\right).$$

Therefore inequality (8.7.32) is shown for any $x, x' \in \mathscr{S}_x$ and t. □

Proof of Lemma 8.2 For any $x \in \bar{\mathscr{S}}_\Lambda$, there exists $x' \in \mathscr{S}_x$ such that $F_{1,x'} > F_{1,x}$. For such an x', one has

$$\lim_{t\to\infty} \exp\left(-\frac{F_{0,x'} - F_{0,x}}{T_0} - \frac{F_{1,x'} - F_{1,x}}{T_1(t)}\right) = 0.$$

Besides, because of $A(u) = uA(1/u) \leq u$ for $u \leq 1$, one has

$$\lim_{t\to\infty} A\left(\frac{Q_{x'}(t)}{Q_x(t)}\right) \leq \lim_{t\to\infty} \frac{Q_{x'}(t)}{Q_x(t)} = 0.$$

Therefore there exists a number t_0 for a given positive number ε_0 such that

$$A\left(\frac{Q_{x'}(t)}{Q_x(t)}\right) \leq \varepsilon_0 \quad \text{for all } t \geq t_0.$$

Using this inequality and $A(u) \leq 1$, it is shown for $t \geq t_0$ that

$$\begin{aligned}
\sum_{x''} p_{x''x} A\left(\frac{Q_{x''}(t)}{Q_x(t)}\right) &= p_{x'x} A\left(\frac{Q_{x'}(t)}{Q_x(t)}\right) + \sum_{x''\neq x'} p_{x''x} A\left(\frac{Q_{x''}(t)}{Q_x(t)}\right) \\
&\leq p_{x'x}\varepsilon_0 + \sum_{x''\neq x'} p_{x''x} = \varepsilon_0 p_{x'x} + 1 - p_{x'x} \\
&= 1 - (1 - \varepsilon_0) p_{x'x}.
\end{aligned}$$

Therefore one obtains for $t \geq t_1$

$$G_{xx}(t) \geq (1 - \varepsilon_0) p_{x',x} \geq (1 - \varepsilon_0) p_0.$$

Noting that

$$\lim_{t\to\infty} p_0 A(1) \exp\left(-\frac{L_0}{T_0} - \frac{L_1}{T_1(t)}\right) = 0,$$

there exists a positive number t_1 for given positive numbers ε_0 and $\varepsilon_1 \leq (1 - \varepsilon_0) p_0$ such that

$$G_{xx}(t) \geq (1 - \varepsilon_0) p_0 \geq \varepsilon_1 \geq p_0 A(1) \exp\left(-\frac{L_0}{T_0} - \frac{L_1}{T_1(t)}\right) \quad \text{for all } t \geq t_1.$$

Thus Lemma 8.2 is proved. □

We next prove that our Markov chain has the weak ergodicity. To this end, we introduce the transition probability $G_{x'x}(t', t)$ from a state x at Monte Carlo time t to x' at $t' > t$. Due to the Chapman-Kolmogorov equation, one can write $G_{x'x}(t+s, t)$ in terms of $G_{x'x}(t)$ as

$$G_{x'x}(t+s, t) = \sum_{x_{s-1},\dots,x_2,x_1} G_{x'x_{s-1}}(t+s-1)\cdots G_{x_2x_1}(t+1)G_{x_1x}(t). \quad (8.7.36)$$

Details of this equation is provided in Sect. 8.A.3.1. Using the matrix notation, $G(t', t)$ and $G(t)$, for $G_{x',x}(t', t)$ and $G_{x',x}(t)$, (8.7.36) is simplified into

$$G_{x'x}(t+s, t) = \big[G(t+s-1)\cdots G(t+1)G(t)\big]_{x'x}.$$

The definition of the weak ergodicity is given by

$$\sup_{P_0, Q_0}\left[\sum_{x'}\left|\sum_{x} G_{x'x}(t+s, t)P_{0,x} - \sum_{x} G_{x'x}(t+s, t)Q_{0,x}\right|\right] \overset{s\to\infty}{\longrightarrow} 0 \quad \text{for any } t. \quad (8.7.37)$$

The weak ergodicity implies that the distribution generated by a Markov chain becomes independent of the initial distribution for long Monte-Carlo times.

Proof (Weak ergodicity) In order to prove the weak ergodicity, we call for the weak ergodic theorem. We define the ergodic coefficient $\alpha(G(t', t))$ by

$$\alpha\big(G(t', t)\big) = \max_{x'',x'}\left[\sum_{x}\max\big\{0, G_{xx''}(t', t) - G_{xx'}(t', t)\big\}\right]. \quad (8.7.38)$$

Observing $\sum_x G_{xx'}(t', t) = 1$ and

$$\begin{aligned}
&\sum_{x}\max\big\{0, G_{xx''}(t', t) - G_{xx'}(t', t)\big\}\\
&\quad = \sum_{x}\big[G_{xx''}(t', t) - \min\big\{G_{xx''}(t', t), G_{xx'}(t', t)\big\}\big]\\
&\quad = 1 - \sum_{x}\min\big\{G_{xx''}(t', t), G_{xx'}(t', t)\big\},
\end{aligned}$$

one can arrange (8.7.38) into

$$\alpha\big(G(t', t)\big) = 1 - \min_{x'',x'}\left[\sum_{x}\min\big\{G_{xx''}(t', t), G_{xx'}(t', t)\big\}\right]. \quad (8.7.39)$$

The weak ergodic theorem is stated as follows.

Let $\{t_i\}$ $(i = 1, 2, \dots)$ be an ascending sequence of time such that $t_i < t_j$ for $i < j$. The necessary and sufficient condition for the weak ergodicity is given by

$$\sum_{i=1}^{\infty}\big\{1 - \alpha\big(G(t_{i+1}, t_i)\big)\big\} = \infty. \quad (8.7.40)$$

Details of this theorem is included in Sect. 8.A.3.3. We shall show that the transition probability defined by (8.7.24) satisfies (8.7.40).

Let x^* be the state which belongs to $\bar{\mathscr{S}}_\Lambda$ and gives $\max_{x'}\{d_{x'x^*}\} = R$. By the definition of R in (8.7.23), the state x^* can be reached from an arbitrary state by at most R Monte Carlo steps. We consider a Markov chain which brings the system to x^* at the Monte Carlo time t from x at $t-R$:

$$G_{x^*x}(t,t-R) = \sum_{x_1,x_2,\ldots,x_{R-1}} G_{x^*x_{R-1}}(t-1)\cdots G_{x_2x_1}(t-R+1)G_{x_1x}(t-R).$$

One finds from the definition of x^* that there exists a number l $(0 \le l \le R)$ and a sequence of transitions from x to x^* such that

$$x = x_0 \neq x_1 \neq x_2 \neq \cdots \neq x_l = x_{l+1} = \cdots = x_R = x^*.$$

For the transition from x_i to x_{i+1} with $0 \le i < l$, Lemma 8.1 gives

$$\begin{aligned} G_{x_{i+1}x_i}(t-R+i) &= p_{x_{i+1}x_i} A\left(\frac{Q_{x_{i+1}}(t-R+i)}{Q_{x_i}(t-R+i)}\right) \\ &\ge p_0 A(1)\exp\left(-\frac{L_0}{T_0} - \frac{L_1}{T_1(t-R+1)}\right). \end{aligned}$$

With $i \ge l$, Lemma 8.2 guarantees that there exists a positive integer t_1' such that for $t \ge t_1'$ one has

$$\begin{aligned} G_{x_{i+1}x_i}(t-R+i) &= G_{x^*x^*}(t-R+i) \\ &\ge p_0 A(1)\exp\left(-\frac{L_0}{T_0} - \frac{L_1}{T_1(t-R+1)}\right). \end{aligned}$$

Therefore one obtains for $t \ge t_1'$

$$\begin{aligned} G_{x^*x}(t,t-R) &\ge G_{x^*x_{R-1}}(t-1)\cdots G_{x_2x_1}(t-R+1)G_{x_1x}(t-R) \\ &\ge \bigl(p_0A(1)\bigr)^R \exp\left(-\frac{RL_0}{T_0} - \sum_{i=0}^{R-1}\frac{L_1}{T_1(t-R+i)}\right) \\ &\ge \bigl(p_0A(1)\bigr)^R \exp\left(-\frac{RL_0}{T_0} - \frac{L_1}{T_1(t-1)}\right), \end{aligned}$$

where we used the monotonicity of $T(t)$ at the last inequality. Now we evaluate the ergodic coefficient. For sufficiently large t, we have

$$\begin{aligned} \sum_x \min\bigl\{G_{xx''}(t,t-R), G_{xx'}(t,t-R)\bigr\} &\ge \min\bigl\{G_{x^*x''}(t,t-R), G_{x^*x'}(t,t-R)\bigr\} \\ &\ge \bigl(p_0A(1)\bigr)^R \exp\left(-\frac{RL_0}{T_0} - \frac{L_1}{T_1(t-1)}\right). \end{aligned}$$

Notice here that the right hand side is independent of x'' and x'. Taking the minimum with respect to x'' and x', this inequality yields

$$\min_{x',x''}\left[\sum_x \min\{G_{xx''}(t,t-R), G_{xx'}(t,t-R)\}\right]$$
$$= 1-\alpha\big(G(t,t-R)\big) \geq \big(p_0 A(1)\big)^R \exp\left(-\frac{RL_0}{T_0} - \frac{L_1}{T_1(t-1)}\right).$$

Therefore there exists a positive integer k_0 such that $k_0 R \geq t_1'$ and the following inequality holds for $k \geq k_0$:

$$1-\alpha\big(G(kR,kR-R)\big) \geq \big(p_0 A(1)\big)^R \exp\left(-\frac{RL_0}{T_0} - \frac{L_1}{T_1(kR-1)}\right).$$

We substitute (8.7.27) for T_1 in the right hand side. Then it is shown

$$\sum_{k=0}^{\infty}\big\{1-\alpha\big(G(kR,kR-1)\big)\big\} \geq \big(p_0 A(1)\big)^R \exp\left(-\frac{RL_0}{T_0}\right)\sum_{k\geq k_0}^{\infty}\frac{1}{kR+1} = \infty. \tag{8.7.41}$$

Therefore, by the weak ergodic theorem, it is proved that a Markov chain generated by the transition probability (8.7.24) has the weak ergodicity. □

We shall finally prove that our Markov chain has the strong ergodicity. The definition of the strong ergodicity is given by

$$\lim_{t\to\infty}\sum_{x'}\left|\sum_x G_{x'x}(t,0)P_{0,x} - Q^{\star}_{x'}\right| = 0, \tag{8.7.42}$$

where $P_{0,x}$ is an arbitrary initial distribution and $Q^{\star}_x = \lim_{t\to\infty} Q_x(t)$ with the equilibrium distribution $Q_x(t)$ of $G_{x'x}(t)$, which satisfies $Q_{x'}(t) = G_{x'x}(t)Q_x(t)$. The strong ergodicity implies that the distribution generated by $G(t)$ from an arbitrary initial distribution converges to a unique distribution Q^* with $t \to \infty$.

Proof (Strong ergodicity) To accomplish the proof of the strong ergodicity, we resort to the strong ergodic theorem. The strong ergodic theorem is stated as follows.

Assume that

(i) *a Markov chain generated by* $G(t)$ *is weakly ergodic,*
(ii) *there exists an equilibrium distribution* $Q(t)$ *such that* $Q_{x'}(t) = \sum_x G_{x'x}(t)Q_x(t)$ *for any* t,
(iii) *the equilibrium distribution* $Q(t)$ *satisfies*

$$\sum_{t=0}^{\infty}\sum_x \big|Q_x(t+1) - Q_x(t)\big| < \infty. \tag{8.7.43}$$

Then the Markov chain has the strong ergodicity.

We refer to Sect. 8.A.3.3 for details of this theorem. Here we are going to prove that the three conditions (i), (ii), and (iii) are satisfied by the transition probability $G(t)$ defined by (8.7.24) and the distribution $Q(t)$ by (8.7.26).

The satisfaction of the condition (i) has been already mentioned. That of condition (ii) is easily shown by noting $p_{x'x} = p_{xx'}$ and $A(Q_{x'}/Q_x) = A(Q_x/Q_{x'})Q_{x'}/Q_x$ in (8.7.24). Here we see that the condition (iii) is satisfied.

Define $F_{1,\min} = \min_x F_{1,x}$ and $\Delta_{1,x} = F_{1,x} - F_{1,\min}$. We write a set of x which provides $F_{1,x} = F_{1,\min}$ as $\mathcal{S}_{1,\min}$ and its complementary set as $\bar{\mathcal{S}}_{1,\min}$. For $x \in \mathcal{S}_{1,\min}$, one can write

$$\begin{aligned} Q_x(t) &= \frac{\exp(-\frac{F_{0,x}}{T_0} - \frac{F_{1,\min}}{T_1(t)})}{\sum_{x'} \exp(-\frac{F_{0,x'}}{T_0} - \frac{F_{1,x'}}{T_1(t)})} \\ &= \frac{\exp(-\frac{F_{0,x}}{T_0})}{\sum_{x' \in \mathcal{S}_{1,\min}} \exp(-\frac{F_{0,x'}}{T_0}) + \sum_{x' \in \bar{\mathcal{S}}_{1,\min}} \exp(-\frac{F_{0,x'}}{T_0} - \frac{\Delta_{1,x'}}{T_1(t)})}. \end{aligned}$$

Since $T_1(t+1) < T_1(t)$ and $\Delta_{1,x'} > 0$ for $x' \in \bar{\mathcal{S}}_{1,\min}$, one has

$$\exp\left(-\frac{\Delta_{1,x'}}{T_1(t+1)}\right) < \exp\left(-\frac{\Delta_{1,x'}}{T_1(t)}\right),$$

which leads us to $Q_x(t+1) > Q_x(t)$. On the other hand, for $x \in \bar{\mathcal{S}}_{1,\min}$, one can write

$$Q_x(t) = \frac{\exp(-\frac{F_{0,x}}{T_0} - \frac{\Delta_{1,x}}{T_1(t)})}{\sum_{x' \in \mathcal{S}_{1,\min}} \exp(-\frac{F_{0,x'}}{T_0}) + \sum_{x' \in \bar{\mathcal{S}}_{1,\min}} \exp(-\frac{F_{0,x'}}{T_0} - \frac{\Delta_{1,x'}}{T_1(t)})}.$$

The derivative of this by t yields

$$\frac{d}{dt} Q_x(t) = \frac{dT_1(t)}{dt} \left\{ \Delta_{1,x} - \sum_{x' \in \bar{\mathcal{S}}_{1,\min}} \Delta_{1,x'} Q_{x'}(t) \right\} \frac{Q_x(t)}{T_1(t)^2}.$$

Notice that $dT_1(t)/dt < 0$ and $\Delta_{1,x} - \sum_{x' \in \bar{\mathcal{S}}_{1,\min}} \Delta_{1,x'} Q_{x'}(t) > 0$ for sufficiently large t since $Q_{x'}(t)$ with $x' \in \bar{\mathcal{S}}_{1,\min}$ vanishes with $t \to \infty$. Hence there exists a number $t_2 < \infty$ such that $dQ_x(t)/dt < 0$ for all $t \geq t_2$. This results in $Q_x(t+1) < Q_x(t)$ for any $t \geq t_2$. Using these relations between $Q(t+1)$ and $Q(t)$, one can show

$$\begin{aligned} \sum_x \left| Q_x(t+1) - Q_x(t) \right| &= \sum_{x \in \mathcal{S}_{1,\min}} \big(Q_x(t+1) - Q_x(t) \big) \\ &\quad + \sum_{x \in \bar{\mathcal{S}}_{1,\min}} \big(Q_x(t) - Q_x(t+1) \big) \\ &= 2 \sum_{x \in \mathcal{S}_{1,\min}} \big(Q_x(t+1) - Q_x(t) \big), \end{aligned}$$

by which one obtains

$$\sum_{t=t_2}^{\infty} \sum_x \left| Q_x(t+1) - Q_x(t) \right| = 2 \sum_{x \in \mathcal{S}_{1,\min}} \big(Q_x(\infty) - Q_x(t_2) \big) \leq 2.$$

Therefore it is shown

$$\begin{aligned}\sum_{t=0}^{\infty}\sum_{x}\big|Q_x(t+1)-Q_x(t)\big| &= \sum_{t=0}^{t_2-1}\sum_{x}\big|Q_x(t+1)-Q_x(t)\big| \\ &\quad + \sum_{t=t_2}^{\infty}\sum_{x}\big|Q_x(t+1)-Q_x(t)\big| \\ &\leq \sum_{t=0}^{t_2-1}\sum_{x}\big|Q_x(t+1)-Q_x(t)\big| + 2 \\ &\leq \sum_{t=0}^{t_2-1}\sum_{x}\big(Q_x(t+1)+Q_x(t)\big) + 2 \\ &= 2t_2 + 2 < \infty.\end{aligned}$$

Thus we verified that the condition (iii) is satisfied. It follows that the Markov chain defined by (8.7.24)–(8.7.27) has the strong ergodicity and the distribution generated by this Markov chain converges to Q^* defined by (8.7.29). □

8.8 Conclusion

Quantum annealing is known as a quantum mechanical algorithm which can be implemented using the quantum computer. We mentioned some studies regarding the size scaling of energy gaps in Sect. 8.5. The size scaling of energy gaps is related to the power of quantum annealing as quantum computation. So far there have been several evidences that the exponential complexity of NP-complete problems cannot be relaxed by quantum annealing. It is probable that quantum annealing is not an almighty algorithm for hard optimisation problems. However, it is too early to be discouraged. There could be unknown paths of quantum annealing which do not encounter exponential closing of the energy gap [351]. In addition, there is still a possibility that NP-complete problems which are solved by quantum annealing efficiently will be found. As for the latter possibility, the Ising spin glass on the cubic lattice might be a candidate. The energy minimisation of this model has been proved to be an NP-hard problem [23]. If the quantum phase transition of this model in the transverse field belongs to an infinite randomness fixed point with $\psi < 1$ [236], then the complexity reduces to a subexponential one (see (8.5.12) in Sect. 8.5.2).

Comparison between quantum annealing and simulated annealing is a significant topic in order to highlight the validity of a quantum annealing. We mentioned in Sect. 8.6 that residual energy after quantum annealing decays faster than simulated annealing in the one-dimensional random Ising model. Apart from this model, no generic theory which shows the convergence of residual errors after quantum annealing has been established. In contrast, the Huse-Fisher law [185] is known for residual energy after simulated annealing in generic systems. It is a future task to make a theory for the convergence of residual errors after quantum annealing in generic systems and show the validity of quantum annealing in comparison to simulated annealing.

In Sect. 8.7, we presented theorems on sufficient conditions of convergence of quantum annealings with the real-time dynamics and the quantum Monte Carlo dynamics. Morita and Nishimori have provided several theorems in Ref. [282], some of which are not included in Sect. 8.7. We list them as follows:

- Sufficient conditions on the coefficient $\Gamma(t)$ of the transverse field to guarantee the convergence of quantum annealing with real-time dynamics (presented in Sect. 8.7.1).
- Sufficient conditions on the coefficient $\Gamma(t)$ of the transverse field and two-body interactions, $-\Gamma(t)(\sum_i S_i^x + \sum_{ij} S_i^x S_j^x)$, to guarantee the convergence of quantum annealing with real-time dynamics.
- Sufficient conditions on the coefficient $\Gamma(t)$ of many-body interactions, $-\Gamma(t) \times \prod_i (1 + S_i^x)$, to guarantee the convergence of quantum annealing with real-time dynamics.
- Sufficient conditions on the coefficient $\Gamma(t)$ of the transverse field to guarantee the convergence of quantum annealing with standard quantum Monte Carlo dynamics (presented in Sect. 8.7.2).
- Sufficient conditions on the coefficient $\Gamma(t)$ of the transverse field to guarantee the weak ergodicity of quantum annealing with a quantum Monte Carlo dynamics using a generalised transition probability [404].
- Sufficient conditions on the time-dependent mass $m(t)$ to guarantee the convergence of quantum annealing with standard quantum Monte Carlo dynamics in a system of distinguishable particles in a finite continuous space.
- Sufficient conditions on the coefficient $\Gamma(t)$ of the transverse field to guarantee the convergence of quantum annealing with Green's function quantum Monte Carlo dynamics.

It is noteworthy that the schedule of the coefficient $\Gamma(t)$ of the transverse field appearing in the sufficient conditions of the convergence of quantum annealing with standard quantum Monte Carlo dynamics depends on a power of t for $t \gg 1$, whereas that of temperature $T(t)$ given by Geman and Geman for simulated annealing depends on t as $1/\ln t$. If Γ and T are equivalent, this fact means that the convergence of quantum annealing is faster than simulated annealing. However, it is not necessarily true. Recalling the Suzuki-Trotter mapping, $1/K_M = 2/\ln\coth(\beta\Gamma/M)$ gives an effective temperature, where β and M are the inverse of real temperature and the number of the Trotter slices. This shows that the coefficient of the transverse field has a non-linear relation with an effective temperature. In order to compare the speed of convergence between quantum annealing and simulated annealing, one needs to investigate the convergence of a quantity which is common to both methods. This remains to be studied.

There are several missing topics related to quantum annealing in this chapter. Among them, probably the most important topic is an equivalence between quantum annealing and the standard model of quantum computation. The standard model means the gate-based circuit model [295]. Their equivalence implies that in principle any gate-based quantum computation can be mapped to quantum annealing and vice versa. It follows that quantum annealing can be seen as one of the most basic model of quantum computation. We refer to Refs. [6, 278] for details.

Quantum annealing is still an active research subject. We expect further development in the study of this subject, which leads us to deeper understanding of quantum computation as well as quantum disordered systems.

Appendix 8.A

8.A.1 Hopf's Theorem

In this appendix, we shall prove the Hopf's theorem [177]. The Hopf's theorem is given by Theorem 8.8. To prove this, we show Theorems 8.5–8.7 in advance.

Suppose that M is a $m \times m$ matrix and $\mathbf{v}$ is a column vector with m components. We use simplified notation $M \geq 0$ and $M > 0$ to represent $M_{ij} \geq 0$ and $M_{ij} > 0$ for any i and j respectively. Also, $\mathbf{v} \geq 0$ and $\mathbf{v} > 0$ imply $v_i \geq 0$ and $v_i > 0$ for any ith component respectively. $\mathbf{v} = 0$ means that all components of $\mathbf{v}$ are zero. As usual, we write the product of a matrix M and a vector $\mathbf{v}$ as $M\mathbf{v}$, namely,

$$[M\mathbf{v}]_i = \sum_j^m M_{ij} v_j.$$

We introduce following notations,

$$\begin{aligned} \operatorname*{osc}_i(v_i) &= \max_i(v_i) - \min_i(v_i) \quad \text{for a real vector } \mathbf{v}, \\ \operatorname*{Osc}_i(v_i) &= \sup_{|\eta|=1} \operatorname*{osc}_i \operatorname{Re}(\eta v_i) = \sup_\phi \operatorname*{osc}_i |v_i| \cos(\phi + \theta_i) \quad \text{for a complex vector } \mathbf{v} \end{aligned}$$

where θ_i is a phase of v_i. For any complex number c and a complex vector v, one has

$$\begin{aligned} \operatorname*{Osc}_i(cv_i) &= \sup_\phi \operatorname*{osc}_i |c||v_i| \cos(\phi + \alpha + \theta_i) = |c| \sup_\psi \operatorname*{osc}_i |v_i| \cos(\psi + \theta_i) \\ &= |c| \operatorname*{Osc}_i(v_i). \end{aligned} \tag{8.A.1}$$

We assume in the following part of this Appendix that M is a positive matrix, namely, $M > 0$. The positivity of M is equivalent to

$$M\mathbf{v} > 0 \quad \text{for any } \mathbf{v} \text{ which is } \mathbf{v} \geq 0 \text{ and } \mathbf{v} \neq 0. \tag{8.A.2}$$

We next assume that the ratio M_{ik}/M_{jk} is bounded, namely, there exists a positive number $\kappa < +\infty$ such that

$$\frac{M_{ik}}{M_{jk}} \leq \kappa \quad \text{for all } i,\ j, \text{ and } k. \tag{8.A.3}$$

One may write this assumption with a vector $\mathbf{v}$ such that $\mathbf{v} \geq 0$ and $\mathbf{v} \neq 0$ as

$$\frac{[M\mathbf{v}]_i}{[M\mathbf{v}]_j} \leq \kappa \quad \text{for all } i,\ j, \text{ and } \mathbf{v}\ (\geq 0,\ \neq 0). \tag{8.A.4}$$

Under assumptions (8.A.2) and (8.A.4), we have the following theorem.

Theorem 8.5 *If M satisfies the conditions* (8.A.2) *and* (8.A.4), *for any vector* $\mathbf{p} > 0$ *and complex-valued vector* $\mathbf{v}$

$$\operatorname*{Osc}_i\left(\frac{[M\mathbf{v}]_i}{[M\mathbf{p}]_i}\right) \leq \frac{\kappa - 1}{\kappa + 1} \operatorname*{Osc}_i\left(\frac{v_i}{p_i}\right). \tag{8.A.5}$$

Proof Consider first a real-valued vector $\mathbf{v}$. Define X_k for fixed i, j, and $\mathbf{p} > 0$ by

$$\frac{[M\mathbf{v}]_i}{[M\mathbf{p}]_i} - \frac{[M\mathbf{v}]_j}{[M\mathbf{p}]_j} = \sum_{k=1}^{m} X_k v_k, \tag{8.A.6}$$

namely, $X_k(i, j, \mathbf{p}) = M_{ik}/[M\mathbf{p}]_i - M_{jk}/[M\mathbf{p}]_j$. It is easy to find that the light hand side of (8.A.6) vanishes when $\mathbf{v} = a\mathbf{p}$ with a constant a. Hence $\sum_k X_k p_k = 0$, and thus

$$\frac{[M\mathbf{v}]_i}{[M\mathbf{p}]_i} - \frac{[M\mathbf{v}]_j}{[M\mathbf{p}]_j} = \sum_{k=1}^{m} X_k (v_k - a p_k). \tag{8.A.7}$$

Suppose now

$$a = \min_i\left(\frac{v_i}{p_i}\right) \quad \text{and} \quad b = \max_i\left(\frac{v_i}{p_i}\right).$$

Because of an identity, $v_k - a p_k = (b - a) p_k - (b p_k - v_k)$, and inequality, $a \leq v_k/p_k \leq b$, $v_k - a p_k$ takes its minimum 0 when $v_k = a p_k$ and its maximum $(b - a) p_k$ when $v_k = b p_k$. Therefore the right hand side of (8.A.6) for a given $\mathbf{p}$ takes the maximum when

$$\mathbf{v} = a\mathbf{p}^- + b\mathbf{p}^+ = a\mathbf{p} + (b - a)\mathbf{p}^+,$$

where we defined

$$p_i^- = \begin{cases} p_i & \text{when } X_i \leq 0 \\ 0 & \text{when } X_i > 0 \end{cases}, \qquad p_i^- = \begin{cases} 0 & \text{when } X_i \leq 0 \\ p_i & \text{when } X_i > 0 \end{cases}.$$

Therefore, one gets from (8.A.6)

$$\begin{aligned} \frac{[M\mathbf{v}]_i}{[M\mathbf{p}]_i} - \frac{[M\mathbf{v}]_j}{[M\mathbf{p}]_j} &\leq \frac{[M\{a\mathbf{p} + (b - a)\mathbf{p}^+\}]_i}{[M\mathbf{p}]_i} - \frac{[M\{a\mathbf{p} + (b - a)\mathbf{p}^+\}]_j}{[M\mathbf{p}]_j} \\ &= (b - a)\left\{\frac{[M\mathbf{p}^+]_i}{[M\mathbf{p}]_i} - \frac{[M\mathbf{p}^+]_j}{[M\mathbf{p}]_j}\right\}. \end{aligned} \tag{8.A.8}$$

Note here $M\mathbf{p} > 0$, $M\mathbf{p}^+ \geq 0$, and $M\mathbf{p}^- \geq 0$, since $M > 0$ and $\mathbf{p} > 0$. If $\mathbf{p}^- = 0$ or $\mathbf{p}^+ = 0$, then $\mathbf{p}^+ = \mathbf{p}$ or $\mathbf{p}^+ = 0$ respectively which implies that the right hand side of (8.A.8) vanishes. Hence we may assume both $\mathbf{p}^- \neq 0$ and $\mathbf{p}^+ \neq 0$, namely,

$M\mathbf{p}^- > 0$ and $M\mathbf{p}^+ > 0$. It follows from this that the terms inside braces in the right hand side of (8.A.8) can be written as

$$\frac{[M\mathbf{p}^+]_i}{[M\mathbf{p}]_i} - \frac{[M\mathbf{p}^+]_j}{[M\mathbf{p}]_j} = \frac{1}{1+t} - \frac{1}{1+t'},$$

where

$$t = \frac{[M\mathbf{p}^-]_i}{[M\mathbf{p}^+]_i} > 0 \quad \text{and} \quad t' = \frac{[M\mathbf{p}^-]_j}{[M\mathbf{p}^+]_j} > 0.$$

Because of the condition (8.A.4), t and t' are bounded from κ^{-1} to κ, namely, $\kappa \le \kappa^2 t \le \kappa^3$ and $t' \le \kappa$. Combining these inequalities, one has $t' \le \kappa^2 t$. Therefore

$$\frac{[M\mathbf{p}^+]_i}{[M\mathbf{p}]_i} - \frac{[M\mathbf{p}^+]_j}{[M\mathbf{p}]_j} \le \left(\frac{1}{1+t} - \frac{1}{1+\kappa^2 t}\right).$$

For $t > 0$ and given $\kappa > 1$, the right hand side takes the maximum $(\kappa - 1)/(\kappa + 1)$ when $t = 1/\kappa$. Therefore one obtains from (8.A.8)

$$\frac{[M\mathbf{v}]_i}{[M\mathbf{p}]_i} - \frac{[M\mathbf{v}]_j}{[M\mathbf{p}]_j} \le \frac{\kappa - 1}{\kappa + 1} \operatorname*{osc}_i \left(\frac{v_i}{p_i}\right).$$

Since this inequality holds for any i and j, we reach

$$\operatorname*{osc}_i \left(\frac{M\mathbf{v}}{M\mathbf{p}}\right) = \max_i \left(\frac{M\mathbf{v}}{M\mathbf{p}}\right) - \min_i \left(\frac{M\mathbf{v}}{M\mathbf{p}}\right) \le \frac{\kappa - 1}{\kappa + 1} \operatorname*{osc}_i \left(\frac{v_i}{p_i}\right).$$

As for a complex-valued vector $\mathbf{v}$, we replace v_i by $\mathrm{Re}(\eta v_i)$ in the above. The same argument as above yields

$$\operatorname*{osc}_i \left(\mathrm{Re}\frac{[\eta M\mathbf{v}]_i}{[M\mathbf{p}]_i}\right) \le \frac{\kappa - 1}{\kappa + 1} \operatorname*{osc}_i \mathrm{Re}\left(\eta \frac{v_i}{p_i}\right),$$

where we note $M\mathrm{Re}(\eta\mathbf{v}) = \mathrm{Re}(\eta M\mathbf{v})$. Taking sup with respect to η on $|\eta| = 1$, we obtain (8.A.5). □

Now we apply Theorem 8.5 to the eigenvalue problem:

$$M\mathbf{v} = \lambda\mathbf{v}. \tag{8.A.9}$$

The Perron-Frobenius theorem states that when $M \ge 0$, the eigenvalue equation (8.A.9) has a non-negative eigenvalue λ_0 that satisfies $\lambda_0 \ge |\lambda|$ for any other eigenvalue λ. This theorem is sharpened for $M > 0$ in the next theorem. We refer to Sect. 8.A.2 for the Perron-Frobenius theorem.

Theorem 8.6 *Under conditions* (8.A.2) *and* (8.A.4), *the eigenvalue equation* (8.A.9) *has a positive eigenvalue* $\lambda_0 > 0$ *and corresponding eigenvector* $\mathbf{q} > 0$. *Moreover, for any vector* $\mathbf{p} \ge 0$ $(\ne 0)$,

$$\mathbf{q}^{(n)} = \frac{M^n\mathbf{p}}{[M^n\mathbf{p}]_k}$$

with a fixed k *converges to* $\mathbf{q}$ *with* $n \to \infty$.

Proof Consider two vectors $\mathbf{p}$ and $\mathbf{p}'$ such that $\mathbf{p} \geq 0$, $\mathbf{p} \neq 0$, $\mathbf{p}' \geq 0$, and $\mathbf{p}' \neq 0$. Define

$$\mathbf{p}^{(n+1)} = M\mathbf{p}^{(n)}, \qquad \mathbf{p}^{(0)} = \mathbf{p}$$
$$\mathbf{p}'^{(n+1)} = M\mathbf{p}'^{(n)}, \qquad \mathbf{p}'^{(0)} = \mathbf{p}'.$$

By the assumption (8.A.2), $\mathbf{p}^{(n)}$ and $\mathbf{p}'^{(n)}$ are positive vectors for $n > 0$. Applying Theorem 8.5, one finds

$$\operatorname*{osc}_i\left(\frac{[\mathbf{p}'^{(n)}]}{[\mathbf{p}^{(n)}]_i}\right) \leq \left(\frac{\kappa-1}{\kappa+1}\right)\operatorname*{osc}_i\left(\frac{[\mathbf{p}'^{(n-1)}]_i}{[\mathbf{p}^{(n-1)}]_i}\right) \leq \left(\frac{\kappa-1}{\kappa+1}\right)^{n-1}\operatorname*{osc}_i\left(\frac{[\mathbf{p}'^{(1)}]_i}{[\mathbf{p}^{(1)}]_i}\right). \tag{8.A.10}$$

It follows that

$$\operatorname*{osc}_i\left(\frac{[\mathbf{p}'^{(n)}]_i}{[\mathbf{p}^{(n)}]_i}\right) \to 0 \quad (n \to \infty),$$

namely, there exists a finite constant λ independent of i such that

$$\frac{[\mathbf{p}'^{(n)}]_i}{[\mathbf{p}^{(n)}]_i} \to \lambda \quad (n \to \infty) \quad \text{for every } i. \tag{8.A.11}$$

We introduce for a fixed k

$$\mathbf{q}^{(n)} = \frac{\mathbf{p}^{(n)}}{[\mathbf{p}^{(n)}]_k} \quad \text{and} \quad \mathbf{q}'^{(n)} = \frac{\mathbf{p}'^{(n)}}{[\mathbf{p}'^{(n)}]_k}.$$

The assumption (8.A.4) leads to

$$\kappa^{-1} \leq \left[\mathbf{q}^{(n)}\right]_i \leq \kappa \quad \text{and} \quad \kappa^{-1} \leq \left[\mathbf{q}'^{(n)}\right]_i \leq \kappa. \tag{8.A.12}$$

It follows that

$$\begin{aligned}
\left|\left[\mathbf{q}'^{(n)}\right]_i - \left[\mathbf{q}^{(n)}\right]_i\right| &= \left|\frac{[\mathbf{p}'^{(n)}]_i}{[\mathbf{p}'^{(n)}]_k} - \frac{[\mathbf{p}^{(n)}]_i}{[\mathbf{p}^{(n)}]_k}\right| = \frac{[\mathbf{p}^{(n)}]_i}{[\mathbf{p}'^{(n)}]_k}\left|\frac{[\mathbf{p}'^{(n)}]_i}{[\mathbf{p}^{(n)}]_i} - \frac{[\mathbf{p}'^{(n)}]_k}{[\mathbf{p}^{(n)}]_k}\right| \\
&\leq \left[\mathbf{q}^{(n)}\right]_i \frac{[\mathbf{p}^{(n)}]_k}{[\mathbf{p}'^{(n)}]_k}\operatorname*{osc}_i\left(\frac{[\mathbf{p}'^{(n)}]_i}{[\mathbf{p}^{(n)}]_i}\right) \leq \kappa\frac{[\mathbf{p}^{(n)}]_k}{[\mathbf{p}'^{(n)}]_k}\operatorname*{osc}_i\left(\frac{[\mathbf{p}'^{(n)}]_i}{[\mathbf{p}^{(n)}]_i}\right).
\end{aligned} \tag{8.A.13}$$

We now choose $\mathbf{p}' = \mathbf{p}^{(1)} = M\mathbf{p}$, namely, $\mathbf{p}'^{(n)} = M\mathbf{p}^{(n)} = \mathbf{p}^{(n+1)}$ and $\mathbf{q}'^{(n)} = \mathbf{q}^{(n+1)}$. From (8.A.11) and (8.A.13), one finds

$$\left|\left[\mathbf{q}^{(n+1)}\right]_i - \left[\mathbf{q}^{(n)}\right]_i\right| \leq \kappa\frac{[\mathbf{p}^{(n)}]_k}{[\mathbf{p}^{(n+1)}]_k}\operatorname*{osc}_i\left(\frac{[\mathbf{p}^{(n+1)}]_i}{[\mathbf{p}^{(n)}]_i}\right) \to 0 \quad (n \to \infty).$$

This implies the sequence $\mathbf{q}^{(n)}$ converges to a vector $\mathbf{q}$, which is positive because of (8.A.12). We now define λ_0 such that

$$\frac{[\mathbf{p}^{(n+1)}]_i}{[\mathbf{p}^{(n)}]_i} = \frac{[M\mathbf{p}^{(n)}]_i}{[\mathbf{p}^{(n)}]_i} = \frac{[M\mathbf{q}^{(n)}]_i}{[\mathbf{q}^{(n)}]_i} \to \lambda_0 \quad (n \to \infty).$$

This implies

$$M\mathbf{q} = \lambda_0 \mathbf{q}.$$

Because of (8.A.10), (8.A.11), and (8.A.13), the sequence $\mathbf{q}^{(n)}$ for any vector $\mathbf{p}$ converges to the same limit $\mathbf{q}$. Thus Theorem 8.6 is proved. □

Theorem 8.6 immediately leads us to the following theorem.

Theorem 8.7 *Under the conditions* (8.A.2) *and* (8.A.4), *Eq.* (8.A.9) *has no other non-negative solution except for* λ_0 *and* $c\mathbf{q}$, *and has no other eigenvectors for the eigenvalue* λ_0 *except for* $c\mathbf{q}$.

Proof Assume that $\mathbf{v} \geq 0$ ($\neq 0$) is a solution of Eq. (8.A.9). Because of the assumption (8.A.2), one has $M\mathbf{v} > 0$ and thus $\lambda > 0$ and $\mathbf{v} > 0$. Now we make $\mathbf{p}_0 = \mathbf{v}$ in Theorem 8.6. One finds

$$\frac{M^n \mathbf{v}}{[M^n \mathbf{v}]_k} = \frac{\lambda^n \mathbf{v}}{[\lambda^n \mathbf{v}]_k} = \frac{\mathbf{v}}{v_k}.$$

Hence the vector $\mathbf{q}$ in Theorem 8.6 is nothing but $\mathbf{v}/v_k$, that is, $\mathbf{v} = c\mathbf{q}$ and $\lambda = \lambda_0$. This proves the first part of Theorem 8.7.

Next, following Theorem 8.6, suppose $\lambda_0 > 0$ and $\mathbf{q} > 0$, and consider a solution λ and $\mathbf{v}$ of Eq. (8.A.9). Application of Theorem 8.5 to $\mathbf{v}$ and $\mathbf{q}$ yields

$$\frac{|\lambda|}{\lambda_0} \operatorname*{Osc}_i \left(\frac{v_i}{q_i} \right) = \operatorname*{Osc}_i \left(\frac{\lambda v_i}{\lambda_0 q_i} \right) = \operatorname*{Osc}_i \left(\frac{[M\mathbf{v}]_i}{[M\mathbf{q}]_i} \right) \leq \frac{\kappa - 1}{\kappa + 1} \operatorname*{Osc}_i \left(\frac{v_i}{q_i} \right), \tag{8.A.14}$$

where we used (8.A.1) in the first equality. If $\lambda = \lambda_0$ in this inequality, one has $\mathrm{Osc}(v_i/q_i) = 0$ since $(\kappa - 1)/(\kappa + 1) < 1$. Hence the eigenvector with respect to $\lambda = \lambda_0$ is $\mathbf{v} = c\mathbf{q}$. This proves the second part of Theorem 8.7. □

Finally, we reach the following theorem.

Theorem 8.8 *Under the conditions* (8.A.2) *and* (8.A.4), *any eigenvalue* $\lambda \neq \lambda_0$ *of Eq.* (8.A.9) *satisfies*

$$|\lambda| \leq \frac{\kappa - 1}{\kappa + 1} \lambda_0. \tag{8.A.15}$$

Proof Recall (8.A.14). If $\lambda \neq \lambda_0$ and $\mathbf{v} \neq 0$, then $\mathbf{v}$ cannot be $c\mathbf{q}$. It follows that $\mathrm{osc}(v_i/q_i) > 0$ and hence (8.A.14) yields (8.A.15). □

8.A.2 Perron-Frobenius Theorem

In this appendix, we provide the statement of the Perron-Frobenius theorem without proof. We refer to textbooks of the matrix analysis (e.g., Ref. [180]) for the proof.

Consider a square matrix A with dimension $N \geq 2$. We define the irreducibility of a matrix A by the property that one cannot make

$$PAP^{-1} = \begin{bmatrix} A_{11} & A_{12} \\ 0 & A_{22} \end{bmatrix}$$

by using any permutation matrix P, where A_{11}, A_{12}, and A_{22} are square matrices. This property is also written as follows: For any i and j with $1 \leq i \neq j \leq N$, one can choose $l_1, l_2, \ldots, l_n$ such that

$$a_{il_1} a_{l_1 l_2} \cdots a_{l_{n-1} l_n} a_{l_n j} \neq 0,$$

where a_{lm} denotes an element of the matrix A.

The Perron-Frobenius theorem is stated as follows.

Theorem 8.9 *If A is a non-negative and irreducible matrix, it has following properties.*

(i) *Maximum eigenvalue λ of A is positive, i.e., $\lambda > 0$, and satisfies $\lambda > |\lambda_i|$ for any other eigenvalue λ_i.*
(ii) *λ is the non-degenerate eigenvalue.*
(iii) *The eigenvector $\mathbf{v}$ to λ can be positive, $\mathbf{v} > 0$, or negative, $\mathbf{v} < 0$.*

Corollary 8.1 *Assume A is a non-negative and irreducible matrix. If $A\mathbf{v} = \lambda\mathbf{v}$ with $\mathbf{v} > 0$, then λ is the maximum eigenvalue of A and $\lambda > 0$.*

8.A.3 Theory of the Markov Chain

In this appendix, we provide a theory of a Markov chain [155]. Starting from the Chapman-Kolmogorov equation for a generic Markov chain, we show the ergodic theorem of a Markov chain with the time-independent transition probability. We then move on to a Markov chain with the time-dependent transition probability. We finally give a proof on the weak and strong ergodic theorems.

We employ an abstract notation x for the microscopic state of a system. We assume that x is a discrete variable. Throughout this appendix, the time is discrete and takes an integral value.

8.A.3.1 Chapman-Kolmogorov Equation

Let $P(X_t = x)$ be the probability that a system is in a state x and time t and $P(X_{t_1} = x_1; X_{t_2} = x_2; \ldots)$ be the joint probability that a system is in x_1 at t_1, x_2 at t_2, and so on. Furthermore, let $P(X_{t'_1} = x'_1; X_{t'_2} = x'_2; \ldots | X_{t_1} = x_1; X_{t_2} = x_2; \ldots)$ be the conditional probability that a system is in x'_1 at t'_1, x'_2 at t'_2, and so on, under the constraint that the system is in x_1 at t_1, x_2 at t_2, and so on, where it is assumed $t'_i > t_j$. These quantities are related through

$$P\big(X_{t_1'}=x_1';\,X_{t_2'}=x_2';\ldots;\,X_{t_1}=x_1;\,X_{t_2}=x_2;\ldots\big)$$
$$=P\big(X_{t_1'}=x_1';\,X_{t_2'}=x_2';\ldots|X_{t_1}=x_1;\,X_{t_2}=x_2;\ldots\big)P(X_{t_1}=x_1;\,X_{t_2}=x_2;\ldots). \tag{8.A.16}$$

Consider the conditional probability $P(X_{t+1}=x'|X_t=x;\,X_{t-s_1}=x_1;\ldots;\,X_{t-s_n}=x_n)$ where $1\le s_1\le s_2\le\cdots\le s_n$. We demand

$$P\big(X_{t+1}=x'|X_t=x;\,X_{t-s_1}=x_1;\ldots;\,X_{t-s_n}=x_n\big)=P\big(X_{t+1}=x'|X_t=x\big). \tag{8.A.17}$$

This is an essential property of the Markov chain. We call the sequence of the stochastic variables, $X_1, X_2, \ldots,$ the Markov chain. Equation (8.A.17) ensures that the probability in the Markov chain that the system is found in x' at $t+1$ depends only on the state at t and does not on the states $t-1$ and before.

We next define the transition probability by $G_{x',x}(t',t)=P(X_{t'}=x'|X_t=x)$ and in particular $G_{x',x}(t)=G_{x',x}(t+1,t)$. The former denotes the transition probability from x at t to x' at t'. These transition probabilities have following properties:

$$0\le G_{x',x}\big(t',t\big)\le 1,\quad \sum_{x'}G_{x',x}\big(t',t\big)=1, \tag{8.A.18}$$

$$0\le G_{x',x}(t)\le 1,\quad \sum_{x'}G_{x',x}(t)=1. \tag{8.A.19}$$

We consider an equation

$$G_{x',x}(t+s,t)=\sum_{x_1,x_2,\ldots,x_{s-1}}G_{x',x_{s-1}}(t+s-1)\cdots G_{x_2,x_1}(t+1)G_{x_1,x}(t), \tag{8.A.20}$$

or

$$G_{x',x}(t+s,t)=\big[G(t+s-1)\cdots G(t+1)G(t)\big]_{x',x} \tag{8.A.21}$$

in the matrix representation. In particular, if $G_{x',x}(t)$ is independent of t, this reduces to $G_{x',x}(t+s,t)=[G^s]_{x',x}$. Equation (8.A.20) or (8.A.21) is called the Chapman-Kolmogorov equation. The Chapman-Kolmogorov equation is proved by the method of induction as follows.

At first, one can verify (8.A.21) with $s=1$ by definition. Next, we assume (8.A.21) and consider $G_{x',x}(t+s+1,t)$. It is easy to see

$$G_{x',x}(t+s+1,t)=P\big(X_{t+s+1}=x'|X_t=x\big)$$
$$=\sum_y P\big(X_{t+s+1}=x';\,X_{t+s}=y|X_t=x\big). \tag{8.A.22}$$

By Eq. (8.A.16), one finds

$$P\big(X_{t+s+1}=x';\,X_{t+s}=y|X_t=x\big)$$
$$=P\big(X_{t+s+1}=x';\,X_{t+s}=y;\,X_t=x\big)/P(X_t=x)$$
$$=P\big(X_{t+s+1}=x'|X_{t+s}=y;\,X_t=x\big)P(X_{t+s}=y;\,X_t=x)/P(X_t=x). \tag{8.A.23}$$

Using the property of the Markov chain (8.A.17), one has

$$P(X_{t+s+1} = x' | X_{t+s} = y; X_t = x) = P(X_{t+s+1} = x' | X_{t+s} = y).$$

With this and $P(X_{t+s} = y; X_t = x)/P(X_t = x) = P(X_{t+s} = y | X_t = x)$, (8.A.23) is arranged as

$$\begin{aligned} &P(X_{t+s+1} = x'; X_{t+s} = y | X_t = x) \\ &\quad = P(X_{t+s+1} = x' | X_{t+s} = y) P(X_{t+s} = y | X_t = x) \\ &\quad = G_{x',y}(t+s) G_{y,x}(t+s,t). \end{aligned}$$

Hence Eq. (8.A.22) is written under the assumption (8.A.21) as

$$\begin{aligned} G_{x',x}(t+s+1,t) &= \sum_y G_{x',y}(t+s) G_{y,x}(t+s,t) \\ &= \left[G(t+s) \cdots G(t+1) G(t)\right]_{x',x}. \end{aligned} \qquad (8.A.24)$$

Therefore, by the method of induction, Eq. (8.A.20) or (8.A.21) are proved for any integer $s \geq 1$.

8.A.3.2 Time-Independent Transition Probability

Let us focus on the time-independent transition probability. We use the notation $G_{x',x} = G_{x',x}(t)$ throughout this subsection. The purpose of this subsection is to prove that the distribution generated by G converges to a unique equilibrium distribution.

The transition probability $G_{x',x}$ has the properties given by (8.A.19). Here we impose additional conditions to $G_{x',x}$.

(i) **(Irreducibility).** For any states x and x', there exists a finite positive integer s such that

$$\sum_{x_1, x_2, \ldots, x_{s-1}} G_{x',x_{s-1}} \cdots G_{x_2,x_1} G_{x_1,x} = \left[G^s\right]_{x',x} > 0. \qquad (8.A.25)$$

This condition implies that any state x' is accessible from any state x by a finite number of application of G.

(ii) **(Aperiodicity).** For any state x and $l = 1, 2, \ldots,$ there is no positive integer $s > 1$ which yields $[G^{ls}]_{x,x} > 0$ and $[G^{n \neq ls}]_{x,x} = 0$. This implies that the system does not return to the initial state with the periodicity larger than 1.

(iii) **(Existence of an equilibrium distribution).** There exists an equilibrium distribution Q_x which satisfies

$$0 < Q_x \leq 1, \quad \sum_x Q_x = 1$$

$$\sum_x G_{x',x} Q_x = Q_{x'} \quad \text{(i.e., } GQ = Q \text{ in the matrix representation).}$$

(8.A.26)

(iv) **(Detailed balance).** $G_{x',x}$ satisfies the detailed balance condition:

$$G_{x',x} Q_x = G_{x,x'} Q_{x'}, \tag{8.A.27}$$

where Q_x is the equilibrium distribution of G.

To tell a truth, the last two conditions, the existence of an equilibrium distribution and the detailed balance, are not necessary for the convergence of the distribution generated by G. However, in practical situations, the transition probability is usually constituted so as to fulfil the last two conditions. Since the proof is simpler with these two conditions, we impose them.

The theorem for the convergence of the Markov chain with the time-independent transition probability is stated as follows.

Theorem 8.10 *The Markov chain generated by the transition probability, which fulfils the conditions* (i)–(iv) *and* (8.A.19), *converges to the unique equilibrium distribution Q.*

Proof Let $P_x(0)$ be an initial distribution. We demand $0 \le P_x(0) \le 1$ and $\sum_x P_x(0) = 1$. Define the distribution after t applications of transition probability as $P_{x'}(t) = \sum_x [G^t]_{x',x} P_x(t)$ or $P(t) = G^t P(t)$ in the matrix notation. Due to the detailed balance condition and $Q_x > 0$, one has

$$Q_{x'}^{-1/2} G_{x',x} Q_x^{1/2} = Q_x^{-1/2} G_{x,x'} Q_{x'}^{1/2}.$$

Introducing a diagonal matrix $D = \mathrm{diag}(Q_x)$, this relation is written as $D^{-1/2} G \times D^{1/2} = D^{1/2} G D^{-1/2}$. This implies that $\bar{G} = D^{-1/2} G D^{1/2}$ is a non-negative symmetric matrix. Hence the eigenvalue of $\bar{G}$ is real and its eigenvectors are mutually orthogonal. In addition, the Perron-Frobenius theorem can be applied to $\bar{G}$. We write the eigenvalue and eigenvector of $\bar{G}$ as λ_n and $\mathbf{v}_n$ respectively. We define the order of λ_n $(n = 0, 1, 2, \ldots)$ such that $\lambda_0 > \lambda_1 \ge \lambda_2 \ge \cdots$. Note that the Perron-Frobenius theorem guarantees $\lambda_0 > 0$ and $\lambda_0 > |\lambda_n|$ $(n = 1, 2, \ldots)$. See Sect. 8.A.2 for the theorem. Here we recall (8.A.26). It follows that

$$\bar{G} D^{-1/2} Q = D^{-1/2} Q. \tag{8.A.28}$$

Hence one finds that $D^{-1/2} Q$ is a positive eigenvector of $\bar{G}$ with the eigenvalue 1. Therefore, by the Perron-Frobenius theorem, the eigenvalue 1 is the maximum eigenvalue of $\bar{G}$, i.e., $\lambda_0 = 1$.

Now we expand the initial distribution $P(0)$ by the eigenvectors of $\bar{G}$ as

$$P(0) = D^{1/2} \left(C_0 \mathbf{v}_0 + \sum_{n \ge 1} C_n \mathbf{v}_n \right).$$

Assuming that $\mathbf{v}_n$ is normalised, one may choose $\mathbf{v}_0 = D^{-1/2} Q$. The coefficient C_0 is easily obtained as $C_0 = Q^T D^{-1/2} D^{-1/2} P(0) = \sum_x Q_x Q_x^{-1} P_x(0) = 1$. The application of G^t yields

$$P(t) = D^{1/2} \left(D^{-1/2} Q + \sum_{n \ge 1} C_n \lambda_n^t \mathbf{v}_n \right).$$

Since $1 > |\lambda_n|$ for $n \ge 1$, it is shown that $P(t) \to Q$ $(t \to \infty)$. □

In practice, one can make $G_{x',x}$ as follows. At first, we suppose that there is a target distribution Q_x we want to realise. An example of Q_x is the Boltzmann distribution, $Q_x = \exp(-H_x/k_B T)/\sum_{x'} \exp(-H_{x'}/k_B T)$, of the Ising model H_x with the fixed temperature T. Next, suppose $p_{x',x}$ the probability that x' is selected in one step from x and $A_{x',x}$ the acceptance function which decides the acceptance of the transition of the state from x to x'. One demands $p_{x',x} \geq 0$, $p_{x,x} = 0$, $\sum_{x'} p_{x',x} = 1$, $p_{x',x} = p_{x,x'}$, and the existence of a positive integer s such that $[p^s]_{x',x} > 0$ for any x and x'. As for the acceptance function $A_{x',x}$, we give it using the ratio $Q_{x'}/Q_x$ of the target distribution function such that $A_{x',x} = A(Q_{x'}/Q_x)$ and

$$0 < A(u) \leq 1, \quad \text{and} \quad A(1/u) = A(u)/u \quad \text{for } u > 0.$$

Usually we choose

$$A(u) = \begin{cases} \min\{1, u\} & \text{in the Metropolis method} \\ \frac{u}{1+u} & \text{in the heat bath method} \end{cases}.$$

Using $p_{x',x}$ and $A_{x',x}$, we define $G_{x',x}$ by

$$G_{x',x} = \begin{cases} p_{x',x} A_{x',x} & \text{for } x' \neq x \\ 1 - \sum_{x''} p_{x'',x} A_{x'',x} & \text{for } x' = x \end{cases}. \tag{8.A.29}$$

The proofs that this transition probability fulfils the conditions (i)–(iv) and (8.A.19) are given as follows.

At first, it is clear that this transition probability satisfies (8.A.19). Next, due to the property of $p_{x',x}$, one finds that for a positive integer s there is a sequence of states, $x = x_0, x_1, x_2, \ldots, x_{s-1}, x_s = x'$, such that $p_{x',x_{s-1}} \cdots p_{x_2,x_1} p_{x_1,x} > 0$. Notice that $x_{i+1} \neq x_i$ since $p_{x_{i+1},x_i} \neq 0$. Therefore, for such a sequence of states,

$$G_{x',x_{s-1}} \cdots G_{x_2,x_1} G_{x_1,x} = p_{x',x_{s-1}} A_{x',x_{s-1}} \cdots p_{x_2,x_1} A_{x_2,x_1} p_{x_1,x} A_{x_1,x} > 0,$$

where we used $A_{x_i,x_j} > 0$. This inequality and $G_{x',x} > 0$ for any x and x' immediately establish (8.A.25), namely, the irreducibility. Third, consider two state, x and x', such that $Q_x > Q_{x'}$. For such x and x', due to the property of $p_{x',x}$, there exists a positive integer s which yields $[p^s]_{x',x} > 0$. This implies that there exists at least a pair of x_i and x_j which yield $Q_{x_i} > Q_{x_j}$ and $p_{x_j,x_i} > 0$. Notice here that if $Q_{x_i} > Q_{x_j}$ then

$$A_{x_j,x_i} = A(Q_{x_j}/Q_{x_i}) = A(Q_{x_i}/Q_{x_j}) Q_{x_j}/Q_{x_i} < A(Q_{x_i}/Q_{x_j}) < 1.$$

With $p_{x_j,x_i} > 0$ and $A_{x_j,x_i} < 1$, one can show

$$\begin{aligned} G_{x_i,x_i} &= 1 - \sum_{x''} p_{x'',x_i} A_{x'',x_i} = 1 - \sum_{x'' \neq x_j,x_i} p_{x'',x_i} A_{x'',x_i} - p_{x_j,x_i} A_{x_j,x_i} \\ &> 1 - \sum_{x'' \neq x_j,x_i} p_{x'',x_i} - p_{x_j,x_i} = 1 - \sum_{x''} p_{x'',x_i} = 0. \end{aligned}$$

Hence there exists a state x which gives $G_{x,x} > 0$. Now suppose $G_{x',x'} > 0$. For an arbitrary x, by the irreducibility of G, there exists a positive integer s

and s' which yield $[G^{s'}]_{x',x} > 0$ and $[G^{s'}]_{x,x'} > 0$. It follows that $[G^{s+s'}]_{x,x} > [G^{s'}]_{x,x'}[G^{s}]_{x',x} > 0$ and $[G^{s+s'+1}]_{x,x} > [G^{s'}]_{x,x'}G_{x',x'}[G^{s}]_{x',x} > 0$. This shows the aperiodicity of G. To show the existence of an equilibrium distribution and the detailed balance, we call for $A(Q_{x'}/Q_x)Q_x = A(Q_x/Q_{x'})Q_{x'}$ and $p_{x,x'} = p_{x',x}$. Using these properties, one can easily show (8.A.27) and (8.A.26).

8.A.3.3 Time-Dependent Transition Probability

In this subsection, we discuss the convergence of a Markov chain generated by a time-dependent transition probability.

Suppose that $P_x(t)$ represents a probability distribution of the state at time t. Note that $P_x(t)$ is identical to $P(X_t = x)$. When $G_{x',x}(t)$ depends on t, the Chapman-Kolmogorov equation (8.A.20) leads to the distribution $P_x(t')$ at t' from a given distribution at t:

$$
\begin{aligned}
P_{x'}(t') &= \sum_x G_{x',x}(t',t)P_x(t) \\
&= \sum_{x,x_1,x_2,\ldots,x_{s-1}} G_{x',x_{s-1}}(t'-1)\cdots G_{x_2,x_1}(t+1)G_{x_1,x}(t)P_x(t).
\end{aligned}
\tag{8.A.30}
$$

This can be written using the matrix representation as

$$
P(t') = G(t'-1)\cdots G(t+1)G(t)P(t). \tag{8.A.31}
$$

For this Markov chain with the time-dependent transition probability $G_{x',x}(t)$, theorems on the condition for the weak and strong ergodicities are known. They are called the weak ergodic theorem and the strong ergodic theorem respectively. We first prove the weak ergodic theorem, and then move on to the strong ergodic theorem.

To make a statement of the weak ergodic theorem, we define the ergodic coefficient by

$$
\alpha\big(G(t_j,t_i)\big) = \max_{x'',x'}\left[\sum_x \max\big\{0, G_{x,x''}(t_j,t_i) - G_{x,x'}(t_j,t_i)\big\}\right]. \tag{8.A.32}
$$

We note that there are several expressions for the ergodic coefficient α. To see this, we define the notations $(\cdot)_+$ and $(\cdot)_-$ by

$$
(A)_+ = \begin{cases} A & (A \geq 0) \\ 0 & (A < 0), \end{cases}
$$

$$
(A)_- = \begin{cases} 0 & (A \geq 0) \\ A & (A < 0). \end{cases}
$$

We next define $\mathscr{S}_+(G, x', x'')$ and $\mathscr{S}_-(G, x', x'')$ by the sets of x which yields $G_{x,x''} - G_{x,x'} \geq 0$ and $G_{x,x''} - G_{x,x'} < 0$ respectively. Note that

$$\sum_x \max\{0, G_{x,x''} - G_{x,x'}\} = \sum_x (G_{x,x''} - G_{x,x'})_+ = \sum_{x \in \mathscr{S}_+(G,x',x'')} (G_{x,x''} - G_{x,x'}). \quad (8.A.33)$$

Recall here $\sum_x G_{x,x''} = \sum_x G_{x,x'} = 1$. Then one can write

$$\sum_x \max\{0, G_{x,x''} - G_{x,x'}\} = \sum_x (G_{x,x''} - G_{x,x'})_+ - \frac{1}{2}\sum_x (G_{x,x''} - G_{x,x'}).$$

The half of the first sum in the right hand side is cancelled by the second sum over $x \in \mathscr{S}_+(G, x', x'')$. Moreover, the second sum over $x \in \mathscr{S}_-(G, x', x'')$ is identical to $\sum_{x \in \mathscr{S}_-(G,x',x'')} |G_{x,x''} - G_{x,x'}|$. Hence one obtains

$$\sum_x \max\{0, G_{x,x''} - G_{x,x'}\} = \frac{1}{2}\sum_x |G_{x,x''} - G_{x,x'}|.$$

Therefore (8.A.32) can be written as

$$\alpha\big(G(t', t)\big) = \max_{x'',x'} \sum_x \big(G_{x,x''}(t', t) - G_{x,x'}(t', t)\big)_+, \quad (8.A.34)$$

$$\alpha\big(G(t', t)\big) = \frac{1}{2} \max_{x'',x'} \left[\sum_x \big|G_{x,x''}(t', t) - G_{x,x'}(t', t)\big|\right]. \quad (8.A.35)$$

Next, observe

$$\sum_x \max\{0, G_{x,x''} - G_{x,x'}\} = \sum_x \big(G_{x,x''} - \min\{G_{x,x''}, G_{x,x'}\}\big) = 1 - \sum_x \min\{G_{x,x''}, G_{x,x'}\}.$$

Due to the non-negativity of G, the second term yields

$$\max_{x'',x'} \left[-\sum_x \min\{G_{x,x''}, G_{x,x'}\}\right] = -\min_{x'',x'} \sum_x \min\{G_{x,x''}, G_{x,x'}\}.$$

Thus another expression of α is obtained as

$$\alpha\big(G(t', t)\big) = 1 - \min_{x'',x'} \sum_x \min\big\{G_{x,x''}(t', t), G_{x,x'}(t', t)\big\}. \quad (8.A.36)$$

In order to simplify expressions in the following argument, we employ the notation for the norm of vector: $\|v\| = \sum_x |v_x|$. Note that $u_x(t) \to v_x$ if $\|u(t) - v\| \to 0$ $(t \to \infty)$.

Now the weak ergodic theorem is stated as follows.

Theorem 8.11 (Condition for the weak ergodicity) *Consider an ascending sequence of time $t_1, t_2, \ldots$, where $t_i < t_j$ for $i < j$. If and only if*

$$\alpha\big(G(t+s), t\big) \stackrel{s\to\infty}{\longrightarrow} 0 \quad \textit{for any } t, \tag{8.A.37}$$

then

$$\max_{P_0, Q_0} \big\| G(t+s,t)P_0 - G(t+s,t)Q_0 \big\| \stackrel{s\to\infty}{\longrightarrow} 0 \quad \textit{for any } t. \tag{8.A.38}$$

Equation (8.A.38) is called the weak ergodicity. The weak ergodicity implies that the distribution given by a Markov chain becomes independent of the initial distribution in the infinite limit of time.

Proof We employ the expression (8.A.35) of α, dropping arguments (t', t) to make formulas concise. Using the Kronecker's δ, (8.A.35) is written as

$$\alpha(G) = \frac{1}{2} \max_{x'', x'} \left[\sum_x \left| \sum_y G_{x,y}(\delta_{y,x''} - \delta_{y,x'}) \right| \right].$$

Let us here consider $\sum_x |\sum_y G_{x,y}(P_y - P'_y)|$, where P and P' are some probability distributions. Using the notations, $(\cdot)_+$ and $(\cdot)_-$, one can write

$$\begin{aligned} &\sum_x \left| \sum_y G_{x,y}(P_y - P'_y) \right| \\ &\quad = \sum_x \sum_y \big(G_{x,y}(P_y - P'_y)\big)_+ - \sum_x \sum_y \big(G_{x,y}(P_y - P'_y)\big)_- . \end{aligned} \tag{8.A.39}$$

Noting that

$$\begin{aligned} &\sum_x \sum_y G_{x,y}(P_y - P'_y) \\ &\quad = \sum_x \left(\sum_y G_{x,y}(P_y - P'_y) \right)_+ + \sum_x \left(\sum_y G_{x,y}(P_y - P'_y) \right)_- = 0, \end{aligned}$$

Eq. (8.A.39) is arranged as

$$\sum_x \left| \sum_y G_{x,y}(P_y - P'_y) \right| = 2 \sum_x \left(\sum_y G_{x,y}(P_y - P'_y) \right)_+ . \tag{8.A.40}$$

Suppose that $\tilde{P}$ and $\tilde{P}'$ are the probability distributions that maximise (8.A.39), namely,

$$\max_{P, P'} \left\{ \sum_x \left| \sum_y G_{x,y}(P_y - P'_y) \right| \right\} = \sum_x \left| \sum_y G_{x,y}(\tilde{P}_y - \tilde{P}'_y) \right|, \tag{8.A.41}$$

or

$$\max_{P, P'} \left\{ \sum_x \left(\sum_y G_{x,y}(P_y - P'_y) \right)_+ \right\} = \left\{ \sum_x \left(\sum_y G_{x,y}(\tilde{P}_y - \tilde{P}'_y) \right)_+ \right\}. \tag{8.A.42}$$

We here define $\mathscr{S}_+(G; \tilde{P}, \tilde{P}')$ as the set of x that yields $\sum_y G_{x,y}(\tilde{P}_y - \tilde{P}'_y) \geq 0$. Notice that

$$\sum_{x \in \mathscr{S}_+(G;\tilde{P},\tilde{P}')} \sum_y G_{x,y}(\tilde{P}_y - \tilde{P}'_y) = \max_{P,P'} \left\{ \sum_{x \in \mathscr{S}_+(G;\tilde{P},\tilde{P}')} \sum_y G_{x,y}(P_y - P'_y) \right\}. \tag{8.A.43}$$

Equations (8.A.42) and (8.A.43) lead to

$$\begin{aligned} &\max_{P,P'} \left\{ \sum_x \left(\sum_y G_{x,y}(P_y - P'_y) \right)_+ \right\} \\ &= \max_{P,P'} \left\{ \sum_{x \in \mathscr{S}_+(G;\tilde{P},\tilde{P}')} \sum_y G_{x,y}(P_y - P'_y) \right\} \\ &= \max_P \left\{ \sum_y \left(\sum_{x \in \mathscr{S}_+(G;\tilde{P},\tilde{P}')} G_{x,y} \right) P_y \right\} - \min_P \left\{ \sum_y \left(\sum_{x \in \mathscr{S}_+(G;\tilde{P},\tilde{P}')} G_{x,y} \right) P_y \right\}. \end{aligned} \tag{8.A.44}$$

We define $y_{\max}$ so as to maximise $\sum_{x \in \mathscr{S}_+(G;\tilde{P},\tilde{P}')} G_{x,y}$ and $y_{\min}$ to minimise it. Then one finds that the maximisation of the first term of (8.A.44) is achieved when $P_y = \tilde{P}_y = \delta_{y,y_{\max}}$. Also one finds that the minimisation of the second term is done when $P_y = \tilde{P}'_y = \delta_{y,y_{\min}}$. From this with (8.A.40) and (8.A.44), one obtains

$$\frac{1}{2} \max_{P,P'} \left\{ \sum_x \left| \sum_y G_{x,y}(P_y - P'_y) \right| \right\} = \sum_{x \in \mathscr{S}_+(G;y_{\max},y_{\min})} (G_{x,y_{\max}} - G_{x,y_{\min}}), \tag{8.A.45}$$

where $\mathscr{S}_+(G; y_{\max}, y_{\min})$ is the set of x that yields $G_{x,y_{\max}} - G_{x,y_{\min}} \geq 0$. By the definition of $y_{\max}$ and $y_{\min}$, one has

$$G_{x,y_{\max}} - G_{x,y_{\min}} = \max_{x'',x'} \{ G_{x,x''} - G_{x,x'} \},$$

and

$$\sum_{x \in \mathscr{S}_+(G;y_{\max},y_{\min})} \max_{x'',x'} \{ G_{x,x''} - G_{x,x'} \} = \max_{x'',x'} \sum_{x \in \mathscr{S}_+(G;x',x'')} \{ G_{x,x''} - G_{x,x'} \}.$$

Therefore one reaches

$$\frac{1}{2} \max_{P,P'} \left\{ \sum_x \left| \sum_y G_{x,y}(P_y - P'_y) \right| \right\} = \max_{x'',x'} \left\{ \sum_x (G_{x,x''} - G_{x,x'})_+ \right\} = \alpha(G), \tag{8.A.46}$$

where we used the expression (8.A.34). Thus it is shown that, if and only if $\alpha(G) \to 0$, one has $\max_{P,P'} \|GP - GP'\| \to 0$. □

The weak ergodic Theorem 8.11 is followed by the next corollary.

Corollary 8.2 *The necessary and sufficient condition for the weak ergodicity is given by*

$$\sum_{i=1}^{\infty}\{1-\alpha(G(t_{i+1},t_i))\}=\infty. \tag{8.A.47}$$

Proof Assume first that there is an ascending sequence of time, $t_1, t_2, \dots$, for which one has

$$\sum_{i=1}^{\infty}\{1-\alpha(G(t_{i+1},t_i))\}=\infty. \tag{8.A.48}$$

Let us observe that, if $\sum_{i=1}^{\infty} a_i = \infty$ for $0 \le a_i \le 1$, then one has $\prod_{i=1}^{\infty}(1-a_i)=0$. This is verified by noting that $0 \le \prod_{i=1}^{\infty}(1-a_i) \le \prod_{i=1}^{\infty} e^{-a_i} = \exp(-\sum_{i=1}^{\infty} a_i)$. Using this, it is shown that (8.A.48) yields

$$\prod_{i=1}^{\infty}\alpha(G(t_{i+1},t_i))=0. \tag{8.A.49}$$

Now we look into $\alpha(G(t_{i+2},t_i))$. By the Chapman-Kolmogorov equation (8.A.20), one has $G(t_{i+2},t_i)=G(t_{i+2},t_{i+1})G(t_{i+1},t_i)$. Hereafter we use the shorthand notation: $G=G(t_{i+2},t_i)$, $G'=G(t_{i+1},t_i)$, and $G''=G(t_{i+2},t_{i+1})$. Observe

$$G_{x,x''}-G_{x,x'}=\sum_y\{G''_{x,y}(G'_{y,x''}-G'_{y,x'})_+ + G''_{x,y}(G'_{y,x''}-G'_{y,x'})_-\}.$$

It follows that

$$\begin{aligned}\sum_x(G_{x,x''}-G_{x,x'})_+ &= \sum_{x\in\mathscr{S}_+(G,x',x'')}(G_{x,x''}-G_{x,x'})\\ &=\sum_y\Bigl(\sum_{x\in\mathscr{S}_+(G,x',x'')}G''_{x,y}\Bigr)\{(G'_{y,x''}-G'_{y,x'})_+\\ &\quad+(G'_{y,x''}-G'_{y,x'})_-\}.\end{aligned} \tag{8.A.50}$$

Note that the second term in the braces are negative. Hence one can bound (8.A.50) by maximising $\sum_{x\in\mathscr{S}_+(G;x',x'')}G''_{x,y}$ with respect to y on the first term and by minimising it on the second term.

$$\begin{aligned}\sum_x(G_{x,x''}-G_{x,x'})_+ &\le \sum_y\Bigl[\max_{y'}\Bigl\{\sum_{x\in\mathscr{S}_+(G,x',x'')}G''_{x,y'}\Bigr\}(G'_{y,x''}-G'_{y,x'})_+\\ &\quad+\min_{y'}\Bigl\{\sum_{x\in\mathscr{S}_+(G,x',x'')}G''_{x,y'}\Bigr\}(G'_{y,x''}-G'_{y,x'})_-\Bigr]\\ &=\max_{y'',y'}\Bigl\{\sum_{x\in\mathscr{S}_+(G,x',x'')}(G''_{x,y''}-G''_{x,y'})\Bigr\}\sum_y(G'_{y,x''}-G'_{y,x'})_+,\end{aligned}$$

where we used $\sum_y (G_{y,x''} - G_{y,x'})_- = -\sum_y (G_{y,x''} - G_{y,x'})_+$. Since

$$\sum_{x\in\mathscr{S}_+(G;x',x'')} \left(G''_{x,y''} - G''_{x,y'}\right) \le \sum x \in \mathscr{S}_+\left(G''; y', y''\right)\left(G''_{x,y''} - G''_{x,y'}\right),$$

one obtains

$$\begin{aligned}\sum_x (G_{x,x''} - G_{x,x'})_+ &\le \max_{y'',y'} \left\{ \sum_{x\in\mathscr{S}_+(G'',y',y'')} \left(G''_{x,y''} - G''_{x,y'}\right)\right\} \sum_y \left(G'_{y,x''} - G'_{y,x'}\right)_+ \\ &= \alpha\left(G''\right) \sum_y \left(G'_{y,x''} - G'_{y,x'}\right)_+. \end{aligned} \tag{8.A.51}$$

Taking the maximum over x'' and x', one reaches

$$\alpha\left(G(t_{i+2}, t_i)\right) \le \alpha\left(G(t_{i+2}, t_{i+1})\right)\alpha\left(G(t_{i+1}, t_i)\right). \tag{8.A.52}$$

With (8.A.49) and (8.A.52), it is shown that

$$\lim_{t'\to\infty} \alpha\left(G\left(t', t_1\right)\right) \le \prod_{i=1}^{\infty} \alpha\left(G(t_{i+1}, t_i)\right) = 0. \tag{8.A.53}$$

Next, we assume that

$$\alpha\left(G\left(t', t\right)\right) \overset{t'\to\infty}{\longrightarrow} 0 \quad \text{for any } t. \tag{8.A.54}$$

This is followed by

$$1 - \alpha\left(G\left(t', t\right)\right) \overset{t'\to\infty}{\longrightarrow} 1 \quad \text{for any } t. \tag{8.A.55}$$

One finds from this that for a fixed t_1, there exists t_2 such that $t_2 > t_1$ and $1 - \alpha(G(t_2, t_1)) > \frac{1}{2}$. In the same way, one can choose $t_3, t_4, \dots, t_n$ such that $t_{i+1} > t_i$ and $1 - \alpha(G(t_{i+1}, t_i)) > \frac{1}{2}$. Then, with an arbitrary positive integer n, one has

$$\sum_{i=1}^{n} \left\{1 - \alpha\left(G(t_{i+1}, t_i)\right)\right\} > \frac{1}{2}n. \tag{8.A.56}$$

Taking the limit of $n \to \infty$, one obtains

$$\sum_{i=1}^{\infty} \left\{1 - \alpha\left(G(t_{i+1}, t_i)\right)\right\} = \infty. \tag{8.A.57}$$

Thus the Corollary 8.2 is proved. □

We finally move on to the strong ergodic theorem.

Theorem 8.12 (Conditions for the strong ergodicity) *Assume the following conditions*:

(i) *The Markov chain is weakly ergodic.*
(ii) *For any t, there exists an equilibrium probability distribution $Q(t)$ such that $Q(t) = G(t)Q(t)$.*

(iii) *The equilibrium distribution* $Q(t)$ *satisfies*

$$\sum_{t=0}^{\infty} \left\| Q(t+1) - Q(t) \right\| < \infty. \tag{8.A.58}$$

Then the probability distribution $P(t)$ *generated by the transition matrix* $G(t) = G(t-1) \cdots G(1)G(0)$ *from an arbitrary initial distribution* P_0 *converges to a unique distribution* $Q^* = \lim_{t\to\infty} Q(t)$ *for* $t \to \infty$, *namely,* $P(t)$ *satisfies*

$$\lim_{t\to\infty} \left\| P(t) - Q^* \right\| = \lim_{t\to\infty} \left\| G(t,0)P_0 - Q^* \right\| = 0. \tag{8.A.59}$$

Proof Because of the condition (i), $P(t) = G(t,0)P_0$ becomes independent of P_0 for $t \to \infty$. Hence one may choose $Q(0)$ as P_0. Define $R(s)$ by

$$R(s) = G(s)Q(s-1) - Q(s). \tag{8.A.60}$$

Noting $G(0)Q(0) = Q(0)$, one can write

$$\begin{aligned} P(t) &= G(t,0)Q(0) = G(t-1)\cdots G(2)G(1)Q(0) \\ &= G(t-1)\cdots G(2)\big(Q(1) + R(1)\big) \\ &= Q(t-1) + R(t-1) + G(t,t-1)R(t-2) + \cdots + G(t,2)R(1). \end{aligned} \tag{8.A.61}$$

Consider here $\|P(t) - Q^*\|$. For this quantity, one has

$$\begin{aligned} \left\| P(t) - Q^* \right\| &\le \left\| P(t) - Q(t-1) \right\| + \left\| Q(t-1) - Q^* \right\| \\ &\le \left\| P(t) - G(t,u)Q(u) \right\| + \left\| G(t,u)Q(u) - Q(t-1) \right\| \\ &\quad + \left\| Q(t-1) - Q^* \right\|, \end{aligned} \tag{8.A.62}$$

where we supposed $0 < u < t$. The first term is estimated as follows:

$$\begin{aligned} \left\| P(t) - G(t,u)Q(u) \right\| &= \left\| G(t,u)G(u,0)Q(0) - G(t,u)Q(u) \right\| \\ &= \left\| G(t,u)\big(G(u,0)Q(0) - Q(u)\big) \right\|. \end{aligned}$$

Due to the weak ergodic theorem, for any given positive number ε one can choose t_1 such that

$$\left\| G(t,u)\big(G(u,0)Q(0) - Q(u)\big) \right\| < \varepsilon \quad \text{for any } t > t_1 \text{ and } u < t,$$

namely,

$$\left\| P(t) - G(t,u)Q(u) \right\| = \varepsilon \quad \text{for any } t > t_1 \text{ and } u < t. \tag{8.A.63}$$

Next, the second term of (8.A.62) is estimated as follows:

$$\left\| G(t,u)Q(u) - Q(t-1) \right\| = \left\| \sum_{s=u+1}^{t-1} G(t,s+1)R(s) \right\| \le \sum_{s=u+1}^{t-1} \left\| G(t,s+1)R(s) \right\|.$$

Using (8.A.60), it follows that

$$\begin{aligned}\left\|G(t,u)Q(u)-Q(t-1)\right\| &\le \sum_{s=u+1}^{t-1}\left\|G(t,s+1)\big(G(s)Q(s-1)-Q(s)\big)\right\|\\ &= \sum_{s=u+1}^{t-1}\left\|G(t,s)\big(Q(s-1)-Q(s)\big)\right\|.\end{aligned}$$

Note that, since $G_{x,y} \ge 0$ and $\sum_x G_{x,y} = 1$, one has $\|GA\| = \sum_x |\sum_y G_{x,y}A_y| \le \sum_x \sum_y G_{x,y}|A_y| = \|A\|$ with an arbitrary vector **A**. Hence one obtains

$$\left\|G(t,u)Q(u)-Q(t-1)\right\| \le \sum_{s=u+1}^{t-1}\left\|Q(s-1)-Q(s)\right\|.$$

Due to the condition (iii), for given ε one can choose u_1 such that $u_1 \ge t_1$ and

$$\left\|G(t,u_1)Q(u_1)-Q(t-1)\right\| \le \sum_{s=u_1+1}^{t-1}\left\|Q(s-1)-Q(s)\right\| < \varepsilon \quad \text{for any } t > u_1+1. \tag{8.A.64}$$

Finally, regarding the third term of (8.A.62), the condition (iii) ensures us that there is a positive integer t_2 for given ε such that $t_2 > u_1 + 1$ and

$$\left\|Q(t-1)-Q^*\right\| < \varepsilon \quad \text{for any } t > t_2. \tag{8.A.65}$$

Thus it is shown that, for an arbitrary $\varepsilon > 0$, there exists a positive integer t_2 and for $t > t_2$ one has

$$\left\|P(t)-Q^*\right\| < 3\varepsilon. \tag{8.A.66}$$

Therefore the strong ergodic Theorem 8.12 is proved. □

Chapter 9
Applications

In the previous chapters, we reviewed several models to explain physical phenomena in which the quantum fluctuation as a transverse field plays important role, especially quantum phase transition in magnetic systems. However, as we saw in the chapter of quantum annealing, transverse Ising model and its variants are useful for various research fields. In this chapter, we introduce several applications of transverse Ising model to the outside area of physics, namely, brain science and information science and technology.

9.1 Hopfield Model in a Transverse Field

Since J.J. Hopfield pointed out, we know that associative memories in artificial neural networks are described by a kind of spin glasses of the Sherrington-Kirkpatrick type. In this section, we explain how we extend the model to the quantum-mechanical variant in terms of the transverse field.

The Hopfield model of neural networks employs the two state (classical) Ising spins $S_i^z = \pm 1$ to represent the McCulloch-Pitts (formal) neurons. The synaptic connections between the neurons are represented by the spin-spin interaction J_{ij}, which are taken to be symmetric. The symmetry of J_{ij} allows one to define an energy function or Hamiltonian for the network. The symmetric connections are constructed following Hebb's rule of learning, which says that for P random (orthogonal) patterns, the synaptic strength J_{ij} for the pair of neurons (i, j) is given by

$$J_{ij} = \frac{1}{N} \sum_{\mu=1}^{P} \xi_i^{\mu} \xi_j^{\mu} \tag{9.1.1}$$

where $\{\xi_i^{\mu}\}$, $i = 1, 2, \ldots, N$, represents the μ-th (random) pattern to be learned by the network. Each ξ_i^{μ} can take values ± 1 and the randomness of the patterns demand $(1/N) \sum_i \xi_i^{\mu} \xi_i^{\nu} = \delta_{\mu\nu}$. N is the total number of neurons (Ising spins) in the

S. Suzuki et al., *Quantum Ising Phases and Transitions in Transverse Ising Models*,
Lecture Notes in Physics 862, DOI 10.1007/978-3-642-33039-1_9,

network; each connected to all others in the network through J_{ij}'s. The Hamiltonian is then defined as

$$H = -\sum_{i>j}^{N} J_{ij} S_i^z S_j^z. \tag{9.1.2}$$

Strictly speaking, about 10 years before of the Hopfield's work, Kaoru Nakano [293] who is a Japanese engineer already proposed almost the same mathematical model as that by Hopfield. However, in his paper, the concept of energy function such as (9.1.2) was not mentioned. As the result, his paper was not drawn attention to physicists.

The idea of Hopfield is that the above choice of J_{ij}'s (9.1.2) will make the energy corresponding to the learned patterns local minima in the free energy landscape. An initial configuration close to a learned pattern will therefore be attracted towards a learned pattern through any energy minimising dynamics. Any pattern evolves following the (zero-temperature Monte Carlo) dynamics

$$S_i^z(t+1) = \mathrm{sgn}\big(h_i(t)\big), \tag{9.1.3}$$

where h_i is the internal field (the active potential) on the neuron i, given by

$$h_i(t) = \sum_{j=1}^{N} J_{ij} S_j^z(t). \tag{9.1.4}$$

Here a fixed point of dynamics or attractor is guaranteed (Hopfield [178]): after a certain number of iterations t^*, the network stabilises and $S_i^z(t^*) = S_i^z(t^* + 1)$.

9.1.1 Statics and Phase Diagrams

Using replica symmetric (mean field) theory for such long range systems (as for the Sherrington-Kirkpatrick model of spin glasses, Amit, Sompolinsky and Gutfreund [13] obtained the following equations for the finite temperature overlap function $m_\mu = (1/N)\sum_i S_i^z(t^*)\xi_i^\mu$ (where t^* refers to the equilibrium state at finite β and to fixed points for $t^* \to \infty$):

$$m_\mu = \int_{-\infty}^{\infty} Dr \tanh\big[\beta(m_\mu + \sqrt{\alpha\tilde{q}}\, r)\big], \tag{9.1.5}$$

$$q = \int_{-\infty}^{\infty} Dr \tanh^2\big[\beta(m_\mu + \sqrt{\alpha\tilde{q}}\, r)\big], \tag{9.1.6}$$

$$\tilde{q} = \frac{q}{[1 - \beta(1-q)]^2}, \tag{9.1.7}$$

where we defined $Dr \equiv dr\, \mathrm{e}^{-r^2/2}/\sqrt{2\pi}$ and $\alpha \equiv P/N$. A self-consistent numerical solution of the equation shows that, at zero temperature ($\beta \to \infty$), the local minima for H in (9.1.2) indeed correspond to the patterns fed to be learned (i.e., $m_\mu = 1$),

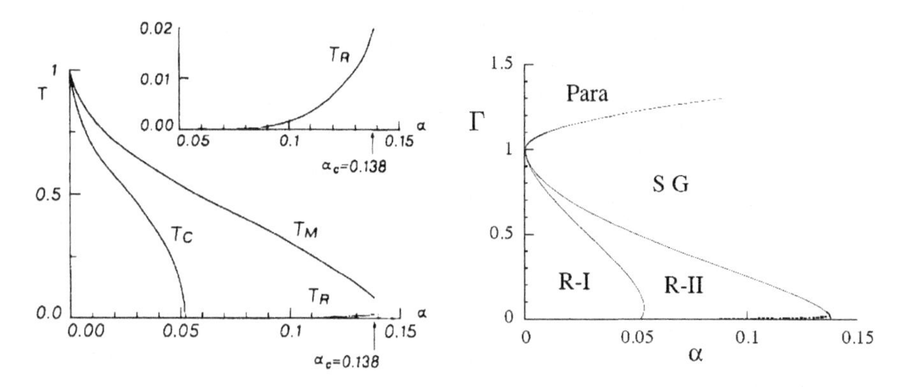

Fig. 9.1 The phase diagram of the classical Hopfield model (*left*: from [13]). *Right panel* is phase diagram for the Hopfield model in a transverse field (from [298]). *R-I* in the *right panel* corresponds to the area where the ferromagnetic memory state is a global minimum, whereas *R-II* denotes the area in which the memory state is a local minimum of the free energy. The areas $0 \leq T \leq T_C$ and $T_C < T \leq T_M$ in the *left panel* correspond to the regions *R-I* and *R-II* in the *right panel*, respectively. The T_R in the *inset* of the *left panel* stands for the de Almeida-Thouless (AT) line [97] and the corresponding line in the *right panel* is shown as a *broken line*

in the limit when memory the loading factor $\alpha(= P/N)$ tends to zero; and they are about 3 % off ($m_\mu \simeq 0.97$) when α is finite but $\alpha \leq \alpha_c \simeq 0.138$ [13] (see the phase diagram in Fig. 9.1 (left)). Above this loading capacity, the network goes to a confused state where the local minima in the energy landscape do not have significant overlap with the patterns fed to be learned. These results are also checked in detailed numerical studies, where the dynamics of relaxation of the distorted patterns, measured quantitatively by the relaxation time t^*, can he studied additionally. These relaxation studies for the Hopfield model show a novel critical slowing down or divergence of relaxation time t^* (Chakrabarti et al. [64]) near the critical loading capacity α_c, where the metastable (memory) overlap states disappear (but no real phase transition occurs in the Hopfield model for which the ground states correspond to spin glass states already at $\alpha > 0.05$).

In order to study the effect of quantum fluctuations on such associative memory neural network models (in connection with the suggestions of nontrivial quantum coherence effects in the brain [307, 414]), one can study the various phases of the Hopfield model in a transverse field [250, 251, 253]. One therefore considers here the Hamiltonian

$$H = -\sum_{i>j}^{N} J_{ij} S_i^z S_j^z - \Gamma \sum_i S_i^x; \quad J_{ij} = \frac{1}{N} \sum_{\mu=1}^{P} \xi_i^\mu \xi_j^\mu, \tag{9.1.8}$$

where ξ_i^μ's represent the same $P = \alpha N$ (random) patterns. The treatment by Ma et al. [250, 251, 253] is based on the so-called 'naive mean field method', and it could not mention the de Almeida-Thouless line [97] at which the replica symmetric solution becomes unstable. Nishimori and Nonomura [298] solve the problem by using a standard approach based on the replica method. Following again the mean field theory of transverse Ising Sherrington-Kirkpatrick model [369], they can generate the self-consistent equations as

$$m = \int Dz Y^{-1} \int Dw\, g u^{-1} \sinh \beta u \tag{9.1.9}$$

$$r = \frac{q}{(1 - \beta \overline{S} + \beta q)^2} \tag{9.1.10}$$

$$t = r + \frac{\overline{S} - q}{1 - \beta \overline{S} + \beta q} \tag{9.1.11}$$

$$q = \int Dz \left(Y^{-1} \int Dw\, q u^{-1} \sinh \beta u \right)^2 \tag{9.1.12}$$

$$S = \int Dz\, Y^{-1} \left(\int Dw\, q^2 u^{-1} \cosh \beta u + T\Gamma \int Dw\, u^{-1} \sinh \beta u \right) \tag{9.1.13}$$

where we defined

$$g = m + \sqrt{\alpha r}\, z + \sqrt{\alpha(t - r)}\, w \tag{9.1.14}$$

$$u = \sqrt{g^2 + \Gamma^2}, \qquad Y = \int Dw \cosh \beta u \tag{9.1.15}$$

where we set $m_\mu = \delta_{\mu,1} m$, namely, we consider the case in which a specific pattern $\xi_1 = (1, \ldots, 1)$ is recalled. The detail of the derivation is given in Appendix 9.A.

We draw the phase diagram in Fig. 9.1 (right). R-I in the panel corresponds to the area where the ferromagnetic memory state is a global minimum, whereas R-II denotes the area in which the memory state is a local minimum of the free energy. The areas $0 \le T \le T_C$ and $T_C < T \le T_M$ in the left panel correspond to the regions R-I and R-II in the right panel, respectively. It appears that the quantum fluctuations destroy the overlap states and the critical loading capacity α_c ($\simeq 0.138$ for $T = \Gamma = 0$) decreases continually with Γ ($\alpha_c = 0$ for $\Gamma = 1$).

9.1.2 Pattern-Recalling Processes

In the above subsections, we discussed equilibrium properties of quantum Hopfield model. However, the dynamical properties of the system, especially, the pattern-recalling process also have a rich behaviour. In the following subsections, we argue such dynamics of the quantum Hopfield model in which of order 1 patterns are embedded [192].

9.1.2.1 The Classical System

We first revisit the conventional Hopfield model described by classical Ising spin. Let us consider the network having N-neurons. Each neuron S_i takes two states, namely, $S_i = +1$ (fire) and $S_i = -1$ (stationary). Neuronal states are given by the set of variables S_i, that is, $\boldsymbol{S} = (S_1, \dots, S_N)$, $S_i \in \{+1, -1\}$. Each neuron is located on a complete graph, namely, graph topology of the network is 'fully-connected'. The synaptic connection between arbitrary two neurons, say, S_i and S_j is defined by the following Hebb rule:

$$J_{ij} = \frac{1}{N} \sum_{\mu,\nu} \xi_i^{\mu} A_{\mu\nu} \xi_j^{\nu} \tag{9.1.16}$$

where $\boldsymbol{\xi}^{\mu} = (\xi_1, \dots, \xi_N)$, $\xi_i^{\mu} \in \{+1, -1\}$ denote the embedded patterns and each of them is specified by a label $\mu = 1, \dots, P$. $A_{\mu\nu}$ denotes $(P \times P)$-size matrix and P stands for the number of built-in patterns. We should keep in mind that there exists an energy function (a Lyapunov function) in the system if the matrix $A_{\mu\nu}$ is symmetric.

Then, the output of the neuron i, that is, S_i is determined by the sign of the local field h_i as

$$h_i = \sum_{\mu,\nu=1}^{p} \xi_i^{\mu} A_{\mu\nu} m^{\nu} + \frac{1}{N} \sum_{a,b=p+1}^{P} \xi_i^{a} A_{ab} \sum_j \xi_j^{\nu'} S_j \tag{9.1.17}$$

where $A_{\mu\nu}$ and A_{ab} are elements of $p \times p$, $(P-p) \times (P-p)$-size matrices, respectively. We also defined the overlap (the direction cosine) between the state of neurons $\boldsymbol{S}$ and one of the built-in patterns $\{\xi_i^{\mu}\}$ by

$$m^{\nu} \equiv \frac{1}{N} \sum_i \xi_i^{\nu} S_i. \tag{9.1.18}$$

Here we should notice that the Hamiltonian of the system is given by $-\sum_i h_i S_i$. The first term appearing in the left hand side of Eq. (9.1.17) is a contribution from $p \sim \mathscr{O}(1)$ what we call 'condensed patterns', whereas the second term stands for the so-called 'cross-talk noise'.

To evaluate the cross-talk noise, let us first consider the case in which the second term is negligibly small in comparison with the first term, namely, the case of $P = p \sim \mathscr{O}(1)$. Then, the cross-talk noise is evaluated as $\sqrt{pN}/N \sim 1/\sqrt{N} = 0$ in the limit of $N \to \infty$. In this sense, we can say that the network is 'far from its saturation'. On the other hand, for the case of $P = \mathscr{O}(N)$, $p = \mathscr{O}(1)$, one can evaluate the cross-talk as $\sqrt{pN}/N \sim \mathscr{O}(1)$, which is the same order as the signal. In this case, we can say that the network is 'near saturation'

9.1.2.2 The Quantum System

To extend the classical system to the quantum-mechanical variant, we rewrite the local field h_i as follows

$$\boldsymbol{\phi}_i = \sum_{\mu,\nu=1}^{p} \xi_i^{\mu} A_{\mu\nu} \left(\frac{1}{N} \sum_i \xi_i^{\nu} \boldsymbol{S}_i^{z} \right) \tag{9.1.19}$$

where S_i^z $(i = 1, \ldots, N)$ stands for the z-component of the Pauli matrix. Thus, the Hamiltonian $\boldsymbol{H}_0 \equiv -\sum_i \phi_i S_i^z$ is a diagonalised $(2^N \times 2^N)$-size matrix and the lowest eigenvalue is identical to the ground state of the classical Hamiltonian $-\sum_i \phi_i S_i$ (S_i is an eigenvalue of the matrix S_i^z).

Then, we introduce quantum-mechanical noise into the Hopfield neural network by adding the transverse field to the Hamiltonian as follows

$$\boldsymbol{H} = \boldsymbol{H}_0 - \Gamma \sum_{i=1}^{N} S_i^x \tag{9.1.20}$$

where S_i^x is the x-component of the Pauli matrix and transitions between eigenvectors of the classical Hamiltonian $\boldsymbol{H}_0$ are induced due to the off-diagonal elements of the matrix $\boldsymbol{H}$ for $\Gamma \neq 0$. In this paper, we mainly consider the system described by (9.1.20).

9.1.2.3 Quantum Monte Carlo Method

The dynamics of the quantum model (9.1.20) follows Schrödinger equation. Thus, we should solve it or investigate the time dependence of the state $|\psi(t)\rangle$ by using the time-evolutionary operator $\mathrm{e}^{-i\boldsymbol{H}\Delta t/\hbar}$ defined for infinitesimal time Δt as

$$\big|\psi(t + \Delta t)\big\rangle = \mathrm{e}^{-i\boldsymbol{H}\Delta t/\hbar}\big|\psi(t)\big\rangle. \tag{9.1.21}$$

However, even if we carry it out numerically, it is very hard for us to do it with reliable precision because $(2^N \times 2^N)$-size Hamilton matrix becomes huge for the number of neurons $N \gg 1$ as in a realistic brain. Hence, here we use the quantum Monte Carlo method to simulate the quantum system in our personal computer and consider the stochastic processes of Glauber-type to discuss the pattern-recalling dynamics of the quantum Hopfield model.

9.1.2.4 The Suzuki-Trotter Decomposition

The difficulty to carry out algebraic calculations in the model system is due to the non-commutation operators appearing in the Hamiltonian (9.1.20), namely, $\boldsymbol{H}_0, \boldsymbol{H}_1 \equiv -\Gamma \sum_i S_i^x$. Thus, we use the following Suzuki-Trotter decomposition [386] in order to deal with the system as a classical spin system

$$\mathrm{tr}\,\mathrm{e}^{\beta(\boldsymbol{H}_0+\boldsymbol{H}_1)} = \lim_{M\to\infty} \mathrm{tr}\left(\exp\left(\frac{\beta \boldsymbol{H}_0}{M}\right)\exp\left(\frac{\beta \boldsymbol{H}_1}{M}\right)\right)^M \tag{9.1.22}$$

where β denotes the 'inverse temperature' and M is the number of the Trotter slices, for which the limit $M \to \infty$ should be taken. Thus, one can deal with d-dimensional quantum system as the corresponding $(d + 1)$-dimensional classical system.

9.1.2.5 Derivation of the Deterministic Flows

In the previous section, we mentioned that we should simulate the quantum Hopfield model by means of the quantum Monte Carlo method to reveal the quantum neuro-dynamics through the time-dependence of the macroscopic quantities such as the overlap. However, in general, it is also very difficult to simulate the quantum-mechanical properties at the ground state by a personal computer even for finite size systems ($N, M < \infty$).

With this fact in mind, in this section, we attempt to derive the macroscopic flow equations from the microscopic master equation for the classical system regarded as the quantum system in terms of the Suzuki-Trotter decomposition. This approach [192] is efficiently possible because the Hopfield model is a fully-connected mean-field model such as the Sherrington-Kirkpatrick model [369] for spin glasses and its equilibrium properties are completely determined by several order parameters.

9.1.2.6 The Master Equation

After the Suzuki-Trotter decomposition, we obtain the local field for the neuron i located on the k-th Trotter slice as follows

$$\begin{aligned}\beta\phi_i\big(\boldsymbol{S}_k : S_i(k \pm 1)\big) &= \frac{\beta}{M}\sum_{\mu,\nu}\xi_i^{\nu}A_{\mu\nu}\left\{\frac{1}{N}\sum_j \xi_j^{\nu}S_j(k)\right\} \\ &\quad + \frac{B}{2}\big\{S_i(k-1) + S_i(k+1)\big\}\end{aligned} \tag{9.1.23}$$

where parameter B is related to the amplitude of the transverse field (the strength of the quantum-mechanical noise) Γ by

$$B = \frac{1}{2}\log\coth\left(\frac{\beta\Gamma}{M}\right). \tag{9.1.24}$$

In the classical limit $\Gamma \to 0$, the parameter B goes to infinity. For the symmetric matrix $A_{\mu\nu}$, the Hamiltonian (scaled by β) of the system is given by $-\sum_i \beta\phi_i(\boldsymbol{S}_k : S(k \pm 1))S_i(k)$.

Then, the transition probability which specifies the Glauber dynamics of the system is given by

$$w_i(\boldsymbol{S}_k) = \frac{1}{2}\big[1 - S_i(k)\tanh\big(\beta\phi_i\big(\boldsymbol{S}_k : S(k \pm 1)\big)\big)\big]. \tag{9.1.25}$$

More explicitly, $w_i(\boldsymbol{S}_k)$ denotes the probability that an arbitrary neuron $S_i(k)$ changes its state as $S_i(k) \to -S_i(k)$ within the time unit. Therefore, the probability that the neuron $S_i(k)$ takes $+1$ is obtained by setting $S_i(k) = -1$ in the above $w_i(\boldsymbol{S}_k)$ and we immediately find $S_i(k) = S_i(k-1) = S_i(k+1)$ with probability 1 in the limit of $B \to \infty$ which implies the classical limit $\Gamma \to 0$.

Hence, the probability that a microscopic state including the M-Trotter slices $\{\boldsymbol{S}_k\} \equiv (\boldsymbol{S}_1, \ldots, \boldsymbol{S}_M)$, $\boldsymbol{S}_k \equiv (S_1(k), \ldots, S_N(k))$ obeys the following master equation:

$$\frac{dp_t(\{\boldsymbol{S}_k\})}{dt} = \sum_{k=1}^{M}\sum_{i=1}^{N}\left[p_t\left(F_i^{(k)}(\boldsymbol{S}_k)\right)w_i\left(F_i^{(k)}(\boldsymbol{S}_k)\right) - p_t(\boldsymbol{S}_k)w_i(\boldsymbol{S}_k)\right] \tag{9.1.26}$$

where $F_i^{(k)}(\cdot)$ denotes a single spin flip operator for neuron i on the Trotter slice k as $S_i(k) \to -S_i(k)$. When we pick up the overlap between neuronal state $\boldsymbol{S}_k$ and one of the built-in patterns $\boldsymbol{\xi}^\nu$, namely,

$$m_k \equiv \frac{1}{N}\left(\boldsymbol{S}_k \cdot \boldsymbol{\xi}^\nu\right) = \frac{1}{N}\sum_i \xi_i^\nu S_i(k) \tag{9.1.27}$$

as a relevant macroscopic quantity, the joint distribution of the set of the overlaps $\{m_1, \ldots, m_M\}$ at time t is written in terms of the probability for realisations of microscopic states $p_t(\{\boldsymbol{S}_k\})$ at the same time t as

$$P_t\left(m_1^\nu, \ldots, m_M^\nu\right) = \sum_{\{\boldsymbol{S}_k\}} p_t(\{\boldsymbol{S}_k\}) \prod_{k=1}^{M} \delta\left(m_k^\nu - m_k^\nu(\boldsymbol{S}_k)\right) \tag{9.1.28}$$

where we defined the sums by

$$\sum_{\{\boldsymbol{S}_k\}}(\cdots) \equiv \sum_{\boldsymbol{S}_1}\cdots\sum_{\boldsymbol{S}_M}(\cdots), \qquad \sum_{\boldsymbol{S}_k}(\cdots) \equiv \sum_{S_1(k)=\pm 1}\cdots\sum_{S_N(k)=\pm 1}(\cdots). \tag{9.1.29}$$

Taking the derivative of Eq. (9.1.28) with respect to t and substituting (9.1.26) into the result, we have the following differential equations for the joint distribution

$$\begin{aligned}
&\frac{dP_t(m_1^\nu, \ldots, m_M^\nu)}{dt} \\
&= \sum_k \frac{\partial}{\partial m_k^\nu}\left\{m_k^\nu P_t\left(m_1^\nu, \ldots, m_k^\nu, \cdots, m_M^\nu\right)\right\} \\
&\quad - \sum_k \frac{\partial}{\partial m_k^\nu}\left\{P_t\left(m_1^\nu, \ldots, m_k^\nu, \ldots, m_M^\nu\right)\int_{-\infty}^{\infty} D[\xi^\nu]\, d\xi^\nu\right. \\
&\quad \times \left.\frac{\sum_{\{\boldsymbol{S}_k\}} p_t(\{\boldsymbol{S}_k\})\xi^\nu \tanh[\beta\phi(k)] \prod_{k,i}\delta(m_k^\nu - m_k^\nu(\boldsymbol{S}_k))}{\sum_{\{\boldsymbol{S}_k\}} p_t(\{\boldsymbol{S}_k\}) \prod_k \delta(m_k^\nu - m_k^\nu(\boldsymbol{S}_k))}\right\} \\
&\quad \times \delta\left(S(k+1) - S_i(k+1)\right)\delta\left(S(k-1) - S_i(k-1)\right)
\end{aligned} \tag{9.1.30}$$

where we introduced several notations

$$D[\xi^\nu] \equiv \frac{1}{N}\sum_i \delta\left(\xi^\nu - \xi_i^\nu\right) \tag{9.1.31}$$

$$\beta\phi(k) \equiv \frac{\beta\sum_{\mu\nu}\xi^\mu A_{\mu\nu}}{M} m_k^\nu + \frac{B}{2}S(k-1) + \frac{B}{2}S(k+1) \tag{9.1.32}$$

for simplicity.

Here we should notice that if the local field $\beta\phi(k)$ is independent of the microscopic variable $\{\boldsymbol{S}_k\}$, one can get around the complicated expectation of the quantity $\tanh[\beta\phi(k)]$ over the time-dependent Gibbs measurement which is defined in the sub-shell: $\prod_k \delta(m_k^\nu - m_k^\nu(\boldsymbol{S}_k))$. As the result, only procedure we should carry out to get the deterministic flow is to calculate the data average (the average over the built-in patterns). However, unfortunately, we clearly find from Eq. (9.1.32) that the local field depends on the $\{\boldsymbol{S}_k\}$. To overcome the difficulty and to carry out the calculation, we assume that the probability $p_t(\{\boldsymbol{S}_k\})$ of realisations for microscopic states during the dynamics is independent of t, namely,

$$p_t(\{\boldsymbol{S}_k\}) = p(\{\boldsymbol{S}_k\}). \tag{9.1.33}$$

Then, our average over the time-dependent Gibbs measurement in the sub-shell is rewritten as

$$\begin{aligned}
&\frac{\sum_{\{\boldsymbol{S}_k\}} p_t(\{\boldsymbol{S}_k\})\xi^\nu \tanh[\beta\phi(k)]\prod_{k,i}\delta(m_k^\nu - m_k^\nu(\boldsymbol{S}_k))}{\sum_{\{\boldsymbol{S}_k\}} p_t(\{\boldsymbol{S}_k\})\prod_k \delta(m_k^\nu - m_k^\nu(\boldsymbol{S}_k))} \\
&\quad \times \delta\big(S(k+1) - S_i(k+1)\big)\delta\big(S(k-1) - S_i(k-1)\big) \\
&\equiv \left\langle \xi^\nu \tanh\big[\beta\phi(k)\big]\prod_i \delta\big(S(k+1) - S_i(k+1)\big)\delta\big(S(k-1) - S_i(k-1)\big)\right\rangle_*
\end{aligned} \tag{9.1.34}$$

where $\langle\cdots\rangle_*$ stands for the average in the sub-shell defined by $m_k^\nu = m_k^\nu(\boldsymbol{S}_k)$ $(\forall_k)$:

$$\langle\cdots\rangle_* \equiv \frac{\sum_{\{\boldsymbol{S}_k\}} p(\{\boldsymbol{S}_k\})(\cdots)\prod_k \delta(m_k^\nu - m_k^\nu(\boldsymbol{S}_k))}{\sum_{\{\boldsymbol{S}_k\}} p(\{\boldsymbol{S}_k\})\prod_k \delta(m_k^\nu - m_k^\nu(\boldsymbol{S}_k))}. \tag{9.1.35}$$

If we notice that the Gibbs measurement in the sub-shell is rewritten as

$$\sum_{\{\boldsymbol{S}_k\}} p(\{\boldsymbol{S}_k\})\prod_k \delta\big(m_k^\nu - m_k^\nu(\boldsymbol{S}_k)\big) = \mathrm{tr}_{\{S\}}\exp\left[\beta\sum_{l=1}^{M}\phi(l)S(l)\right] \tag{9.1.36}$$

($\mathrm{tr}_{\{S\}}(\cdots) \equiv \prod_k \sum_{S_k}(\cdots)$), and the quantity

$$\tanh\big[\beta\phi(k)\big] = \frac{\sum_{S(k)=\pm 1} S(k)\exp[\beta\phi(k)S(k)]}{\sum_{S(k)=\pm 1}\exp[\beta\phi(k)S(k)]} \tag{9.1.37}$$

is independent of $S(k)$, the average appearing in (9.1.34) leads to

$$\begin{aligned}
&\left\langle \xi^\nu \tanh\big[\beta\phi(k)\big]\prod_i \delta\big(S(k\pm 1) - S_i(k\pm 1)\big)\right\rangle_* \\
&\quad = \frac{\mathrm{tr}_{\{S\}}\,\xi^\nu\{\frac{1}{M}\sum_{l=1}^{M} S(l)\}\exp[\beta\phi(k)S(k)]}{\mathrm{tr}\{S\}\exp[\beta\phi(k)S(k)]} \\
&\quad \equiv \xi^\nu \langle S\rangle_{path}^{(\xi^\nu)}
\end{aligned} \tag{9.1.38}$$

in the limit of $M \to \infty$. This is nothing but a path integral for the effective single neuron problem in which the neuron updates its state along the imaginary time axis: $\mathrm{tr}_{\{S\}}(\cdots) \equiv \sum_{S(1)=\pm 1} \cdots \sum_{S(M)=\pm 1}(\cdots)$ with weights $\exp[\beta\phi(k)S(k)]$ $(k = 1, \ldots, M)$.

Then, the differential equation (9.1.30) leads to

$$\frac{dP_t(m_1^\nu, \ldots, m_M^\nu)}{dt} = \sum_k \frac{\partial}{\partial m_k^\nu}\{m_k^\nu P_t(m_1^\nu, \ldots, m_k^\nu, \ldots, m_M^\nu)\} - \sum_k \frac{\partial}{\partial m_k^\nu}\left\{P_t(m_1^\nu, \ldots, m_k^\nu, \ldots, m_M^\nu)\int_{-\infty}^{\infty} D[\xi^\nu]\,d\xi^\nu \xi^\nu \langle S\rangle_{path}^{(\xi^\nu)}\right\}. \tag{9.1.39}$$

In order to derive the compact form of the differential equations with respect to the overlaps, we substitute $P_t(m_1^\nu, \ldots, m_M^\nu) = \prod_{k=1}^M \delta(m_k^\nu - m_k^\nu(t))$ into the above (9.1.39) and multiplying m_l^ν by both sides of the equation and carrying out the integral with respect to $dm_1^\nu \cdots dm_M^\nu$ by part, we have for $l = 1, \ldots, M$ as

$$\frac{dm_l^\nu}{dt} = -m_l^\nu + \int_{-\infty}^{\infty} D[\xi^\nu]\,d\xi^\nu \xi^\nu \langle S\rangle_{path}^{(\xi^\nu)}. \tag{9.1.40}$$

Here we should notice that the path integral $\xi^\nu \langle S\rangle_{path}^{(\xi^\nu)}$ depends on the embedded patterns $\boldsymbol{\xi}^\nu$. In the next subsection, we carry out the quenched average explicitly under the so-called static approximation.

9.1.2.7 Static Approximation

In order to obtain the final form of the deterministic flow, we assume that macroscopic quantities such as the overlap are independent of the Trotter slices k during the dynamics. Namely, we must use the so-called static approximation:

$$m_k^\nu = m^\nu \ (\forall k). \tag{9.1.41}$$

Under the static approximation, let us use the following inverse process of the Suzuki-Trotter decomposition (9.1.22):

$$\lim_{M\to\infty} Z_M = \mathrm{tr}\exp\left[\beta\sum_{\mu\nu} \xi^\mu A_{\mu\nu} m^\nu S_z + \beta\Gamma S_x\right] \tag{9.1.42}$$

$$Z_M \equiv \mathrm{tr}_{\{S\}}\exp\left[\frac{\beta\sum_{\mu\nu}\xi^\mu A_{\mu\nu}m^\nu}{M}\sum_k S(k) + B\sum_k S(k)S(k+1)\right]. \tag{9.1.43}$$

Then, one can calculate the path integral immediately as

$$\langle S\rangle_{path}^{(\xi^\nu)} = \frac{\sum_{\mu\nu}\xi^\mu A_{\mu\nu}m^\nu}{\sqrt{(\sum_{\mu\nu}\xi^\mu A_{\mu\nu}m^\nu)^2 + \Gamma^2}} \tanh\beta\sqrt{\left(\sum_{\mu\nu}\xi^\mu A_{\mu\nu}m^\nu\right)^2 + \Gamma^2}. \tag{9.1.44}$$

Inserting this result into (9.1.40), we obtain

$$\frac{dm^\nu}{dt} = -m^\nu + \ll \frac{\xi^\nu \sum_{\mu\nu} \xi^\mu A_{\mu\nu} m^\nu}{\sqrt{(\sum_{\mu\nu} \xi^\mu A_{\mu\nu} m^\nu)^2 + \Gamma^2}} \tanh \beta \sqrt{\left(\sum_{\mu\nu} \xi^\mu A_{\mu\nu} m^\nu\right)^2 + \Gamma^2} \gg \tag{9.1.45}$$

where we should bear in mind that the empirical distribution $D[\xi^\nu]$ in (9.1.40) was replaced by the built-in pattern distribution $\mathscr{P}(\xi^\nu)$ as

$$\lim_{N\to\infty} \frac{1}{N} \sum_i \delta(\xi_i^\nu - \xi^\nu) = \mathscr{P}(\xi^\nu) \tag{9.1.46}$$

in the limit of $N \to \infty$ and the average is now carried out explicitly as

$$\int D[\xi^\nu] d\xi^\nu (\cdots) = \int \mathscr{P}(\xi^\nu)\, d\xi^\nu (\cdots) \equiv \ll \cdots \gg. \tag{9.1.47}$$

Equation (9.1.45) is a general solution for the problem.

9.1.2.8 The Classical and Zero-Temperature Limits

It is easy for us to take the classical limit $\Gamma \to 0$ in the result (9.1.45). Actually, we have immediately

$$\frac{dm^\nu}{dt} = -m^\nu + \ll \xi^\nu \tanh\left(\beta \sum_{\mu\nu} \xi^\mu A_{\mu\nu} m^\nu\right) \gg. \tag{9.1.48}$$

The above equation is identical to the result by Coolen and Ruijgrok [87] who considered the retrieval process of the conventional Hopfield model under thermal noise.

We can also take the zero-temperature limit $\beta \to \infty$ in Eq. (9.1.45) as

$$\frac{dm^\nu}{dt} = -m^\nu + \ll \frac{\xi^\nu \sum_{\mu\nu} \xi^\mu A_{\mu\nu} m^\nu}{\sqrt{(\sum_{\mu\nu} \xi^\mu A_{\mu\nu} m^\nu)^2 + \Gamma^2}} \gg. \tag{9.1.49}$$

Thus, Eq. (9.1.45) including the above two limiting cases is our general solution for the neuro-dynamics of the quantum Hopfield model in which $\mathscr{O}(1)$ patterns are embedded. Thus, we can discuss any kind of situations for such pattern-recalling processes and the solution is always derived from (9.1.45) explicitly.

9.1.2.9 Limit Cycle Solution for Asymmetric Connections

In this section, we discuss a special case of the general solution (9.1.45). Namely, we investigate the pattern-recalling processes of the quantum Hopfield model with asymmetric connections $\boldsymbol{A} \equiv \{A_{\mu\nu}\}$.

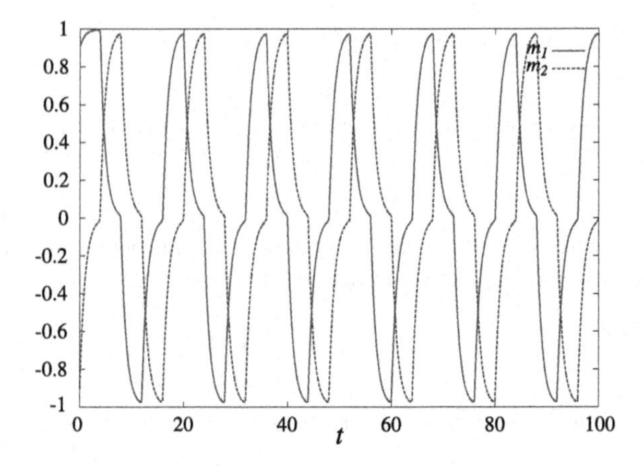

Fig. 9.2 Time evolutions of m_1 and m_2 for the case of $\Gamma = 0.01$

9.1.2.10 Result for Two-Patterns

Let us consider the case in which just only two patterns are embedded via the following matrix:

$$A = \begin{pmatrix} 1 & -1 \\ 1 & 1 \end{pmatrix}. \tag{9.1.50}$$

Then, from the general solution (9.1.45), the differential equations with respect to the two overlaps m_1 and m_2 are written as

$$\frac{dm_1}{dt} = -m_1 + \frac{m_1}{\sqrt{(2m_1)^2 + \Gamma^2}} - \frac{m_2}{\sqrt{(2m_2)^2 + \Gamma^2}}$$

$$\frac{dm_2}{dt} = -m_2 + \frac{m_1}{\sqrt{(2m_1)^2 + \Gamma^2}} + \frac{m_2}{\sqrt{(2m_2)^2 + \Gamma^2}}.$$

In Fig. 9.2, we show the time evolutions of the overlaps m_1 and m_2 for the case of the amplitude $\Gamma = 0.01$. From this figure, we clearly find that the neuronal state evolves as $A \to B \to \overline{A} \to \overline{B} \to A \to B \to \cdots$ ($\overline{A}, \overline{B}$ denote the 'mirror images' of A and B, respectively), namely, the network behaves as a limit cycle.

To compare the effects of thermal and quantum noises on the pattern-recalling processes, we plot the trajectories m_1–m_2 for $(T \equiv \beta^{-1}, \Gamma) = (0, 0.01)$, $(0.01, 0)$ (left panel), $(T, \Gamma) = (0, 0.8)$, $(0.8, 0)$ (right panel) in Fig. 9.3. From these panels, we find that the limit cycles are getting collapsed as the strength of the noise level is increasing for both thermal and quantum-mechanical noises, and eventually the trajectories shrink to the origin $(m_1, m_2) = (0, 0)$ in the limit of $T, \Gamma \to \infty$.

9.2 Statistical Mechanics of Information

In the previous section, we consider the quantum-mechanical variant of the Hopfield model in terms of transverse Ising model. We could apply the knowledge of

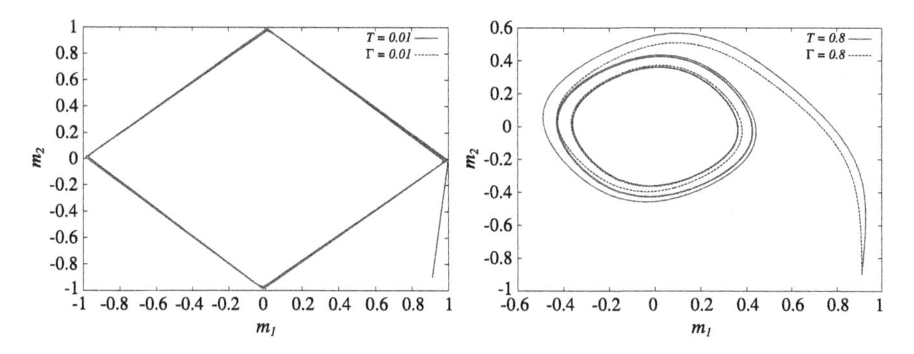

Fig. 9.3 Trajectories m_1–m_2 for $(T, \Gamma) = (0, 0.01), (0.01, 0)$ (*left panel*), $(T, \Gamma) = (0, 0.8), (0.8, 0)$ (*right panel*)

transverse Ising model to a specific problem in the outside of physics. However, application of the transverse field Ising model is not limited to the Hopfield model. As we saw, the quantum fluctuation in the Hopfield model works on the retrieval properties as 'quantum-mechanical noise' to prevent the network from recalling an embedded pattern. In this sense, the use of the transverse field is rather 'negative' effect on the information processing in artificial brain. However, we use the quantum fluctuation induced by the transverse field to construct useful algorithms for massive information processing such as image restoration or error-correcting codes.

Recently, problems of information processing were investigated from statistical mechanical point of view [297]. Among them, image restoration (see [320, 396, 422] and references there in) and error-correcting codes [374] are most suitable subjects. In the field of the error-correcting codes, Sourlas [374] showed that the convolution codes can be constructed by infinite range spin-glasses Hamiltonian and the decoded message should correspond to the zero temperature spin configuration of the Hamiltonian. Ruján [337] suggested that the error of each bit can be suppressed if one uses finite temperature equilibrium states (sign of the local magnetisation) as the decoding result, what we call the *MPM* (*Maximizer of Posterior Marginal*) estimate, instead of zero temperature spin configurations, and this optimality of the retrieval quality at a specific decoding temperature (this temperature is well known as the *Nishimori temperature* in the field of spin glasses) is proved by Nishimori [296].

The next remarkable progress in this direction was done by Nishimori and Wong [299]. They succeeded in giving a new procedure in order to compare the performance of the zero temperature decoding (statisticians call this strategy the *MAP* (*Maximum A Posteriori*) estimation) with that of the finite temperature decoding, the MPM estimation. They introduced an infinite range model of spin-glasses like the Sherrington-Kirkpatrick (SK) model [369] as an exactly solvable example. Kabashima and Saad [210] succeeded in constructing more practical codes, namely, low density parity check (LDPC) codes by using the spin glass model with finite connectivities. In these decoding process, one of the most important problems is how one obtains the minimum energy states of the effective Hamiltonian as quickly as possible. Geman and Geman [155] used simulated annealing [228] in the context of image restoration to obtain good recovering of the original image from its corrupted version. Recently, Tanaka and Horiguchi [396, 397] introduced a quantum fluctuation, instead of the thermal one, into the mean-field annealing algorithm and showed that performance of the image recovery is improved by controlling the quantum fluctuation appropriately during its annealing process.

As we saw in Chap. 10, the attempt to use the quantum fluctuation to search the lowest energy states in the context of annealings by Markov chain Monte Carlo methods, what we call *quantum annealing*, is originally introduced by [11, 135] and its application to the combinatorial optimisation problems including the ground state searching for several spin glass models was done by Kadowaki and Nishimori [211] and Santoro et al. [342]. However, these results are restricted to research aided by computer simulations, although there exist some extensive studies on the Landau-Zener's model for the single spin problems [276, 277, 443].

Recently, the averaged case performance of the both MPM and MAP estimations for image restoration with quantum fluctuation was investigated by Inoue [190] for the mean-field model. He also carried out the quantum Monte Carlo method to evaluate the performance for two dimensional pictures and found that the quantum fluctuation suppress the error due to failing to set the hyperparameters effectively, however, the best possible value of the bit-error rate does not increases by the quantum fluctuation. In this result the quantum and the thermal fluctuations are combined in the MPM estimation (the effective temperature is unity). Therefore, it is important for us to revisit this problem and investigate to what extent the MPM estimation, which is based on pure quantum fluctuation and without any thermal one, works effectively. From this direction of studies, the MPM estimation by using quantum fluctuation at zero temperature was discussed in [93] (see a chapter given by Inoue, pp. 259–296). In this reference, the decoding performance of the Sourlas codes by MPM estimation with quantum fluctuation was also given and the relation ship between the phase transition and the so-called Shannon's bound was clarified.

In this subsection according to [93], we make this point clear and show that the best possible performance obtained by the MPM estimation, which is purely induced by quantum fluctuations, is exactly the same as the results by the thermal MPM estimation. The Nishimori-Wong condition [296, 299] for the quantum fluctuation, on which the best possible performance is achieved, is also discussed. Moreover, we extend the Sourlas codes [374] by means of the spin glass model with p-spin interaction in a transverse field [65, 158] and discuss the tolerance of error-less (or quite low-error) state to the quantum uncertainties in the prior distribution. In last part of this section, we check the performance of the MAP and MPM image restorations predicted by the analysis of the mean-field infinite range model by using the quantum Markov chain Monte Carlo method [386] and the quantum annealing [11, 135, 211, 342].

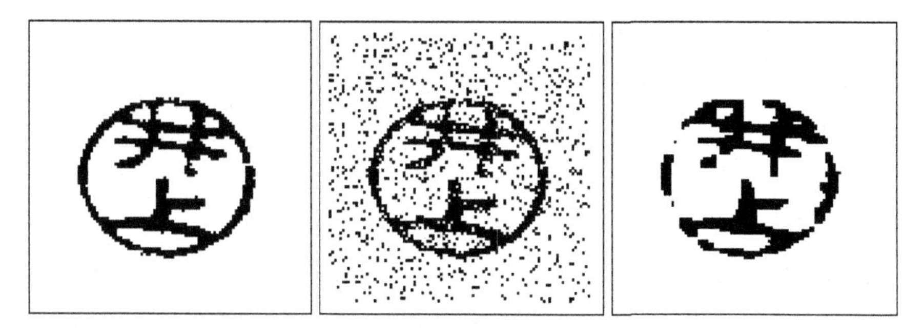

Fig. 9.4 A typical example of image data retrieval. From the *left* to the *right*, the original $\{\xi\}$, the degraded $\{\tau\}$ and the recovering $\{S\}$ images. The above restored image was obtained by quantum annealing. The detailed account of this method will be explained and discussed in the last section. The original image (*left*) is a Japanese seal (what we call 'inkan') for surname 'Inoue' in kanji character

9.2.1 Bayesian Statistics and Information Processing

In the field of signal processing or information science, we need to estimate the original message which is sent via email or fax. Usually, these massages are degraded by noise and we should retrieve the original messages (Fig. 9.4). Then, if possible, we can send these messages not only as sequence of information bits but also as some redundant information such as *parity check*. In such problems, noise channels or statistical properties of the original message are specified by some appropriate probabilistic models (see [254] for the basics of such probabilistic models). In this subsection, we explain the general definitions of our problems and how these problems link to statistical physics.

9.2.1.1 General Definition of the Model System

Let us suppose that the original information is represented by a configuration of Ising spins $\{\xi\} \equiv (\xi_1, \xi_2, \ldots, \xi_N)$ (each spin takes $\xi_i = \pm 1, i = 1, \ldots, N$) with probability $P(\{\xi\})$. Of course, if each message/pixel ξ_i is generated from independent identical distribution (i.i.d.), the probability of the configuration $\{\xi\}$ is written by the product of the probability $P(\xi_i)$, namely, $P(\{\xi\}) = \prod_{i=1}^{N} P(\xi_i)$.

These messages/pixels $\{\xi\}$ are sent through the noisy channel by not only the form of the sequence of the original messages/pixels $\{\xi\}$, but also as 'spin products' $\{\xi_{i1} \cdots \xi_{ip}\} \equiv \{J^0_{i1\cdots ip}\}$ for appropriately chosen set of indexes $\{i1, \ldots, ip\}$ (what we call *parity check* in the context of error-correcting codes). Therefore, the outputs of the noisy channel are exchange interactions $\{J_{i1\cdots ip}\}$ and fields $\{\tau\}$. Namely,

$$\{\xi\} \xrightarrow{\text{noise}} \{\tau\}$$
$$\{J^0_{i1\cdots ip}\} \xrightarrow{\text{noise}} \{J_{i1\cdots ip}\}.$$

In the field of information theory (see [254] for example), the noisy channel is specified by the conditional probability like $P(\{\tau\}|\{\xi\})$ or $P(\{J\}|\{J^0\})$. If each message/pixel ξ_i and parity check $J^0_{i1 \cdot ip}$ are affected by the channel noise independently, the probabilities of output sequences $\{\tau\} \equiv (\tau_1, \dots, \tau_2)$ or $\{J\} \equiv \{J_{i1\cdots ip}\}\forall_{(i1\cdots ip)}$ for given input sequences $\{\xi\} = (\xi_1, \dots, \xi_N)$ or $\{J^0\} = \{J^0_{i1\cdots ip}\}\forall_{(i1\cdots ip)}$ are written by

$$P(\{\tau\}|\{\xi\}) = \prod_{i=1}^{N} P(\tau_i|\xi_i), \quad P(\{J\}|\{J^0\}) = \prod_{(i1\cdots ip)} P(J_{i1\cdots ip}|J^0_{i1\cdots ip}), \tag{9.2.1}$$

respectively, where $\prod_{(i1\cdots ip)}(\cdots)$ stands for the product of all possible combinations of p indices $(i1\cdots ip)$.

In this book, we use the following two kinds of the noisy channel. The first one is referred to as *binary symmetric channel* (*BSC*). In this channel, each message/pixel ξ_i and parity check $J_{i1\dots jp}$ change their signs with probabilities p_τ and p_r, respectively. By introducing the parameters

$$\beta_\tau \equiv \frac{1}{2}\log\left(\frac{1-p_\tau}{p_\tau}\right), \qquad \beta_r \equiv \frac{1}{2}\log\left(\frac{1-p_r}{p_r}\right), \tag{9.2.2}$$

the conditional probabilities (9.2.1) are given by

$$P(\{\tau\}|\{\xi\}) = \prod_{i=1}^{N} P(\tau_i|\xi_i) = \prod_{i=1}^{N}\left\{\frac{e^{\beta_\tau \tau_i \xi_i}}{\sum_{\tau_i=\pm\xi_i} e^{\beta_\tau \tau_i \xi_i}}\right\} = \frac{e^{\beta_\tau \sum_i \tau_i \xi_i}}{(2\cosh\beta_\tau)^N} \tag{9.2.3}$$

$$\begin{aligned} P(\{J\}|\{J^0\}) &= \prod_{(i1\cdots ip)} P(J_{i1\cdots ip}|J^0_{i1\cdots ip}) = \prod_{(i1\cdots ip)}\left\{\frac{e^{\beta_r J_{i1\cdots ip} J^0_{i1\cdots ip}}}{\sum_{J_{i1\cdots ip}=\pm J^0_{i1\cdots ip}} e^{\beta_r J_{i1\cdots ip} J^0_{i1\cdots ip}}}\right\} \\ &= \frac{e^{\beta_r \sum_{(i1\cdots ip)} J_{i1\cdots ip} J^0_{i1\cdots ip}}}{(2\cosh\beta_r)^{N_B}} \end{aligned} \tag{9.2.4}$$

where we defined $N \equiv \sum_i 1$, $N_B \equiv \sum_{(i1\cdots ip)} 1$.

Thus, the probability of the output sequences $\{J\}$, $\{\tau\}$ provided that the corresponding input sequence of the original messages/pixels is $\{\xi\}$ is obtained by $\sum_{\{J_0\}} P(\{J\}|\{J^0\})P(\{J^0\}|\{\xi\})P(\{\tau\}|\{\xi\})$, that is to say,

$$P(\{J\},\{\tau\}|\{\xi\}) = \frac{\exp(\beta_r \sum_{(i1,\cdots,ip)} J_{i1\cdots ip}\,\xi_{i1}\cdots\xi_{ip} + \beta_\tau \sum_i \tau_i \xi_i)}{(2\cosh\beta_r)^{N_B}(2\cosh\beta_\tau)^N} \tag{9.2.5}$$

where we used the following condition:

$$P(\{J^0\}|\{\xi\}) = \prod_{(i1\cdots ip)} \delta_{J^0_{i1\cdots ip},\,\xi_{i1}\cdots\xi_{ip}}. \tag{9.2.6}$$

The second type of the noisy channel is called as *Gaussian channel* (*GC*). The above BSC (9.2.5) is simply extended to the GC as follows.

$$P(\{J\},\{\tau\}|\{\xi\}) = \frac{e^{-\frac{1}{2J^2}\sum_{(i1,\dots,ip)}(J_{i1\cdots ip} - J_0\xi_{i1}\cdots\xi_{ip})^2 - \frac{1}{2a^2}\sum_i(\tau_i - a_0\xi_i)^2}}{(\sqrt{2\pi}\,J)^{N_B}(\sqrt{2\pi}\,a)^N}. \tag{9.2.7}$$

We should notice that these two channels can be treated within a single form:

$$P(\{J\},\{\tau\}|\{\xi\}) = \prod_{(i1\cdots ip)} F_r(J_{i1\cdots ip}) \prod_i F_\tau(\tau_i) \times \exp\left(\beta_r \sum_{i1\cdots ip} J_{i1\cdots ip}\xi_{i1}\cdots\xi_{ip} + \beta_\tau \sum_i \tau_i \xi_i\right) \quad (9.2.8)$$

with

$$F_r(J_{i1\ldots ip}) = \frac{\sum_{j=\pm 1}\delta(J_{i1\cdots ip} - j)}{2\cosh\beta_r}, \qquad F_\tau(\tau_i) = \frac{\sum_{j=\pm 1}\delta(\tau_i - j)}{2\cosh\beta_\tau} \quad (9.2.9)$$

for the BSC and

$$F_r(J_{i1\cdots ip}) = \frac{\exp[-\frac{1}{2J^2}(J_{i1\cdots ip}^2 + J_0^2)]}{\sqrt{2\pi J^2}}, \qquad F_\tau(\tau_i) = \frac{\exp[-\frac{1}{2a^2}(\tau_i^2 + a_0^2)]}{\sqrt{2\pi a^2}} \quad (9.2.10)$$

for the GC. Therefore, it must be noted that there exist relationship between the parameters for both channels as

$$\beta_r = \frac{J_0}{J^2}, \qquad \beta_\tau = \frac{a_0}{a^2}. \quad (9.2.11)$$

Main purpose of signal processing we are dealing with in this subsection is to estimate the original sequence of messages/pixels $\{\xi\}$ from the outputs $\{J\}$, $\{\tau\}$ of the noisy channel. For this aim, it might be convenient for us to construct the probability of the estimate $\{S\}$ for the original messages/pixels sequence $\{\xi\}$ provided that the outputs of the noisy channel are $\{J\}$ and $\{\tau\}$.

From the Bayes formula:

$$P(B|A) = \frac{P(A|B)P(B)}{\sum_B P(A|B)P(B)}, \quad (9.2.12)$$

where $P(B)$, $P(A|B)$, $P(B|A)$ are referred to as *prior*, *likelihood* and *posterior*, respectively. Thus, in our model system, the probability $P(\{S\}|\{J\},\{\tau\})$ is written in terms of the likelihood: $P(\{J\},\{\tau\}|\{S\})$ and the prior: $P_m(\{S\})$ as follows

$$P(\{S\}|\{J\},\{\tau\}) = \frac{P(\{J\},\{\tau\}|\{S\})P_m(\{S\})}{\sum_{\{S\}} P(\{J\},\{\tau\}|\{S\})P_m(\{S\})}. \quad (9.2.13)$$

As the likelihood has a meaning of the probabilistic model of the noisy channel, we might choose it naturally as

$$P(\{J\},\{\tau\}|\{S\}) = \frac{\exp(\beta_J \sum_{(i1\cdots ip)} J_{i1\cdots ip} S_{i1}\cdots S_{ip} + h\sum_i \tau_i S_i)}{(2\cosh\beta_J)^{N_B}(2\cosh h)^N} \quad (9.2.14)$$

for the BSC and

$$P(\{J\},\{\tau\}|\{S\}) = \frac{\mathrm{e}^{-\frac{\beta_J}{2}\sum_{(i1\cdots ip)}(J_{i1\cdots ip} - S_{i1}\cdots S_{ip})^2 - h\sum_i(\tau_i - S_i)^2}}{(2\pi/\beta_J)^{N_B/2}(\pi/h)^{N/2}} \quad (9.2.15)$$

for the GC.

Therefore, the posterior $P(\{S\}|\{J\},\{\tau\})$ which is defined by (9.2.12), (9.2.13) is rewritten in terms of the above likelihood as follows.

$$P\big(\{S\}|\{J\},\{\tau\}\big)=\frac{\mathrm{e}^{-\beta\mathscr{H}_{\mathrm{eff}}}}{\sum_{\{S\}}\mathrm{e}^{-\beta\mathscr{H}_{\mathrm{eff}}}} \tag{9.2.16}$$

where we defined the inverse temperature $\beta=1/T$ and set $T=1$ in the above case. The effective Hamiltonian $\mathscr{H}_{\mathrm{eff}}$ is also defined by

$$\mathscr{H}_{\mathrm{eff}}=-\beta_J\sum_{(i1\cdots ip)}J_{i1\cdots jp}S_{i1}\cdots S_{ip}-h\sum_i\tau_i S_i-\log P_m\big(\{S\}\big) \tag{9.2.17}$$

for the BSC and

$$\mathscr{H}_{\mathrm{eff}}=-\frac{\beta_J}{2}\sum_{(i1\cdots ip)}(J_{i1\cdots ip}-S_{i1}\cdots S_{ip})^2-h\sum_i(\tau_i-S_i)^2-\log P_m\big(\{S\}\big) \tag{9.2.18}$$

for the GC.

9.2.1.2 MAP Estimation and Simulated Annealing

As we mentioned, the posterior $P(\{S\}|\{J\},\{\tau\})$ is a useful quantity in order to determine the estimate $\{S\}$ of the original messages/pixels sequence. As the estimate of the original message/pixel sequence, we might choose a $\{S\}$ which maximises the posterior for a given set of the output sequence $\{J\}$, $\{\tau\}$. Apparently, this estimate $\{S\}$ corresponds to the ground state of the effective Hamiltonian $\mathscr{H}_{\mathrm{eff}}$. In the context of Bayesian statistics, this type of estimate $\{S\}$ is referred to as *Maximum A posteriori* (*MAP*) estimate.

As we saw in Chap. 10, from the view point of important sampling from the posterior as a Gibbs distribution (Gibbs sampler), such a MAP estimate is obtained by controlling the temperature T as $T\to 0$ during the Markov chain Monte Carlo steps. This kind of optimisation method is well-known and is widely used as *simulated annealing* (*SA*) [155, 228]. As the optimal scheduling of the temperature T is $T(t)=c/\log(1+t)$, which was proved by using mathematically rigorous arguments [155].

9.2.1.3 MPM Estimation and a Link to Statistical Mechanics

From the posterior $P(\{S\}|\{J\},\{\tau\})$, we can attempt to make another kind of estimations. For this estimation, we construct the following marginal distribution for each pixel S_i:

$$P\big(S_i|\{J\},\{\tau\}\big)=\sum_{\{S\}\neq S_i}P\big(\{S\}|\{J\},\{\tau\}\big). \tag{9.2.19}$$

Then, we might choose the sign of the difference between the probabilities $P(1|\{J\},\{\tau\})$ and $P(-1|\{J\},\{\tau\})$ as the estimate of the i-th message/pixel, to put it another way, we assume that the original pixel $\hat{\xi}_i$ is $+1$ if $P(1|\{J\},\{\tau\}) > P(-1|\{J\},\{\tau\})$ and $\hat{\xi}_i = -1$ vice versa. Thus we have

$$\hat{\xi}_i \equiv \mathrm{sgn}\left[\sum_{S_i=\pm 1} S_i P\big(S_i|\{J\},\{\tau\}\big)\right] = \mathrm{sgn}\left(\frac{\sum_{\{S\}} S_i\, \mathrm{e}^{-\mathcal{H}_{\mathrm{eff}}}}{\sum_{\{S\}} \mathrm{e}^{-\mathcal{H}_{\mathrm{eff}}}}\right) \equiv \mathrm{sgn}\big(\langle S_i\rangle_1\big) \quad (9.2.20)$$

where we defined the bracket $\langle\cdots\rangle_\beta$ as

$$\langle\cdots\rangle_\beta \equiv \frac{\sum_{\{S\}}(\cdots)\,\mathrm{e}^{-\beta\mathcal{H}_{\mathrm{eff}}}}{\sum_{\{S\}} \mathrm{e}^{-\beta\mathcal{H}_{\mathrm{eff}}}}. \quad (9.2.21)$$

In this sense, the estimate $\hat{\xi}_i$ is regarded as the result of 'majority rule' in which each voting $S_i = \pm 1$ fluctuates due to the 'thermal' fluctuation. Therefore, the above estimate $\hat{\xi}_i$ has a link to statistical mechanics through the local magnetisation $\langle S_i\rangle_1$ for the spin system described by $\mathcal{H}_{\mathrm{eff}}$ at temperature $T = 1$. This estimate $\hat{\xi}_i = \mathrm{sgn}(\langle S_i\rangle_1)$ is referred to as *Maximizer of Posterior Marginal (MPM) estimate* or *Finite Temperature (FT) estimate* [337].

It is well-known that this estimate minimises the following *bit-error rate* [296, 299, 337]:

$$p_b^{(MPM)} = P_b^{(1)}(\beta_J, h : P_m) = \frac{1}{2}\big[1 - R^{(1)}(\beta_J, h : P_m)\big] \quad (9.2.22)$$

with the overlap between the original message/pixel ξ_i and its MPM estimate $\hat{\xi}_i = \mathrm{sgn}(\langle S_i\rangle)$:

$$R^{(1)}(\beta_J, h : P_m) = \sum_{\{\xi,J,\tau\}} P\big(\{J\},\{\tau\},\{\xi\}\big)\xi_i\hat{\xi}_i. \quad (9.2.23)$$

Obviously, the bit-error rate for the MAP estimate is given by

$$p_b^{(MAP)} = \lim_{\beta\to\infty} P^{(\beta)}(\beta_J, h : P_m) = \frac{1}{2}\Big[1 - \lim_{\beta\to\infty} R^{(\beta)}(\beta_J, h : P_m)\Big] \quad (9.2.24)$$

with

$$R^{(\beta)}(\beta_J, h : P_m) = \sum_{\{\xi,J,\tau\}} P\big(\{J\},\{\tau\},\{\xi\}\big)\xi_i\, \mathrm{sgn}\big(\langle S_i\rangle_\beta\big). \quad (9.2.25)$$

In the next section, we compare $p_b^{(MPM)}$ with $p_b^{(MAP)}$ by using replica method and show the former is smaller than the later.

9.2.2 The Priors and Corresponding Spin Systems

In the previous two subsections, we showed the close relationship between Bayesian inference of the original messages/pixels under some noises and statistical physics

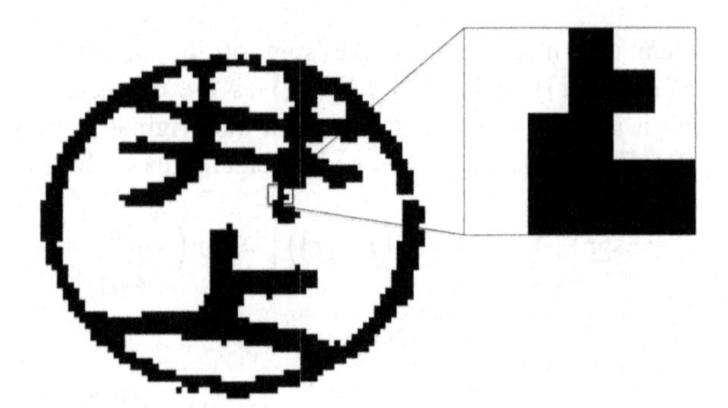

Fig. 9.5 We assume that the local structure in binary images is *smooth* so as to be well-described by a snapshot from the Gibbs distribution for the ferromagnetic Ising model

[297]. However, we do not yet mention about the choice of the prior distribution $P_m(\{S\})$ in the effective Hamiltonian $\mathcal{H}_{\mathrm{eff}}$. In the framework of the Bayesian statistics, the choice of the prior is arbitrary, however, the quality of the estimation for a given problem strongly depends on the choice (what we call 'model selection').

9.2.2.1 Image Restoration and Random Field Ising Model

In image restoration, we might have an assumption that in realistic two dimensional pictures, the nearest neighbouring sites should be inclined to be the same value (the same colour), in other words, we assume that real picture should be locally smooth (see Fig. 9.5). Taking this smoothness into account, then, it is quite reasonable for us to choose the prior for image restoration as

$$P_m(\{S\}) = \frac{\mathrm{e}^{\beta_m \sum_{\langle ij\rangle} S_i S_j}}{Z(\beta_m)}, \qquad Z(\beta_m) = \sum_{\{S\}} \mathrm{e}^{\beta_m \sum_{\langle ij\rangle} S_i S_j}, \tag{9.2.26}$$

where $\sum_{\langle i,j\rangle}(\cdots)$ implies that the sum should be carried out for all pairs locating as the nearest neighbours in the lattice. In conventional image restoration, we do not send any parity check and only available information is the degraded sequence of the pixels $\{\tau\}$. Hence, we set $\beta_J = 0$ for this problem. Ultimately, we obtain the effective Hamiltonian for our image restoration as

$$\mathcal{H}_{\mathrm{eff}} = -\beta_m \sum_{\langle ij\rangle} S_i S_j - h \sum_i \tau_i S_i. \tag{9.2.27}$$

This Hamiltonian is identical to that of the *random field Ising model* in which random field on each cite corresponds to each degraded pixel τ_i.

9.2.2.2 Error-Correcting Codes and Spin Glasses with *p*-Body Interaction

In error-correcting codes, we usually use so-called *uniform prior distribution* because we do not have any idea about the properties of the original message sequence $\{\xi\}$ as we assumed smoothness for images. Thus, we set the prior as $P_m(\{S\}) = 2^{-N}$

and substituting $-\log P_m(\{S\}) = N \log 2 = \text{const.}$ into $\mathcal{H}_{\text{eff}}$ (usually, we neglect the constant term).

In this case, we do not use any a priori information to estimate the original message, however, in error-correcting codes, we compensate this lack of information with extra redundant information as a form of $\xi_{i1} \cdots \xi_{ip}$ (a parity), besides the original message sequence $\{\xi\}$. In information theory, it is well-known that we can decode the original message $\{\xi\}$ without any error when the transmission rate R, which is defined by $R = N/N_B$ (N original message length, N_B: redundant message length), is smaller than the channel capacity C (see for example [19, 254]). The channel capacity is given by

$$C = \begin{cases} 1 + p \log_2 p + (1-p) \log_2(1-p) & \text{(BSC)} \\ \frac{1}{2} \log_2(1 + \frac{J_0^2}{J^2}) & \text{(GC)} \end{cases}. \tag{9.2.28}$$

As we will mention in the next subsection, when we send ${}_N C_r$ combinations of p bits among the original image $\{\xi\}$, as products $\xi_{i1} \cdots \xi_{ip}$, error-less decoding might be achieved in the limit of $p \to \infty$. We call this type of code as *Sourlas codes* [374]. For this Sourlas codes, we obtain the following effective Hamiltonian.

$$\mathcal{H}_{\text{eff}} = -\beta_J \sum_{(i1\cdots ip)} J_{i1\cdots ip} S_{i1} \cdots S_{ip} - h \sum_i \tau_i S_i. \tag{9.2.29}$$

It is clear that this Hamiltonian is identical to that of the *Ising spin glass model with p-body interaction* under some random fields on cites.

9.2.3 Quantum Version of Models

In the previous subsections, we explained the relationship between the Bayesian statistics and statistical mechanics. We found that there exists the effective Hamiltonian for each problem of image restoration and error-correcting codes. In order to extend the model systems to their quantum version, we add the transverse field term: $-\Gamma \sum_i S_i^x$ into the effective Hamiltonian. In this expression, $\{S^x\}$ means the x-component of the Pauli matrix and Γ controls the strength of *quantum fluctuation*. Each term ΓS_i^x appearing in the sum might be understood as tunnelling probability between the 'up-state' $|+\rangle_i$ ($S_i^z|+\rangle_i = +1|+\rangle_i$) and the 'down-state' $|-\rangle_i$ ($S_i^z|-\rangle_i = -1|-\rangle_i$), namely,

$$\Gamma S_i^x|+\rangle_i = \Gamma|-\rangle_i, \qquad \Gamma S_i^x|-\rangle_i = \Gamma|+\rangle_i, \tag{9.2.30}$$

intuitively. The eigenvalues $S_i = \pm 1$ of the matrix S_i^z correspond to the classical Ising spin (information message/pixel). As the result, the quantum version of image restoration is reduced to that of statistical mechanics for the following effective Hamiltonian

$$\mathcal{H}_{\text{eff}}^{quantum} = -\beta_m \sum_{\langle ij \rangle} S_i^z S_j^z - h \sum_i \tau_i S_i^z - \Gamma \sum_i S_i^x. \tag{9.2.31}$$

We also obtain the quantum version of the effective Hamiltonian for error-correcting codes as

$$\mathscr{H}_{\mathrm{eff}}^{quantum} = -\beta_J \sum_{(i1\cdots ip)} J_{i1\cdots ip} S_{i1}^z \cdots S_{ip}^z - h \sum_i \tau_i S_i^z - \Gamma \sum_i S_i^x. \tag{9.2.32}$$

We should keep in mind that in the context of the MAP estimation, it might be useful for us to controlling the strength of the quantum fluctuation, namely, the amplitude of the transverse field Γ as $\Gamma \to 0$ during the quantum Markov chain Monte Carlo steps. If this annealing process of Γ is slow enough, at the end $\Gamma = 0$, we might obtain the ground states of the classical spin systems described by the following Hamiltonian

$$\mathscr{H}_{\mathrm{eff}}^{classical} = -\beta_m \sum_{\langle ij \rangle} S_i S_j - h \sum_i \tau_i S_i \tag{9.2.33}$$

for image restoration and

$$\mathscr{H}_{\mathrm{eff}}^{classical} = -\beta_J \sum_{(i1\cdots ip)} J_{(i1\cdots ip)} S_{i1} \cdots S_{ip} - h \sum_i \tau_i S_i \tag{9.2.34}$$

for error-correcting codes, where S_i stands for the eigenvalue of the matrix S_i^z. This is an essential idea of the *quantum annealing*. We will revisit this problem in the last section. In this section, we investigate its averaged case performance by analysis of the infinite range model and by caring out quantum Markov chain Monte Carlo simulations.

9.2.4 Analysis of the Infinite Range Model

In the previous subsections, we completely defined our two problems of information processing, that is to say, image restoration and error-correcting codes as random spin systems in a transverse field. We found that there exist two possible candidates to determine the original sequence of the messages/pixels. The first one is the MAP estimation and the estimate is regarded as ground states of the effective Hamiltonian that is defined as a minus of logarithm of the posterior distribution. As we mentioned, to carry out the optimisation of the Hamiltonian, both the simulated annealing and the quantum annealing are applicable. In order to construct the quantum annealing, we should add the transverse field to the effective Hamiltonian and control the amplitude of the field Γ during the quantum Markov chain Monte Carlo steps. Therefore, the possible extension of the classical spin systems to the corresponding quantum spin systems in terms of the transverse field is an essential idea.

Besides the MAP estimate as a solution of the optimisation problems, the MPM estimate, which is given by the sign of the local magnetisation of the spin system, is also available. This estimate is well-known as the estimate that minimises the bit-error rate. Performances of both the MAP and the MPM estimations are evaluated through this bit-error rate.

In order to evaluate the performance, we first attempt to calculate the bit-error rate analytically by using the mean-field infinite range model. As the most famous example of solvable model, Sherrington-Kirkpatrick model [369] in spin glasses, we also introduce the solvable models for both image restoration and error-correcting codes. In this section, according to the previous work by the present author [190], we first investigate the performance of image restoration.

It is important to bear in mind that in our Hamiltonian, there exists two types of terms, namely, $\mathscr{A}_0 = -\mathscr{H}_{\mathrm{eff}}^{classical}$ and $\mathscr{A}_1 = -\Gamma \sum_i S_i^x$, and they do not commute with each other. Therefore, it is impossible to calculate the partition function directly. Hence, here we use the *Suzuki-Trotter (ST) decomposition* [386, 403]

$$Z_{\mathrm{eff}} = \lim_{M\to\infty} \mathrm{tr}\left\{\exp\left(\frac{\mathscr{A}_0}{M}\right)\exp\left(\frac{\mathscr{A}_1}{M}\right)\right\}^M \tag{9.2.35}$$

to cast the problem into an equivalent classical spin system. In following, we calculate the macroscopic behaviour of the model system with the assistance of the ST formula [386, 403] and replica method [369] for the data $\{\xi, J, \tau\}$ average $\ll\cdots\gg$:

$$\ll \log Z_{\mathrm{eff}} \gg = \lim_{n\to 0} \frac{\ll Z_{\mathrm{eff}}^n \gg - 1}{n} \tag{9.2.36}$$

of the infinite range model.

9.2.4.1 Image Restoration

In order to analyse the performance of the MAP and the MPM estimation in image restoration, we suppose that the original image is generated by the next probability distribution,

$$P\big(\{\xi\}\big) = \frac{\exp(\frac{\beta_s}{N}\sum_{ij}\xi_i\xi_j)}{Z(\beta_s)}, \qquad Z(\beta_s) = \sum_{\{\xi\}} \exp\left(\frac{\beta_s}{N}\sum_{ij}\xi_i\xi_j\right), \tag{9.2.37}$$

namely, the Gibbs distribution of the ferromagnetic Ising model at the temperature $T_s = \beta_s^{-1}$. For this original image and under the Gaussian channel, the macroscopic properties of the system like a bit-error rate are derived from the data-averaged free energy $\ll \log Z_{\mathrm{eff}} \gg$. Using the ST formula and the replica method, we write down the replicated partition function as follows

$$\begin{aligned}
\ll Z_{\mathrm{eff}}^n \gg &= \sum_{\{\xi\}} \int_{-\infty}^{\infty} \prod_{ij} \frac{dJ_{ij}}{\sqrt{2\pi J^2/N}} \, \mathrm{e}^{-\frac{N}{2J^2}\sum_{ij}(J_{ij}-\frac{J_0}{N}\xi_i\xi_j)^2} \\
&\quad \times \int_{-\infty}^{\infty} \prod_i \frac{d\tau_i}{\sqrt{2\pi}a} \, \mathrm{e}^{-\frac{1}{2a^2}\sum_i(\tau_i - a_0\xi_i)^2} \times \frac{\mathrm{e}^{(\beta_s/N)\sum_{ij}\xi_i\xi_j}}{Z(\beta_s)} \\
&\quad \times \mathrm{tr}_{\{S\}} \prod_{\alpha=1}^{n}\prod_{K=1}^{M} \exp\left[\frac{\beta_J}{M}\sum_{ij} J_{ij} S_{iK}^{\alpha} S_{jK}^{\alpha} + \frac{\beta_m}{MN}\sum_{ij} S_{iK}^{\alpha} S_{jK}^{\alpha} \right. \\
&\quad \left. + \frac{h}{M}\sum_i \tau_i S_{iK}^{\alpha} + B\sum_i S_{iK}^{\alpha} S_{i,K+1}^{\alpha}\right]
\end{aligned} \tag{9.2.38}$$

where $\ll \cdots \gg$ means average over the quenched randomness, namely, over the joint probability $P(\{J\},\{\tau\},\{\xi\})$. We should keep in mind that these quantities $\{\xi\}$ and $\{J\},\{\tau\}$ mean the data we send to the receiver and the outputs of the channel the receiver obtain, respectively. Therefore, by calculating these averages $\ll \cdots \gg$, we can evaluate the *data-averaged case performance* of the image restoration [190]. We also defined the partition function $Z(\beta_s)$ for the original images and B as $Z(\beta_s) \equiv \sum_{\{\xi\}} \mathrm{e}^{(\beta_s/N)\sum_{ij}\xi_i\xi_j}$, $B \equiv (1/2)\log\coth(\Gamma/M)$. The standard replica calculation leads to the following expressions of the free energy density:

$$\ll \log \mathrm{Z}_{\mathrm{eff}} \gg = \frac{\ll Z_{\mathrm{eff}}^n \gg -1}{nN} = -\frac{f_0^{RS}}{n} - f^{RS} \tag{9.2.39}$$

with

$$f_0^{RS} = \frac{1}{2}\beta_s m_0^2 - \log 2\cosh(\beta_s m_0) \tag{9.2.40}$$

$$\begin{aligned} f^{RS} &= -\frac{(\beta_J J)^2}{2}Q^2 + \frac{(\beta_J J)^2}{2}S^2 + \frac{\beta_m}{2}m^2 + \frac{\beta_J J_0}{2}t^2 \\ &\quad - \sum_{\xi}\mathcal{M}(\xi)\int_{-\infty}^{\infty} Du \log \int_{-\infty}^{\infty} Dw\, 2\cosh\sqrt{\Phi^2+\Gamma^2} \end{aligned} \tag{9.2.41}$$

and the saddle point equations with respect to the order parameters as follows

$$\ll \langle S_{iK}^{\alpha}\rangle \gg = m = \sum_{\xi}\mathcal{M}(\xi)\int_{-\infty}^{\infty} Du \int_{-\infty}^{\infty} D\omega \left(\frac{\Phi \sinh \Xi}{\Xi \Omega}\right) \tag{9.2.42}$$

$$\ll \xi_i \langle S_{iK}^{\alpha}\rangle \gg = t = \sum_{\xi}\xi\mathcal{M}(\xi)\int_{-\infty}^{\infty} Du \int_{-\infty}^{\infty} D\omega \left(\frac{\Phi \sinh \Xi}{\Xi \Omega}\right) \tag{9.2.43}$$

$$\ll \langle (S_{iK}^{\alpha})^2\rangle \gg = Q = \sum_{\xi}\mathcal{M}(\xi)\int_{-\infty}^{\infty} Du \left[\int_{-\infty}^{\infty} D\omega \left(\frac{\Phi \sinh \Xi}{\Xi \Omega}\right)\right]^2 \tag{9.2.44}$$

$$\ll \langle S_{iK}^{\alpha} S_{iL}^{\alpha}\rangle \gg = S = \sum_{\xi}\mathcal{M}(\xi)\int_{-\infty}^{\infty} \frac{Du}{\Omega}\int_{-\infty}^{\infty}\left[\left(\frac{\Phi}{\Xi}\right)^2 \cosh \Xi + \Gamma^2\left(\frac{\sinh \Xi}{\Xi^3}\right)\right] \tag{9.2.45}$$

with $\ll \xi_i \gg = m_0 = \tanh(\beta_s m_0)$ and $\mathcal{M}(\xi) = \mathrm{e}^{\beta_s m_0 \xi}/2\cosh(\beta_s m_0)$, where we used the replica symmetric and the static approximation, that is,

$$t_K = t, \qquad S_\alpha(KL) = \begin{cases} S & (K \neq L) \\ 1 & (K = L) \end{cases}, \qquad Q_{\alpha\beta} = Q \tag{9.2.46}$$

and $\langle \cdots \rangle$ denotes the average over the posterior distribution and Φ, y and Ω are defined as

$$\Phi \equiv u\sqrt{(ah)^2 + (J\beta_J)^2 Q} + J\beta_J \omega\sqrt{S-Q} + (a_0 h + J_0\beta_J t)\xi + \beta_m m \tag{9.2.47}$$

$$\Xi \equiv \sqrt{\Phi^2 + \Gamma^2}, \qquad \Omega \equiv \int_{-\infty}^{\infty} D\omega \cosh \Xi. \tag{9.2.48}$$

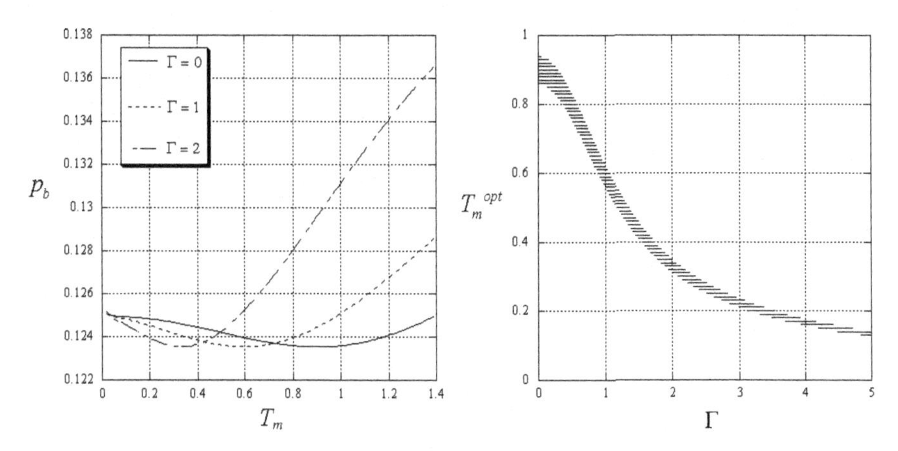

Fig. 9.6 The bit-error rate $p_b = (1-R)/2$ without exchange term ($\beta_J = 0$) as a function of temperature $T_m = \beta_m^{-1}$ (*left*: from [190]). Keeping the ratio to $h/\beta_m = \beta_\tau/\beta_s = a_0/a^2\beta_s = 0.9$ (we set $a_0 = a = 1$), we change the value of T_m. For the case of $\Gamma = 0$, p_b takes its minimum at $T_m = T_s = 0.9$. For finite Γ, the optimal temperature T_m is not T_s, however, the minimum of p_b does not change. The *right panel* shows the optimal temperature T_m^{opt} as a function of Γ

Then, the overlap R which is a measure of retrieval quality is calculated explicitly as

$$\langle\!\langle \xi_i \,\mathrm{sgn}\big(\langle S_{iK}^{\alpha}\rangle\big)\rangle\!\rangle = R = \sum_{\xi} \xi \mathscr{M}(\xi) \int_{-\infty}^{\infty} Du \int_{-\infty}^{\infty} Dw \,\mathrm{sgn}(\Phi), \tag{9.2.49}$$

then, of course, the bit-error rate is given by $p_b = (1-R)/2$.

9.2.4.2 Image Restoration at Finite Temperature

We first investigate the image restoration without parity check term $\beta_J = 0$. For this case, the saddle point equations lead to the following much simpler coupled equations:

$$m_0 = \tanh(\beta_s m_0) \tag{9.2.50}$$

$$m = \sum_{\xi} \mathscr{M}(\xi) \int_{-\infty}^{\infty} Du \, \frac{\Phi_0 \tanh\sqrt{\Phi_0^2 + \Gamma^2}}{\sqrt{\Phi_0^2 + \Gamma^2}} \tag{9.2.51}$$

with $\Phi_0 \equiv m\beta_m + a_0 h\xi + ahu$. Then, the overlap R is also reduced to

$$R = \sum_{\xi} \xi \mathscr{M}(\xi) \int_{-\infty}^{\infty} Du \,\mathrm{sgn}(\Phi_0) = 1 - 2p_b \tag{9.2.52}$$

where R depends on Γ through m. In Fig. 9.6 (left), we plot the bit-error rate p_b as a function of $T_m = \beta_m^{-1}$ for the case of no parity check $\beta_J = 0$. We choose the temperature of the original image $T_s^{-1} = \beta_s = 0.9$ and noise rate $\beta_\tau = a_0/a^2 = 1$. We

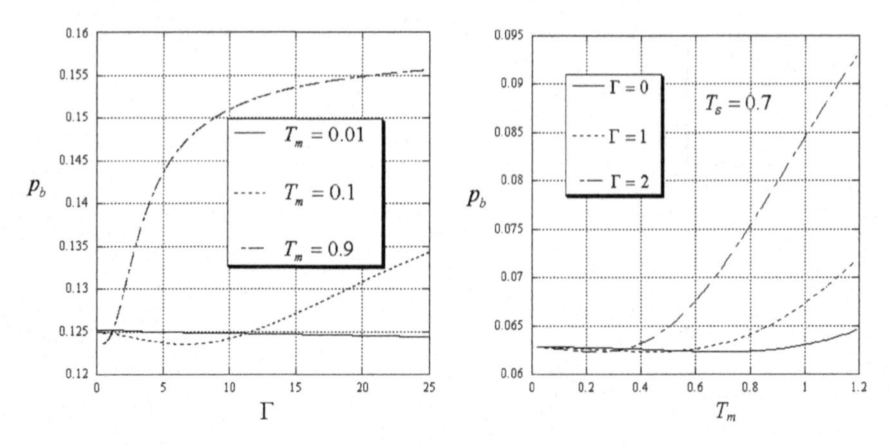

Fig. 9.7 The bit-error rate p_b is drawn for $T_m = 0.01, 0.1$ and $T_m = 0.9$ as a function of Γ (*left*). The *right panel* is the same type of the plot as the *right panel* in Fig. 9.6 for the case of $T_s = 0.7$

keep the ratio h/β_m to its optimal value $\beta_\tau/\beta_s = 0.9$ and investigate T_m-dependence of p_b. Then, the parameter T_m has a meaning of 'temperature' for simulated annealing. Obviously, $p_b^{(MAP)} = \lim_{T_m \to 0} p_b$ and p_b at $T_m = T_s$ is the lowest value of $p_b^{(MPM)}$ for $\Gamma = 0$.

Let us stress again that in practice, the infinite range model is not relevant for realistic two dimensional image restoration because all pixels are neighbour each other. In order to restore these two dimensional images, we should use the prior $P(\{\xi\})$ for two dimension. In fact, let us think about the overlap r between an original pixel ξ_i and corresponding degraded pixel τ_i, namely,

$$r = \langle\langle \xi_i \tau_i \rangle\rangle = \frac{\sum_{\tau,\xi=\pm 1} \mathrm{e}^{\beta_\tau \xi \tau + \beta_s m_0 \xi} (\xi \tau)}{4\cosh(\beta_\tau)\cosh(\beta_s m_0)} = \tanh(\beta_\tau). \tag{9.2.53}$$

From this relation, the error probability p_τ is given as

$$p_\tau = (1-r)/2 = 1/\big(1+\mathrm{e}^{2\beta_\tau}\big) = 0.119 < p_b^{(MPM)} \tag{9.2.54}$$

for $\beta_\tau = 1$, and unfortunately, the restored image becomes much worse than the degraded (see Fig. 9.6 (left)). This is because any spacial structure is ignored in this artificial model. This result might be understood as a situation in which we try to restore the finite dimensional image with some structures by using the infinite range prior without any structure (namely, the correlation length between pixels is also infinite).

However, the infinite range model is useful to predict the qualitative behaviour of macroscopic quantities like bit-error rate and we can grasp the details of its hyper-parameters (namely, T_m, h or Γ) dependence and can also compare the MAP with the MPM estimations. This is a reason why we introduce this model to the analysis of image restoration problems. Of course, if we use two dimensional structural priors, the both the MAP estimations via simulated and quantum annealing and the MPM estimation by using thermal and quantum fluctuations work well for realistic

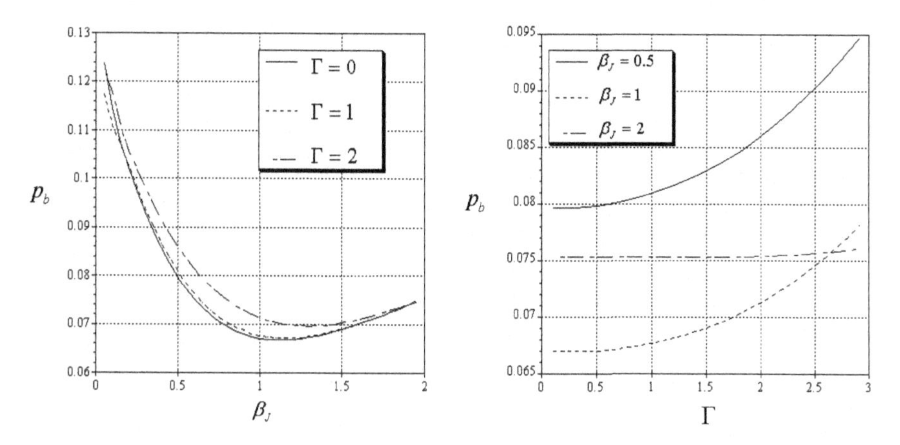

Fig. 9.8 The bit-error rate p_b as a function of β_J for $\Gamma = 0, 1, 2$ keeping the ration constant $h/\beta_m = \beta_\tau/\beta_s$ (*left*). In *right panel*, p_b as a function of Γ is plotted for the case of $\beta_J = 0.5, 1, 2.0$

two dimensional image restoration. In the following subsections, we will revisit this problem and find it. It is also important for us to bear in mind that the quality of the restoration depends on the macroscopic properties of the original image.

In our choice of the original image, its macroscopic qualities are determined by the temperature T_s and magnetisation m_0 as a solution of $m_0 = \tanh(\beta_s m_0)$. Although we chose the temperature $T_s = 0.9$ in Fig. 9.6 (left), it is important to check the retrieval quality for different temperatures T_s. In Fig. 9.7 (right), we plot the bit-error rate for the case of $T_s = 0.7$. From this panel, we find $p_b < p_\tau$ and the MPM estimation improves the quality of the restoration.

For $\Gamma > 0$, the optimal temperature which gives the minimum of p_b is not T_s. In the right panel of Fig. 9.6, we plot the T_m^{opt} as a function of Γ. In Fig. 9.7 (left), we plot the bit-error rate as a function of Γ for $T_m = T_s = 0.9$ setting the ratio to its optimal value $h/\beta_m = \beta_\tau/\beta_s = 0.9$. From this figure, we find that the MPM optimal estimate no longer exists by adding the transverse field $\Gamma > 0$ and the bit-error rate p_b increases as the amplitude of the transverse field Γ becomes much stronger.

On the other hand, when we set the temperature $T_m = 0.01$, the Γ-dependence of the bit-error rate is almost flat (see Fig. 9.7 (right)). We should notice that p_b at $\Gamma = 0$ for $T_m = 0$ corresponds to the performance of the MAP estimation by quantum annealing. We discuss the performance of the quantum annealing in the following subsections.

We next consider the performance for the MAP and the MPM estimations with parity check term ($\beta_J \neq 0$). We plot the result in Fig. 9.8. As we mentioned before, two body parity check term works very well to decrease the bit-error rate p_b. However, in this case, there does not exist the optimal β_J which minimises the bit-error rate for any finite values of Γ. As we see the left panel in Fig. 9.8, for small value of β_J, the restoration by a finite Γ is superior to that of absence of the transverse field ($\Gamma = 0$).

9.2.4.3 Hyperparameter Estimation

In this subsection, we evaluated the performance of the MAP and the MPM estimations in image restoration through the bit-error rate. In these results, we found that the macroscopic parameters, β_m, h and Γ-dependence of the bit-error rate have important information to retrieve the original image. However, from the definition, (9.2.49), (9.2.52), as the bit-error rate contains the original image $\{\xi\}$, it is impossible for us to use p_b as a cost function to determine the best choice of these parameters. In statistics, we usually use the *marginal likelihood* [193] which is defined by the logarithm of the normalisation constant of $\mathrm{tr}_{\{S\}} P(\{S\}|\{\tau\})P_m(\{S\})$, that is,

$$K\left(\beta_m, h, \Gamma : \{\tau\}\right) \equiv \log Z_{Pos.} - \log Z_{Pri.} - \log Z_L \qquad (9.2.55)$$

where $Z_{Pos.}$, $Z_{Pri.}$ and Z_L are normalisation constants for the posterior, the prior and the likelihood, and which are given by

$$Z_{Pos.} = \mathrm{tr}_{\{\sigma\}}\, \mathrm{e}^{\beta_m \sum_{ij} S_i^z S_j^z + h\sum_i \tau_i S_i^z + \Gamma \sum_i S_i^x} \qquad (9.2.56)$$

$$Z_{Pri.} = \mathrm{tr}_{\{\sigma\}}\, \mathrm{e}^{\beta_m \sum_{ij} S_i^z S_j^z + \Gamma \sum_i S_i^x}, \qquad Z_L = \mathrm{tr}_{\{\tau\}}\, \mathrm{e}^{h\sum_i \tau_i S_i^z}, \qquad (9.2.57)$$

respectively. For simplicity, let us concentrate ourselves to the case of no parity check $\beta_J = 0$.

It must be noted that the marginal likelihood (9.2.55) is constructed by using the observables $\{\tau\}$ and does not contain the original image $\{\xi\}$ at all. Therefore, in practice, the marginal likelihood has a lot of information to determine the macroscopic parameters, what we call *hyperparameters*, before we calculate the MAP and the MPM estimates.

In the infinite range model, it is possible for us to derive the data-averaged marginal likelihood per pixel $K(\beta_J, h, \Gamma) = \langle\!\langle K(\beta_J, h, \Gamma : \{\tau\})\rangle\!\rangle / N$ explicitly. Here we first investigate the hyperparameter dependence of the marginal likelihood. $\log Z_{Pri}$ and $\langle\!\langle \log Z_L \rangle\!\rangle = \langle\!\langle \log \int_{-\infty}^{\infty} \prod_i d\tau_i\, F_\tau(\tau_i) \mathrm{e}^{h\tau_i S_i^z} \rangle\!\rangle$ per pixel can be calculated as

$$\frac{\log Z_{Pri}}{N} = -\frac{\beta_m m_1^2}{2} + \log 2\cosh\sqrt{(\beta_m m_1)^2 + \Gamma^2} \qquad (9.2.58)$$

$$\frac{\langle\!\langle \log Z_L \rangle\!\rangle}{N} = -\frac{h^2}{2}\left[\left(\frac{a_0}{ah}\right)^2 - a^2\right] \qquad (9.2.59)$$

and the data average of the first term of the right hand side of (9.2.55) is identical to the free energy density for $\beta_J = 0$. Thus, we obtain the data-averaged marginal likelihood as follows.

$$K(\beta_m, h, \Gamma) = -\frac{\beta_m m^2}{2} + \sum_{\xi} \mathscr{M}(\xi) \int_{-\infty}^{\infty} Du \log 2\cosh\sqrt{\Phi_0^2 + \Gamma^2} + \frac{\beta_m m_1^2}{2} - \log 2\cosh\sqrt{(\beta_m m_1)^2 + \Gamma^2} + \frac{h^2}{2}\left[\left(\frac{a_0}{ah}\right)^2 - a^2\right] \qquad (9.2.60)$$

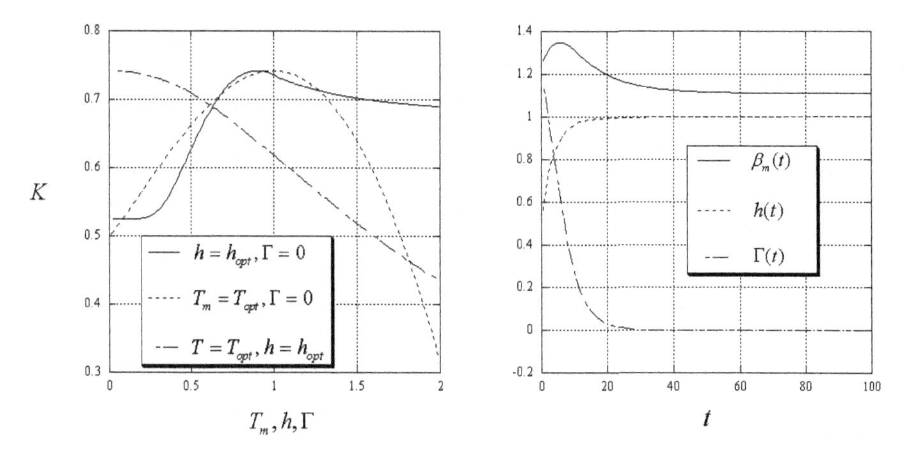

Fig. 9.9 The data-averaged marginal likelihood as a function of hyperparameters, β_m, h and Γ (*left*). The *right panel* shows the time development of the hyperparameter β_m, h and Γ via gradient descent of the marginal likelihood. We set the time constants $c_{\beta_m} = c_h = c_\Gamma = 1$ and the values of true hyperparameters as $T_s = \beta_s^{-1} = 0.9$, $\beta_\tau = 1$

where m_1, m mean the magnetisations of the prior and the posterior, are given by

$$m_1 = \frac{\beta_m m_1 \tanh \sqrt{(\beta_m m_1)^2 + \Gamma^2}}{\sqrt{(\beta_m m_1)^2 + \Gamma^2}} \tag{9.2.61}$$

and (9.2.51), respectively. In Fig. 9.9 (left), we plot $K(\beta_m, h, \Gamma)$. In this figure, we set $T_s = 0.9$, $\beta_\tau = 1$. We found that the data-averaged marginal likelihood takes its maximum at $T_m = T_s$, $h = \beta_\tau$ and $\Gamma = 0$. This result might be naturally understood because the performance of both the MAP and MPM estimation should be the best for setting the probabilistic models of the noise channel and the distribution of the original image to the corresponding true probabilities. Therefore, it seems that the transverse field Γ has no meaning for restoration.

When we attempt to maximise the marginal likelihood via gradient descent, we need to solve the following coupled equations

$$c_{\beta_m} \frac{d\beta_m}{dt} = \frac{\partial K}{\partial \beta_m} = \left\langle \sum_{ij} S_i^z S_j^z \right\rangle_{Pos.} - \left\langle \sum_{ij} S_i^z S_j^z \right\rangle_{Pri.} \tag{9.2.62}$$

$$c_h \frac{dh}{dt} = \frac{\partial K}{\partial h} = \left\langle \sum_i \tau_i S_i^z \right\rangle_{Pos.} - \left\langle \sum_i \tau_i S_i^z \right\rangle_{Pri.} - \left\langle \sum_i \tau_i S_i^z \right\rangle_L \tag{9.2.63}$$

$$c_\Gamma \frac{d\Gamma}{dt} = \frac{\partial K}{\partial \Gamma} = \left\langle \sum_i S_i^x \right\rangle_{Pos.} - \left\langle \sum_i S_i^x \right\rangle_{Pri.} \tag{9.2.64}$$

with the definitions of the brackets

$$\langle \cdots \rangle_{Pos.} = \frac{\mathrm{tr}_{\{S\}} (\cdots) \, \mathrm{e}^{\beta_m \sum_{ij} S_i^z S_j^z + h \sum_i \tau_i S_i^z + \Gamma \sum_i S_i^x}}{\mathrm{tr}_{\{\sigma\}} \, \mathrm{e}^{\beta_m \sum_{ij} S_i^z S_j^z + h \sum_i \tau_i S_i^z + \Gamma \sum_i S_i^x}} \tag{9.2.65}$$

$$\langle\cdots\rangle_{Pri.} = \frac{\mathrm{tr}_{\{S\}}(\cdots)\,\mathrm{e}^{\beta_m \sum_{ij} S_i^z S_j^z + \Gamma \sum_i S_i^x}}{\mathrm{tr}_{\{S\}}\,\mathrm{e}^{\beta_m \sum_{ij} S_i^z S_j^z + \Gamma \sum_i S_i^x}}, \qquad \langle\cdots\rangle_L = \frac{\mathrm{tr}_{\{\tau\}}(\cdots)\,\mathrm{e}^{h \sum_i \tau_i S_i^x}}{\mathrm{tr}_{\{\tau\}}\,\mathrm{e}^{h \sum_i \tau_i S_i^x}} \tag{9.2.66}$$

and time constants c_{β_m}, c_h and c_Γ.

For the infinite range model, we easily calculate these expectation explicitly. The results are given by

$$c_{\beta_m}\frac{d\beta_m}{dt} = \frac{m_1^2 - m^2}{2} - \frac{\beta_m m_1^2 \tanh\sqrt{(\beta_m m_1)^2 + \Gamma^2}}{\sqrt{(\beta_m m_1)^2 + \Gamma^2}} + m\sum_\xi \mathscr{M}(\xi)\int_{-\infty}^{\infty} Du\,\frac{\Phi_0 \tanh\sqrt{\Phi_0^2 + \Gamma^2}}{\sqrt{\Phi_0^2 + \Gamma^2}} \tag{9.2.67}$$

$$c_h\frac{dh}{dt} = -a^2 h + \sum_\xi \mathscr{M}(\xi)\int_{-\infty}^{\infty} Du\,\frac{\Phi_0(a_0\xi + au)\tanh\sqrt{\Phi_0^2 + \Gamma^2}}{\sqrt{\Phi_0^2 + \Gamma^2}} \tag{9.2.68}$$

$$c_\Gamma\frac{d\Gamma}{dt} = -\frac{\Gamma\tanh\sqrt{(\beta_m m_1)^2 + \Gamma^2}}{\sqrt{(\beta_m m_1)^2 + \Gamma^2}} + \Gamma\sum_\xi \mathscr{M}(\xi)\int_{-\infty}^{\infty} Du\,\frac{\tanh\sqrt{\Phi_0^2 + \Gamma^2}}{\sqrt{\Phi_0^2 + \Gamma^2}} \tag{9.2.69}$$

where m_1 and m satisfy (9.2.51) and (9.2.61). We plot the results by solving the differential equations with respect to the hyperparameters, namely, (9.2.67), (9.2.68), (9.2.69) numerically in Fig. 9.9 (right). From this figure, we find that the β_m and h converge to the true values, whereas the Γ drops to zero. Hence, the gradient learning algorithm actually works well and we find the solution that maximises the marginal likelihood, however, it is impossible to obtain the optimal Γ which minimises the bit-error rate. As we mentioned, this result comes from the fact that our probabilistic model with transverse field is inconsistent with the true model.

In practice for the realistic two-dimensional image, we should solve the above equations, and then, we need to evaluate these expectations for every time steps by using the quantum Markov chain Monte Carlo method. It is obvious that it takes quite long time to obtain the solutions. From reasons mentioned above, it is convenient for us to suppress the error of hyperparameter estimation by introducing the transverse field. From Fig. 9.6 (left), Fig. 9.7 (right), we actually find these desirable properties.

9.2.4.4 Image Restoration Driven by Pure Quantum Fluctuation

In the above discussion, we investigated mainly the MPM estimation at finite temperature $T_m > 0$ according to Ref. [190]. However, it is worth while for us to check the following limit: $\beta_m \to \infty$ keeping the *effective amplitude of transverse field*

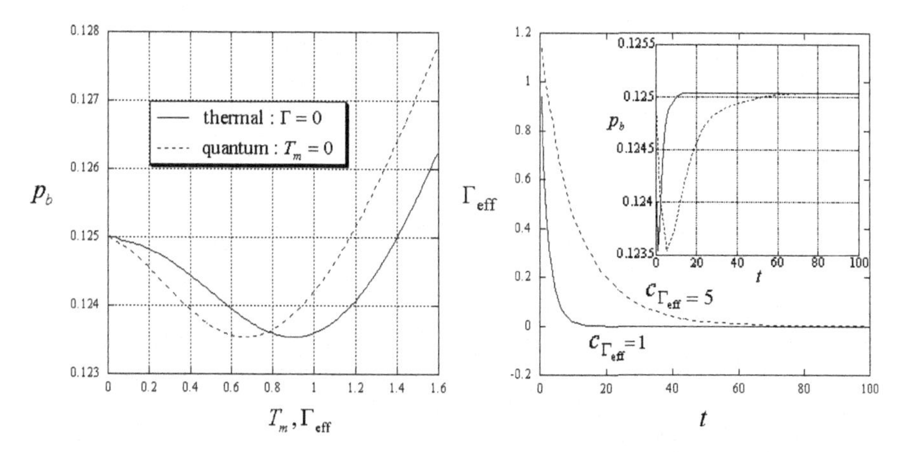

Fig. 9.10 The bit-error rate for the quantum ($T_m = 0$) and the thermal (classical) ($\Gamma = 0$) estimation (*left*). The *right panel* shows the time development of the effective amplitude of the transverse field $\Gamma_{\rm eff} = \Gamma/\beta_m$. The *inset* means the time dependence of the bit-error rate

$\Gamma_{\rm eff} = \Gamma/\beta_m$ finite. In this limit, we investigate pure effect of the quantum fluctuation without any thermal one. To evaluate the performances of the MAP and the MPM estimations for this zero temperature case, we set $\Phi_0 = \beta_m(m + h_* a_0 \xi + h_* au) = \beta_m \phi_0$, where h_* is its optimal value $h_* = \beta_s/\beta_\tau$, and consider the asymptotic form of the saddle point equations with respect to m and m_1 in the limit of $\beta_m \to \infty$. We easily find

$$m_1 = \sqrt{1 - \Gamma_{\rm eff}^2}, \qquad m = \sum_\xi \mathscr{M}(\xi) \int_{-\infty}^{\infty} \frac{\phi_0 \, Du}{\sqrt{\phi_0^2 + \Gamma_{\rm eff}^2}} \tag{9.2.70}$$

and the time evolution of $\Gamma_{\rm eff}$ as follows

$$c_{\Gamma_{\rm eff}} \frac{d\Gamma_{\rm eff}}{dt} = -\frac{\Gamma_{\rm eff}}{\sqrt{m_1^2 + \Gamma_{\rm eff}^2}} + \sum_\xi \mathscr{M}(\xi) \int_{-\infty}^{\infty} \frac{\Gamma_{\rm eff} \, Du}{\sqrt{\phi_0^2 + \Gamma_{\rm eff}^2}} \tag{9.2.71}$$

where $c_{\Gamma_{\rm eff}} = \beta_m c_\Gamma$. The bit-error rate is given by $p_b = (1 - m_0)/2 + \sum_\xi \mathscr{M}(\xi) \times \xi H(u_*)$, where $u_* = (a_0 h_* \xi + m)/a h_*$. We fist plot the $\Gamma_{\rm eff}$-dependence of the bit-error rate at $T_m = 0$ in Fig. 9.10. In this figure, the value at $\Gamma_{\rm eff} = 0$ corresponds to the quantum MAP estimation which might be realised by the quantum annealing. From this figure, we find that the performance of the quantum MPM estimation is superior to the MAP estimation and there exists some finite value of the amplitude Γ at which the bit-error rate takes its minimum. In the same figure, we also plot the T_m-dependence of the bit-error rate for $\Gamma = 0$. We find that, for both the quantum and the thermal cases, the best possible values of both the MAP and the MPM estimation is exactly the same. In Fig. 9.10 (right), we plot the time development of the effective amplitude of transverse field and the resultant bit-error rate. From this figure, we notice that at the beginning of the gradient descent the bit-error rate decreases but as Γ decreases to zero, the error converges to the best possible value for the quantum MAP estimation. The speed of the convergence is exponentially fast. Actually, in the asymptotic limit $t \to \infty$, $\Gamma_{\rm eff} \to 0$, Eq. (9.2.71) is solved as

$$\Gamma_{\rm eff} = \Gamma_{\rm eff}(0)\,{\rm e}^{-\theta_{\Gamma_{\rm eff}} t} \tag{9.2.72}$$

where

$$\theta_{\Gamma_{\rm eff}} \equiv (1/c_{\Gamma_{\rm eff}})\left(1 - \sum_{\xi} \mathscr{M}(\xi) \int_{-\infty}^{\infty} Du/|\phi_0|\right). \tag{9.2.73}$$

However, this fact does not mean that it is possible for us to decrease the effective amplitude of the transverse field to zero by using exponentially fast scheduling to realise the best possible performance of the quantum MAP estimation. This is because the time unit t appearing in (9.2.71) does not corresponds to the quantum Monte Carlo step and the dynamics (9.2.71) requires the (equilibrium) magnetisation $m(\Gamma_{\rm eff})$ at each time step in the differential equation. As the result, we need the information about m near $\Gamma_{\rm eff} \to 0$, namely, the asymptotic form: $m(t \to \infty, \Gamma_{\rm eff} \to 0)$ to discuss the annealing schedule to obtain the MAP estimation. Although we assume that each time step in (9.2.71), the system obeys the equilibrium condition:

$$m = \sum_{\xi} \mathscr{M}(\xi) \int_{-\infty}^{\infty} \phi_0\, Du/\sqrt{\phi_0^2 + \Gamma_{\rm eff}^2}, \tag{9.2.74}$$

we need the dynamics of m to discuss the optimal annealing scheduling about $\Gamma_{\rm eff}$. This point will be discussed in last section by means of the quantum Markov chain Monte Carlo method.

9.2.4.5 The Nishimori-Wong Condition on the Effective Transverse Field

From Fig. 9.10 (left), we find that the lowest value of the bit-error rate is the same both for the thermal and the quantum MPM estimations. In the thermal MPM estimation, Nishimori and Wong [299] found that the condition on which the best performance is obtained, namely, what we call *Nishimori-Wong condition*. They showed that the condition:

$$m/m_0 = (h/\beta_\tau)(\beta_s/\beta_m) \tag{9.2.75}$$

should hold in order to obtain the lowest value of the bit-error rate. When we set the hyperparameter h to its true value $h = \beta_\tau$, the condition is reduced to the simple form:

$$T_m^{\rm opt} = T_s. \tag{9.2.76}$$

Therefore, it is important for us to derive the same kind of condition which gives the best performance of the quantum MPM estimation. Here we derive the condition and show the lowest values of the p_b for the thermal and the quantum MPM estimations are exactly the same.

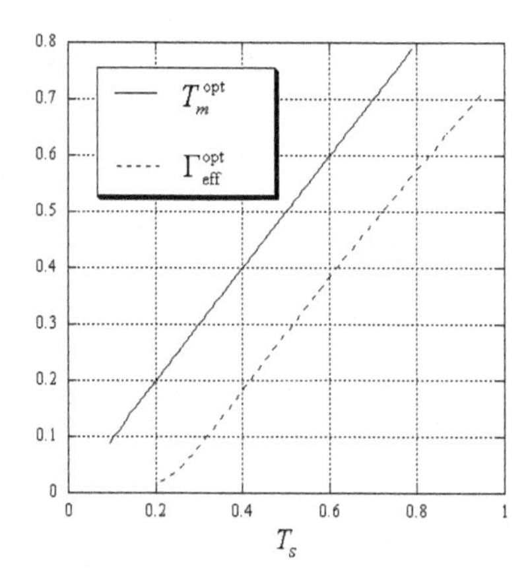

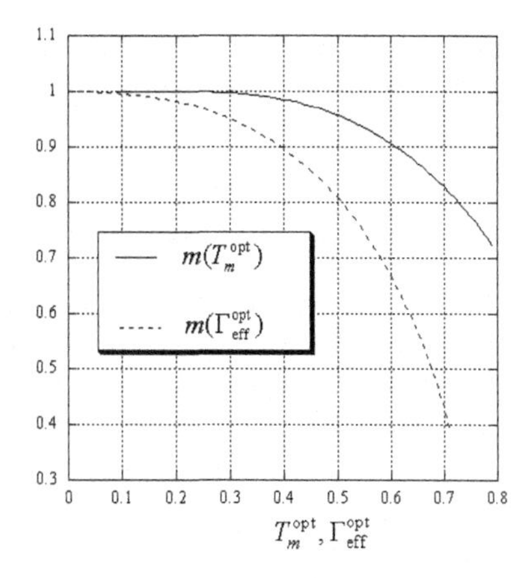

Fig. 9.11 The optimal temperature $T_m^{\rm opt}$ and the optimal transverse field $\Gamma_{\rm eff}^{\rm opt}$ as a function of the temperature T_s of the original image (*left*), respectively. The optimal temperature for the thermal MPM estimation $T_m^{\rm opt}$ is simply given by $T_m^{\rm opt} = T_s$ (Nishimori temperature). The *right panel* shows the magnetisations $m(T_m^{\rm opt})$ and $m(\Gamma_{\rm eff}^{\rm opt})$

To look for the condition, we first investigate the equation, $(\partial p_b/\partial \Gamma_{\rm eff}) = 0$ for $p_b = (1-m_0)/2 + \sum_\xi \xi \mathcal{M}(\xi) H(u_*)$. After some simple algebra, we obtain

$$m\big(\Gamma_{\rm eff}^{\rm opt}\big) \sum_\xi \xi \mathcal{M}(\xi) \exp\left[-\frac{\{a_0 h_* \xi + m(\Gamma_{\rm eff}^{\rm opt})\}^2}{2a^2 h_*^2}\right] = 0. \tag{9.2.77}$$

Taking into account that $m(\Gamma_{\rm eff}) \neq 0$ is needed for meaningful image restorations, the Nishimori-Wong condition for the quantum MPM estimation is written by

$$\frac{m_0(\beta_s)}{m(\Gamma_{\rm eff}^{\rm opt})} = \frac{a_0}{a^2 h_* \beta_s}. \tag{9.2.78}$$

As we chose $h_* = \beta_\tau/\beta_s$, $\beta_\tau = a_0/a^2$, this condition is simply rewritten as

$$m_0(\beta_s) = m(\Gamma_{\rm eff}). \tag{9.2.79}$$

Let us summarise the Nishimori-Wong condition for the MPM estimation:

$$\textbf{Thermal:} \quad T_m^{\rm opt} = T_s \quad \text{(Nishimori and Wong 1999)}$$

$$\textbf{Quantum:} \quad m_0(\beta_s) = \sum_\xi \mathcal{M}(\xi) \int_{-\infty}^{\infty} \frac{\phi_0\, Du}{\sqrt{\phi_0^2 + (\Gamma_{\rm eff}^{\rm opt})^2}}$$

In Fig. 9.11, we plot the temperature of the original image T_s-dependence of the optimal temperature $T_m^{\rm opt}$ and the optimal amplitude of the transverse field $\Gamma_{\rm eff}^{\rm opt}$. In the right panel of this figure, the magnetisations $m(T_m^{\rm opt})$ and $m(\Gamma_{\rm eff}^{\rm opt})$ are plotted. The effective amplitude of the transverse field $\Gamma_{\rm eff}$ at which the bit-error rate takes its minimum in Fig. 9.10 is consistent with the $\Gamma_{\rm eff}^{\rm opt}(T_s = 0.9) \simeq 0.66$ as shown in Fig. 9.11 (left).

From these results, it is shown that the lowest values of the of the bit-error rate for both the thermal and the quantum MPM estimations are exactly the same and the value is given by

$$p_b = \frac{1-m_0}{2} + \sum_{\xi} \xi \mathscr{M}(\xi) H\left(\frac{a_0 h_* \xi + m_0}{a h_*}\right). \tag{9.2.80}$$

Therefore, we conclude that it is possible for us to construct the MPM estimation purely induced by the quantum fluctuation (without any thermal fluctuation) and the best possible performance is exactly the same as that of the thermal MPM estimation.

9.2.4.6 Error-Correcting Codes

Here we introduce our model system of error-correcting codes and mention the Shannon's bound. In our error-correcting codes, in order to transmit the original message $\{\xi\} \equiv (\xi_1, \ldots, \xi_N), \xi_i \in \{-1, 1\}$ through some noisy channel, we send all possible combinations ${}_N C_p$ of the products of p-components in the N-dimensional vector $\{\xi\}$ such as

$$J^0_{i1,\ldots,ip} = \xi_{i1}\xi_{i2}\cdots\xi_{ip} \tag{9.2.81}$$

as 'parity'. Therefore, the rate of the transmission is now evaluated as

$$R = \frac{N}{{}_N C_p} \simeq \frac{p!}{N^{p-1}} \tag{9.2.82}$$

in the limit of $N \to \infty$ keeping the p finite.

On the other hand, when we assume the additive white Gaussian noise (AWGN) channel with mean $(J_0 p!/N^{p-1})J^0_{i1i2\cdots ip}$ and variance $\{J\sqrt{p!/2N^{p-1}}\}^2$, that is, when the output of the channel $J_{i1i2\cdots ip}$ is given by

$$J_{i1i2\cdots ip} = \left(\frac{J_0 p!}{N^{p-1}}\right) J^0_{i1i2\cdots ip} + J\sqrt{\frac{p!}{2N^{p-1}}}\,\eta, \quad \eta = \mathscr{N}(0,1), \tag{9.2.83}$$

the channel capacity C [19, 254] leads to

$$C = \frac{1}{2}\log_2\left(1 + \frac{\{(J_0 p!/N^{p-1})J_{i1\cdots ip}\}^2}{J^2 p!/2N^{p-1}}\right) \simeq \frac{J_0^2 p!}{J^2 N^{p-1}\log 2} \tag{9.2.84}$$

in the same limit as in the derivation (9.2.82) (we also used the fact $(J_{i1\cdots ip})^2 = 1$). The factors $p!/N^{p-1}$ or $\sqrt{p!/2N^{p-1}}$ appearing in (9.2.83) are needed to take a proper thermodynamic limit (to make the energy of order 1 object) as will be explained in the next section.

Then, the channel coding theorem tells us that zero-error transmission is achieved if the condition $R \le C$ is satisfied. For the above case, we have $R/C = (J/J_0)^2 \log 2 \le 1$, that is,

$$\frac{J_0}{J} \ge \left(\frac{J_0}{J}\right)_c \equiv \sqrt{\log 2}. \tag{9.2.85}$$

The above inequality means that if the signal-to-noise ratio J_0/J is greater than or equal to $\sqrt{\log 2}$, the error probability of decoding behaves as $P_e \simeq 2^{-N(C-R)} \to 0$ in the thermodynamic limit $N \to \infty$. In this sense, we might say that the zero-error transmission is achieved asymptotically in the limit $N \to \infty$, $C, R \to 0$ keeping $R/C = \mathcal{O}(1) \le 1$ for the above what we call *Sourlas codes* [374].

In the following subsections, we investigate the performance of the decoding in the so-called Sourlas codes [374], in which uncertainties in the prior are introduced as the quantum transverse field. As we saw, we usually choose the prior in the Sourlas codes as $P(\{\sigma\}) = 2^{-N}$ (the uniform prior).

To introduce the quantum fluctuation into the system, we shall multiply the posterior by the factor $\prod_i \mathrm{e}^{-\Gamma S_i^x}$ and rewrite the effective Hamiltonian. The resulting Hamiltonian of the Sourlas codes extended by means of quantum fluctuation leads to

$$\mathcal{H}_{\mathrm{eff}} = -\beta_J \sum_{i1,\cdots,ip} J_{i1\cdots ip} S_{i1}^z S_{i2}^z \cdots S_{ip}^z - h \sum_i \tau_i S_i^z - \Gamma \sum_i S_i^x. \tag{9.2.86}$$

Hereafter, we call this type of error-correcting codes as *Quantum Sourlas codes*. We first derive the Γ-dependence of the bit-error rate for a given p (the number of bits in a parity). Then, the Gaussian channel noise is specified by the next output distribution:

$$P\big(\{J\},\{\tau\}|\{\xi\}\big) = \frac{\mathrm{e}^{-\frac{N^{p-1}}{J^2 p!}\sum_{i1,\dots,ip}(J_{i1\cdots ip} - \frac{J_0 p!}{N^{p-1}}\xi_{i1}\cdots\xi_{ip})^2 - \frac{1}{2\tau^2}(\tau_i - a_0\xi_i)^2}}{(J^2\pi p!/N^{p-1})^{1/2}\sqrt{2\pi}\,a}. \tag{9.2.87}$$

For a simplicity, we treat the case in which the original message sequence $\{\xi\}$ is generated by the following uniform distribution $P(\{\xi\}) = 2^{-N}$. Then, the moment of the effective partition function Z_{eff} leads to

$$\begin{aligned} Z_{\mathrm{eff}}^n = \exp\Bigg[& \frac{\beta_J}{M} \sum_{i1,\cdots,ip} \sum_{\alpha=1}^{n} \sum_{t=1}^{M} J_{i1\cdots ip} S_{i1}^{\alpha}(t) S_{i2}^{\alpha}(t) \cdots S_{ip}^{\alpha}(t) \\ & + \frac{h}{M} \sum_i \sum_{\alpha=1}^{n} \sum_{t=1}^{M} \tau_i S_i^{\alpha}(t) + B \sum_i \sum_{t=1}^{M} S_i(t) S_i(t+1) \Bigg] \end{aligned} \tag{9.2.88}$$

where α and t mean the indexes of the replica number and the Trotter slice, respectively. We set $B \equiv (1/2)\log\coth(\Gamma/M)$ and used the gauge transform: $J_{i1\cdots ip} \to J_{i1\cdots ip}\xi_{i1}\cdots\xi_{ip}$, $S_{ip} \to \xi_{ip} S_{ip}^z$.

After averaging Z_{eff}^n over the quenched randomness $\ll\cdots\gg$, namely, over the joint distribution $P(\{J\},\{\tau\},\{\xi\})$, we obtain the following data averaged effective partition function:

$$\begin{aligned} \ll Z_{\mathrm{eff}}^n \gg = & \prod_{t'} \prod_{\alpha\beta} \int_{-\infty}^{\infty} dQ_{\alpha\beta}\big(t,t'\big) \int_{-\infty}^{\infty} d\lambda_{\alpha\beta}\big(t,t'\big) \int_{-\infty}^{\infty} dm_{\alpha}(t) \int_{-\infty}^{\infty} d\hat{m}_{\alpha}(t) \\ & \times \exp\big[-Nf(m,\hat{m},\boldsymbol{Q},\boldsymbol{\lambda})\big] \end{aligned} \tag{9.2.89}$$

with

$$\begin{aligned}f(m,\hat{m},\boldsymbol{Q},\boldsymbol{\lambda}) &= -\frac{\beta_J J_0}{M}\sum_{t,\alpha} m_\alpha^p(t) - \frac{h\tau_0}{M}\sum_{t,\alpha} m_\alpha(t)\\ &\quad - \frac{(\beta_J J)^2}{4M^2}\sum_{tt',\alpha\beta} Q_{\alpha\beta}^p(t,t') - \frac{(h\tau)^2}{2M^2}\sum_{tt',\alpha\beta} Q_{\alpha\beta}(t,t')\\ &\quad + \frac{1}{M}\sum_{t,\alpha}\hat{m}_\alpha(t)m_\alpha(t) + \frac{1}{M^2}\sum_{tt',\alpha\beta}\lambda_{\alpha\beta}(t,t')Q_{\alpha\beta}(t,t')\\ &\quad - \frac{1}{M}\sum_{t,\alpha}\hat{m}_\alpha(t)S^\alpha(t) - \frac{1}{M^2}\sum_{tt',\alpha\beta}\lambda_{\alpha\beta}(t,t')S^\alpha(t)S^\beta(t')\\ &\quad - B\sum_t S(t)S(t+1)\end{aligned} \tag{9.2.90}$$

where we labelled each Trotter slice by index t. Using the replica symmetric and the static approximations, namely,

$$m_\alpha(t) = m, \qquad \hat{m}_\alpha(t) = \hat{m} \tag{9.2.91}$$

$$Q_{\alpha\beta}(t,t') = \begin{cases} \chi & (\alpha=\beta)\\ q & (\alpha\neq\beta)\end{cases}, \qquad \lambda_{\alpha\beta}(t,t') = \begin{cases}\lambda_1 & (\alpha=\beta)\\ \lambda_2 & (\alpha\neq\beta)\end{cases}, \tag{9.2.92}$$

we obtain the free energy density f^{RS}:

$$\begin{aligned}\beta_J f^{RS}(m,\chi,q) &= (p-1)\beta_J J_0 m^p + \frac{1}{4}(p-1)(\beta_J J)^2\left(\chi^p - q^p\right)\\ &\quad - \int_{-\infty}^{\infty} Dw \log\int_{-\infty}^{\infty} Dz\, 2\cosh\Xi\end{aligned} \tag{9.2.93}$$

where we used the saddle point equations with respect to $\hat{m}$, λ_1, λ_2, namely, $\hat{m} = p\beta_J J_0 m^{p-1} + a_0 h$ and $\lambda_1 = \frac{p}{2}(\beta_J J)^2\chi^{p-1} + (ah)^2$, $\lambda_2 = \frac{p}{2}(\beta_J J)^2 q^{p-1} + (ah)^2$. Then, the saddle point equations are derived as follows:

$$m = \int_{-\infty}^{\infty} D\omega \int_{-\infty}^{\infty} Dz\left(\frac{\Phi\sinh\Xi}{\Xi\Omega}\right) \tag{9.2.94}$$

$$\chi = \int_{-\infty}^{\infty}\frac{D\omega}{\Omega}\int_{-\infty}^{\infty} Dz\left[\left(\frac{\Phi}{\Xi}\right)^2\cosh\Xi + \Gamma^2\left(\frac{\sinh\Xi}{\Xi^3}\right)\right] \tag{9.2.95}$$

$$q = \int_{-\infty}^{\infty} D\omega\left[\int_{-\infty}^{\infty} Dz\left(\frac{\Phi\sinh\Xi}{\Xi\Omega}\right)\right]^2 \tag{9.2.96}$$

where we defined

$$\begin{aligned}\Phi &= \omega\sqrt{\frac{p}{2}(\beta_J J)^2 q^{p-1} + (ah)^2} + z\sqrt{\frac{p}{2}(\beta_J J)^2\left(\chi^{p-1} - q^{p-1}\right)}\\ &\quad + p\beta_J J_0 m^{p-1} + a_0 h\end{aligned} \tag{9.2.97}$$

and $\Xi = \sqrt{\Phi^2 + \Gamma^2}$, $\Omega = \int_{-\infty}^{\infty} Dz\cosh\Xi$. The resultant overlap leads to

$$R = \int_{-\infty}^{\infty} D\omega\int_{-\infty}^{\infty} Dz\,\mathrm{sgn}(\Phi) = 1 - 2\int_{-\infty}^{\infty} Dw\, H\left(-z_p^*\right) \tag{9.2.98}$$

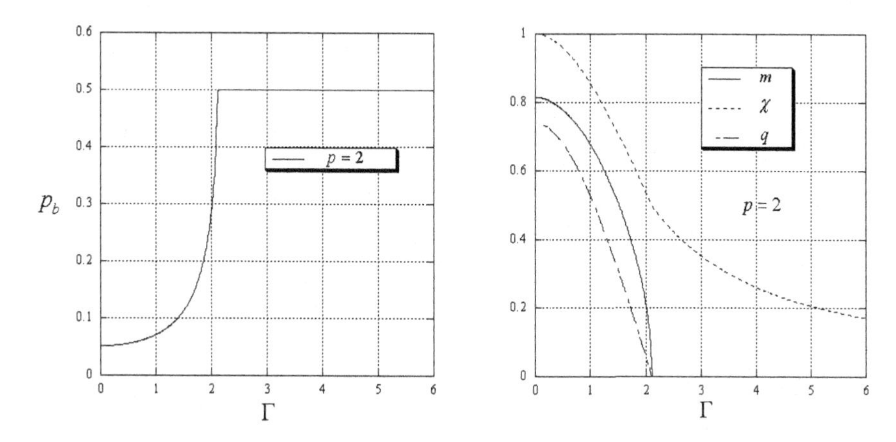

Fig. 9.12 The Γ-dependence of the bit error-rate p_b for the case of $p = 2$ without magnetic field $h = 0$ (*left*) and order parameters m, χ and q as a function of Γ (*right*). We set $\beta_J = 1$, $J = J_0 = 1$

where we defined z_p^* by

$$z_p^* = -\frac{(p\beta_J J_0 m^{p-1} + a_0 h) + w\sqrt{\frac{p}{2}(\beta_J J)^2 q^{p-1} + (ah)^2}}{\sqrt{\frac{p}{2}(\beta_J J)^2(\chi^{p-1} - q^{p-1})}} \tag{9.2.99}$$

and the error function $H(x)$ defined as $H(x) = \int_x^\infty Dz$. Thus, the bit-error rate for the problem of error-correcting codes is given by $p_b = (1 - R)/2 = \int_{-\infty}^{\infty} Dw\, H(-z_p^*)$, where the above bit-error rate p_b depends on Γ through the order parameters χ, q and m.

9.2.4.7 Analysis for Finite p

We first evaluate the performance of the quantum Sourlas codes for the case of finite p by solving the saddle point equations numerically.

In Fig. 9.12 (left), we first plot the Γ-dependence of the bit-error rate p_b for the case of $p = 2$ without magnetic field $h = 0$. In this plot, we choose $J = J_0 = 1$ and set $\beta_J = 1$. It must be noted that J_0/J corresponds to the signal to noise ratio (SN ratio). From this figure, we find that the bit error rate gradually approaches to the random guess limit $p_b = 0.5$ as Γ increases. This transition is regarded as a second order phase transition between the ferromagnetic and the paramagnetic phases. We plot the Γ-dependence of the order parameters m, χ and q in the right panel of Fig. 9.12. We should notice that in the classical limit $\Gamma \to 0$, the order parameter χ should takes 1 and both magnetisation m and spin glass order parameter q continuously becomes zero at the transition point. Therefore, for the case of $p = 2$, the increase of the quantum fluctuation breaks the error-less state gradually.

On the other hand, in Fig. 9.13, we plot the Γ-dependence of the bit-error rate p_b for the case of $p = 3$. In this figure, we find that the bit-error rate suddenly increases

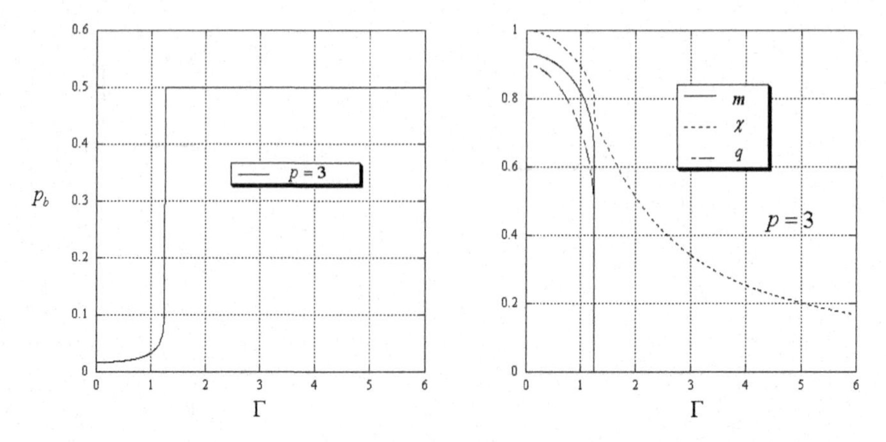

Fig. 9.13 The Γ-dependence of the bit error-rate p_b for the case of $p = 3$ without magnetic field $h = 0$ (*left*) and order parameters m, χ and q as a function of Γ (*right*). We set $\beta_J = 1$, $J = J_0 = 1$

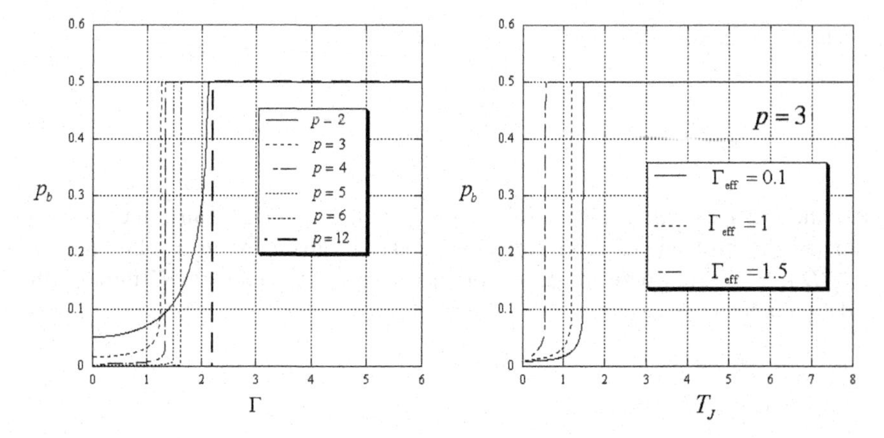

Fig. 9.14 The Γ-dependence of the bit error-rate p_b for $p = 2, \dots, 6$ and $p = 12$ without magnetic field $h = 0$ (*left*). We set $\beta_J = 1$, $J = J_0 = 1$. The *right panel* shows the $T_J = \beta_J^{-1}$ dependence of the bit-error rate for keeping the ratio: $\Gamma/\beta_J \equiv \Gamma_{\text{eff}}$ to the values $\Gamma_{\text{eff}} = 0.1, 1$ and 1.5

to 0.5 at the transition point $\Gamma = \Gamma_c$ and the quality of the message-retrieval becomes the same performance as the random guess. This first order phase transition from the ferromagnetic error-less phase to the paramagnetic random guess phase is observed in the right panel of Fig. 9.13.

We find that the system undergoes the first order phase transition for $p \geq 3$. In Fig. 9.14, we plot the Γ-dependence of the bit-error rate for $p = 2, 3, \dots, 6$ and $p = 12$. From this figure, we find that the transition for $p \geq 3$ is first order and the bit-error rate changes its state from the ferromagnetic almost perfect information retrieval phase to the paramagnetic random guess phase at $\Gamma = \Gamma_c$. The tolerance to the quantum fluctuation increases as the number of degree p of the interaction increases.

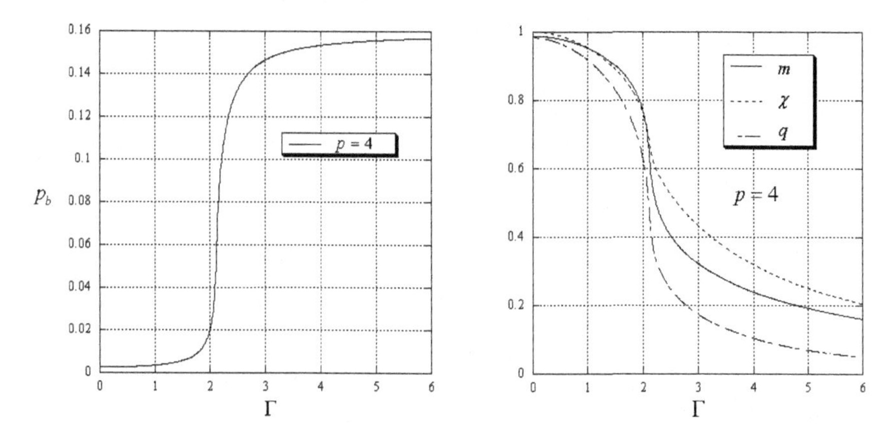

Fig. 9.15 The Γ-dependence of the bit error-rate p_b for the case of $p = 4$ with magnetic field $h = 1$ (*left*) and order parameters m, χ and q as a function of Γ (*right*). We set $\beta_J = 1$, $J = J_0 = 1$ and $a_0 = a = 1$

We next consider the case of $h \neq 0$. This means that we send not only the parity check $\{J_{i1 \cdots ip}\}$ but also bit sequence $\{\xi\}$ itself. We plot the bit-error rate as a function of Γ for $p = 4$ in Fig. 9.15 (left). From this figure, we find that the bit-error rate goes to some finite value which is below the random guess limit gradually. The right panel of this figure tells us that in this case there is no sharp phase transition induced by the quantum fluctuation. In Fig. 9.16 (left), we plot the bit-error rate and corresponding order parameters as a function of Γ for $p = 5$. This figure tells us that the bit-error rate suddenly increases at some critical length of the transverse field Γ_c. At the same critical point, order parameters suddenly changes their values (see Fig. 9.16 (right)). As we add the external field h, this is not a ferro-para magnetic phase transition, however, there exist two stable states, namely good retrieval phase and poor retrieval phase.

In Fig. 9.17, we plot the Γ-dependence of the bit-error rate for $p = 3, \ldots, 6$ and $p = 12$ (left) and for $p = 6$ and $\beta_J = 0.2, \ldots, 12$ (right). From this right panel, interesting properties are observed. For small Γ, the bit-error rate becomes small as we increases p. On the other hand, for large Γ, the bit error rate becomes large as p increases. Moreover, the bit-error rate for $p = 6$ takes its maximum at some finite value of Γ.

9.2.4.8 Phase Diagrams for $p \to \infty$ and Replica Symmetry Breaking

In this subsection, we investigate properties of the quantum Sourlas codes in the limit of $p \to \infty$. In this limit, we easily obtain several phase boundaries analytically and draw the phase diagrams.

First of all, we consider the simplest case, namely, the case of $J_0 = 0$, $h = 0$. For this choice of parameters, the ferromagnetic phase does not appear and possible

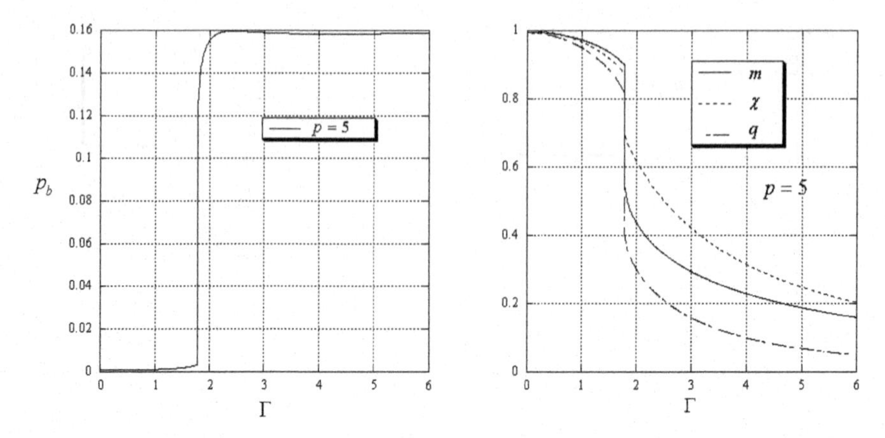

Fig. 9.16 The Γ-dependence of the bit error-rate p_b for the case of $p = 5$ with magnetic field $h = 1$ (*left*) and order parameters m, χ and q as a function of Γ (*right*). We set $\beta_J = 1$, $J = J_0 = 1$ and $a_0 = a = 1$

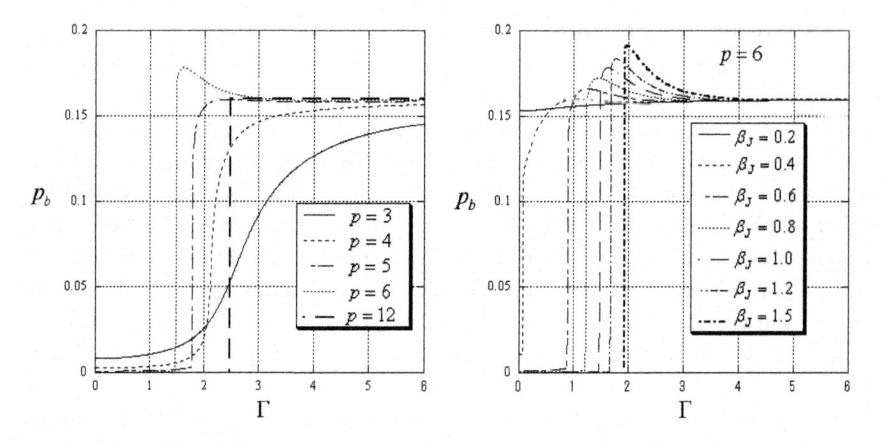

Fig. 9.17 The Γ-dependence of the bit-error rate for $p = 3, \ldots, 6$ and 12 (*left*). For $p = 6$, the Γ-dependences of the bit-error rate for $\beta_J = 0.2, \ldots, 1.2$ are shown in the *right panel*

phases are paramagnetic phase and spin glass phase. The free energy density we evaluate is now rewritten by

$$
\begin{aligned}
f^{RS} &= -\frac{1}{4}(p-1)\beta_J J^2 \big(q^p - \chi^p\big) \\
&\quad - T_J \int_{-\infty}^{\infty} Dw \log \int_{-\infty}^{\infty} Dz\, 2\cosh \beta_J \sqrt{\phi_0^2 + \Gamma_{\mathrm{eff}}^2} \qquad (9.2.100)
\end{aligned}
$$

with $\phi_0 = w\sqrt{pJ^2 q^{p-1}/2} + z\sqrt{pJ^2(\chi^{p-1} - q^{p-1})/2}$, where we defined $\Gamma_{\mathrm{eff}} = \Gamma/\beta_J$. In the paramagnetic phase, there is no spin glass ordering, namely, $q = 0$.

Thus, the free energy density in the paramagnetic phase leads to

$$f_{para}^{RS} = \frac{J^2\beta_J}{4}(p-1)\chi^p - T_J \log \int_{-\infty}^{\infty} Dz\, 2\cosh \beta_J \sqrt{\Gamma_{\rm eff}^2 + \frac{p}{2}J^2\chi^{p-1}z^2}. \tag{9.2.101}$$

The saddle point equation with respect to χ is given by

$$\chi = \frac{\int_{-\infty}^{\infty} Dz \left\{ \left(\frac{\phi_{00}}{\Gamma_{\rm eff}^2+\phi_{00}}\right)\cosh \beta_J \sqrt{\Gamma_{\rm eff}^2+\phi_{00}} + \Gamma_{\rm eff}^2 T_J \frac{\sinh \beta_J \sqrt{\Gamma_{\rm eff}^2+\phi_{00}}}{\sqrt{\Gamma_{\rm eff}^2+\phi_{00}}^{\,3}} \right\}}{\int_{-\infty}^{\infty} Dz \cosh \beta_J \sqrt{\Gamma_{\rm eff}^2+\phi_{00}^2}} \tag{9.2.102}$$

with $\phi_{00} = pJ^2\chi^{p-1}z^2/2$. In the limit of $p \to \infty$, there are two possible solutions of χ, that, is $\chi^p = 1$ and $\chi^p = 0$. The former is explicitly given from (9.2.102) as $\chi \simeq 1 - 4\Gamma_{\rm eff}^2 T_J^2/p^2 J$. Then, we obtain the free energy density for this solution as $f_I = -J^2/4T_J - T_J \log 2$ by substituting this χ into (9.2.101) and evaluating the integral with respect to z at the saddle point in the limit of $p \to \infty$. Let us call this phase as *PI*. The later solution is explicitly evaluated as $\chi = (T_J/\Gamma_{\rm eff})\tanh(\Gamma_{\rm eff}/T_J)$ (< 1, thus, $\chi^p = 0$) and corresponding free energy density leads to $f_{II} = -T_J \log 2 - T_J \log\cosh(\Gamma_{\rm eff}/T_J)$. We call this phase as *PII*.

Here we should not overlook the entropy in *PI*, namely, $S = -(\partial f_I/\partial T) = -J^2/4T_J^2 + \log 2$. Obviously, S becomes negative for $T < (J/2\sqrt{\log 2})^{-1}$ and in this region, the replica symmetry of the order parameters might be broken. Therefore, in this low temperature region, we should construct the replica symmetry breaking (RSB) solution. To obtain the RSB solution, we break the symmetry of the matrices $\boldsymbol{q}$ and $\boldsymbol{\lambda}$ as

$$q_{l\delta,l'\delta'} = \begin{cases} q_0 & (l=l') \\ q_1 & (l \neq l') \end{cases}, \qquad \lambda_{l\delta,l'\delta'} = \begin{cases} \hat{\lambda}_0 & (l=l') \\ \hat{\lambda}_1 & (l \neq l') \end{cases} \tag{9.2.103}$$

for $l = 1, \dots, n/x$, $\delta = 1, \dots, x$. Then, we obtain the free energy density for one step RSB solution as

$$\begin{aligned} f^{1RSB} &= (p-1)J_0 m^p + \frac{\beta_J J^2}{4}\left[xq_1^p + (1-x)q_0^p\right] + \frac{\beta_J J^2}{4}(p-1)\chi^p \\ &\quad - \frac{\beta_J}{2}\left[xq_1\hat{\lambda}_1 + (1-x)q_0\hat{\lambda}_0\right] \\ &\quad - \frac{T_J}{x}\int_{-\infty}^{\infty} Dw \log \int_{-\infty}^{\infty} Dz \left(\int_{-\infty}^{\infty} Dy\, 2\cosh \beta_J \sqrt{\hat{\phi}^2 + \Gamma_{\rm eff}^2}\right)^x \end{aligned} \tag{9.2.104}$$

with $\hat{\phi} = w\sqrt{\hat{\lambda}_1} + z\sqrt{\hat{\lambda}_0 - \hat{\lambda}_1} + y\sqrt{pJ^2\chi^{p-1}/2 - \hat{\lambda}_0} + p\beta_J J_0 m^{p-1} + a_0 h$. By taking $(\partial f^{1RSB}/\partial q_0) = (\partial f^{1RSB}/\partial q_1) = 0$, we obtain $\hat{\lambda}_1 = pJ^2 q_1^{p-1}/2, \hat{\lambda}_0 = pJ^2 q_0^{p-1}/2$.

Here we set the parameters J_0, h again to $J_0 = h = 0$. At low temperature, we naturally assume $q_1 < 0$ ($\hat{\lambda}_1 = 0$), $q_0 = 1$ ($\hat{\lambda}_0 = pJ^2/2$) and $\chi = 1$. Substituting

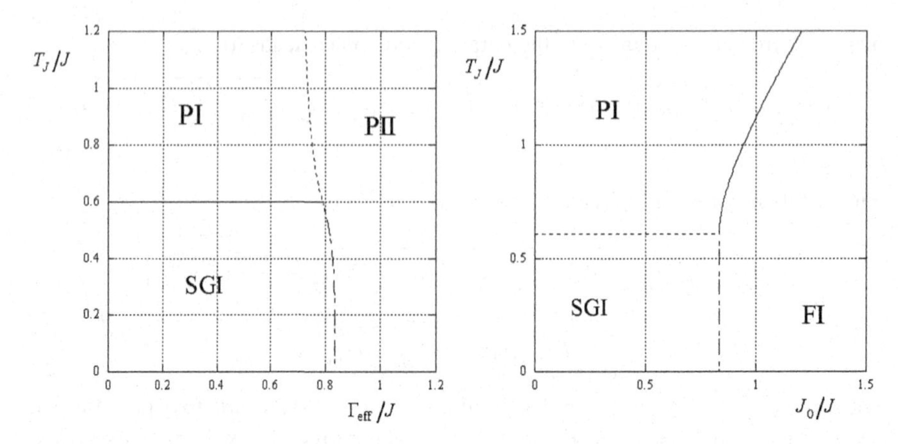

Fig. 9.18 The phase diagrams in the limit of $p \to \infty$. In the case of $J_0 = 0$, there exist three phases, namely, *PI*, *PII* and *SGI*. Below the critical point $(T_J/J)_c = (1/2J\sqrt{\log 2})$, the replica symmetry is broken (the *left panel*). The *right panel* shows the critical SN ratio $(J_0/J)_c$, above which (labelled *FI* in the panel) decoding without errors is achieved, is given by $\sqrt{\log 2}$

these conditions into (9.2.104) and evaluating the integral with respect to y at the saddle point in the limit of $p \to \infty$, we obtain the free energy density in this phase, which will be referred to as *SGI*, as $f_{SGI} = -\beta_J J^2 x/4 - \log 2/(\beta_J x)$. Substituting the solution of $(\partial f_{SGI}/\partial x) = 0$, namely, $x = 2\sqrt{\log 2}/(JT_J)$ into f_{SG}, we obtain the free energy density which specifies *SGI* as $f_{SGI} = -J\sqrt{\log 2}$.

Let us summarise:

$$PI \text{ (para)}: \quad f_I = -\frac{J^2}{4T_J} - T_J \log 2 \quad (\chi = 1,\ q = 0)$$

$$PII \text{ (para)}: \quad f_{II} = -T_J \log 2 - T_J \log\cosh\left(\frac{\Gamma_{\text{eff}}}{T_J}\right) \quad (\chi^p = q = 0)$$

$$SGI \text{ (spin glass)}: \quad f_{SGI} = -J\sqrt{\log 2} \quad (\chi = q = 1)$$

We illustrate the phase diagram in Fig. 9.18 (left).

As the phase transitions between arbitrary two phases among these three (*PI*, *PII*, *SGI*) are all first order, each phase boundary is obtained by balancing of the free energy density. Namely,

$$\Gamma_{\text{eff}} = \begin{cases} T_J \cosh^{-1}(\mathrm{e}^{J^2/4T_J^2}) \quad (T_J > T_c) & \text{(for } PI\text{–}PII) \\ T_J \cosh^{-1}(\mathrm{e}^{J\sqrt{\log 2}/T_J}/2) \quad (T_J < T_c) & \text{(for } PII\text{–}SGI) \\ T_J = J/2\sqrt{\log 2} = T_c & \text{(for } SGI\text{–}PI) \end{cases} \cdot \tag{9.2.105}$$

9.2.4.9 The Shannon's Bound and Phase Boundaries

We next consider the case of $J_0 \neq 0$. This case is much more important in the context of error-correcting codes. For the case of absence of the external field

$h = 0$, the phase transition between the error-less phase and the random guess phase is specified as the ferro-paramagnetic (or spin glass) phase transition. From reasons we mentioned above, our main purpose here is to determine the transition point $(J_0/J)_c$ below which the ferromagnetic phase is stable. The critical SN ratio $(J_0/J)_c$ is important because as we mentioned before, the error-less decoding is possible when the channel capacity C and the transmission rate R satisfy the inequality $R \le C$. The channel capacity for the Gaussian channel we are dealing with is given by $C = (1/2)\log_2(1 + J_0^2/J^2)$ with $J_0 = J_0 p!/N^{p-1}$, $J = J^2 p!/2N^{p-1}$, that is, $C \simeq J_0^2 p!/(J^2 N^{p-1} \log 2)$ in the limit of $N \to \infty$ for a given p. On the other hand, the transmission rate R is given as $R = N/N_B = N/{}_N C_p \simeq p!/N^{p-1}$. Therefore, the error-less decoding is possible when the following inequality:

$$\frac{R}{C} = \left(\frac{J}{J_0}\right)^2 \log 2 \le 1 \tag{9.2.106}$$

holds and the question now arises, namely, it is important to ask whether the above inequality is satisfied or not at the critical point $(J/J_0)_c$. In following, we make this point clear.

We start from the saddle point equations which are derived from the free energy density of the one step RSB (9.2.104). These equations are given explicitly as

$$m = \int_{-\infty}^{\infty} Dw \frac{\int_{-\infty}^{\infty} Dz\, (\int_{-\infty}^{\infty} Dy\, 2\cosh \beta_J \hat{\Xi})^{x-1} \int_{-\infty}^{\infty} Dy\, (\frac{\hat{\phi}}{\hat{\Xi}}) 2\sinh \beta_J \hat{\Xi}}{\int_{-\infty}^{\infty} Dz\, (\int_{-\infty}^{\infty} Dy\, 2\cosh \beta_J \hat{\Xi})^x}$$

$$q_0 = \int_{-\infty}^{\infty} Dw \frac{\int_{-\infty}^{\infty} Dz\, (\int_{-\infty}^{\infty} Dy\, 2\cosh \beta_J \hat{\Xi})^{x-2} (\int_{-\infty}^{\infty} Dy\, (\frac{\hat{\phi}}{\hat{\Xi}}) 2\sinh \beta_J \hat{\Xi})^2}{\int_{-\infty}^{\infty} Dz\, (\int_{-\infty}^{\infty} Dy\, 2\cosh \beta_J \hat{\Xi})^x}$$

$$q_1 = \int_{-\infty}^{\infty} Dw \left\{ \frac{\int_{-\infty}^{\infty} Dz\, (\int_{-\infty}^{\infty} Dy\, (\frac{\hat{\phi}}{\hat{\Xi}}) 2\sinh \beta_J \hat{\Xi})^x}{\int_{-\infty}^{\infty} Dz\, (\int_{-\infty}^{\infty} Dy\, 2\cosh \beta_J \hat{\Xi})^x} \right\}^2 \tag{9.2.107}$$

$$\begin{aligned} \chi &= \int_{-\infty}^{\infty} Dw \frac{\int_{-\infty}^{\infty} Dz\, (\int_{-\infty}^{\infty} Dy\, 2\cosh \beta_J \hat{\Xi})^{x-1} \int_{-\infty}^{\infty} Dy\, (\frac{\hat{\phi}}{\hat{\Xi}})^2 2\sinh \beta_J \hat{\Xi}}{\int_{-\infty}^{\infty} Dz\, (\int_{-\infty}^{\infty} Dy\, 2\cosh \beta_J \hat{\Xi})^x} \\ &\quad + \Gamma_{\mathrm{eff}}^2 T_J \int_{-\infty}^{\infty} Dw \frac{\int_{-\infty}^{\infty} Dz\, (\int_{-\infty}^{\infty} Dy\, 2\cosh \beta_J \hat{\Xi})^{x-1} \int_{-\infty}^{\infty} Dy\, (\frac{2\sinh \beta_J \hat{\Xi}}{\hat{\Xi}^3})}{\int_{-\infty}^{\infty} Dz\, (\int_{-\infty}^{\infty} Dy\, 2\cosh \beta_J \hat{\Xi})^x} \end{aligned} \tag{9.2.108}$$

with

$$\hat{\phi} = Jw\sqrt{\frac{p}{2} q_1^{p-1}} + Jz\sqrt{\frac{p}{2}(q_0^{p-1} - q_1^{p-1})} + Jy\sqrt{\frac{p}{2}(\chi^{p-1} - q_0^{p-1})} + pJ_0 m^{p-1} \tag{9.2.109}$$

and $\hat{\Xi} = \sqrt{\hat{\phi}^2 + \Gamma_{\mathrm{eff}}^2}$.

When the number of product p of the estimate of the original bits is extremely large and $J/J_0, m$ is positive, $\hat{\phi} = pJ_0m^{p-1}$ and the solutions of the above saddle point equations lead to $m = q_0 = q_1 = 1$ and $\chi = 1$. Thus, the system is in the ferromagnetic phase and the replica symmetry is not broken ($q_0 = q_1$). Substituting the replica symmetric solution $m = q = 1$ into (9.2.93) and evaluating the integral with respect to w at the saddle point in the limit of $p \to \infty$, we obtain the free energy density in this phase (let us call *FI*) as $f_{FI} = -J_0$. We should notice that this free energy density does not depend on the effective amplitude of the transverse field Γ_{eff} at all. From the argument of $J_0 = 0$ case, the phase specified $\chi = 1, T_J < T_c = (2\sqrt{\log 2})$ is spin glass phase. Therefore, the condition (9.2.106) is satisfied and the ferromagnetic error-less phase exists for $(J_0/J) \geq (J_0/J)_c = \sqrt{\log 2}$, where (J_0/J) is determined by balancing of the free energy densities $f_{FI} = f_{SGI}$. As the result, we conclude that the error-less decoding is achieved if the SN ratio (J_0/J) is greater than the critical value $(J_0/J)_c = \sqrt{\log 2}$ and the condition is independent of Γ_{eff}. To put it into another word, the Shannon's bound is not violated by the quantum uncertainties in the prior distribution in the limit of $p \to \infty$.

9.2.5 Mean Field Algorithms

In the previous subsections, we evaluated the performance of Bayesian inference for the original image or original bit sequence by using the solvable model. In this subsection, we mention concrete algorithms to achieve the original information retrieval. For such algorithms, mean field approaches are useful to carry out.

9.2.5.1 Naive Mean Field Algorithm for Image Restoration

Tanaka and Horiguchi [397] proposed mean field algorithm for image restoration. For the image defined on the two dimensional square lattice, the effective Hamiltonian $\hat{\boldsymbol{H}}$ for image restoration is given by the matrix form:

$$\boldsymbol{H} = -\sum_{i,j=1}^{L} \left(J\boldsymbol{S}_{i,j}^z \boldsymbol{S}_{i+1,j}^z + J\boldsymbol{S}_{i,j}^z \boldsymbol{S}_{i,j+1}^z + h\tau_{i,j}\boldsymbol{S}_{i,j}^z + \Gamma \boldsymbol{S}_{i,j}^x \right) \tag{9.2.110}$$

where τ_{ij} is degraded pixel on the site (i,j) and the dimension of the matrix is $2^N \times 2^N$, $N = L^2$. We also defined the following tensor products:

$$\begin{cases} \boldsymbol{S}_{i,j}^x \equiv \overset{(1,1)}{\boldsymbol{I}} \otimes \cdots \otimes \boldsymbol{I} \otimes \overset{(i,j)}{\boldsymbol{S}^x} \otimes \boldsymbol{I} \otimes \cdots \otimes \overset{(L,L)}{\boldsymbol{I}} \\ \boldsymbol{S}_{i,j}^z \equiv \overset{(1,1)}{\boldsymbol{I}} \otimes \cdots \otimes \boldsymbol{I} \otimes \overset{(i,j)}{\boldsymbol{S}^z} \otimes \boldsymbol{I} \otimes \cdots \otimes \overset{(L,L)}{\boldsymbol{I}} \end{cases} \tag{9.2.111}$$

with

$$\boldsymbol{S}^x = \begin{pmatrix} 0 & 1 \\ 1 & 0 \end{pmatrix}, \quad \boldsymbol{S}^z = \begin{pmatrix} 1 & 0 \\ 0 & -1 \end{pmatrix}, \quad \boldsymbol{I} = \begin{pmatrix} 1 & 0 \\ 0 & 1 \end{pmatrix}. \tag{9.2.112}$$

For this Hamiltonian (9.2.110), the density matrix is given by

$$\boldsymbol{\rho} = \frac{\exp(-\boldsymbol{H}/T)}{\mathrm{Tr}\exp(-\boldsymbol{H}/T)} \tag{9.2.113}$$

where we used the formula

$$\exp(\boldsymbol{A}) = \sum_{n=0}^{\infty} \frac{\boldsymbol{A}^n}{n!}. \tag{9.2.114}$$

Under the mean field approximation, the density matrix is rewritten by

$$\boldsymbol{\rho} \simeq \prod_{i,j=1}^{L} \bigotimes \rho_{ij}, \qquad \rho_{ij} = \frac{\exp(-\boldsymbol{H}_{ij}/T)}{\mathrm{Tr}\exp(-\boldsymbol{H}_{ij}/T)} \tag{9.2.115}$$

with

$$\boldsymbol{H}_{ij} = \begin{pmatrix} \hat{H}_{ij}(+1) & -\Gamma \\ -\Gamma & \hat{H}_{ij}(-1) \end{pmatrix} \tag{9.2.116}$$

$$\hat{H}_{ij}(s) = -\{J(m_{i+1,j} + m_{i-1,j} + m_{i,j+1} + m_{i,j-1}) + h\tau_{ij}\}s \tag{9.2.117}$$

$$m_{i,j} = \mathrm{Tr}[\boldsymbol{S}^z \rho_{ij}]. \tag{9.2.118}$$

For the eigenvalues $\lambda_{ij}(n)$, $n = 1, 2$ of the matrix $\boldsymbol{H}_{ij}$, the marginal density matrix ρ_{ij} is expressed as

$$\rho_{ij} = \frac{1}{Z_{ij}} \sum_{n=1}^{2} |\phi_{ij}(n)\rangle \mathrm{e}^{-\lambda_{ij}(n)/T} \langle \phi_{ij}(n)|, \quad Z_{ij} = \sum_{n=1}^{2} \mathrm{e}^{-\lambda_{ij}(n)/T}. \tag{9.2.119}$$

In the zero temperature limit $T \to 0$, the marginal density matrix is reduced to

$$\rho_{ij} = |\phi_{ij}(1)\rangle\langle\phi_{ij}(1)| \tag{9.2.120}$$

and the local field at site (i, j) is rewritten as

$$m_{i,j} = \mathrm{Tr}[\boldsymbol{S}^z \rho_{ij}] = \mathrm{Tr}\{\boldsymbol{S}^z |\phi_{ij}(1)\rangle\langle\phi_{ij}(1)| = \langle\phi_{ij}(1)|\boldsymbol{S}^z|\phi_{ij}(1)\rangle \tag{9.2.121}$$

were we defined $|\phi_{ij}(1)\rangle$ as eigenvector of the minimum eigenvalue $\lambda_{ij}(1)$ of the matrix $\boldsymbol{H}_{ij}$.

Thus, from Eqs. (9.2.118), (9.2.121), we obtain the mean field equation for each site (i, j) as

$$m_{i,j} = \frac{J(m_{i+1,j} + m_{i-1,j} + m_{i,j+1} + m_{i,j-1}) + h\tau_{ij}}{\sqrt{\Gamma^2 + \{J(m_{i+1,j} + m_{i-1,j} + m_{i,j+1} + m_{i,j-1}) + h\tau_{ij}\}^2}}. \tag{9.2.122}$$

Obviously, the solution of the above equation $m_{i,j}$ corresponds to the local magnetisation at site (i, j) obtained by mean field approximation. To solve the non-linear coupled equations, we usually use the following recursion relations (maps) by introducing the 'time step' t as

$$m_{i,j}^{(t+1)} = \frac{J(m_{i+1,j}^{(t)} + m_{i-1,j}^{(t)} + m_{i,j+1}^{(t)} + m_{i,j-1}^{(t)}) + h\tau_{ij}}{\sqrt{\Gamma^2 + \{J(m_{i+1,j}^{(t)} + m_{i-1,j}^{(t)} + m_{i,j+1}^{(t)} + m_{i,j-1}^{(t)}) + h\tau_{ij}\}^2}}. \tag{9.2.123}$$

Fig. 9.19 Original image for the demonstration of the mean field algorithm. The size is $L \times L = 32 \times 32$

By solving the above equations until the time dependent error:

$$\varepsilon_t = \frac{1}{N} \sum_{i,j=1}^{L} \left| m_{i,j}^{(t+1)} - m_{i,j}^{(t)} \right| \tag{9.2.124}$$

becomes smaller than some appropriate precision, say, $\varepsilon_{t*} \simeq 10^{-8}$, the MPM estimate is given by

$$\hat{\xi}_{i,j} = \text{sgn}\big(m_{i,j}^{(t*)}\big) = \text{sgn}(m_{i,j}) \tag{9.2.125}$$

for each pixel (i, j).

As a benchmark test, we shall apply this mean field algorithm for two dimensional image with size $N = L \times L = 32 \times 32$ given in Fig. 9.19. The original image is degraded by noise probability (flip probability) $p = 0.2$. The parameters are set to $J = 1.1$ and $h = 1$. The schedule of the transverse field Γ is selected as in Table 9.1. As the performance measurements, we choose the distance

$$d(\boldsymbol{a}, \boldsymbol{b}) = \frac{1}{N} \sum_{i,j=1}^{L} (1 - \delta_{a_{i,j}, b_{i,j}}) \tag{9.2.126}$$

and the number of edges

$$\sigma_2(\boldsymbol{a}) = \frac{1}{2N} \sum_{i,j=1}^{L} (1 - \delta_{a_{i,i}, a_{i,j+1}}) + \frac{1}{2N} \sum_{i,j=1}^{L} (1 - \delta_{a_{i,j}, a_{i+1,j}}). \tag{9.2.127}$$

From the table, we find that these quantities converge to the finite values as the transverse field goes to zero. Although there exists a gap between the resulting values of d, σ_2 and the true values, the quantum fluctuation works well on the original image retrieval on the contest of the mean field annealing.

In Fig. 9.20, we show the resultant restore images.

9.2.5.2 Mean Field Decoding via the TAP-Like Equation for Sourlas Codes

In the previous subsections, we introduced a quantum variant of the Sourlas codes and investigated the averaged-case performance by means of the replica method. In practice, we must use some concrete algorithm to decode the original message. As we saw in the previous subsections, zero-error decoding is achieved for $(J/J_0) \geq$

Table 9.1 The schedule of the transverse field (Γ, # Iteration) and several measurement to quantify the performance

Γ	$d(\boldsymbol{\xi}, \hat{\boldsymbol{\xi}})$	$d(\boldsymbol{\tau}, \hat{\boldsymbol{\xi}})$	$\sigma_2(\hat{\boldsymbol{\xi}})$	# iteration
8.0	0.190430	0	0.341797	18
7.0	0.156250	0.034180	0.273438	22
6.0	0.125977	0.066406	0.213867	27
5.0	0.088867	0.103516	0.160156	36
4.0	0.062500	0.133789	0.125977	62
3.0	0.044922	0.159180	0.099609	72
2.0	0.039063	0.172852	0.086914	66
1.0	0.038086	0.175781	0.084961	32
0.0	0.038086	0.175781	0.084961	2
True Value	0	0.190430	0.083008	

Fig. 9.20 From *left* to *right*, the snapshot at $\Gamma = 7, 6, 5$ and $\Gamma = 0$

$\sqrt{\log 2}$ in the limit of $p \to \infty$ (see Fig. 9.21). However, for finite p, it is not obvious to clarify to what extent the bit-error rate becomes small for a given algorithm. To make this point clear, we introduce a mean field algorithm to decode the original message for finite p.

In practical decoding, for a given set of the parity $\{J\}$, we should calculate the estimate

$$\hat{\xi}_i = \lim_{\beta \to \infty} \mathrm{sgn}\big[\mathrm{tr}\big(\hat{S}_i^z \hat{\rho}_\beta\big)\big], \qquad \hat{\rho}_\beta = \frac{\exp[-\beta \hat{H}]}{\mathrm{tr}\exp[-\beta \hat{H}]} \tag{9.2.128}$$

for each bit. To calculate the trace effectively by sampling, the quantum Monte Carlo method (QMCM) might be applicable and useful [386]. However, unfortunately, the QMCM approach encounters several crucial difficulties. First, it takes quite long time for us to simulate the quantum states for large number of the Trotter slices. Second, in general, it is technically quite hard to simulate the quantum states at zero temperature. Thus, we are now stuck for the computational cost problem.

Nevertheless, as an alternative to decode the original message practically, we here examine a TAP (Thouless-Anderson-Palmer)-like mean field algorithm which has a lot of the variants applying to various information processing [203, 302].

According to [194], here we shall provide a simple attempt to apply the mean field equations to the Sourlas error-correcting codes for the case of $p = 2$. In following, the derivation of the equations is briefly explained.

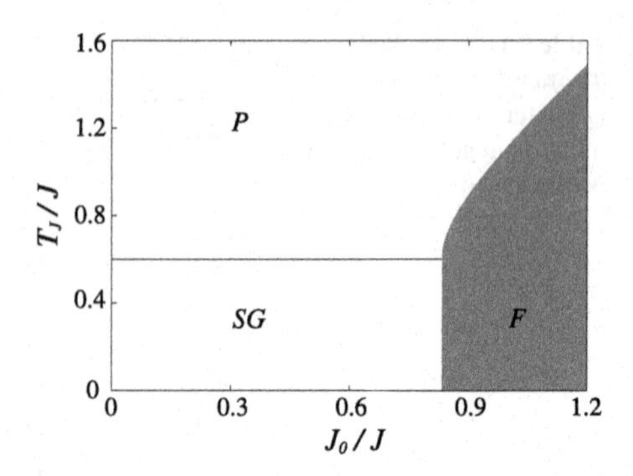

Fig. 9.21 Phase diagram of the Sourlas codes for $p \to \infty$. In the *shaded area* (F), zero-error transmission is achieved. The area P denotes the para-magnetic phase and the area SG is the spin glass phase. For instance, at the ground state, the critical signal-to-noise ratio is $(J_0/J)_c = \sqrt{\log 2} = 0.8326$. We set $T_J \equiv \beta_J^{-1}$

We shall start the Hamiltonian:

$$\hat{H} = -\sum_{ij} J_{ij} \hat{S}_i^z \hat{S}_j^z - \Gamma \sum_i \hat{S}_i^x \tag{9.2.129}$$

$$J_{ij} = \left(\frac{2J_0}{N}\right) J_{ij}^0 + \frac{J}{\sqrt{N}} \eta, \quad J_{ij}^0 = \xi_i \xi_j, \quad \eta = \mathcal{N}(0,1) \tag{9.2.130}$$

Then, we rewrite the above Hamiltonian as follows

$$\begin{aligned}\hat{H} &= -\sum_i \left(\Gamma \hat{S}_i^x + h_i \hat{S}_i^z\right) + \sum_{ij} J_{ij} (m_i \hat{I}_i)(m_j \hat{I}_j) \\ &\quad - \sum_{ij} J_{ij} \left(\hat{S}_i^z - m_i \hat{I}_i\right)\left(\hat{S}_j^z - m_j \hat{I}_j\right) \equiv \hat{H}^{(0)} + \hat{V}\end{aligned} \tag{9.2.131}$$

$$\hat{H}^{(0)} \equiv -\sum_i \left(\Gamma \hat{S}_i^x + h_i \hat{S}_i^z\right) + \sum_{ij} J_{ij} (m_i \hat{I}_i)(m_j \hat{I}_j) \tag{9.2.132}$$

$$\hat{V} \equiv -\sum_{ij} J_{ij} \left(\hat{S}_i^z - m_i \hat{I}_i\right)\left(\hat{S}_j^z - m_j \hat{I}_j\right), \qquad h_i \equiv 2\sum_j J_{ij} m_j \tag{9.2.133}$$

where we defined the $2^N \times 2^N$ identity matrix $\hat{I}_i$, which is formally defined by $\hat{I}_i \equiv \boldsymbol{I}_{(1)} \otimes \cdots \otimes \boldsymbol{I}_{(i)} \otimes \cdots \otimes \boldsymbol{I}_{(N)}$. We also use the definitions in terms of tensor product:

$$\hat{S}_i^z = \boldsymbol{I}_{(1)} \otimes \cdots \otimes \boldsymbol{S}_{(i)}^z \otimes \cdots \otimes \boldsymbol{I}_{(N)} \tag{9.2.134}$$

$$\hat{S}_i^x = \boldsymbol{I}_{(1)} \otimes \cdots \otimes \boldsymbol{S}_{(i)}^x \otimes \cdots \otimes \boldsymbol{I}_{(N)} \tag{9.2.135}$$

with

$$\boldsymbol{I}_{(i)} = \begin{pmatrix} 1 & 0 \\ 0 & 1 \end{pmatrix}, \qquad \boldsymbol{S}_{(i)}^x = \begin{pmatrix} 0 & 1 \\ 1 & 0 \end{pmatrix}, \qquad \boldsymbol{S}_{(i)}^z = \begin{pmatrix} 1 & 0 \\ 0 & -1 \end{pmatrix}. \tag{9.2.136}$$

m_i is the local magnetisation for the system described by the mean field Hamiltonian $\hat{H}^{(0)}$, that is,

$$m_i \equiv m_i^z = \lim_{\beta \to \infty} \mathrm{tr}\left(\hat{S}_i^z \hat{\rho}_\beta^{(0)}\right), \quad \hat{\rho}_\beta^{(0)} \equiv \frac{\exp(-\beta \hat{H}^{(0)})}{\mathrm{tr} \exp(-\beta \hat{H}^{(0)})}. \tag{9.2.137}$$

Shortly, we derive closed equations to determine m_i. It is very tough problem for us to diagonalise the $2^N \times 2^N$ matrix $\hat{H}$, whereas it is rather easy to diagonalise the mean field Hamiltonian $\hat{H}^{(0)}$. Actually, we immediately obtain the ground state internal energy as

$$E^{(0)} = -\sum_i E_i + \frac{1}{2}\sum_i h_i m_i, \quad E_i \equiv \sqrt{\Gamma^2 + h_i^2}. \tag{9.2.138}$$

Then, taking the derivative of the $E^{(0)}$ with respect to m_i and setting it to zero, namely, $\partial E^{(0)}/\partial m_i = \sum_k (\partial h_k/\partial m_i)\{h_k/\sqrt{\Gamma^2 + h_k^2} - m_k\} = 0$, we have

$$(\forall_i) \quad m_i = \frac{h_i}{\sqrt{\Gamma^2 + h_i^2}}, \quad h_i = 2\sum_j J_{ij} m_j. \tag{9.2.139}$$

The above equations are nothing but the so-called *naive mean field equations* for the Ising spin glass (the Sherrington-Kirkpatrick model [369]) in a transverse field. It should be noted that the equations are reduced to $(\forall_i)\ m_i = h_i/|h_i| = \mathrm{sgn}(h_i) = \lim_{\beta\to\infty}\tanh(\beta h_i)$ which is naive mean field equations at the ground state for the corresponding classical system.

To improve the approximation, according to [194, 195, 433], we introduce the reaction term R_i for each pixel i and rewrite the local field h_i such as $2\sum_j J_{ij} m_j - R_i$. Then, the naive mean field equations (9.2.139) are rewritten as

$$m_i = \frac{2\sum_j J_{ij} m_j - R_i}{\sqrt{\Gamma^2 + (2\sum_j J_{ij} m_j - R_i)^2}} \simeq \frac{h_i}{(\Gamma^2 + h_i^2)^{3/2}}\left[1 - \frac{\Gamma^2}{\Gamma^2 + h_i^2}\left(\frac{R_i}{h_i}\right)\right] \tag{9.2.140}$$

for all indices i. In the last line of the above equation, we expanded the equation with respect to R_i up to the first order. We next evaluate the expectation of the Hamiltonian $\hat{H}$ by using the eigenvector that diagonalises the mean field Hamiltonian $\hat{H}^{(0)} = -\sum_i (\Gamma \hat{S}_i^x + h_i \hat{S}_i^z) + \sum_{ij} J_{ij}(m_i \hat{I}_i)(m_i \hat{I}_j)$. We obtain

$$E_g = E^{(0)} - \Gamma^4 \sum_{ij}\left(\frac{J_{ij}^2}{2E_i^2 E_j^2 (E_i + E_j)}\right). \tag{9.2.141}$$

Then, $(\partial E_g/\partial m_i) = 0$ gives

$$m_i = \frac{h_i}{(\Gamma^2 + h_i^2)^{3/2}}\left[1 - \frac{\Gamma^2}{\Gamma^2 + h_i^2}\left(\frac{1}{h_i}\right)\right. \\ \left. \times \sum_j \frac{J_{ij}^2 m_i [2(1-m_i^2)(1-m_j^2)^{\frac{3}{2}} + 3(1-m_i^2)^{\frac{1}{2}}(1-m_j^2)^2]}{2\Gamma[(1-m_i^2)^{\frac{1}{2}} + (1-m_j^2)^{\frac{1}{2}}]^2}\right]. \tag{9.2.142}$$

By comparing (9.2.140) and (9.2.142), we might choose the reaction term R_i for each bit i consistently as

$$R_i = \sum_j \frac{J_{ij}^2 m_i[2(1-m_i^2)(1-m_j^2)^{\frac{3}{2}} + 3(1-m_i^2)^{\frac{1}{2}}(1-m_j^2)^2]}{2\Gamma[(1-m_i^2)^{\frac{1}{2}} + (1-m_j^2)^{\frac{1}{2}}]^2}. \quad (9.2.143)$$

Therefore, we now have a decoding dynamics described by

$$m_i(t+1) = \frac{2\sum_j J_{ij} m_j(t) - R_i(t)}{\sqrt{\Gamma^2 + \{2\sum_j J_{ij} m_j(t) - R_i(t)\}^2}} \quad (9.2.144)$$

$$R_i(t) = \sum_j \frac{J_{ij}^2 m_i(t)[2(1-m_i(t)^2)(1-m_j(t)^2)^{\frac{3}{2}} + 3(1-m_i(t)^2)^{\frac{1}{2}}(1-m_j(t)^2)^2]}{2\Gamma[(1-m_i(t)^2)^{\frac{1}{2}} + (1-m_j(t)^2)^{\frac{1}{2}}]^2}$$
(9.2.145)

for each bit i. Then, the MPM estimate is given as a function of time t as $\overline{\xi}_i(t) = \mathrm{sgn}[m_i(t)]$ and the bit-error rate (BER) is evaluated at each time step through the following expression

$$p_B(t) = \frac{1}{2}\left(1 - \frac{1}{N}\sum_i \xi_i \overline{\xi}_i(t)\right). \quad (9.2.146)$$

We should notice that the naive mean field equations are always retrieved by setting $R_i = 0$ for all i. The naive mean field equations were applied to image restoration by Tanaka and Horiguchi [397] in the previous subsections. We plot several results in Fig. 9.22. In the left panel of this figure, we plot the dynamics of mean field decoding. We plot them for several cases of the signal-to-noise ratio. During the decoding dynamics, we control the Γ by means of

$$\Gamma(t) = \Gamma_0\left(1 + \frac{c}{t+1}\right) \quad (9.2.147)$$

where t denotes the number of time step in the TAP-like update described by Eqs. (9.2.144) and (9.2.145). In Fig. 9.22, we set $\Gamma_0 = 0.5$. From this figure, we find that the BER drops monotonically as the number of iterations increases. We also find in the right panel that the BER drops beyond the SN ratio $J_0/J \simeq 1$. From these limited results, we might confirm that TAP-like mean-field approach examined here is useful to decode the original message with a low BER for relatively large SN ratio.

9.2.6 *Quantum Monte Carlo Method for Information Processing*

In the previous subsections, we investigated the performance of the MAP and the MPM estimations for the problems of image restoration and error-correcting codes

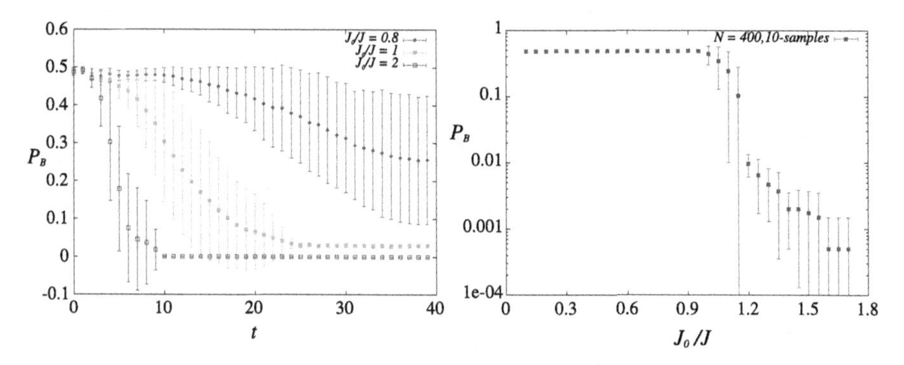

Fig. 9.22 The dynamics of the TAP-like mean field decoding (*left*, $N = 1000$, the *error-bars* are evaluated by independent 10-samples). We set $p = 2$, $\Gamma_0 = 0.5$ and $J/J_0 = 0.8, 2$ and 1. The horizontal axis t in the *left panel* denotes the number of time step in the TAP-like update described by Eqs. (9.2.144) and (9.2.145). The *right panel* shows the signal-to-noise ratio dependence of the BER. We set $p = 2$ and $\Gamma_0 = 0.5$. These panels are borrowed from [194]

by using analysis of the mean-field infinite range model. We also provide several concrete algorithms based on the mean field approach to obtain the MPM estimations. In Sourlas codes, the infinite range model is naturally accepted because we do not have to consider any structure in the bit sequence $\{\xi\}$, and in that sense, the range of interactions in the parity check $\{\xi_{i1} \cdots \xi_{ip}\}$ is infinite.

On the other hand, in image restoration, there should exist some geometrical structures in each pair in the sequence of the original image $\{\xi\}$. Then, we should introduce appropriate two dimensional lattice on which each pixel is located. Therefore, here we carry out computer simulations for the two dimensional model system to investigate the qualities of the MAP and the MPM image restorations quantitatively.

9.2.6.1 Application of Quantum Monte Carlo Method

Let us remind of readers that our effective Hamiltonian for image restoration is described by $\mathscr{H}_{\rm eff} = -\beta_m \sum_{\langle ij\rangle} S_i^z S_j^z - h\sum_i \tau_i S_i^z - \Gamma \sum_i S_i^x$. In this section, we suppose that each pixel σ_i^z is located on the two dimensional square lattice. To evaluate the expectation value of arbitrary quantity A in the quantum spin system

$$\langle A\rangle = \frac{{\rm tr}_{\{S\}}\, A\, {\rm e}^{-\beta \mathscr{H}_{\rm eff}}}{{\rm tr}_{\{S\}}\, {\rm e}^{-\beta \mathscr{H}_{\rm eff}}}, \tag{9.2.148}$$

we use the following Suzuki-Trotter formula [386] to carry out the above trace in practice as

$$\exp(-\beta \mathscr{H}_{\rm eff}) = \lim_{M\to\infty} \left({\rm e}^{\frac{\mathscr{A}}{M}}\, {\rm e}^{\frac{\mathscr{B}}{M}}\right) \tag{9.2.149}$$

where we defined

$$\mathscr{A} = \beta\left(\beta_m \sum_{\langle ij\rangle} S_i^z S_j^z + h\sum_i \tau_i S_i^z\right) = -\beta \mathscr{H}_{\rm eff}^{classical}, \qquad \mathscr{B} = \beta\Gamma\sum_i S_i^x. \tag{9.2.150}$$

We should keep in mind that these two terms $\mathscr{A}$ and $\mathscr{B}$ are easily diagonalised.

Then, by inserting the complete set: $\sum_{\{S_{jk}\}} |\{S_{jk}\}\rangle\langle\{S_{jk}\}| = 1$, the partition function Z_M for a fixed Trotter size M leads to

$$Z_M = \mathrm{tr}_{\{\sigma\}}\left(\mathrm{e}^{\frac{\mathscr{A}}{M}}\mathrm{e}^{\frac{\mathscr{B}}{M}}\right) = \sum_{\{S_{jk}=\pm 1\}} \langle\{S_{j1}\}|\mathrm{e}^{\frac{\mathscr{A}}{M}}|\{S'_{j1}\}\rangle\langle\{S'_{j1}\}|\mathrm{e}^{\frac{\mathscr{B}}{M}}|\{S_{j2}\}$$

$$\times\cdots\times\langle\{S_{jM}\}|\mathrm{e}^{\frac{\mathscr{A}}{M}}|\{S'_{jM}\}\rangle\langle\{S'_{jM}\}|\mathrm{e}^{\frac{\mathscr{B}}{M}}|\{S_{j1}\}\rangle \tag{9.2.151}$$

where $|\{S_{jk}\}\rangle$ is M-th product of eigenvectors $\{S\}$ and is explicitly given by $|\{S_{jk}\}\rangle = |S_{j1}\rangle \otimes |S_{j2}\rangle \otimes \cdots \otimes |S_{jM}\rangle$.

By taking the limit of $M \to \infty$, we obtain the *effective partition function* $Z_{\rm eff}$ of the quantum spin system with $B = (1/2)\log\coth(\beta\Gamma/M)$ as follows

$$\begin{aligned}
Z_{\rm eff} &\equiv \lim_{M\to\infty} Z_M \\
&= \lim_{M\to\infty} (a_M)^N \sum_{\{S_{jk}\}=\pm 1} \mathrm{e}^{\frac{\beta\beta_m}{M}\sum_{ij,k} S_i^k S_j^k + \frac{\beta h}{M}\sum_{i,k}\tau_i S_i^k + B\sum_{i,k} S_i^k S_i^{k+1}} \\
&= \lim_{M\to\infty} (a_M)^N \\
&\quad\times \sum_{\{S_{jk}=\pm 1\}} \exp\left[\beta_{\rm eff}\left\{\beta_m \sum_{ij,k} S_i^k S_j^k + h\sum_{i,k}\tau_i S_i^k + B_M \sum_{i,k} S_i^k S_i^{k+1}\right\}\right]
\end{aligned} \tag{9.2.152}$$

where we defined a_M and B_M as

$$a_M \equiv \left\{(1/2)\sinh(2\beta_{\rm eff}\Gamma)\right\}^{1/2}, \quad B_M \equiv (1/2\beta_{\rm eff})\log\coth(\beta_{\rm eff}\Gamma) \tag{9.2.153}$$

and introduced the following *effective inverse temperature*: $\beta_{\rm eff} = \beta/M$. Thus, this is the partition function of a $(d+1)$-dimensional classical system at the effective temperature $T_{\rm eff} = \beta_{\rm eff}^{-1}$.

Let us think about the limit of $\Gamma \to 0$ in this expression. Then, the coupling constant of the last term appearing in the argument of the exponential becomes strong. As the result, copies of the original system, which are described by the $\mathscr{H}_{\rm eff}^{classical}$ and located in the Trotter direction labelled by k, have almost the same spin configurations. Thus, the partition function is now reduced to that of the classical system at temperature $T = \beta^{-1}$.

We should not overlook that when we describe the same quantum system at $T = 0$ of the effective Hamiltonian $\mathscr{H}_{\rm eff}^{Quantum}$ by analysis of Schrödinger equation: $i\hbar(\partial|\psi(t)\rangle/\partial t) = \mathscr{H}(t)|\psi(t)\rangle$ for the time dependent Hamiltonian: $\mathscr{H}(t) =$

$-\beta_m \sum_{\langle ij \rangle} S_i^z S_j^z - h \sum_i \tau_i S_i^z - \Gamma(t) \sum_i S_i^x$, the inverse temperature β does not appear in the above expression. Therefore, we can not use β in the quantum Monte Carlo method to simulate the quantum system at $T = 0$.

To realise the equilibrium state at the ground state $T = 0$ for a finite amplitude of the quantum fluctuation $\Gamma \neq 0$, we take the limit $\beta \to \infty$, $M \to \infty$ keeping the effective inverse temperature $\beta_{\text{eff}} = \mathcal{O}(1)$. Namely, effective parameters to simulate the pure quantum system by the quantum Monte Carlo method are β_{eff} and M, instead of β and M. This choice is quite essential especially in the procedure of quantum annealing [211] because the quantum annealing searches the globally minimum energy states by using only the quantum fluctuation without any thermal fluctuation. Therefore, if we set the effective inverse temperature β_{eff} as of order 1 object in the limit of $M \to \infty$ (we can take into account the quantum effect correctly in this limit) and $\beta \to \infty$ (the thermal fluctuation is completely suppressed in this limit), we simulate the quantum spin system at the ground state $T = 0$.

9.2.6.2 Quantum Annealing and Simulated Annealing

According to the argument in the previous subsection, we construct the quantum annealing algorithm to obtain the globally minimum energy states of our effective Hamiltonian $\mathcal{H}_{\text{eff}}^{classical}$. To realise the algorithm, we control the amplitude of the transverse field as

$$\textbf{Quantum Annealing (QA):} \quad \Gamma \to 0 \quad \text{for } \beta_{\text{eff}} = 1,\ M \to \infty.$$

We should notice that the simulated annealing (thermal annealing) is achieved by controlling the parameter β as

$$\textbf{Simulated Annealing (SA):} \quad \beta \to \infty \quad \text{for finite } M \text{ and } \Gamma = 0.$$

As we mentioned, the scheduling of $T(t)$ and $\Gamma(t)$ might be essential in the simulated annealing and the quantum annealing. Although we know the optimal temperature scheduling $T(t) \sim (\log t)^{-1}$, whereas the schedule of the $\Gamma(t)$ is given as $\Gamma(t) \sim (1+t)^{-c}$ from the argument of the convergence theorem [282]. Nevertheless, in this section, we use the same scheduling for $\Gamma(t)$ as that of the simulated annealing, namely, $T(t) = \Gamma(t)$ to compare the accuracy of these two algorithms. The justification of identification of $\Gamma(t)$ and $T(t)$ comes from the results we obtained in the previous section, that is, the shape of the bit-error rate, especially for image restoration at $T = 0$ as a function of Γ is almost same as the bit-error rate for the thermal one. Thus, we assume that Γ and T might have the same kind of role to generate the equilibrium states for a given Γ and T.

9.2.6.3 Application to Image Restoration

We investigate the MAP and MPM estimations by the quantum Monte Carlo method and the quantum annealing for the two dimensional pictures which are generated by

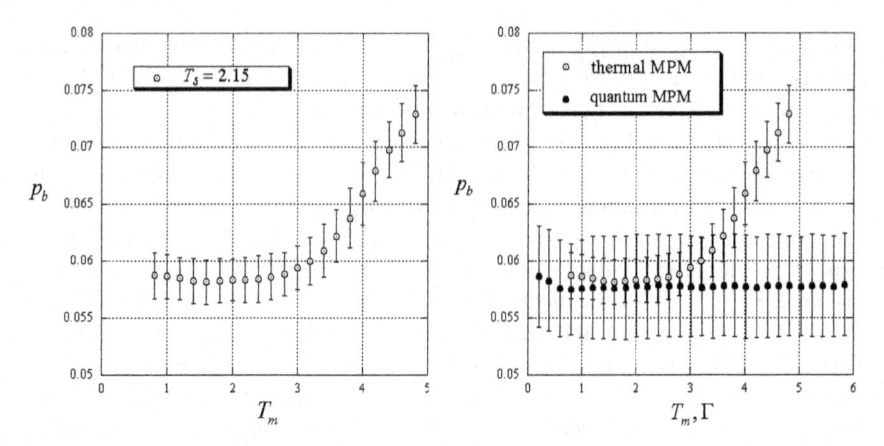

Fig. 9.23 The bit-error rate p_b for the thermal MPM estimation as a function of the temperature T_m (*left*). The plots were obtained from 30-independent runs for the system size 100×100. We set the temperature of the original image $T_s = 2.15$ and the noise rate $p_\tau = 0.1$. The *right panel* shows the bit-error rate for the quantum MPM estimation for the system size 50×50, and the Trotter number $M = 200$ for the same noise level $p_\tau = 0.1$ as the *left panel*. The *error-bars* are obtained from 50-independent runs

the Gibbs distribution: $P(\{\xi\}) = e^{\beta_s \sum_{\langle ij \rangle} \xi_i \xi_j} / Z(\beta_s)$. It must be noted that in the above sum $\sum_{\langle ij \rangle}(\cdots)$ should be carried out for the nearest neighbouring pixels located on the two dimensional square lattice. A typical snapshot from this distribution is shown in Fig. 9.24 (upper left).

9.2.6.4 Thermal MPM Estimation Versus Quantum MPM Estimation

Before we investigate the performance of the simulated annealing and the quantum annealing, as a simple check for our simulations, we demonstrate the thermal MPM estimation for the degraded image with $p_\tau = 0.1$ of the original image generated at $T_s = 2.15$ by using the thermal and the quantum Markov chain Monte Carlo methods. We show the result of the T_m-dependence of the bit-error rate in Fig. 9.23. We carried out 30-independent runs for system size 100×100. We set $h/\beta_m = T_s \beta_\tau = (T_s/2)\log(1 - p_\tau/p_\tau)$. From this figure, we find that the best performance is achieved around the temperature $T_m = T_s = 2.15$. In Fig. 9.24, we show the original, the degraded and restored images. From this figure, we found that the restored image at relatively low temperature $T_m = 0.6$ is pained in even for the local structure of the original images. On the other hand, at the optimal temperature $T_m = 2.15$, the local structures of the original image are also restored.

We next investigate the quantum MPM estimation. In Fig. 9.23, we plot the bit-error rate for the quantum MPM estimation of the original image generated by the Gibbs distribution for the two dimensional ferromagnetic Ising model. We control the effective transverse field $\Gamma_{\rm eff}$ on condition that the inverse temperature β is setting to $\beta = \beta_{\rm eff} M$, namely, the effective inverse temperature $\beta_{\rm eff} = 1$. The hyper-parameter $\beta_m^{-1} = T_m$ and h are fixed to their optimal values $T_m = T_s = 2.15$ and

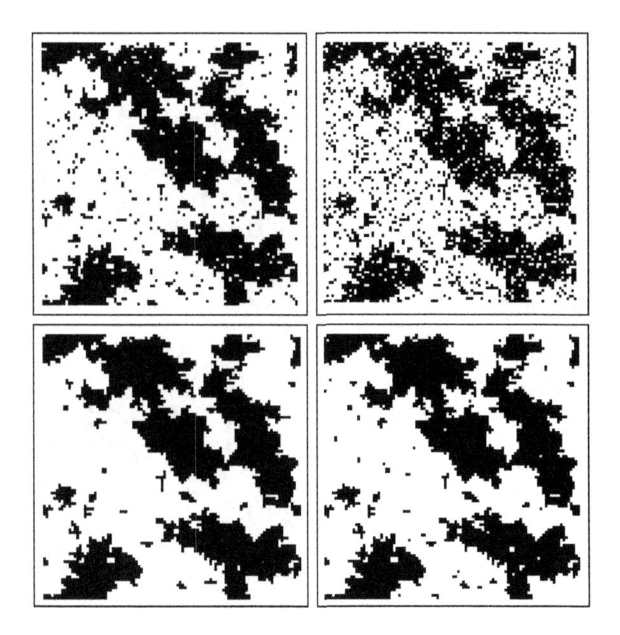

Fig. 9.24 From the *upper left* to the *lower right*, the original, the degraded ($p_\tau = 1$), and the restored at $T_m = 0.6$ and $T_m = T_s = 2.15$ pictures are displayed

$h = \beta_\tau = (1/2)\log(1 - p_\tau/p_\tau)$. To draw this figure, we carry out 50-independent runs for the system size 50×50 for the Trotter size $M = 200$. The Monte Carlo Step (MCS) needed to obtain the equilibrium state is chosen as $t' = Mt$, where $t = 10^5$ is the MCS for the thermal MPM estimation. One Monte Carlo step in calculation the quantum MPM estimate takes M times evaluations of spin flips than the calculation of the thermal MPM estimate. Thus, we provide a reasonable definition of the time t' of which the quantity is plotted and compared as a function as $t' = t$ (thermal) and $t' = Mt$ (quantum).

From this figure, we find that the lowest values of the bit-error rate for the quantum and the thermal MPM estimations are almost the same value as our analysis of the mean-field infinite range model predicted, however, the Γ-dependence of the bit-error rate is almost flat. We display several typical examples of restored images by the thermal and quantum MPM estimations in Fig. 9.25. From this figure, we find that the performance of the quantum MPM estimation is slightly superior to the thermal MPM.

9.2.6.5 Simulated Annealing Versus Quantum Annealing

In last part of this section, we investigate how effectively the quantum tunnelling process possibly leads to the global minimum of the effective Hamiltonian for the image restoration problem in comparison to temperature-driven process used in the simulated annealing. It is important for us to bear in mind that the observables we should check in the problem of image restoration are not only the energy on time E but also the bit-error rate p_b. As we mentioned, the globally minimum energy state

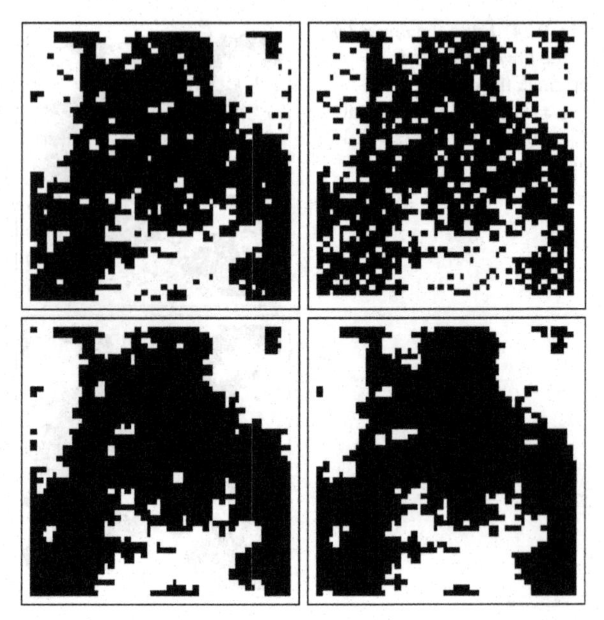

Fig. 9.25 From the *upper left* to the *lower right*, 50 × 50 original image generated at $T_s = 2.15$, degraded images ($p_\tau = 0.1$), and restored image by the thermal MPM estimation, and the restored image by the quantum MPM estimation. Each bit-error rate is $p_b = 0.06120$ for the thermal MPM at $T_m = T_s = 2.15$ and $p_b = 0.06040$ for the quantum MPM estimation with $\Gamma = 0.8$ (at the nearest point form the solution of $m_0 = m(\Gamma)$), respectively

of the classical Hamiltonian does not always minimise the bit-error rate. Therefore, from the view point of image restoration, the dynamics of the bit-error rate is also relevant quantity, although, to evaluate the performance of the annealing procedure, the energy on time is much more important measure. In this section, we investigate both of these two measures.

In our simulations discussed below, we choose the temperature and the amplitude of transverse scheduling as $\Gamma(t) = T(t) = 3/\sqrt{t}$ according to Kadowaki and Nishimori [211]. To suppress the thermal and the quantum fluctuation at the final stage of the annealing procedure, we set $\Gamma = T = 0$ in last 10 % of the MCS.

In Fig. 9.26, we plot the time development of the bit-error rate and the energy on time, namely, $E_t = -\beta_m \sum_{\langle ij\rangle} S_i^z S_j^z - h\sum_i \tau_i S_i^z$, where we defined $S_i^z = (1/M)\sum_k S_i^k$ for the quantum annealing. As the MCS t' for the quantum annealing is defined by $t' = Mt$ for the MCS, where t is the MCS for the SA, we should not overlook that the initial behaviour of the first M-th MCS in the quantum annealing is not shown in this figure. We carried out this simulation for system size 50 × 50 with Trotter size $M = 200$. The noise rate is $p_\tau = 0.1$. We set $\beta_m^{-1} = T_s = 2.15$ and $h = (1/2)\log(1 - p_\tau/p_\tau) = 1.1$. From this figure, we find that the mean value of the bit-error rate calculated by the quantum annealing is smaller than that of the simulated annealing. However, the energy on time of the simulated annealing is slightly lower than that of the quantum annealing. Although this result is not enough to decide which annealing is superior, the simulated annealing with temperature scheduling $T(t) = 3/\sqrt{t}$ seems to be much more effective than the quantum annealing with the same scheduling of the amplitude of the transverse field for finding the minimum energy state. Of course, we should check more carefully to choose the optimal or much more effective scheduling of Γ. This might be one of the im-

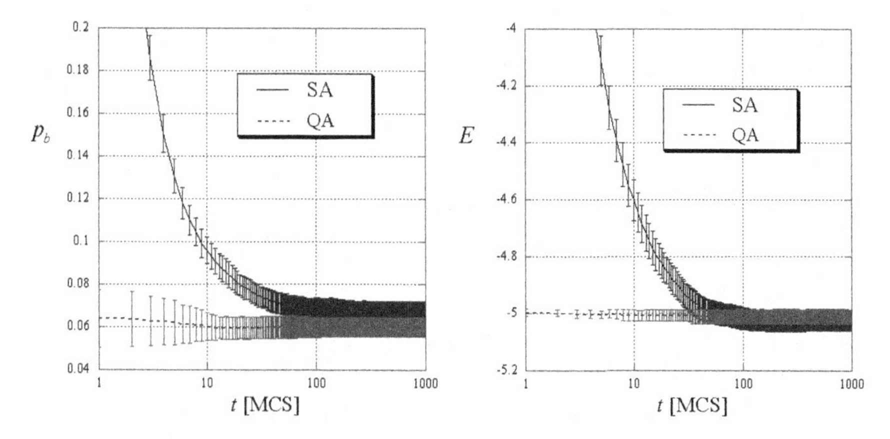

Fig. 9.26 The time dependence of the bit-error rate for the simulated annealing (SA) and the quantum annealing (QA). The Monte Carlo step (MCS) t' for the quantum annealing is defined by $t' = Mt$, where t is the MCS for the SA. The *right panel* indicates the dynamical process of the energy function by the SA and the QA. We carried out this simulation for system size 50×50 with the Trotter size $M = 200$. The noise rate is $p_\tau = 0.1$. The *error-bars* are calculated by 50-independent runs

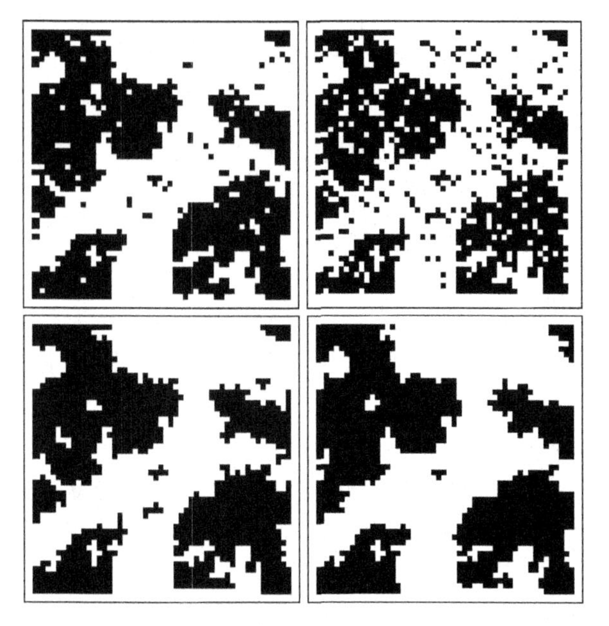

Fig. 9.27 From the *upper left* to the *lower right*, the original image ($T_s = 2.15$), the degraded image ($p_\tau = 0.1$), and typical restored images by the simulated annealing and the quantum annealing. The resultant bit-error rates are $p_b = 0.066400$ for the SA and $p_b = 0.058000$ for the QA

portant future problems. In Fig. 9.27, we display the resultant restored images by the simulated annealing and the quantum annealing. For this typical example, the performance of the quantum annealing restoration measured by the bit-error rate is better than that of the simulated annealing. The difference of the correct pixels is estimated as $\Delta n = 50 \times 50 \times \Delta p_b = 2500 \times 0.0084 = 21$ (pixels), where

$\Delta p_b = p_b(\mathrm{SA}) - p_b(\mathrm{QA})$. From reasons we mentioned above, the MAP estimate obtained by the quantum annealing is not a correct MAP estimate, however, the quality of the restoration is really fine.

Appendix 9.A: Derivation of Saddle Point Equations for the Quantum Hopfield Model

In this appendix, according to Nishimori and Nonomura [298], we derive the saddle point equations (9.1.9)–(9.1.15) for the Hopfield model at zero temperature. We starts our argument from the Suzuki-Trotter decomposition:

$$Z = \lim_{M\to\infty} \mathrm{Tr}\left(\mathrm{e}^{-\beta H_0/M}\,\mathrm{e}^{-\beta H_1/M}\right)^M = \lim_{M\to\infty} Z_M \tag{9.A.1}$$

where we defined the two parts of the Hamiltonian as

$$H_0 = -\frac{1}{N}\sum_{ij}\sum_{\mu=1}^{P}\xi_i^\mu \xi_j^\mu S_i^z S_j^z, \qquad H_1 = -\Gamma\sum_i S_i^x. \tag{9.A.2}$$

Z_M is given by

$$\begin{aligned} Z_M &= \sum_{\{\sigma=\pm1\}} \exp\left[\frac{\beta}{MN}\sum_{K=1}^{M}\sum_{ij}\sum_{\mu=1}^{P}\xi_i^\mu\xi_j^\mu S_{iK}S_{jK} + B\sum_{K=1}^{M}\sum_{i=1}^{N} S_{iK}S_{i,k+1}\right] \\ &= \int\prod_{K,\mu} dm_{K\mu}\sum_\sigma \exp\left[-\frac{N\beta}{2M}\sum_{K\mu}m_{K\mu}^2 + \frac{\beta}{M}\sum_{K,\mu}\sum_i m_{K\mu}\xi_i^\mu S_{iK}\right. \\ &\quad \left. + B\sum_{K,i} S_{ik}S_{i,K+1}\right] \end{aligned} \tag{9.A.3}$$

with

$$B = \frac{1}{2}\log\operatorname{cosec}\left(\frac{\beta\Gamma}{M}\right). \tag{9.A.4}$$

To average out the pattern-dependence in the self-averaging quantity such as free energy, we use the replica trick:

$$\ll \log Z_M \gg = \lim_{n\to 0}\frac{\ll Z_M \gg - 1}{n}. \tag{9.A.5}$$

Then, we have

$$\begin{aligned} &\ll Z_M \gg \\ &= \ll \int\prod_{K\mu\rho} dm_{K\mu\rho}\sum_\sigma \exp\left(-\frac{N\beta}{2M}\sum_{K\mu\rho}m_{K\mu\rho}^2 + \frac{\beta}{M}\sum_{K\rho}\sum_{\mu=1}^{s} m_{K\mu\rho}\sum_i \xi_i^\mu S_{i\sigma K}\right. \\ &\quad \left. + \frac{\beta}{M}\sum_{K\rho}\sum_{\mu=s+1}^{P}\sum_i \xi_i^\mu S_{i\rho K} + B\sum_{Ki\rho} S_{i\rho K}S_{i\rho K+1}\right) \gg . \end{aligned} \tag{9.A.6}$$

Here we assume that for the first s patterns, namely, for the so-called 'condensed patterns', the quantity $(\beta/M)\sum_{K\rho}\sum_{\mu=1}^{s} m_{K\mu\rho}\sum_i \xi_i^{\mu} S_{i\sigma K}$ is of order 1 object, whereas for the un-condensed patterns $\mu = s,\dots,p$, the quantity $(\beta/M)\sum_{K\rho}\sum_{\mu=s+1}^{P}\sum_i \xi_i^{\mu} S_{i\rho K}$ is of order $1/\sqrt{N}$. Hence, we can rewrite (9.A.6) as

$$
\begin{aligned}
&\ll \exp\left[\frac{\beta}{M}\sum_{K\rho}\sum_{\mu=s+1}^{P}\sum_i \xi_i^{\mu} S_{i\rho K}\right]\gg \\
&=\prod_{i\mu}\ll \exp\left[\frac{\beta}{M}\sum_{K\rho} m_{K\mu\rho}\xi_i^{\mu} S_{i\rho K}\right]\gg \\
&=\prod_{\mu i}\cosh\left(\frac{\beta}{M}\sum_{K\rho} m_{K\mu\rho} S_{i\rho K}\right) \\
&=\prod_{\mu i}\exp\left\{\log\left(1-\frac{1}{2!}\frac{\beta^2}{M^2}\sum_{K\rho}\sum_{L\sigma} m_{K\mu\rho} m_{L\mu\sigma} S_{i\rho K} S_{i\sigma L}+\cdots\right)\right\} \\
&=\prod_{\mu}\exp\left[\frac{\beta^2}{2M^2}\sum_{K\rho}\sum_{L\sigma} m_{K\mu\rho} m_{L\mu\sigma}\sum_i S_{i\rho K} S_{i\sigma L}+\mathcal{O}(1/N)\right]. \qquad (9.A.7)
\end{aligned}
$$

Collecting the quadratic part in (9.A.6), we obtain

$$
\prod_{\mu}\exp\left[-\frac{N\beta}{2M^2}\sum_{K\rho}\sum_{L\sigma} m_{K\mu\rho}^2+\frac{\beta^2}{2M^2}\sum_{K\rho}\sum_{L\sigma} m_{K\mu\rho} m_{L\mu\sigma}\sum_i S_{i\rho K} S_{i\sigma L}\right]=\prod_{\mu} E_{\mu} \qquad (9.A.8)
$$

where we defined

$$
E_{\mu}=\exp\left[-\frac{N\beta}{2M}\sum_{\mu K\rho} \Lambda_{K\rho,L\sigma} m_{K\mu\rho} m_{L\mu\sigma}\right] \qquad (9.A.9)
$$

with $(Mn\times Mn)$-matrix:

$$
\begin{aligned}
\Lambda_{K\rho,L\sigma} &= \delta_{K\rho,L\sigma}-\frac{\beta}{MN}\sum_i S_{i\rho K} S_{i\sigma L} \\
&= \delta_{K\rho,L\sigma}-\frac{\beta}{M} q_{\rho\sigma}(KL)-\frac{\beta}{M}\overline{S}_{\rho}(KL) \qquad (9.A.10)
\end{aligned}
$$

and

$$
q_{\rho\sigma}(KL)=\frac{1}{N}\sum_i S_{i\rho K} S_{i\sigma L}, \qquad \overline{S}_{q}(KL)=\frac{1}{N}\sum_i S_{i\rho K} S_{i\sigma L}. \qquad (9.A.11)
$$

Thus, we immediately obtain

$$\prod_{\mu} \int \prod_{K\rho} dm_{K\mu\rho}\, E_\mu = \left\{\text{const.}(\det \Lambda)^{-1/2}\right\}^{-(p-s)/2}$$
$$\simeq (\det \Lambda)^{-p/2} = \left(\prod_{K\rho} \lambda_{K\rho}\right)^{-p/2}$$
$$= \exp\left(-\frac{p}{2}\sum_{K\rho} \log \lambda_{K\rho}\right) \tag{9.A.12}$$

where $\lambda_{K\rho}$ are the eigenvalues of the $\Lambda_{K\rho,L\sigma}$. Inserting the definitions (9.A.11) by using the identities:

$$\int \prod_{(K\rho,L\sigma)} dq_{\rho\sigma}(KL)\delta\left(q_{\rho\sigma}(KL) - \frac{1}{N}\sum_i S_{i\rho K} S_{i\sigma L}\right) = 1 \tag{9.A.13}$$

$$\int \prod_{\rho,(KL)} d\overline{S}_\rho(KL)\,\delta\left(\overline{S}_\rho(KL) - \frac{1}{N}\sum_i S_{i\rho K} S_{i\rho L}\right) = 1. \tag{9.A.14}$$

We write the factor of unity $G\{\lambda_{K\rho}\} = 1$ as

$$G\{\lambda_{K\rho}\} = \int \prod_{(K\rho,L\sigma)} dq_{\rho\sigma}(KL) \prod_{\rho,(KL)} d\overline{S}_\rho(KL)$$
$$\times\, \delta\left(q_{\rho\sigma}(KL) - \frac{1}{N}\sum_i S_{i\rho K} S_{i\sigma L}\right)\delta\left(\overline{S}_\rho(KL) - \frac{1}{N}\sum_i S_{i\rho K} S_{i\rho L}\right)$$
$$= \int \prod_{(K\rho,L\sigma)} dq_{\rho\sigma}(KL)\,dr_{\rho\sigma}(KL) \prod_{\rho,(KL)} d\overline{S}_\rho(KL)\,dt_\rho(KL)$$
$$\times \exp\left[-\frac{N\alpha\beta^2}{M^2} r_{\rho\sigma}(KL) q_{\rho\sigma}(KL) + \frac{\alpha\beta^2}{M^2} r_{\rho\sigma}(KL)\sum_i S_{i\rho K} S_{i\rho L}\right]$$
$$\times \exp\left[-\frac{N\alpha\beta^2}{M^2} t_\rho(KL)\overline{S}_\rho(KL) + \frac{\alpha\beta^2}{M^2} t_\rho(KL)\sum_i S_{i\rho K} S_{i\rho L}\right] = 1$$
$$\tag{9.A.15}$$

where we used the Fourier transform of the delta function. Multiplying the $G\{\lambda_{K\rho}\} = 1$ by $\ll Z^n \gg$, one obtains

$$\ll Z^n \gg = \int \prod_{K\mu\rho} dm_{K\mu\rho} \prod_{(K\rho,L\sigma)} dq_{\rho\sigma}(KL)\,dr_{\rho\sigma}(KL) \prod_{\rho,(KL)} d\overline{S}_\rho(KL)\,dt_\rho(KL)$$
$$\times \exp\left[-\frac{N\beta}{2M}\sum_{K\rho}\sum_{\mu=1}^{s} m_{K\rho}^2 - \frac{N\alpha}{2}\sum_{K\rho}\log\hat{\lambda}_{K\rho}\right.$$
$$\left. - \frac{N\alpha\beta^2}{2M^2}\sum_{(K\rho,L\sigma)} r_{\rho\sigma}(KL) q_{\rho\sigma}(KL) - \frac{N\alpha\beta^2}{2M^2}\sum_{\rho,(KL)} t_\rho(KL)\overline{S}_\rho(KL)\right]$$

$$
\begin{aligned}
&\times \ll \sum_{\sigma} \exp\left[\frac{\beta}{M}\sum_{K\rho}\sum_{\mu=1}^{s} m_{K\mu\rho}\sum_{i}\xi_i^{\mu}S_{i\rho K} + B\sum_{i}\sum_{K\rho}S_{i\rho K}S_{i\rho,K+1}\right.\\
&+\frac{\alpha\beta^2}{2M^2}\sum_{i}\sum_{(K\rho,L\sigma)} r_{\rho\sigma}(KL)S_{i\rho K}S_{i\sigma L}\\
&\left.+\frac{\alpha\beta^2}{2M^2}\sum_{i}\sum_{\rho,(KL)} t_{\rho}(KL)S_{i\rho K}S_{i\rho L}\right]\gg\\
&=\exp[-N\beta f]
\end{aligned}
\tag{9.A.16}
$$

where we used self-averaging properties in the limit of $N \to \infty$ and dropped the site i dependence. f denotes the free energy per spin, which is given by

$$
\begin{aligned}
f &= \frac{1}{2M}\sum_{K\rho}\sum_{\mu=1}^{s} m_{K\mu\rho}^2 + \frac{\alpha}{2\beta}\sum_{K\rho}\log\hat{\lambda}_{K\rho} + \frac{\alpha\beta}{2M^2}\sum_{(K\rho,L\sigma)} r_{\rho\sigma}(KL)q_{\rho\sigma}(KL)\\
&+\frac{\alpha\beta}{2M^2}\sum_{\rho,(KL)} t_{\rho}(KL)\overline{S}_{\rho}(KL) - T\ll\log\sum_{S}\exp\left[\frac{\beta}{M}\sum_{K\rho}\sum_{\mu=1}^{s} m_{K\mu\rho}\xi^{\mu}S_{\sigma K}\right.\\
&+B\sum_{K\rho}S_{\rho K}S_{\rho,K+1} + \frac{\alpha\beta^2}{2M^2}\sum_{(K\rho,L\sigma)} r_{\rho\sigma}(KL)S_{\rho K}S_{\sigma L}\\
&\left.+\frac{\alpha\beta^2}{2M^2}\sum_{\rho,(KL)} t_{\rho}(KL)S_{\rho K}S_{\rho L}\right]\gg
\end{aligned}
\tag{9.A.17}
$$

where T is temperature $T = \beta^{-1}$. Extremum conditions for the free energy give the following saddle points:

$$
m_{K\mu\rho} = \ll\xi^{\mu}\langle S_{\rho K}\rangle\gg \tag{9.A.18}
$$

$$
q_{\sigma\rho}(KL) = \ll\langle S_{\rho K}\rangle\langle S_{\sigma L}\rangle\gg \tag{9.A.19}
$$

$$
r_{\rho\sigma}(KL) = \frac{1}{\alpha}\sum_{\mu=1}^{s}\ll m_{K\mu\rho}m_{L\mu\sigma}\gg \tag{9.A.20}
$$

$$
\overline{S}_{\rho}(KL) = \ll\langle S_{\rho K}S_{\rho L}\rangle\gg \tag{9.A.21}
$$

$$
t_{\rho}(KL) = \frac{1}{\alpha}\sum_{\mu=1}^{s}\ll m_{K\mu\rho}m_{L\mu\rho}\gg \tag{9.A.22}
$$

where we should notice that $m_{K\mu\rho}$, $q_{\sigma\rho}(KL)$ and $r_{\rho\sigma}(KL)$ are overlap, spin glass order parameter and noise from un-condensed patterns, respectively. These are the same as in the classical Hopfield model. On the other hand, $\overline{S}_{\rho}(KL)$ and $t_{\rho}(KL)$ represent quantum effects. The former represents the degree of quantum fluctuation by the deviation from unity and the latter denotes the effect of un-correlated patterns in the same replica.

9.A.1 Replica Symmetric and Static Approximation

To proceed the calculations, we assume the replica symmetric and static approximations:

$$m_{K\mu\rho} = m_{\mu} \tag{9.A.23}$$

$$q_{\rho\sigma}(KL) = q \tag{9.A.24}$$

$$r_{\rho\sigma}(KL) = r \tag{9.A.25}$$

$$t_{\rho}(KL) = t \tag{9.A.26}$$

$$\overline{S}_{\rho}(KL) = \begin{cases} \overline{S} & (K \neq L) \\ 1 & (K \neq L) \end{cases} \tag{9.A.27}$$

namely, we consider the case in which these order parameters are independent on the replica and Trotter indexes. Then, we have the free energy per replica as follows

$$\begin{aligned} \frac{f}{n} &= \frac{1}{2}\boldsymbol{m}^2 + \frac{\alpha}{2\beta n}\sum_{K\rho}\log\lambda_{K\rho} + \frac{\alpha\beta}{2M^2}M^2(n-1)rq + \frac{\alpha\beta}{2M^2}M^2 t\overline{S} \\ &\quad - \frac{T}{n}\ll \log\sum_{S}\exp\left[\frac{\beta}{M}\sum_{K\rho}S_{\rho K}\sum_{\mu=1}^{s}m_{\mu}\xi^{\mu} + B\sum_{K\rho}S_{\rho K}S_{\rho,K+1}\right. \\ &\quad \left. + \frac{\alpha\beta^2}{2M^2}r\left\{\left(\sum_{\rho,K}S_{\sigma K}\right)^2 - \sum_{\rho}\left(\sum_{K}S_{\rho K}\right)^2\right\} + \frac{\alpha\beta^2}{2M^2}t\left(\sum_{K}S_{\rho K}\right)^2\right]\gg. \end{aligned} \tag{9.A.28}$$

By using the Hubbard-Stratonovich transformation, we obtain in the limit of $n \to 0$ as

$$\begin{aligned} &\log\sum_{S}\exp\left[\frac{\beta}{M}\sum_{K\rho}S_{\rho K}\sum_{\mu=1}^{s}m_{\mu}\xi^{\mu} + B\sum_{K\rho}S_{\rho K}S_{\rho,K+1}\right. \\ &\quad \left. + \frac{\alpha\beta^2}{2M^2}r\left\{\left(\sum_{\rho,K}S_{\sigma K}\right)^2 - \sum_{\rho}\left(\sum_{K}S_{\rho K}\right)^2\right\} + \frac{\alpha\beta^2}{2M^2}t\left(\sum_{K}S_{\rho K}\right)^2\right] \\ &= \log\int Dz\sum_{S}\exp\left[\frac{\beta}{M}\sum_{K\rho}S_{\rho K}\sum_{\mu}m_{\mu}\xi^{\mu} + B\sum_{K\rho}S_{\rho K}S_{\rho,K+1}\right. \\ &\quad \left. + \frac{\beta\sqrt{\alpha r}}{M}z\sum_{K\rho}S_{\rho K} + \frac{\alpha\beta^2}{2M^2}(t-r)\sum_{\rho}\left(\sum_{K}S_{\rho K}\right)^2\right] \\ &= \log\int Dz\left\{\sum_{S}\exp\left[\frac{\beta}{M}(\boldsymbol{m}\cdot\boldsymbol{\xi})\sum_{K}S_{K} + B\sum_{K}S_{K}S_{K+1} + \frac{\beta\sqrt{\alpha r}}{M}z\sum_{K}S_{K}\right.\right. \\ &\quad \left.\left. + \frac{\alpha\beta^2}{2M^2}(t-r)\left(\sum_{K}S_{K}\right)^2\right]\right\}^n \end{aligned}$$

$$
\begin{aligned}
&\simeq n \int Dz \log \sum_{S} \exp\left[\frac{\beta}{M}\left((\boldsymbol{m}\cdot\boldsymbol{\xi}) + \sqrt{\alpha r}\, z\right)\sum_{K} S_K + B\sum_{K} S_K S_{K+1}\right.\\
&\quad \left. + \frac{\alpha\beta^2}{2M^2}(t-r)\left(\sum_{K} S_K\right)^2\right]\\
&= n \int Dz \log \int Dw \sum_{S} \exp\left[\frac{\beta}{M}\left(\boldsymbol{m}\cdot\boldsymbol{\xi} + \sqrt{\alpha r}\, z + \sqrt{\alpha(t-r)}\, w\right)\sum_{K} S_K\right.\\
&\quad \left. + B\sum_{K} S_K S_{K+1}\right]\\
&= n \int Dz \log \int Dw \,\mathrm{Tr} \exp\left[\left(\boldsymbol{m}\cdot\boldsymbol{\xi} + \sqrt{\alpha r}\, z + \sqrt{\alpha(t-r)}\, w\right) S_z + \Gamma S_x\right]\\
&= n \int Dz \log \int Dz\, 2\cosh \beta\sqrt{\left(\boldsymbol{m}\cdot\boldsymbol{\xi} + \sqrt{\alpha r}\, z + \sqrt{\alpha(t-r)}\, w\right)^2 + \Gamma^2}.
\end{aligned}
\tag{9.A.29}
$$

We next consider the term $(1/n)\log\lambda_{K\rho}$ in (9.A.28). The matrix elements $\{\Lambda_{K\rho,L\sigma}\}$ are $-\beta/M$ ($\rho \neq \sigma$), $-\beta\overline{S}/M$ ($\rho = \sigma$, $K \neq L$) and $1 - \beta/M$ ($\rho = \sigma$, $K = L$). Then, the eigenvalues are given by

$$
\lambda_1 = 1 - \frac{\beta}{M} - \frac{(M-1)}{M}\beta\overline{S} - (n-1)\beta q \quad \text{(with 1-degeneracy)} \tag{9.A.30}
$$

$$
\lambda_2 = 1 - \frac{\beta}{M} - \frac{(M-1)}{M}\beta\overline{S} + \beta q \quad \bigl(\text{with } (n-1)\text{-degeneracy}\bigr) \tag{9.A.31}
$$

$$
\lambda_3 = 1 - \frac{\beta}{M} + \frac{\beta}{M}\overline{S} \quad \bigl(\text{with } (M-1)n\text{-degeneracy}\bigr). \tag{9.A.32}
$$

Thus, one can evaluate

$$
\begin{aligned}
\frac{1}{n}\sum_{K\rho}\lambda_{K\rho} &= \frac{1}{n}\left(\log \lambda_1^1 \cdot \lambda_2^{(n-1)} \cdot \lambda_3^{(M-1)n}\right)\\
&= \frac{1}{n}\log\frac{\lambda_1}{\lambda_2} + \log\lambda_2 + (M-1)\log\lambda_3\\
&= \frac{1}{n}\left(1 - \frac{\beta q n}{1 - \frac{\beta}{M} - \frac{(M-1)}{M}\beta\overline{S} + \beta q}\right)\\
&\quad + \log\left(1 - \frac{\beta}{M} - \frac{(M-1)}{M}\beta\overline{S} + \beta q\right)\\
&\quad + (M-1)\log\left(1 - \frac{\beta}{M} + \frac{\beta}{M}\overline{S}\right)\\
&= -\frac{\beta q}{1 - \beta\overline{S} + \beta q} + \log(1 - \beta\overline{S} + \beta q) - \beta + \beta\overline{S} \quad (n \to 0,\ M \to \infty).
\end{aligned}
\tag{9.A.33}
$$

Collecting therms in (9.A.28), (9.A.29) and (9.A.33), we obtain

$$\frac{f}{n} = \frac{1}{2}\boldsymbol{m}^2 + \frac{\alpha}{2\beta}\left[\log(1-\beta\overline{S}+\beta q) - \frac{\beta q}{1-\beta\overline{S}+\beta q} - \beta(1-\overline{S})\right] + \frac{\alpha\beta}{2}(t\overline{S}-rq)$$
$$- T\ll \int Dz \log \int Dz 2\cosh \beta\sqrt{\left(\boldsymbol{m}\cdot\boldsymbol{\xi} + \sqrt{\alpha r}\,z + \sqrt{\alpha(t-r)}\,w\right)^2 + \Gamma^2} \gg. \tag{9.A.34}$$

For simplicity, we now set $m_\mu = \delta_{\mu,1}m, \xi_\mu = \delta_{\mu,1}$, namely, we consider the case in which a single pattern is recalled. The extremum conditions gives the following saddle point equations.

$$m = \int Dz\, Y^{-1} \int Dw\, gu^{-1} \sinh \beta u \tag{9.A.35}$$

$$r = \frac{q}{(1-\beta\overline{S}+\beta q)^2} \tag{9.A.36}$$

$$t = r + \frac{\overline{S}-q}{1-\beta\overline{S}+\beta q} \tag{9.A.37}$$

$$q = \int Dz \left(Y^{-1} \int Dw\, qu^{-1} \sinh \beta u\right)^2 \tag{9.A.38}$$

$$S = \int Dz Y^{-1} \left(\int Dw\, q^2 u^{-1} \cosh \beta u + T\Gamma \int Dw\, u^{-1} \sinh \beta u\right) \tag{9.A.39}$$

where we defined

$$g = m + \sqrt{\alpha r}\,z + \sqrt{\alpha(t-r)}\,w \tag{9.A.40}$$

$$u = \sqrt{g^2+\Gamma^2}, \qquad Y = \int Dw \cosh \beta u. \tag{9.A.41}$$

9.A.2 Zero Temperature Limit

At the ground state $T \to 0$, we naturally expect that $\overline{S} = q$. For convenience, we introduce the parameter C as

$$C \equiv \lim_{\beta\to\infty} \beta(\overline{S}-q) = \Gamma^2 \int \frac{Dz}{[(m+\sqrt{\alpha r}\,z)^2+\Gamma^2]^{3/2}}. \tag{9.A.42}$$

Then, we can rewrite the saddle point equations in terms of C as

$$m = \int Dz \frac{m+\sqrt{\alpha r}\,z}{\sqrt{(m+\sqrt{\alpha r}\,z)^2+\Gamma^2}} \tag{9.A.43}$$

$$q = \int Dz \frac{(m+\sqrt{\alpha r}\,z)^2}{(m+\sqrt{\alpha r}\,z)^2+\Gamma^2} \tag{9.A.44}$$

$$r = \frac{q}{(1-C)^2}. \tag{9.A.45}$$

Chapter 10
Related Models

10.1 XY Model in a Transverse Field

The XY model in a transverse field is another simple model, which, apart from the transverse Ising model, exhibits a zero-temperature quantum phase transition. This model also appears in the pseudo-spin representation of the BCS Hamiltonian and its mean field treatment yields exactly the BCS gap equation [16]. The spin-1/2 transverse XY chain can be diagonalised using the Jordan-Wigner transformation (see Sect. 2.2).

10.1.1 Mean Field Theory and BCS Equations

The reduced BCS Hamiltonian [26], operating only within the pair subspace (which includes only the configurations having both the Bloch states with opposing spin and momenta occupied or empty), can be written as

$$H=\sum_k \varepsilon_k\left(C_k C_k^\dagger + C_{-k}^\dagger C_{-k}\right) - V\sum_{k,k'} C_{k'}^\dagger C_{-k'}^\dagger C_{-k} C_k, \tag{10.1.1}$$

where $C_k^\dagger(C_k)$ are the creation (annihilation) operator for electrons in Bloch state $(k, 1/2)$ and the interaction V is nonzero only if ε_k lies within a small shell of thickness $\hbar\omega_D$ (ω_D is the Debye frequency), centred around the Fermi energy. One can rewrite the Hamiltonian (10.1.1) in terms of the occupation number operator n_k and n_{-k} of the electrons

$$H=-\sum_k \varepsilon_k(1-n_k-n_{-k}) - V\sum_{k,k'} C_{k'}^\dagger C_{-k'}^\dagger C_{-k} C_k. \tag{10.1.2}$$

In the pair subspace, eigenvalues of n_k and n_{-k} are the same (1 or 0) and defining the basis states as $\phi(1_k, 1_{-k})$ (both states with momenta and spin $(k, 1/2)$ and $(-k, -1/2)$ are occupied) and $\phi(0_k, 0_k)$ (both states are empty), one can readily check that

S. Suzuki et al., *Quantum Ising Phases and Transitions in Transverse Ising Models*, Lecture Notes in Physics 862, DOI 10.1007/978-3-642-33039-1_10,

$$(1 - n_k - n_{-k})\phi(1_k, 1_{-k}) = -\phi(1_k, 1_{-k}), \tag{10.1.3}$$

$$(1 - n_k - n_{-k})\phi(0_k, 0_{-k}) = \phi(0_k, 0_{-k}). \tag{10.1.4}$$

The above relations imply that in the pair subspace the operator $(1 - n_k - n_{-k})$ can be written as the z-component of a Pauli spin matrix

$$(1 - n_k - n_{-k}) = S_z^k. \tag{10.1.5}$$

Similarly one can check that

$$C_k^\dagger C_{-k}^\dagger = \frac{1}{2} S_k^-; \qquad C_k C_{-k} = \frac{1}{2} S_k^+. \tag{10.1.6}$$

In terms of the pseudo-spin operators S_z, S^+ and S^-, defined above, one can readily rewrite the reduced BCS Hamiltonian as

$$H = -\sum_k \varepsilon_k S_k^z - \frac{V}{4} \sum_{k,k'} \left(S_{k'}^x S_k^x + S_{k'}^y S_k^y\right), \tag{10.1.7}$$

which turns out to be an XY model in a transverse field, written in terms of the pseudo-spin operators.

The above Hamiltonian (10.1.7) can be studied using the molecular field theory (as in Sect. 1.2), where the many-particle Hamiltonian is written as the collection of isolated spins in a fictitious magnetic field [16]

$$H = -\sum_k \boldsymbol{h}_k \cdot \boldsymbol{S}_k, \tag{10.1.8}$$

where $\boldsymbol{h}_k$ and $\boldsymbol{S}_k$ are treated as classical vectors in the pseudo-spin space and are written as

$$\boldsymbol{h}_k = \varepsilon_k \hat{z} + \frac{V}{2} \sum_{k'} \left(\langle S_{k'}^x \rangle \hat{x} + \langle S_{k'}^y \rangle \hat{y}\right), \tag{10.1.9}$$

$$\boldsymbol{S}_k = S_k^x \hat{x} + S_k^y \hat{y} + S_k^z \hat{z}. \tag{10.1.10}$$

Since, at zero temperature, the energy is minimised when $\boldsymbol{S}_k$ is polarised in the direction of the effective molecular field $\boldsymbol{h}_k$, one can write (with $\langle S_{k'}^y \rangle = 0$),

$$\frac{h_k^x}{h_k^z} = \frac{S_k^x}{S_k^z} = \frac{\frac{V}{2}\sum_{k'} \langle S_{k'}^x \rangle}{\varepsilon_k} = \tan\theta_k. \tag{10.1.11}$$

Now using $S_{k'}^x = \sin\theta_{k'}$ we get the BCS integral equation (at $T = 0$)

$$\tan\theta_k = (V/2\varepsilon_k) \sum_{k'} \sin\theta_{k'}. \tag{10.1.12}$$

To solve Eq. (10.1.12), we set $\Delta = (V/2) \sum_{k'} \sin\theta_{k'} = \varepsilon_k \tan\theta_k$, so that

$$\Delta = \frac{V}{2} \sum_{k'} \frac{\Delta}{(\Delta^2 + \varepsilon_{k'}^2)^{1/2}}. \tag{10.1.13}$$

Replacing the above summation by an integral over f from $-\omega_D$ to $+\omega_D$, one arrives at the BCS gap equation

$$\Delta = \frac{\omega_D}{\sinh(1/\rho_F V)} \simeq 2\omega_D \exp(-1/\rho_F V), \tag{10.1.14}$$

where ρ_F is the density of states at the Fermi level and it is assumed that $V\rho_F \ll 1$. It is clear that the energy gap is positive if the interaction V is positive. At a finite temperature T the average magnetisation can be written as (cf. (1.5))

$$\langle S_k \rangle = \tanh\bigl(\beta |h_k|\bigr); \quad |h_k| = \bigl(\varepsilon_k^2 + \Delta^2(T)\bigr)^{1/2}. \tag{10.1.15}$$

The BCS integral equation (10.1.12) is then modified to

$$\tanh\theta_k = (V/2\varepsilon_k)\sum_{k'} \tanh\bigl(\beta|h_k|\bigr)\sin\theta_{k'} = \frac{\Delta}{\varepsilon_k}. \tag{10.1.16}$$

The phase transition from the superconducting state to the normal state occurs at a temperature T_c given by $\Delta(T_c) = 0$, or

$$1 = V\sum_{k'} \frac{1}{2\varepsilon_{k'}\tanh(\varepsilon_{k'}/T_c)}. \tag{10.1.17}$$

Replacing the summation once again by an integral and performing graphical integration, one obtains

$$T_c = 1.14\omega_D \exp(-1/\rho_F), \tag{10.1.18}$$

for $V\rho_F \ll 1$. One can use (10.1.14) in (10.1.18) to obtain the BCS gap equation

$$2\Delta = 3.5T_c. \tag{10.1.19}$$

Thus, transforming the reduced BCS Hamiltonian into a XY model in a transverse field, one can obtain the BCS results using simple mean field theory.

10.1.2 Exact Solution of Transverse XY Chain

The transverse XY chain Hamiltonian was first studied in details by Katsura [216] as a limiting case of anisotropic Heisenberg chain in a magnetic field. The Hamiltonian can be exactly diagonalised [322, 385] using Jordan-Wigner transformation to rewrite the spin Hamiltonian in terms of spinless Jordan-Wigner fermions (cf. Sect. 2.2). We start from the nearest neighbour XY chain Hamiltonian in a transverse (tunnelling) field, with periodic boundary condition

$$\begin{aligned} H &= -\sum_{i=1}^{N}\bigl(J_x S_i^x S_{i+1}^x + J_y S_i^y S_{i+1}^y\bigr) - \Gamma\sum_{i=1}^{N} S_i^z \\ &= -\frac{J}{2}\sum_{i=1}^{N}\bigl[(1+\gamma)S_i^x S_{i+1}^x + (1-\gamma)S_i^y S_{i+1}^y\bigr] - \Gamma\sum_{i=1}^{N} S_i^z \end{aligned} \tag{10.1.20}$$

where $\gamma J = J_x - J_y$ measures anisotropy in the in-plane interaction and for $\gamma = 0$ and $\gamma = 1$ the model reduces to the isotropic XY chain in a transverse field and transverse Ising chain, respectively. Using Jordan-Wigner transformation (cf. Sect. 2.2, Sect. 2.A.1) one can readily rewrite the Hamiltonian in terms of spin-less lattice fermions

$$H = -J\sum_{i=1}^{N}\left[\left(c_i^\dagger c_{i+1} - c_i c_{i+1}^\dagger\right) + \gamma\left(c_i^\dagger c_{i+1}^\dagger - c_i c_{i+1}\right)\right] - \Gamma\sum_{i=1}^{N}\left(2c_i^\dagger c_i - 1\right). \tag{10.1.21}$$

The above Hamiltonian can be diagonalised by transforming the fermion operators in momentum space and then using Bogoliubov transformation [245] (cf. Sect. 2.2 and Sect. 2.A.1) and can be written in the form

$$H = \sum_q \omega_q \eta_q^\dagger \eta_q - \frac{1}{2}\sum_q \omega_q, \tag{10.1.22}$$

where the dispersion for the elementary excitations ω_q is given as

$$\omega_q = 2\sqrt{J^2\cos^2 q + \gamma^2 J^2 \sin^2 q + \Gamma^2 + 2\Gamma J\cos q}. \tag{10.1.23}$$

For $\gamma = 1$, the dispersion relation reduces to that of a transverse Ising chain [312]. The normal modes η_q and $\eta_q^\dagger$ are as given in the Sect. 2.2. In the isotropic limit, $\gamma = 0$, the dispersion relation writes as

$$\omega_q = 2(\Gamma + J\cos q) = 2J(\lambda + \cos q); \quad \lambda = \frac{\Gamma}{J}. \tag{10.1.24}$$

Defining $q' = \pi - q$ as the pseudo-wavevector for the excitations, it is readily seen that the energy gap ω_q vanishes at $q' = 0$ for a critical value of transverse field given by $\lambda = \Gamma_c/J = 1$; $\Gamma_c = J$, indicating a quantum phase transition at that value of the transverse field. One thus finds using (10.1.23) that, for the anisotropic transverse XY chain, the quantum phase transition occurs (energy gap vanishes) at a value of transverse field given by

$$\Gamma_c = J = J_x + J_y; \quad \lambda = 1. \tag{10.1.25}$$

Near the critical point ($\lambda = 1$), the long wavelength ($q \to 0$) characteristic energy dispersion and its gap Δ (for $q = 0$) have the scaling form (see also [110])

$$\omega_q \sim q^z\left[1 + (q\xi)^{-z}\right], \quad \Delta = \omega_0 = |\Gamma - \Gamma_c|^s, \quad \xi = |\Gamma - \Gamma_c|^{-\nu}. \tag{10.1.26}$$

From Eqs. (10.1.23) and (10.1.24), one gets the exponents $z = 1$, $s = 1$ and $\nu = 1$ for $0 < \gamma \leq 1$ and $z = 2$, $s = 1$ and $\nu = 1/2$ for $\gamma = 0$. One should note at this point that for the isotropic case, the excitations (and hence z and s) are not defined below the critical point ($\lambda < 1$).

At this point we should mention that a two dimensional periodic Ising model ($N \times \infty$) in the absence of a magnetic field

$$H^I = -J_1\sum_i S_{i,k}^x S_{i+1,k}^x - J_2\sum_i S_{i,k}^x S_{i,k+1}^x \tag{10.1.27}$$

at a temperature T is equivalent to the ground state of a linear periodic anisotropic XY chain in a transverse field [385] with the Hamiltonian (10.1.20) through the relations,

$$\frac{J_y}{J_x} = \tanh^2 K_1^*, \qquad \frac{\Gamma}{J_x} = 2\tanh K_1^* \coth K_2, \tag{10.1.28}$$

where $K_1^* = \exp(-2K_1)$, $K_1 = \beta J_1$, or in terms of γ the equivalence holds when

$$\cosh\left(2K_1^*\right) = \gamma^{-1}, \qquad \tanh K_2 = \frac{(1-\gamma^2)^{1/2}}{\lambda}. \tag{10.1.29}$$

The above equivalence can be analytically established by exact analytic diagonalisation of both the Hamiltonians in terms of fermions. The critical temperature of the Ising model T_c, given by the relation [237, 349]

$$\sinh\left(\frac{2J_1}{k_B T_c}\right)\sinh\left(\frac{2J_2}{k_B T_c}\right) = 1, \tag{10.1.30}$$

corresponds to the critical transverse field of the XY-Hamiltonian given by $\Gamma_c = J_x + J_y$, through the relations (10.1.28)–(10.1.29). The high (low) temperature region with $T > T_c$ ($T < T_c$), of the Ising model corresponds to the high (low) transverse field, $\Gamma > \Gamma_c$ ($\Gamma < \Gamma_c$), region of the transverse XY chain. It can also be analytically established that the exponents associated with the finite-temperature thermal phase transition in the Ising model are identical to those associated with the zero temperature quantum transition in the transverse XY chain. From the relation (10.1.28), one can verify that the above mentioned equivalence between the two dimensional Ising model and the one dimensional transverse XY system holds outside the unit circle ($\gamma^2 + \lambda^2 \geq 1$), in the γ–λ plane (Fig. 10.1). From the elaborate calculation [27] it has been established that spin-correlation functions of the transverse XY chain have non-oscillatory asymptotic behaviour outside this unit circle, whereas they show oscillatory asymptotic behaviour (mass gap is always zero) inside the circle. This can be understood realising that the behaviour of the model outside the circle is classical whereas inside the circle the nature is quantum (where it may be related to a two-dimensional Ising model with complicated interactions using the Suzuki-Trotter formalism (cf. Sect. 3.1)).

10.1.2.1 Slow Quench Dynamics in Transverse XY Chain

Due to the richness in the phase diagram as shown in Fig. 10.1, the transverse XY chain involves several properties with respect to the quench dynamics [74, 99–101, 106, 107, 287, 364], which are absent in the transverse Ising chain.

Let us return to the Hamiltonian of the transverse XY chain in the fermion representation (10.1.21). Using the Fourier transformation $c_q = (1/\sqrt{N})\sum_j c_j e^{-iqR_j}$, the Hamiltonian is arranged into

$$H = \sum_{q>0} H_q, \tag{10.1.31}$$

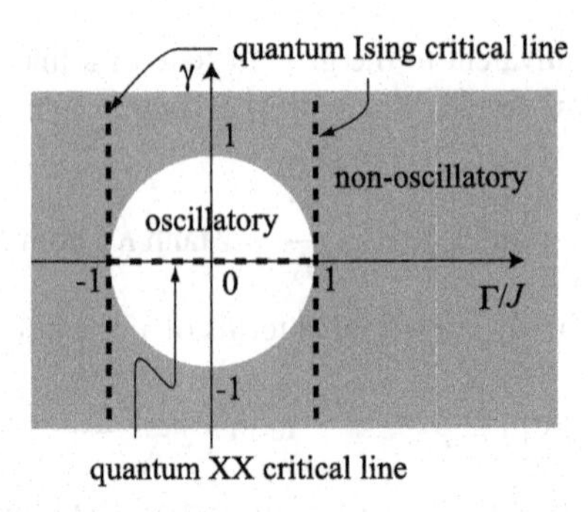

Fig. 10.1 Zero temperature phase diagram of the transverse XY chain [27]. The equivalence between the 2-dimensional Ising model and the transverse XY chain holds outside the unit circle in the γ–λ plane ($\lambda = \Gamma/J$), where spin-correlation functions of the transverse XY chain have non-oscillatory asymptotic behaviour. The spin correlation functions show oscillatory behaviour inside the *circle*. The *dashed lines* show critical lines where excitation gap vanishes. When one crosses a vertical critical line ($\Gamma/J = 1$ or -1), the ground state undergoes a quantum phase transition which shares the nature of transition with the pure transverse Ising chain. On the critical line of the isotropic $\gamma = 0$ case, the ground state is quantum critical. The transition across this line brings about a change of the anisotropic character in the ground state, such as a change from a J_x dominant state to a J_y dominant state

$$\begin{aligned} H_q &= -2\begin{pmatrix} c_q^\dagger & c_{-q} \end{pmatrix}\left\{(\lambda + \cos q)\sigma^z - \gamma \sin q \sigma^y\right\}\begin{pmatrix} c_q \\ c_{-q}^\dagger \end{pmatrix} \\ &= -2\begin{pmatrix} \tilde{c}_q^\dagger & \tilde{c}_{-q} \end{pmatrix}\begin{pmatrix} \lambda + \cos q & -\gamma \sin q \\ -\gamma \sin q & -\lambda - \cos q \end{pmatrix}\begin{pmatrix} \tilde{c}_q \\ \tilde{c}_{-q}^\dagger \end{pmatrix}, \end{aligned} \tag{10.1.32}$$

where we made $\lambda = \Gamma/J$ and chose $J = 1$ as the unit of energy. σ^x, σ^y and σ^z denote the Pauli matrices. Note that we applied a unitary transformation $U = \exp(i\sigma^z\pi/4)$ such that $U\sigma^y U^\dagger = \sigma^x$ and defined $(\tilde{c}_q\ \tilde{c}_{-q}^\dagger)^T = U(c_q\ c_{-q}^\dagger)^T$ to have the last line. Since each mode q is decoupled from the other in the Hamiltonian, the time-dependent Schrödinger equation reduces to that of a two-level system, which is given by

$$i\frac{d}{dt}\left|\psi_q(t)\right\rangle = H_q\left|\psi_q(t)\right\rangle. \tag{10.1.33}$$

We first consider a quench of the transverse field as $\lambda = t/\tau$ ($-\infty < t < \infty$) with a rate $1/\tau$ and assume that $|\psi_q(t = -\infty)\rangle$ is the ground state of the Hamiltonian $H_q(t)$ at $t = -\infty$ [74, 287]. We inquire how much the system is excited at $t = +\infty$. Defining $|\tilde{0}_q\rangle$ as the vacuum of $\tilde{c}_q$ and $\tilde{c}_{-q}$ and $|\tilde{1}_q\rangle = \tilde{c}_q^\dagger \tilde{c}_{-q}^\dagger |\tilde{0}_q\rangle$, we expand the state vector as $|\psi_q(t)\rangle = \tilde{\psi}_{q,0}(t)|\tilde{0}_q\rangle + \tilde{\psi}_{q,1}(t)|\tilde{1}_q\rangle$. Using $\tilde{\psi}_{q,0}(t)$ and $\tilde{\psi}_{q,1}(t)$, the Schrödinger equation is expressed as

$$i\frac{d}{dt}\begin{pmatrix} \tilde{\psi}_{q,1}(t) \\ \tilde{\psi}_{q,0}(t) \end{pmatrix} = -2\begin{pmatrix} \lambda + \cos q & -\gamma \sin q \\ -\gamma \sin q & -\lambda - \cos q \end{pmatrix}\begin{pmatrix} \tilde{\psi}_{q,1}(t) \\ \tilde{\psi}_{q,0}(t) \end{pmatrix}, \tag{10.1.34}$$

and the initial condition is given by $(\tilde{\psi}_{q,1}(-\infty), \tilde{\psi}_{q,0}(-\infty)) = (0, 1)$. This equation is immediately solved using the Landau-Zener solution (cf. Sect. 7.2.2.2 and Sect. 7.A.2). The excitation probability at $t = +\infty$ is

$$p_q = \left|\tilde{\psi}_{q,0}(+\infty)\right|^2 = e^{-2\pi\tau\gamma^2 \sin^2 q}. \tag{10.1.35}$$

Thus the density of defects (excitations) is obtained as

$$\begin{aligned} n = \int_0^\pi \frac{dq}{\pi} p_q &\approx \int_0^\infty \frac{dq}{\pi} e^{-2\pi\tau\gamma^2 q^2} + \int_{-\infty}^\pi \frac{dq}{\pi} e^{-2\pi\tau\gamma^2(\pi-q)^2} \\ &= \frac{1}{\pi\gamma\sqrt{2\tau}}. \end{aligned} \tag{10.1.36}$$

This result is consistent with the Kibble-Zurek scaling $n \propto \tau^{-d\nu/(z\nu+1)}$ (cf. (7.2.111) in Sect. 7.2.2.2), which reduces to $\tau^{-1/2}$ using the values $d = z = \nu = 1$ of the quantum critical point on the quantum Ising critical line.

Note that the scaling (10.1.36) of the density of defects does not make sense when $\gamma = 0$, namely, the system passes through the quantum XX critical line. In this case, one gets $p_q = 1$ for all q's and hence $n = 1$.

Let us next consider a quench in which the transverse field λ and the anisotropy γ are changed as $\lambda + 1 = \gamma = t/\tau$. We assume that the system is in the ground state at $t = -\infty$. Using this schedule of λ and γ, the system passes the multicritical point $(\lambda, \gamma) = (-1, 0)$ at $t = 0$. Note that this multicritical point is characterised by exponents $z = 2$ and $\nu = 1/2$. The Hamiltonian (10.1.32) with time-dependent λ and γ is written as

$$H_q(t) = -2\begin{pmatrix}\tilde{c}_q^\dagger & \tilde{c}_{-q}\end{pmatrix}\begin{pmatrix} t/\tau - 1 + \cos q & -(t/\tau)\sin q \\ -(t/\tau)\sin q & -(t/\tau) + 1 - \cos q \end{pmatrix}\begin{pmatrix}\tilde{c}_q \\ \tilde{c}_{-q}^\dagger\end{pmatrix}. \tag{10.1.37}$$

We introduce a unitary matrix $\overline{U}$ such that

$$\overline{U}\begin{pmatrix} 1 & -\sin q \\ -\sin q & -1 \end{pmatrix}\overline{U}^\dagger = \begin{pmatrix} \sqrt{1+\sin^2 q} & 0 \\ 0 & -\sqrt{1+\sin^2 q} \end{pmatrix}, \tag{10.1.38}$$

and define $(\overline{c}_q\ \overline{c}_{-q}^\dagger)^T = \overline{U}(\tilde{c}_q\ \tilde{c}_{-q}^\dagger)^T$. One can see that, for $|t|/\tau \to \infty$, the Hamiltonian $H_q(t)$ is diagonalised by $\overline{U}$, so that $H_q(t) \to -2(t/\tau)(1+\sin^2 q)^{1/2}(\overline{c}_q^\dagger \overline{c}_q - \overline{c}_{-q}\overline{c}_{-q}^\dagger)$. Therefore the ground state at $t = -\infty$ is found to be the vacuum of $\overline{c}_q$ and $\overline{c}_{-q}$. Writing this vacuum as $|\overline{0}_q\rangle$ and defining $|\overline{1}_q\rangle = \overline{c}_q^\dagger \overline{c}_{-q}^\dagger |\overline{0}_q\rangle$, we express the state vector as $|\psi_q(t)\rangle = \overline{\psi}_{q,0}(t)|\overline{0}_q\rangle + \overline{\psi}_{q,1}(t)|\overline{1}_q\rangle$. Now a little calculation shows that for $q \ll 1$

$$\overline{U}\begin{pmatrix} -1+\cos q & 0 \\ 0 & 1-\cos q \end{pmatrix}\overline{U}^\dagger \approx -\frac{1}{2}\begin{pmatrix} q^2 & q^3 \\ q^3 & -q^2 \end{pmatrix}. \tag{10.1.39}$$

Using this and (10.1.38), the Schrödinger equation with $q \ll 1$ is written as

$$i\frac{d}{dt}\begin{pmatrix}\overline{\psi}_{q,1}(t)\\ \overline{\psi}_{q,0}(t)\end{pmatrix}=\begin{pmatrix}-2t/\tau+q^2 & q^3\\ q^3 & 2t/\tau-q^2\end{pmatrix}\begin{pmatrix}\overline{\psi}_{q,1}(t)\\ \overline{\psi}_{q,0}(t)\end{pmatrix}. \tag{10.1.40}$$

The initial condition for $\overline{\psi}_{q,1}(t)$ and $\overline{\psi}_{q,0}(t)$ is given by $(\overline{\psi}_{q,1}(-\infty),\overline{\psi}_{q,0}(-\infty))=(0,1)$. Applying the Landau-Zener solution, the excitation probability at $t=+\infty$ is obtained as

$$p_q=\left|\overline{\psi}_{q,0}(+\infty)\right|^2=e^{-\pi\tau q^6/2} \tag{10.1.41}$$

for $q \ll 1$, and the density of defects as

$$n=\int_0^{\pi}\frac{dq}{\pi}p_q\approx\int_0^{\infty}\frac{dq}{\pi}e^{-\pi\tau q^6/2}\sim\frac{1}{\tau^{1/6}}. \tag{10.1.42}$$

This result conflict with the Kibble-Zurek scaling $\tau^{-d\nu/(z\nu+1)}$ which with $d=1$, $z=2$ and $\nu=1/2$ predicts $\tau^{-1/4}$. To resolve this conflict, Divakaran et al. [106] proposed an extension of the Kibble-Zurek scaling. According to them, when the off diagonal term of the two-level Hamiltonian H_q has an exponent z' and one has the critical exponent ν' such that $z'\nu'=1$, then the scaling of n follows $n\sim\tau^{-d\nu'/(z'\nu'+1)}=\tau^{-d/2z'}$.

One may also consider a quench along a gapless line. Let us assume $\gamma=t/\tau$ and λ fixed at $\lambda=1$. We choose the initial condition that the system is in the ground state at $t=-\infty$. The Hamiltonian of the mode q with this quench schedule is written as

$$H_q(t)=-2\left(\tilde{c}_q^\dagger\ \ \tilde{c}_{-q}\right)\left\{(1+\cos q)\sigma^z-(t/\tau)\sin q\sigma^x\right\}\begin{pmatrix}\tilde{c}_q\\ \tilde{c}_{-q}^\dagger\end{pmatrix}. \tag{10.1.43}$$

Applying a unitary transformation $U'=\exp(-i\sigma^y\pi/4)$, the Hamiltonian is arranged as

$$\begin{aligned}H_q(t)&=-2\left(\tilde{c}_q^\dagger\ \ \tilde{c}_{-q}\right)U'^\dagger U'\left\{(1+\cos q)\sigma^z-(t/\tau)\sin q\sigma^x\right\}U'^\dagger U'\begin{pmatrix}\tilde{c}_q\\ \tilde{c}_{-q}^\dagger\end{pmatrix}\\ &=-2\left(c_q'^\dagger\ \ c'_{-q}\right)\left\{(1+\cos q)\sigma^x+(t/\tau)\sin q\sigma^z\right\}\begin{pmatrix}c'_q\\ c'^\dagger_{-q}\end{pmatrix},\end{aligned} \tag{10.1.44}$$

where $(c'_q\ c'^\dagger_{-q})^T=U'(\tilde{c}_q\ \tilde{c}^\dagger_{-q})^T$. Letting $|\psi_q(t)\rangle=\psi'_{q,0}(t)|0'_q\rangle+\psi'_{q,1}(t)|1'_q\rangle$ where $|0'_q\rangle$ denotes the vacuum of c'_q and c'_{-q} and $|1'_q\rangle=c_q'^\dagger c'^\dagger_{-q}|0'_q\rangle$, the Schrödinger equation with $\pi-q\ll 1$ is written as

$$i\frac{d}{dt}\begin{pmatrix}\psi'_{q,1}(t)\\ \psi'_{q,0}(t)\end{pmatrix}=-\begin{pmatrix}2(t/\tau)(\pi-q) & (\pi-q)^2\\ (\pi-q)^2 & -2(t/\tau)(\pi-q)\end{pmatrix}\begin{pmatrix}\psi'_{q,1}(t)\\ \psi'_{q,0}(t)\end{pmatrix}. \tag{10.1.45}$$

The initial condition for $\psi'_{q,1}(t)$ and $\psi'_{q,0}(t)$ is given by $(\psi'_{q,1}(-\infty),\psi'_{q,0}(-\infty))=(0,1)$. The Landau-Zener solution for this equation yields the excitation probability at $t=+\infty$

$$p_q=\left|\psi'_{q,0}(+\infty)\right|^2=e^{-\pi\tau(\pi-q)^3/2}. \tag{10.1.46}$$

The density of defects is thus obtained as

$$n = \int_0^{\pi} \frac{dq}{\pi} p_q \approx \int_{-\infty}^{\pi} \frac{dq}{\pi} p_q \sim \frac{1}{\tau^{1/3}}. \qquad (10.1.47)$$

Divakaran et al. generalised this result to a quench across a critical line in a d dimensional system with the energy spectrum of the quasiparticle of the form $\omega_q = |\gamma| q^z + \delta q^{z'}$, where δ is a constant. They proposed $n \sim \tau^{d/(2z'-z)}$ when $2z' > z$ and γ is quenched as $\gamma = t/\tau$ with fixed δ.

Before closing this subsection, we comment that Sengupta and Sen [364] have discussed entanglement production due to a quench of the transverse field in the XY chain. They have shown that, when transverse field is quenched from $\Gamma = -\infty$ to $+\infty$ with rate $1/\tau$, the concurrence (cf. Sect. 2.2.1) acquires a finite value only between even neighbour sites and there is a critical rate $1/\tau_c$ such that for $1/\tau > 1/\tau_c$ no concurrence is generated due to a quench.

10.1.3 Transverse XY Chain and Harper Model

The studies on metal-insulator transition induced by quenched (random) disorder had been naturally extended to systems with quasi-periodic disorder (see e.g. [373]), in order to find the link with and to compare with the quantum transition from extended to localised electronic wavefunctions in such systems. The one dimensional Harper model

$$\psi_{i+1} + \psi_{i-1} + \Gamma_0 \cos(2\pi\sigma i) = E\psi_i \qquad (10.1.48)$$

is considered a paradigm in the study of quasi-periodic systems exhibiting (global) transitions from metallic or Bloch-type extended states (for $\Gamma_0 < 2$) to the insulating or exponentially localised states (for $\Gamma_0 > 2$) with irrational σ [373]. At the metal-insulator transition point, the states are critical, having power law localisation characterised by fractal spectrum and wave functions. The energy spectrum here becomes self-similar (the butterfly spectrum). The quasi-periodic systems being intermediate between periodic and random systems, provide useful link for understanding the crossover, and the Harper model has been extensively studied in this context, as well as in the context of the quantum Hall effect and the mean field theory of Hubbard model.

The above Harper model has been approximately mapped [344–346] to a general isotropic XY chain in quasi-periodic transverse field, suggesting intriguing possibilities of identifying the various states in both the models. Let us consider the anisotropic XY chain in a transverse field

$$H = -\sum_{i=1}^{N} \left(J_x S_i^x S_{i+1}^x + J_y S_i^y S_{i+1}^y + \Gamma_i S_i^z \right) \qquad (10.1.49)$$

with the transverse field having a quasi-periodic variation

$$\Gamma_i = \Gamma_0 \cos(2\pi\sigma i) \qquad (10.1.50)$$

along the chain length i, controlled by the parameter σ and amplitude Γ_0. Using the Jordan-Wigner transformation (see Sect. 2.2) the above spin chain Hamiltonian may be related to a quadratic (free) fermion Hamiltonian

$$H = -\sum_i \left[(J_x + J_y) c_i^\dagger c_{i+1} + (J_x - J_y) c_i^\dagger c_{i+1}^\dagger + \Gamma_i c_i^\dagger c_i + h.c. \right] \tag{10.1.51}$$

where c_i's are the anti-commuting fermion operators. The Hamiltonian (10.1.51) is bi-linear in fermion operators and can be readily recast to the general form (cf. Sect. 2.A.2)

$$H = \sum_{ij} \left[c_i^\dagger A_{ij} c_j + c_i B_{ij} c_j + h.c. \right] \tag{10.1.52}$$

with

$$A_{ij} = (J_x + J_y)\delta_{i+1,j} + \Gamma_i \delta_{ij}; \quad B_{ij} = (J_x - J_y)\delta_{i+1,j}. \tag{10.1.53}$$

Using Bogoliubov transformations (cf. Sect. 2.A.2) one can readily obtain the matrix equations

$$\Phi_k (A - B)(A + B) = E_k^2 \Phi_k; \qquad \Psi_k (A + B)(A - B) = E_k^2 \Psi_k \tag{10.1.54}$$

where$(A + B)^T = A - B$ because A is symmetric and B is antisymmetric. Following Lieb et al. [245], the diagonalisation of the above Hamiltonian can be reduced to the diagonalisation of the tight-binding Hamiltonian [344]

$$\begin{aligned} E^2 \psi_i = {} & 16 J_x J_y (\psi_{i-2} + \psi_{i+2}) + 4(J_x \Gamma_{i-1} + J_x \Gamma_i)\psi_{i-1} \\ & + 4(J_y \Gamma_i + J_x \Gamma_{i+1})\psi_{i+1} + 4\left(J_x + J_y + \Gamma_i^2\right)\psi_i. \end{aligned} \tag{10.1.55}$$

In the isotropic limit ($J_x = J_y$), the above equation reduces to the squared Harper equation (10.1.48). This mapping can be employed to extract and compare the generalised Harper butterfly energy spectrum with the various correlations in the equivalent transverse XY model [346]. In the anisotropic case ($J_x \neq J_y$), the localised to extended state transition (at $\Gamma_0 = 2$) gets split with a critical region in between (localised to critical transition and critical to extended transition).

10.1.4 Infinite Range XY Spin Glass in a Transverse Field

In this section, we shall briefly discuss the infinite range XY model in the presence of a transverse field. The Hamiltonian of the N interacting spins in the presence of a transverse field can be put in the form

$$H = -\sum_{i<j} J_{ij} \left(S_i^x S_j^x + S_i^y S_j^y \right) - \Gamma \sum_{i=1}^{N} S_i^x, \tag{10.1.56}$$

where S_i^α's are Pauli spin operators and random interactions J_{ij}'s are distributed according to the Gaussian distribution with zero mean and variance $(J/\sqrt{N})$,

$$P(J_{ij}) = \left(\frac{N}{2\pi J^2}\right)^{1/2} \exp\left(-\frac{NJ_{ij}^2}{2J^2}\right). \tag{10.1.57}$$

One can now apply the Suzuki-Trotter method (cf. Sect. 3.1) to the above quantum Hamiltonian to derive the partition function of the equivalent classical system. The partition function (in the M-th Trotter approximation) one arrives at, after a few steps of algebra, can be written in the form [407]

$$Z = \mathrm{Tr}_{S,\tau} \prod_{k=1}^{M} \exp(h_x + h_y) \prod_{i=1}^{N} A \exp\left[i\left(\frac{\pi}{4} S_{ik}\tau_{ik} + B\tau_{ik}S_{i,k+1}\right)\right], \tag{10.1.58}$$

where $\{|S_{ik}\rangle\}$ and $\{|\tau_{jk}\rangle\}$ denote the complete set of eigenvectors of the operators S_{ik}^x and S_{ik}^y respectively and

$$h_x = \frac{\beta}{M} \sum_{i<j} J_{ij} S_{ik} S_{jk}, \tag{10.1.59}$$

$$h_y = \frac{\beta}{M} \sum_{i<j} J_{ij} \tau_{ik} \tau_{jk}, \tag{10.1.60}$$

$$A = \frac{1}{2}\sqrt{\cosh\left(\frac{2\Gamma\beta}{M}\right)}, \tag{10.1.61}$$

$$B = -\frac{1}{2} \tan^{-1}\left(\frac{1}{\sinh(2\beta\Gamma/M)}\right). \tag{10.1.62}$$

As mentioned in the previous sections, the partition function may be considered as $\mathrm{Tr}\exp(-\beta H_{\mathrm{eff}}^0)$, where H_{eff}^0 is the effective classical Hamiltonian on a $N \times M$ lattice, corresponding to the quantum model in the $M \to \infty$ limit.

At this point one has to consider the random interactions and use the n-replicated partition function Z^n. Performing the configuration average and the Hubbard-Stratonovich transformation (to linearise the quadratic terms appearing in the expression of the partition function) and using the saddle-point method for a thermodynamically large system, one obtains for the expression of the free energy [51]

$$f = \frac{1}{2nM^2} \sum_{k,\alpha} \sum_{k',\alpha'} q_{k\alpha k'\alpha'}^2 - \frac{1}{n} \ln \mathrm{Tr} \exp(H_{\mathrm{eff}}), \tag{10.1.63}$$

where $t = 1/(\beta J)$ and the effective replicated Hamiltonian is given by

$$H_{\mathrm{eff}} = \frac{1}{2M^2 t^2} \sum_{k\alpha k'\alpha'} q_{k\alpha k'\alpha'} (S_{\alpha k} S_{\alpha' k'} + \tau_{\alpha k} \tau_{\alpha' k'}) + \sum_{k,\alpha} i\left(\frac{\pi}{4} S_{\alpha k} \tau_{\alpha k} + B \tau_{\alpha k} S_{\alpha, k+1}\right), \tag{10.1.64}$$

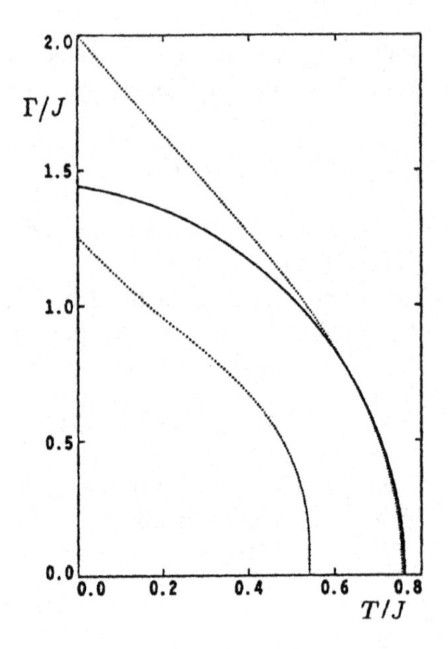

Fig. 10.2 Phase diagram of transverse XY spin glass (infinite range). The *solid line* represents exact result. The *lower dotted line* represents the thermofield dynamical result and the *upper one* is the static approximation result [51]

and order parameters are given by the self-consistency condition

$$q_{k\alpha k'\alpha'} = \langle S_{\alpha k} S_{\alpha' k'} \rangle = \langle \tau_{\alpha k} \tau_{\alpha' k'} \rangle \tag{10.1.65}$$

(the angular bracket denotes the thermal average with respect to the Hamiltonian H_{eff}). There are two types of order parameters: the order parameters $R_{kk'}$ for $\alpha = \alpha'$ are the spin self-interaction terms (cf. Sect. 6.3) (independent of the replica index) and the spin glass order parameter ($\alpha \neq \alpha'$) $q_{\alpha\beta}$ which vanishes in the paramagnetic phase. Similar considerations as in Sect. 6.3 (with the only exception that in this calculation discretised Trotter indices are used) lead to the relation

$$\frac{1}{M} \sum_k R_{kk'} = \frac{k_B T_c}{J}. \tag{10.1.66}$$

Büttner and Usadel [51], numerically solved the self-consistency condition (10.1.65), up to $M = 14$, by direct spin summation. Thermodynamic quantities obtained for finite systems can be extrapolated to the limit $M \to \infty$, using the M^{-2}-extrapolation law proposed by Suzuki (cf. Sect. 6.3). The phase boundary was calculated exactly [51] up to $(k_B/J) = 0.18$ and extrapolated for the lower temperatures (Fig. 10.2). The value of Γ_c ($T = 0$), thus obtained is around $1.44J$ and the value of the critical temperature, in the limit of vanishing Γ is $0.763J$. The thermofield dynamical treatment [52] gives the lower value of T_c whereas the phase boundary, obtained from the static approximation deviates visibly from the above result in the low temperature region (Fig. 10.2).

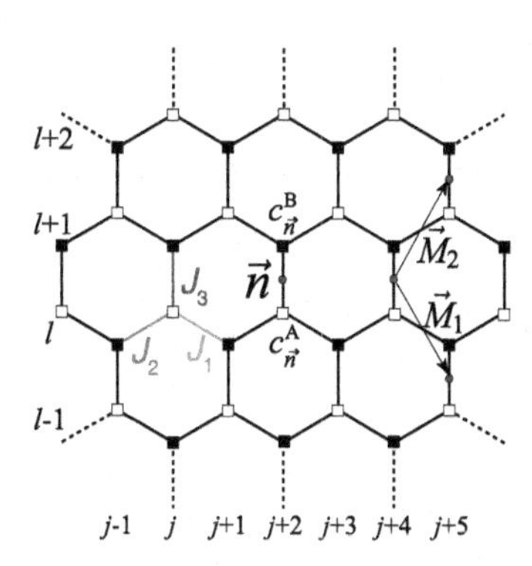

Fig. 10.3 A honeycomb lattice is decomposed into an A-sublattice and a B-sublattice denoted by *open* and *closed squares* respectively. The horizontal bonds with an A-site on the left are x-links, those with a B-site on the left are y-links, and the vertical bonds are z-links. The interaction $S^{\alpha}_i S^{\alpha}_j$ is defined on an α-link. J_1, J_2, and J_3 are the coupling constants of $S^x_i S^x_j$, $S^y_i S^y_j$, and $S^z_i S^z_j$ respectively. $\mathbf{n}$ represents the position vector of a unit cell. $c^{\mathrm{A}}_{\mathbf{n}}$ ($c^{\mathrm{B}}_{\mathbf{n}}$) is the Jordan-Wigner fermion operator on the A(B)-site of the unit cell $\mathbf{n}$

10.2 Kitaev Model

10.2.1 Fermion Representation and Diagonalisation

The Kitaev model [229] is a quantum spin model defined on the honeycomb lattice in two dimension. Its Hamiltonian is given by

$$H = \sum_{j+l:\text{even}} \left(J_1 S^x_{j,l} S^x_{j+1,l} + J_2 S^y_{j-1,l} S^y_{j,l} + J_3 S^z_{j,l} S^z_{j,l+1}\right), \tag{10.2.1}$$

where j and l respectively specify the horizontal and vertical coordinates of a site. The honeycomb lattice can be decomposed into two sublattices. We name the sublattice composed of the sites (j,l) with even $j+l$ A-sublattice and that composed of the odd $j+l$ sites B-sublattice (Fig. 10.3). The bonds connecting two sites of the A-sublattice and the B-sublattice are classified into three types, x-link, y-link, and z-link, according to their positioning. We define the x-link between an A-sublattice site on the left and a B-sublattice site on the right, y-link between a B-sublattice site on the left and an A-sublattice site on the right, and z-link between a B-sublattice site on the top and an A-sublattice on the bottom. In the Kitaev model, neighbouring spins connected by the α-link interact only through α component out of x, y and z. Therefore one can write the Hamiltonian (10.2.1) as

$$H = J_1 \sum_{x\text{-link}} S^x_{\mathbf{r}} S^x_{\mathbf{r}'} + J_2 \sum_{y\text{-link}} S^y_{\mathbf{r}} S^y_{\mathbf{r}'} + J_3 \sum_{z\text{-link}} S^z_{\mathbf{r}} S^z_{\mathbf{r}'}, \tag{10.2.2}$$

where $\mathbf{r}$ indicates the position a site, and $\sum_{\alpha\text{-link}}$ stands for the summation with respect to all α-links.

A prominent feature of the Kitaev model is that it can be mapped to a free-fermion model by means of the Jordan-Wigner transformation. Following the procedure in Sect. 2.2 or Sect. 2.A.1, we define fermion operators as

$$c_{j,l} = \Theta_{j,l} S^-_{j,l}, \tag{10.2.3}$$

$$\Theta_{j,l} = \prod_{m<l} \prod_{k} \exp\left[i\pi S^+_{k,m} S^-_{k,m}\right] \prod_{k<j} \exp\left[i\pi S^+_{k,l} S^-_{k,l}\right]. \tag{10.2.4}$$

Note that the operator $\Theta_{j,l}$ includes spin operators at all the sites which are on the left to or below (j,l). Suppose now that a site (j,l) belongs to the A-sublattice. We define the unit cell specified by the position vector $\mathbf{n}(j,l)$ such that it includes the site (j,l) of the A-sublattice and the site $(j,l+1)$ of the B-sublattice. We then write the fermion operator at site (j,l) as $c^{\rm A}_{\mathbf{n}(j,l)} = c_{j,l}$ and that at $(j,l+1)$ as $c^{\rm B}_{\mathbf{n}(j,l)} = c_{j,l+1}$. Employing this notation, each term in the Hamiltonian (10.2.1) is expressed as follows

$$S^x_{j,l} S^x_{j+1,l} = \left(c^{\rm A\dagger}_{\mathbf{n}(j,l)} - c^{\rm A}_{\mathbf{n}(j,l)}\right)\left(c^{\rm B\dagger}_{\mathbf{n}(j+1,l-1)} + c^{\rm B}_{\mathbf{n}(j+1,l-1)}\right), \tag{10.2.5}$$

$$S^y_{j-1,l} S^y_{j,l} = -\left(c^{\rm B\dagger}_{\mathbf{n}(j-1,l-1)} + c^{\rm B}_{\mathbf{n}(j-1,l-1)}\right)\left(c^{\rm A\dagger}_{\mathbf{n}(j,l)} - c^{\rm A}_{\mathbf{n}(j,l)}\right), \tag{10.2.6}$$

$$S^z_{j,l} S^z_{j,l+1} = \left(c^{\rm A\dagger}_{\mathbf{n}(j,l)} - c^{\rm A}_{\mathbf{n}(j,l)}\right)\left(c^{\rm B\dagger}_{\mathbf{n}(j,l)} + c^{\rm B}_{\mathbf{n}(j,l)}\right)\left(c^{\rm A\dagger}_{\mathbf{n}(j,l)} + c^{\rm A}_{\mathbf{n}(j,l)}\right)\left(c^{\rm B\dagger}_{\mathbf{n}(j,l)} - c^{\rm B}_{\mathbf{n}(j,l)}\right). \tag{10.2.7}$$

We next define new fermion operators by a unitary transformation:

$$\begin{pmatrix} \beta_{\mathbf{n}} \\ \beta^\dagger_{\mathbf{n}} \\ \alpha_{\mathbf{n}} \\ \alpha^\dagger_{\mathbf{n}} \end{pmatrix} = \frac{1}{2} \begin{pmatrix} 1 & 1 & -1 & 1 \\ 1 & 1 & 1 & -1 \\ -1 & 1 & 1 & 1 \\ 1 & -1 & 1 & 1 \end{pmatrix} \begin{pmatrix} c^{\rm A}_{\mathbf{n}} \\ c^{\rm A\dagger}_{\mathbf{n}} \\ c^{\rm B}_{\mathbf{n}} \\ c^{\rm B\dagger}_{\mathbf{n}} \end{pmatrix}. \tag{10.2.8}$$

In terms of these operators, (10.2.5)–(10.2.7) are written as

$$S^x_{j,l} S^x_{j+1,l} = \left(\alpha_{\mathbf{n}(j,l)} - \alpha^\dagger_{\mathbf{n}(j,l)}\right)\left(\alpha_{\mathbf{n}(j,l)+\mathbf{M}_1} + \alpha^\dagger_{\mathbf{n}(j,l)+\mathbf{M}_1}\right), \tag{10.2.9}$$

$$S^y_{j-1,l} S^y_{j,l} = -\left(\alpha_{\mathbf{n}(j,l)-\mathbf{M}_2} + \alpha^\dagger_{\mathbf{n}(j,l)-\mathbf{M}_2}\right)\left(\alpha_{\mathbf{n}(j,l)} - \alpha^\dagger_{\mathbf{n}(j,l)}\right), \tag{10.2.10}$$

$$S^z_{j,l} S^z_{j,l+1} = \left(1 - 2\alpha^\dagger_{\mathbf{n}(j,l)} \alpha_{\mathbf{n}(j,l)}\right) D_{\mathbf{n}(j,l)}, \tag{10.2.11}$$

where

$$D_{\mathbf{n}} = \left(c^{\rm A\dagger}_{\mathbf{n}} + c^{\rm A}_{\mathbf{n}}\right)\left(c^{\rm B\dagger}_{\mathbf{n}} - c^{\rm B}_{\mathbf{n}}\right) = 2\beta^\dagger_{\mathbf{n}}\beta_{\mathbf{n}} - 1. \tag{10.2.12}$$

Therefore the Hamiltonian (10.2.1) is expressed as [73]

$$\begin{aligned} H = \sum_{\mathbf{n}} \Big\{ & J_1\left(c^{\rm A\dagger}_{\mathbf{n}} - c^{\rm A}_{\mathbf{n}}\right)\left(c^{\rm B\dagger}_{\mathbf{n}+\mathbf{M}_1} + c^{\rm B}_{\mathbf{n}+\mathbf{M}_1}\right) + J_2\left(c^{\rm A\dagger}_{\mathbf{n}} - c^{\rm A}_{\mathbf{n}}\right)\left(c^{\rm B\dagger}_{\mathbf{n}-\mathbf{M}_2} + c^{\rm B}_{\mathbf{n}-\mathbf{M}_2}\right) \\ & + J_3\left(c^{\rm A\dagger}_{\mathbf{n}} - c^{\rm A}_{\mathbf{n}}\right)\left(c^{\rm B\dagger}_{\mathbf{n}} + c^{\rm B}_{\mathbf{n}}\right) D_{\mathbf{n}} \Big\}, \qquad (10.2.13) \\ = \sum_{\mathbf{n}} \Big\{ & J_1\left(\alpha_{\mathbf{n}} - \alpha^\dagger_{\mathbf{n}}\right)\left(\alpha_{\mathbf{n}+\mathbf{M}_1} + \alpha^\dagger_{\mathbf{n}+\mathbf{M}_1}\right) + J_2\left(\alpha_{\mathbf{n}} - \alpha^\dagger_{\mathbf{n}}\right)\left(\alpha_{\mathbf{n}-\mathbf{M}_2} + \alpha^\dagger_{\mathbf{n}-\mathbf{M}_2}\right) \\ & + J_3\left(1 - 2\alpha^\dagger_{\mathbf{n}}\alpha_{\mathbf{n}}\right) D_{\mathbf{n}} \Big\}. \qquad (10.2.14) \end{aligned}$$

As one can see from (10.2.12), the operator $D_{\mathbf{n}}$ has eigenvalue ± 1, and it has a remarkable property:

$$[D_{\mathbf{n}}, H] = 0. \tag{10.2.15}$$

Therefore, $D_{\mathbf{n}}$ is a constant of motion. It turns out that the Kitaev model possesses an extensive number of constants of motion.

Let us fix $D_{\mathbf{n}}$ at $D_{\mathbf{n}} = (-1)^{\zeta_{\mathbf{n}}} = e^{i\pi\zeta_{\mathbf{n}}}$, where $\zeta_{\mathbf{n}}$ takes the value 0 or 1. The Hamiltonian is written as

$$\begin{aligned} H = &\sum_{\mathbf{n}} \big\{ J_1 \big(c_{\mathbf{n}}^{\mathrm{A}\dagger} - c_{\mathbf{n}}^{\mathrm{A}}\big)\big(c_{\mathbf{n}+\mathbf{M}_1}^{\mathrm{B}\dagger} + c_{\mathbf{n}+\mathbf{M}_1}^{\mathrm{B}}\big) + J_2 \big(c_{\mathbf{n}}^{\mathrm{A}\dagger} - c_{\mathbf{n}}^{\mathrm{A}}\big)\big(c_{\mathbf{n}-\mathbf{M}_2}^{\mathrm{B}\dagger} + c_{\mathbf{n}-\mathbf{M}_2}^{\mathrm{B}}\big) \\ &+ J_3 \big(e^{i\pi\zeta_{\mathbf{n}}} c^{\mathrm{A}\dagger} - e^{-i\pi\zeta_{\mathbf{n}}} c_{\mathbf{n}}^{\mathrm{A}}\big)\big(c_{\mathbf{n}}^{\mathrm{A}\dagger} + c_{\mathbf{n}}^{\mathrm{A}}\big)\big\} \\ = &\sum_{\mathbf{n}} \big\{ J_1 \big(e^{-i\pi\zeta_{\mathbf{n}}} \tilde{c}_{\mathbf{n}}^{\mathrm{A}\dagger} - e^{i\pi\zeta_{\mathbf{n}}} \tilde{c}_{\mathbf{n}}^{\mathrm{A}}\big)\big(\tilde{c}_{\mathbf{n}+\mathbf{M}_1}^{\mathrm{B}\dagger} + \tilde{c}_{\mathbf{n}+\mathbf{M}_1}^{\mathrm{B}}\big) \\ &+ J_2 \big(e^{-i\pi\zeta_{\mathbf{n}}} \tilde{c}_{\mathbf{n}}^{\mathrm{A}\dagger} - e^{i\pi\zeta_{\mathbf{n}}} \tilde{c}_{\mathbf{n}}^{\mathrm{A}}\big)\big(\tilde{c}_{\mathbf{n}-\mathbf{M}_2}^{\mathrm{B}\dagger} + \tilde{c}_{\mathbf{n}-\mathbf{M}_2}^{\mathrm{B}}\big) \\ &+ J_3 \big(\tilde{c}_{\mathbf{n}}^{\mathrm{A}\dagger} - \tilde{c}_{\mathbf{n}}^{\mathrm{A}}\big)\big(\tilde{c}_{\mathbf{n}}^{\mathrm{B}\dagger} + \tilde{c}_{\mathbf{n}}^{\mathrm{B}}\big)\big\}, \end{aligned} \tag{10.2.16}$$

where we applied a unitary transformation $\tilde{c}_{\mathbf{n}}^{\mathrm{A}} = e^{-\pi\zeta_{\mathbf{n}}} c_{\mathbf{n}}^{\mathrm{A}}$ and $\tilde{c}_{\mathbf{n}}^{\mathrm{B}} = c_{\mathbf{n}}^{\mathrm{B}}$. This Hamiltonian represents the tight-binding model with pair creations and annihilations in the presence of the flux phase.

The flux phase is defined on each hexagonal plaquette. Let us consider a round trip starting from a B-sublattice site of the cell at $\mathbf{n}_1$ counterclockwise to the original site, passing A-site at $\mathbf{n}_2$, B-site at $\mathbf{n}_2$, A-site at $\mathbf{n}_3$, B-site at $\mathbf{n}_4$, and A-site at $\mathbf{n}_4$ (see Fig. 10.4(a)). According to the Hamiltonian (10.2.16), hopping of a fermion from the B-site of the cell $\mathbf{n}_1$ to A-site of cell $\mathbf{n}_2$ is represented by $e^{-i\pi\zeta_{\mathbf{n}_2}} c_{\mathbf{n}_2}^{\mathrm{A}\dagger} c_{\mathbf{n}_1}^{\mathrm{B}}$, where note $\mathbf{n}_2 = \mathbf{n}_1 + \mathbf{M}_2$. A fermion acquires the phase $e^{-i\pi\zeta_{\mathbf{n}_2}}$ by this motion. Applying this consideration to other hoppings, one finds that the total flux phase of the round trip is given

$$e^{i\pi\zeta_{\mathbf{n}_4}} \times e^{i\pi\zeta_{\mathbf{n}_3}} \times e^{-i\pi\zeta_{-\mathbf{n}_3}} \times e^{-i\pi\zeta_{\mathbf{n}_2}} = e^{-i\pi(\zeta_{\mathbf{n}_2} - \zeta_{\mathbf{n}_4})}.$$

According to the Lieb's flux phase theorem [246], the ground state of this system is realised when the flux phase on any plaquette is zero, namely, $\zeta_{\mathbf{n}_2} = \zeta_{\mathbf{n}_4}$. This means that the ground state of the Kitaev model belongs to the subspace with $D_{\mathbf{n}} = 1$ for all $\mathbf{n}$ or -1 for all $\mathbf{n}$.

We next consider a change $D_{\mathbf{n}} \Rightarrow -D_{\mathbf{n}}$ for all $\mathbf{n}$. Note that this is equivalent to the change $J_3 \Rightarrow -J_3$. Then the Hamiltonian (10.2.13) changes into

$$\begin{aligned} H \to H' = &\sum_{\mathbf{n}} \big\{ J_1 \big(c_{\mathbf{n}}^{\mathrm{A}\dagger} - c_{\mathbf{n}}^{\mathrm{A}}\big)\big(c_{\mathbf{n}+\mathbf{M}_1}^{\mathrm{B}\dagger} + c_{\mathbf{n}+\mathbf{M}_1}^{\mathrm{B}}\big) + J_2 \big(c_{\mathbf{n}}^{\mathrm{A}\dagger} - c_{\mathbf{n}}^{\mathrm{A}}\big)\big(c_{\mathbf{n}-\mathbf{M}_2}^{\mathrm{B}\dagger} + c_{\mathbf{n}-\mathbf{M}_2}^{\mathrm{B}}\big) \\ &- J_3 \big(c_{\mathbf{n}}^{\mathrm{A}\dagger} - c_{\mathbf{n}}^{\mathrm{A}}\big)\big(c_{\mathbf{n}}^{\mathrm{B}\dagger} + c_{\mathbf{n}}^{\mathrm{B}}\big) D_{\mathbf{n}} \big\}. \end{aligned} \tag{10.2.17}$$

We here apply a unitary transformation for $c_{\mathbf{n}}^{\mathrm{A}}$ and $c_{\mathbf{n}}^{\mathrm{B}}$ on the l-th row with even l:

$$c_{\mathbf{n}}^{\mathrm{A}} \quad \Rightarrow \quad \bar{c}_{\mathbf{n}}^{\mathrm{A}} = -c_{\mathbf{n}}^{\mathrm{A}} \quad \text{for even } j, \tag{10.2.18a}$$

$$c_{\mathbf{n}}^{\mathrm{B}} \quad \Rightarrow \quad \bar{c}_{\mathbf{n}}^{\mathrm{B}} = -c_{\mathbf{n}}^{\mathrm{B}} \quad \text{for odd } j, \tag{10.2.18b}$$

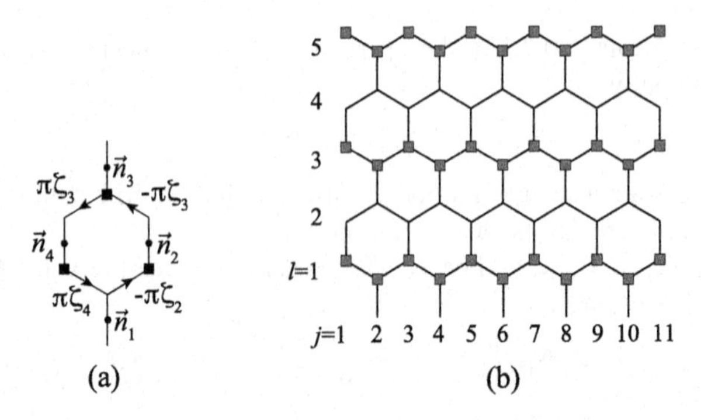

Fig. 10.4 (**a**) A plaquette made by unit cells $\mathbf{n}_1$, $\mathbf{n}_2$, $\mathbf{n}_3$, and $\mathbf{n}_4$. *Squares* denote A-sublattice sites. When a fermion moves around this plaquette counterclockwise, it acquires a flux phase $\exp\{-i\pi(\zeta_2-\zeta_4)\}$. Due to the Lieb's theorem, this flux phase is unity in the ground state. It follows that the ground state belongs to the subspace of uniform $\zeta_{\mathbf{n}}$. (**b**) A unitary transformation that leads to a change in sign of $D_{\mathbf{n}}$. The fermion operators at the sites marked by the *square* are transformed as $c_{\mathbf{n}}^{\mathrm{A,B}} \Rightarrow -c_{\mathbf{n}}^{\mathrm{A,B}}$

whereas those on the odd l-th row are unchanged. This transformation changes the sign of the operators $C_{\mathbf{n}}^{\mathrm{A}}$ and $C_{\mathbf{n}}^{\mathrm{B}}$ at sites marked by squares in Fig. 10.4. As a result of (10.2.18a), (10.2.18b), the negative sign in front of J_3 is cancelled to yield

$$H' = \sum_{\mathbf{n}} \big\{ J_1\big(\bar{c}_{\mathbf{n}}^{\mathrm{A}\dagger} - \bar{c}_{\mathbf{n}}^{\mathrm{A}}\big)\big(\bar{c}_{\mathbf{n}+\mathbf{M}_1}^{\mathrm{B}\dagger} + \bar{c}_{\mathbf{n}+\mathbf{M}_1}^{\mathrm{B}}\big) + J_2\big(\bar{c}_{\mathbf{n}}^{\mathrm{A}\dagger} - \bar{c}_{\mathbf{n}}^{\mathrm{A}}\big)\big(\bar{c}_{\mathbf{n}-\mathbf{M}_2}^{\mathrm{B}\dagger} + \bar{c}_{\mathbf{n}-\mathbf{M}_2}^{\mathrm{B}}\big)$$
$$+ J_3\big(\bar{c}_{\mathbf{n}}^{\mathrm{A}\dagger} - \bar{c}_{\mathbf{n}}^{\mathrm{A}}\big)\big(\bar{c}_{\mathbf{n}}^{\mathrm{B}\dagger} + \bar{c}_{\mathbf{n}}^{\mathrm{B}}\big) D_{\mathbf{n}} \big\}. \qquad (10.2.19)$$

Therefore the eigenenergy remains the same under the change of $D_{\mathbf{n}} \to -D_{\mathbf{n}}$ for all $\mathbf{n}$. This consequence and the Lieb's flux phase theorem show that the ground state is realised when $D_{\mathbf{n}} = 1$ for all $\mathbf{n}$ and when $D_{\mathbf{n}} = -1$ for all $\mathbf{n}$. We hereafter fix $D_{\mathbf{n}} = 1$ for all $\mathbf{n}$ and study the property of the ground state of the Kitaev model.

The Hamiltonian (10.2.14) with $D_{\mathbf{n}} = 1$ is diagonalised by successive application of the Fourier transformation and the Bogoliubov transformation. Let us first define the Fourier transformation:

$$\alpha_{\mathbf{k}} = \frac{1}{\sqrt{N}} \sum_{\mathbf{n}} e^{-i\mathbf{k}\cdot\mathbf{n}} \alpha_{\mathbf{n}}. \qquad (10.2.20)$$

In terms of this, the Hamiltonian is written as

$$H = \sum_{\mathbf{k}\in\mathrm{BZ}} \big\{ J_1\big(e^{-i\mathbf{k}\cdot\mathbf{M}_1}\alpha_{\mathbf{k}}\alpha_{-\mathbf{k}} + e^{-i\mathbf{k}\cdot\mathbf{M}_1}\alpha_{\mathbf{k}}\alpha_{\mathbf{k}}^\dagger - e^{i\mathbf{k}\cdot\mathbf{M}_1}\alpha_{\mathbf{k}}^\dagger\alpha_{\mathbf{k}} - e^{i\mathbf{k}\cdot\mathbf{M}_1}\alpha_{\mathbf{k}}^\dagger\alpha_{-\mathbf{k}}^\dagger\big)$$
$$\times J_2\big(e^{i\mathbf{k}\cdot\mathbf{M}_2}\alpha_{\mathbf{k}}\alpha_{-\mathbf{k}} + e^{i\mathbf{k}\cdot\mathbf{M}_2}\alpha_{\mathbf{k}}\alpha_{\mathbf{k}}^\dagger - e^{-i\mathbf{k}\cdot\mathbf{M}_2}\alpha_{\mathbf{k}}^\dagger\alpha_{\mathbf{k}} - e^{-i\mathbf{k}\cdot\mathbf{M}_2}\alpha_{\mathbf{k}}^\dagger\alpha_{-\mathbf{k}}^\dagger\big)$$
$$+ J_3\big(\alpha_{\mathbf{k}}\alpha_{\mathbf{k}}^\dagger - \alpha_{\mathbf{k}}^\dagger\alpha_{\mathbf{k}}\big)\big\}$$
$$= \sum_{\mathbf{k}\in\mathrm{BZ}/2} \big(\alpha_{\mathbf{k}}^\dagger \quad \alpha_{-\mathbf{k}}\big) H_{\mathbf{k}} \begin{pmatrix} \alpha_{\mathbf{k}} \\ \alpha_{-\mathbf{k}}^\dagger \end{pmatrix}, \qquad (10.2.21)$$

where

$$H_{\mathbf{k}} = -2(J_3 + J_1 \cos \mathbf{k} \cdot \mathbf{M}_1 + J_2 \cos \mathbf{k} \cdot \mathbf{M}_2)\sigma^z + 2(J_1 \sin \mathbf{k} \cdot \mathbf{M}_1 - J_2 \sin \mathbf{k} \cdot \mathbf{M}_2)\sigma^y, \tag{10.2.22}$$

with the Pauli matrices, σ^y and σ^z. Note that BZ stands for the Brillouin zone and BZ/2 is the half of BZ. This Hamiltonian is diagonalised to yield

$$H = \sum_{\mathbf{k} \in \mathrm{BZ}/2} E_{\mathbf{k}} \left(\eta_{\mathbf{k}}^{\dagger} \eta_{\mathbf{k}} + \eta_{-\mathbf{k}}^{\dagger} \eta_{-\mathbf{k}} - 1 \right), \tag{10.2.23}$$

where

$$\begin{pmatrix} \eta_{\mathbf{k}} \\ \eta_{-\mathbf{k}}^{\dagger} \end{pmatrix} = \begin{pmatrix} u_{\mathbf{k}}^* & v_{\mathbf{k}}^* \\ -v_{\mathbf{k}} & u_{\mathbf{k}} \end{pmatrix} \begin{pmatrix} \alpha_{\mathbf{k}} \\ \alpha_{-\mathbf{k}}^{\dagger} \end{pmatrix}, \tag{10.2.24}$$

and

$$u_{\mathbf{k}} = b_{\mathbf{k}} / \left\{ 2\left(a_{\mathbf{k}}^2 + b_{\mathbf{k}}^2\right) - 2a_{\mathbf{k}} \sqrt{a_{\mathbf{k}}^2 + b_{\mathbf{k}}^2} \right\}^{1/2},$$
$$v_{\mathbf{k}} = -i \left(\sqrt{a_{\mathbf{k}}^2 + b_{\mathbf{k}}^2} - a_{\mathbf{k}} \right) / \left\{ 2\left(a_{\mathbf{k}}^2 + b_{\mathbf{k}}^2\right) - 2a_{\mathbf{k}} \sqrt{a_{\mathbf{k}}^2 + b_{\mathbf{k}}^2} \right\}^{1/2}$$

with

$$a_{\mathbf{k}} = -2(J_3 + J_1 \cos \mathbf{k} \cdot \mathbf{M}_1 + J_2 \cos \mathbf{k} \cdot \mathbf{M}_2),$$
$$b_{\mathbf{k}} = -2(J_1 \sin \mathbf{k} \cdot \mathbf{M}_1 - J_2 \sin \mathbf{k} \cdot \mathbf{M}_2).$$

The energy spectrum of the Bogoliubov quasiparticle is given by

$$E_{\mathbf{k}} = \sqrt{a_{\mathbf{k}}^2 + b_{\mathbf{k}}^2}. \tag{10.2.25}$$

The eigenenergy per site of the ground state is given by

$$\varepsilon_{\mathrm{GS}} = -\frac{1}{2N} \sum_{\mathbf{k} \in \mathrm{BZ}/2} E_{\mathbf{k}} \overset{N \to \infty}{\longrightarrow} -\frac{1}{2A} \int_{\mathrm{BZ}/2} d\mathbf{k}\, E_{\mathbf{k}}, \tag{10.2.26}$$

where $A = 4\pi^2/3\sqrt{3}$ is the area of the half Brillouin zone. The energy spectrum (10.2.25) has a zero when

$$a_{\mathbf{k}^*} = -2\left(J_3 + J_1 \cos \mathbf{k}^* \cdot \mathbf{M}_1 + J_2 \cos \mathbf{k}^* \cdot \mathbf{M}_2\right) = 0, \tag{10.2.27}$$
$$b_{\mathbf{k}^*} = -2\left(J_1 \sin \mathbf{k}^* \cdot \mathbf{M}_1 - J_2 \sin \mathbf{k}^* \cdot \mathbf{M}_2\right) = 0, \tag{10.2.28}$$

where $\mathbf{k}^*$ denotes the solution of these equations. Equations (10.2.27) and (10.2.28) are arranged into

$$J_3^2 = J_1^2 + J_2^2 + 2J_1 J_2 \cos \mathbf{k}^* \cdot (\mathbf{M}_1 - \mathbf{M}_2), \tag{10.2.29}$$

or equivalently

$$\cos \mathbf{k}^* \cdot (\mathbf{M}_1 - \mathbf{M}_2) = \frac{J_3^2 - J_1^2 - J_2^2}{2J_1 J_2}. \tag{10.2.30}$$

A condition that (10.2.30) has a real solution is given by

$$-1 \le \left(J_3^2 - J_1^2 - J_2^2\right)/2J_1 J_2 \le 1.$$

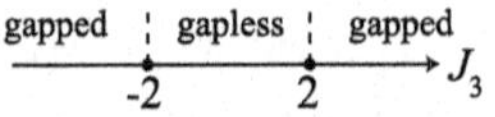

gapped ¦ gapless ¦ gapped
-2 2 J_3

Fig. 10.5 The ground-state phase diagram of the Kitaev model with $J_1 = J_2 = 1$. We have a gapped ground state for $|J_3| > 2$ and a gapless ground state for $-2 < J_3 < 2$

Therefore $E_{\mathbf{k}}$ vanishes when

$$|J_1 - J_2| \le |J_3| \le |J_1 + J_2|. \tag{10.2.31}$$

This implies that the gapless excitation is present when J_3 satisfies (10.2.31), while a energy gap opens when $|J_3| < |J_1 - J_2|$ or $|J_3| > |J_1 + J_2|$ [73, 229]. In particular, when $J_1 = J_2 = 1$, one has the gapped phase for $|J_3| > 2$ and the gapless phase for $-2 < J_3 < 2$, which are separated by a quantum phase transition at $J_3 = \pm 2$. We show the phase diagram of the ground state of the Kitaev model with $J_1 = J_2 = 1$ in Fig. 10.5.

Let us look into the spectrum $E_{\mathbf{k}}$ of the gapless phase in detail. We expand $E_{\mathbf{k}}$ around its zero point, supposing $\mathbf{k} = \mathbf{k}^* + \delta\mathbf{k}$ and $|\delta\mathbf{k}| \ll 1$.

$$\begin{aligned} E_{\mathbf{k}} \approx \left.\frac{\partial E_{\mathbf{k}}}{\partial \mathbf{k}}\right|_{\mathbf{k}^*} \cdot \delta\mathbf{k} = \left(\left[\frac{a_{\mathbf{k}}}{E_{\mathbf{k}}}\frac{\partial a_{\mathbf{k}}}{\partial k_x}\right]_{\mathbf{k}^*} + \left[\frac{b_{\mathbf{k}}}{E_{\mathbf{k}}}\frac{\partial b_{\mathbf{k}}}{\partial k_x}\right]_{\mathbf{k}^*}\right)\delta k_x \\ + \left(\left[\frac{a_{\mathbf{k}}}{E_{\mathbf{k}}}\frac{\partial a_{\mathbf{k}}}{\partial k_y}\right]_{\mathbf{k}^*} + \left[\frac{b_{\mathbf{k}}}{E_{\mathbf{k}}}\frac{\partial b_{\mathbf{k}}}{\partial k_y}\right]_{\mathbf{k}^*}\right)\delta k_y, \end{aligned} \tag{10.2.32}$$

where we note

$$\begin{aligned} \frac{\partial a_{\mathbf{k}}}{\partial \mathbf{k}} &= 2\big(J_1 \sin(\mathbf{k}\cdot\mathbf{M}_1)\mathbf{M}_1 + J_2 \sin(\mathbf{k}\cdot\mathbf{M}_2)\mathbf{M}_2\big), \\ \frac{\partial b_{\mathbf{k}}}{\partial \mathbf{k}} &= 2\big(-J_1 \cos(\mathbf{k}\cdot\mathbf{M}_1)\mathbf{M}_1 + J_2 \cos(\mathbf{k}\cdot\mathbf{M}_2)\mathbf{M}_2\big). \end{aligned}$$

Equation (10.2.32) implies that the quasiparticle has a linear dispersion around the zero point. It turns out that the present system has the dynamical exponent $z = 1$. We remark that the solution $\mathbf{k}^*$ of (10.2.30) moves with J_3 for fixed J_1 and J_2. This shows that the quantum criticality maintains inside the gapless phase. This fact plays a significant role in the slow quench dynamics across the critical region (Sect. 10.2.3).

We finally mention that the eigenstate of the Kitaev model is composed of the "particle" sector and the "spin" sector. The ground state denoted by $|\Psi_{\mathrm{GS}}\rangle = |\psi_{\mathrm{GS}}; \mathscr{G}\rangle|\mathscr{G}\rangle$ is the Bogoliubov vacuum with respect to the particle sector, which satisfies

$$\eta_{\mathbf{k}}|\psi_{\mathrm{GS}}; \mathscr{G}\rangle = 0 \quad \text{for all } \mathbf{k}, \tag{10.2.33}$$

and the *fully polarised* state with respect to the "spin" sector,

$$D_{\mathbf{n}}|\mathscr{G}\rangle = |\mathscr{G}\rangle \text{ for all } \mathbf{n} \quad \text{or} \quad D_{\mathbf{n}}|\mathscr{G}\rangle = -|\mathscr{G}\rangle \text{ for all } \mathbf{n}. \tag{10.2.34}$$

10.2.2 Correlation Functions

Let us next consider the spin-spin correlation function given by [29]

$$
\begin{aligned}
g^{\alpha\beta}_{(j,l)(j',l')}(t) &= \langle\Psi_{\rm GS}|S^{\alpha}_{j,l}(t)S^{\beta}_{j',l'}(0)|\Psi_{\rm GS}\rangle \\
&= \langle\Psi_{\rm GS}|S^{\alpha}_{j,l}e^{-i(H-E_{\rm GS})t}S^{\beta}_{j',l'}|\Psi_{\rm GS}\rangle,
\end{aligned} \tag{10.2.35}
$$

where $S^{\alpha}_{j,l}(t)=e^{iHt}S^{\alpha}_{j,l}e^{-iHt}$ is the Heisenberg operator of $S^{\alpha}_{j,l}$ and $E_{\rm GS}$ denotes the ground-state energy. For the meantime we suppose that (j,l) and (j',l') belong to A- and B-sublattice respectively. We first investigate $S^{\alpha}_{j,l}|\Psi_{\rm GS}\rangle$ $(\alpha=x,y,z)$ with (j,l) belonging to the A-sublattice. Using the fermion representation, they are written as

$$
\begin{aligned}
S^{x}_{j,l}|\Psi_{\rm GS}\rangle &= \Theta_{j,l}\big(c^{\rm A\dagger}_{\mathbf{n}(j,l)}+c^{\rm A}_{\mathbf{n}(j,l)}\big)|\Psi_{\rm GS}\rangle = \Theta_{j,l}\big(\beta^{\dagger}_{\mathbf{n}(j,l)}+\beta_{\mathbf{n}(j,l)}\big)|\Psi_{\rm GS}\rangle, \\
S^{y}_{j,l}|\Psi_{\rm GS}\rangle &= i\Theta_{j,l}\big(c^{\rm A}_{\mathbf{n}(j,l)}-c^{\rm A\dagger}_{\mathbf{n}(j,l)}\big)|\Psi_{\rm GS}\rangle = i\Theta_{j,l}\big(\alpha^{\dagger}_{\mathbf{n}(j,l)}-\alpha_{\mathbf{n}(j,l)}\big)|\Psi_{\rm GS}\rangle, \\
S^{z}_{j,l}|\Psi_{\rm GS}\rangle &= \big(c^{\rm A\dagger}_{\mathbf{n}(j,l)}+c^{\rm A}_{\mathbf{n}(j,l)}\big)\big(c^{\rm A}_{\mathbf{n}(j,l)}-c^{\rm A\dagger}_{\mathbf{n}(j,l)}\big)|\Psi_{\rm GS}\rangle \\
&= \big(\beta^{\dagger}_{\mathbf{n}(j,l)}+\beta_{\mathbf{n}(j,l)}\big)\big(\alpha^{\dagger}_{\mathbf{n}(j,l)}-\alpha_{\mathbf{n}(j,l)}\big)|\Psi_{\rm GS}\rangle,
\end{aligned}
$$

where the string operator $\Theta_{j,l}$ is defined by (10.2.4). Note that, since

$$
\begin{aligned}
&\exp\big[i\pi S^{+}_{k,m}S^{-}_{k,m}\big] \\
&= \begin{cases} (\alpha^{\dagger}_{\mathbf{n}(k,l)}-\alpha_{\mathbf{n}(k,l)})(\beta^{\dagger}_{\mathbf{n}(k,l)}+\beta_{\mathbf{n}(k,l)}) & \text{for } (k,l)\in\text{A-sublattice} \\ (\beta^{\dagger}_{\mathbf{n}(k,l)}-\beta_{\mathbf{n}(k,l)})(\alpha^{\dagger}_{\mathbf{n}(k,l)}+\alpha_{\mathbf{n}(k,l)}) & \text{for } (k,l)\in\text{B-sublattice} \end{cases},
\end{aligned} \tag{10.2.36}
$$

the application of $S^{\alpha}_{j,l}$ to $|\Psi_{\rm GS}\rangle$ causes a change in the "spin" sector. It turns out that the spin operators $S^{\alpha}_{j',l'}$ with $(j',l')\in$ B-sublattice which cancel this change are only $S^{x}_{j+1,l}$, $S^{y}_{j-1,l}$, and $S^{z}_{j,l+1}$ for $\alpha=x$, y, and z respectively. Moreover, since the "spin" sector of $|\Psi_{\rm GS}\rangle$ is unchanged during the time evolution, the correlation function vanishes unless it is between α-components of the spins connected by the α-link. Therefore, as far as the spin-spin correlation between sites of the A-sublattice and B-sublattice is concerned, the non-zero correlation functions are $g^{xx}_{(j,l)(j+1,l)}$, $g^{yy}_{(j,l)(j-1,l)}$, and $g^{zz}_{(j,l)(j,l+1)}$ [29].

10.2.3 Slow Quench Dynamics

The slow quench dynamics of the Kitaev model was studied by Sengupta et al. [366]. We recall the Hamiltonian (10.2.21). Applying a unitary transformation $U=\exp(i\sigma^{z}\pi/4)$, the Hamiltonian can be written as

$$
H=\sum_{\mathbf{k}\in{\rm BZ}/2}\begin{pmatrix}\alpha'^{\dagger}_{\mathbf{k}} & \alpha'_{-\mathbf{k}}\end{pmatrix}H'_{\mathbf{k}}\begin{pmatrix}\alpha'_{\mathbf{k}} \\ \alpha'^{\dagger}_{-\mathbf{k}}\end{pmatrix}, \tag{10.2.37}
$$

where

$$\begin{pmatrix} \alpha'_{\mathbf{k}} \\ \alpha'^{\dagger}_{-\mathbf{k}} \end{pmatrix} = U \begin{pmatrix} \alpha_{\mathbf{k}} \\ \alpha^{\dagger}_{-\mathbf{k}} \end{pmatrix} = \begin{pmatrix} e^{i\pi/4}\alpha_{\mathbf{k}} \\ e^{-i\pi/4}\alpha^{\dagger}_{-\mathbf{k}} \end{pmatrix}, \tag{10.2.38}$$

$$\begin{aligned} H'_{\mathbf{k}} = U H_{\mathbf{k}} U^{\dagger} &= -2(J_3 + J_1 \cos \mathbf{k} \cdot \mathbf{M}_1 + J_2 \cos \mathbf{k} \cdot \mathbf{M}_2)\sigma^z \\ &\quad + 2(J_1 \sin \mathbf{k} \cdot \mathbf{M}_1 - J_2 \sin \mathbf{k} \cdot \mathbf{M}_2)\sigma^x. \end{aligned} \tag{10.2.39}$$

Since each mode is decoupled from the others in the Hamiltonian, the state of a mode $\mathbf{k}$ does not mix with other modes during the time evolution. Supposing that $|\psi_{\mathbf{k}(t)}\rangle$ denotes the wave function with mode $\mathbf{k}$, the time-dependent Schrödinger equation is written as

$$i\frac{d}{dt}\big|\psi_{\mathbf{k}}(t)\big\rangle = H'_{\mathbf{k}}\big|\psi_{\mathbf{k}}(t)\big\rangle. \tag{10.2.40}$$

We choose the vacuum $|0_{\mathbf{k}}\rangle$ and the occupied state $|1_{\mathbf{k}}\rangle$ of $\alpha'_{\mathbf{k}}$ and $\alpha'_{-\mathbf{k}}$ as the basis, and expand $|\psi_{\mathbf{k}}(t)\rangle$ as

$$\big|\psi_{\mathbf{k}}(t)\big\rangle = \psi_{\mathbf{k},0}(t)|0_{\mathbf{k}}\rangle + \psi_{\mathbf{k},1}(t)|1_{\mathbf{k}}\rangle. \tag{10.2.41}$$

Then the Schrödinger equation (10.2.40) reduces to

$$i\frac{d}{dt}\begin{pmatrix} \psi_{\mathbf{k},1}(t) \\ \psi_{\mathbf{k},0}(t) \end{pmatrix} = H'_{\mathbf{k}}\begin{pmatrix} \psi_{\mathbf{k},1}(t) \\ \psi_{\mathbf{k},0}(t) \end{pmatrix}. \tag{10.2.42}$$

Let us now assume $J_1 = J_2 = 1$ for simplicity and consider a quench of J_3 as

$$J_3 = -2 + t/\tau, \tag{10.2.43}$$

with the quench rate $1/\tau$. We fix the initial state at the ground state, i.e., $|\psi_{\mathbf{k}}(-\infty)\rangle = |0_{\mathbf{k}}\rangle$ or equivalently $\psi_{\mathbf{k}}(-\infty) = 1$ and $\psi_{\mathbf{k},0}(-\infty) = 1$. Define now

$$g_{\mathbf{k}} = 2 - \cos \mathbf{k} \cdot \mathbf{M}_1 - \cos \mathbf{k} \cdot \mathbf{M}_2, \tag{10.2.44}$$

$$h_{\mathbf{k}} = (\sin \mathbf{k} \cdot \mathbf{M}_1 - \sin \mathbf{k} \cdot \mathbf{M}_2). \tag{10.2.45}$$

The Schrödinger equation is then written as

$$i\frac{d}{dt}\begin{pmatrix} \psi_{\mathbf{k},1}(t) \\ \psi_{\mathbf{k},0}(t) \end{pmatrix} = 2\begin{pmatrix} g_{\mathbf{k}} - t/\tau & h_{\mathbf{k}} \\ h_{\mathbf{k}} & -(g_{\mathbf{k}} - t/\tau) \end{pmatrix}\begin{pmatrix} \psi_{\mathbf{k},1}(t) \\ \psi_{\mathbf{k},0}(t) \end{pmatrix}, \tag{10.2.46}$$

with the initial condition

$$\begin{pmatrix} \psi_{\mathbf{k},1}(-\infty) \\ \psi_{\mathbf{k},0}(-\infty) \end{pmatrix} = \begin{pmatrix} 0 \\ 1 \end{pmatrix}. \tag{10.2.47}$$

This is nothing but the Landau-Zener problem. The excitation probability at $t = +\infty$ is given by (see Sect. 7.A.2)

$$p_{\mathbf{k}} = \big|\psi_{\mathbf{k},0}(+\infty)\big|^2 = e^{-2\pi h_{\mathbf{k}}^2 \tau}. \tag{10.2.48}$$

Integrating $p_{\mathbf{k}}$ with respect to $\mathbf{k}$ over the half Brillouin zone, we obtain the density of excitation as

$$n = \frac{1}{A}\int_{\mathrm{BZ}/2} d\mathbf{k}\, p_{\mathbf{k}} = \frac{1}{A}\int_{\mathrm{BZ}/2} d\mathbf{k}\, e^{-2\pi h_{\mathbf{k}}^2 \tau}, \tag{10.2.49}$$

where A is the area of the half Brillouin zone. $h_{\mathbf{k}}$ has zero lines in the $\mathbf{k}$-space along $(k_x, k_y) = (\pm\pi/\sqrt{3}, k_y)$, $(k_x, 0)$, and $(k_x, \pm 2\pi/\sqrt{3})$. The excitation probability is $p_{\mathbf{k}} = 1$ along these lines. For sufficiently large τ, only narrow areas around these zero lines have a contribution to n and hence one can safely expand $h_{\mathbf{k}}$ up to square terms of the variation from the zero lines. We introduce the wave number $k_\parallel$ and $k_\perp$, which are respectively parallel and perpendicular to the zero line, such that $h_{\mathbf{k}}$ is expanded as $h_{\mathbf{k}} \approx a k_\perp^2$. Then one finds that the density of excitation results in

$$n \approx \frac{1}{A} \int dk_\parallel \int_{-\infty}^{\infty} dk_\perp \, e^{-2\pi a \tau k_\perp^2} \sim \frac{1}{\sqrt{\tau}}. \tag{10.2.50}$$

We recall here that the Kibble-Zurek scaling: $n \sim \tau^{-d\nu/(z\nu+1)}$. In the case of the Kitaev model, one has $z = \nu = 1$ and $d = 2$ which leads to $n \sim t^{-1}$ in conflict with (10.2.50). This conflict originates from the presence of the zero line of $h_{\mathbf{k}}$ in the $\mathbf{k}$-space. $h_{\mathbf{k}}$ is dispersionless along the zero line and only the component of the momentum perpendicular to the zero line is the relevant variable of $h_{\mathbf{k}}$. It follows that the dimension of the integral in n reduces to $d' = 1$ effectively. Thus one obtains $n \sim \tau^{-d'\nu/(z\nu+1)} = \tau^{-1/2}$. We consider a d-dimensional system with exponents ν and z, and a critical region where a quasiparticle has a dispersion only in m-direction. If one considers a quench across the critical region of this system, then the modes which contribute to the excitation during the quench amount to $\hat{k}^m$, where $\hat{k}$ is the largest mode which can excite during the quench. Since the energy scale of excitation $\hat{\Delta}$ scales as $\Delta \sim \tau^{-z\nu/(z\nu+1)}$ and $\hat{\Delta} \sim \hat{k}^z$, one obtains $\hat{k} \sim \tau^{-\nu/(z\nu+1)}$. This leads to a generic scaling relation between the density of excitation of the quench rate: $n \sim \hat{k}^m \sim \tau^{-m\nu/(z\nu+1)}$ [366]. Hikichi et al. [174] studied quench dynamics of the Kitaev model in the presence of a slow quench which ends at an onset of the critical region. In such a case, the excitation probability (10.2.48) is not applicable to obtain the density of excitation n, but instead one can write n in terms of the exact solution of the Landau-Zener equation (cf. Sect. 7.A.2). What is interesting is that the energy spectrum (10.2.25) at the critical point (e.g., $J = 2$) has a dispersion along $k_\parallel$ such as $E_{\mathbf{k}} \sim k_\perp$ and $E_{\mathbf{k}} \sim k_\parallel^2$. The excitation probability is then a function of $\tau k_\perp^2$ and $\tau k_\parallel^4$. This makes the scaling the density of excitation modified from (10.2.50) into $n \sim \int dk_\parallel \, dk_\perp \, p_{\mathbf{k}}(\tau k_\perp^2, \tau k_\parallel^4) \sim \tau^{-3/4}$. If one considers d-dimensional anisotropic critical points, where the energy spectrum $E_{\mathbf{k}} \sim k_i^z$ for m directions and $\sim k_i^{z'}$ for $d - m$ directions, the density of excitation scales as $n \sim \tau^{-(m+(d-m)z/z')\nu/(z\nu+1)}$ [174].

Chapter 11
Brief Summary and Outlook

The transverse Ising models have occupied an outstanding position in the study of quantum statistical physics and quantum magnetism for a half century. Sometimes they were useful to reveal fundamental properties of a quantum phase transition, or sometimes they were studied to understand experimental findings. Impressively, the study of the transverse Ising models are still developing. This is because they are not only simple but rich in their physical content. In this book, we tried to present the rich dispositions of various transverse Ising models.

After a general introduction to the Transverse Ising models and their relevance in the context of modelling various quantum order-disorder transitions, including quantum glass transitions in condensed matter systems, in the introductory chapter, we discussed in Chap. 2 the energy spectrum of the pure transverse Ising chains. One of the peculiarity of the transverse Ising models is the solvability of the pure transverse Ising chain. We presented the exact solution using the Jordan-Wigner transformation described in this chapter. This chapter also includes several numerical and analytic methods including numerical diagonalisation methods, renormalisation group methods as well as a duality argument and a perturbative method. These theories and calculation reveal the nature of a quantum phase transition resulting from the cooperative interaction and the quantum tunnelling effect in the pure transverse Ising chains. What is found here is quite important to understand the universal properties of a quantum phase transition in quantum spin systems with Ising anisotropy. Remarkably, in spite of its rather mathematical character, theoretical results on this model has also been confirmed in experiments. We have mentioned some recent experimental results in this context. We also discussed there the finite size scaling behaviour and the applications of real space renormalisation group etc. techniques. In the next chapter, we discussed the Suzuki-Trotter mapping of the d-dimensional quantum Ising systems to $(d+1)$-dimensional classical Ising systems and presented various classical Monte Carlo results for the pure transverse Ising systems in 2 and 3 dimensions. We also discussed there various approximate analytical tricks to study the behaviour of the transverse Ising and other generalised models, employing scaling theory and both real space and field theoretic renormalisation group techniques.

S. Suzuki et al., *Quantum Ising Phases and Transitions in Transverse Ising Models*, Lecture Notes in Physics 862, DOI 10.1007/978-3-642-33039-1_11,

In Chap. 4, we discussed the behaviour of regularly frustrated transverse Ising systems; more specifically for the ANNNI models in transverse fields, both in one and higher dimensions, using approximate Fermionic representation (in one dimension), field theoretic real space renormalisations, Monte Carlo etc. techniques. Some recent results for intriguing behaviour of the phase diagram for one dimensional ANNNI chains in transverse fields, using the Bosonisation and a renormalisation group analysis has also been given here. In the next chapter, we discussed the behaviour of randomly dilute Ising systems in transverse fields, in particular near the percolation point, as well as random ferromagnetic Ising system in random transverse fields. Detailed analytical and numerical results revealing the nature of the possible disordered Griffiths phase and its singularities are presented here.

Chapter 6 presents in some details the behaviour of randomly frustrated transverse Ising spin glasses (in particular for both Edwards-Anderson and Sherrington-Kirkpatrick models in transverse fields) and also of random field transverse Ising models, employing both approximate analytic and extensive Monte Carlo etc. techniques. The intriguing nature of the quantum glass phases and the dynamics of such systems are now of extreme importance, both epistemologically as well as in the context of designing important applications. In particular, the indications of a possible restoration of replica symmetry in such quantum spin glasses have been discussed extensively in this chapter. The significance comes from the fact that if these indications are true, it can have very important consequences in the search for solutions (ground states) of computationally hard problems and therefore for designing analog quantum computers, employing quantum annealing techniques (discussed in details in Chap. 8). The next chapter discusses the dynamics of quantum transverse Ising models (without disorder or frustration) and in particular for quenching (tuning the transverse field) through the critical point. Detailed analytical (and also numerical) analyses of slow and sudden quantum quenches were also discussed. The dynamic hysteresis in such transverse Ising models when the external (longitudinal) field is periodically driven has been investigated and some exact results in the cases of pulsed fields were presented.

In Chap. 8, we presented in great details the quantum annealing techniques and results. Like the simulated annealing technique, where the classical noise helps the glassy system (of the equivalent computationally hard problem) to come out of local minima in energy or cost function as the "temperature" is tuned, the quantum annealing technique helps the system to come out of the local minima, utilising the quantum fluctuations. The difficulties of the classical annealing procedures for extremely high barriers (energy or cost function depending on the macroscopic size of the system or problem) can indeed be avoided by quantum tunnelling if the effective width of the barrier is small. These advantages of quantum fluctuations can be appointed efficiently in the searches for energy or cost function minima in computationally hard and challenging problems. Several analytical results for the bounds on search times and numerical techniques and results for annealing were presented here in details. The next chapter discusses some applications of tuning quantum fluctuations in some memory models (like the Hopfield model) or in retrieving the patterns with randomly lost information. Statistical analysis of Information has been presented in this context. The last chapter (Chap. 10) discusses how various techniques presented earlier in the book can be easily translated for studying the behaviour of other quantum mechanical many-body systems (including how the BCS theory corresponds to the mean field theory for transverse Ising model discussed earlier). It also discusses the connections of transverse Ising or XY models with several other well known old and new models (for example the Harper model, Kitaev Model, etc.), developed and studied in other contexts.

The advantage of the separability of the cooperative interactions and quantum non-commutability in the transverse field Ising models help designing appropriate models and developing several analytical techniques to explore their properties. The very recent studies of dynamical properties of quantum glasses, quantum annealing studies in transverse Ising models in particular, have been helping enormously in recent times in making advances towards the development of analog quantum computers (see for example discussions in Chap. 8). All these exciting possibilities make the studies on this intriguing class of transverse Ising models a truly inspiring one of our time.

References[1]

1. Abou-Chacra, R., Thouless, D.J., Anderson, P.W.: A selfconsistent theory of localization. J. Phys. C, Solid State Phys. **6**(10), 1734 (1973). [8.5.4]
2. Acharyya, M., Chakrabarti, B.K.: Ising system in oscillating field: hysteretic response. In: Stauffer, D. (ed.) Annual Reviews of Computational Physics, vol. 1, p. 107. World Scientific, Singapore (1994). [7.2.3]
3. Acharyya, M., Chakrabarti, B.K.: Response of Ising systems to oscillating and pulsed fields: hysteresis, ac, and pulse susceptibility. Phys. Rev. B **52**, 6550–6568 (1995). [1.1, 1.3, 7.2.3]
4. Acharyya, M., Chakrabarti, B.K., Stinchcombe, R.B.: Hysteresis in Ising model in transverse field. J. Phys. A, Math. Gen. **27**(5), 1533 (1994). [1.1, 1.3, 7.2.3]
5. Achlioptas, D., Naor, A., Peres, Y.: Rigorous location of phase transitions in hard optimization problems. Nature **435**, 759–764 (2005). [8.4.1]
6. Aharonov, D., van Dam, W., Kempe, J., Landau, Z., Lloyd, S., Regev, O.: Adiabatic quantum computation is equivalent to standard quantum computation. In: Proc. 45th FOCS, pp. 42–51 (2004). arXiv:quant-ph/0405098. [8.8]
7. Aharony, A.: Tricritical points in systems with random fields. Phys. Rev. B **18**, 3318–3327 (1978). [6.7.1, 6.7.2]
8. Akhiezer, I.A., Spol'nik, A.I.: Sov. Phys., Solid State **25**, 81 (1983). [1.3]
9. Allen, D., Azaria, P., Lecheminant, P.: A two-leg quantum Ising ladder: a bosonization study of the ANNNI model. J. Phys. A, Math. Gen. **34**(21), L305 (2001). [1.1, 1.3, 4.3.7]
10. Altshuler, B., Krovi, H., Roland, J.: Anderson localization makes adiabatic quantum optimization fail. Proc. Natl. Acad. Sci. **107**(28), 12446–12450 (2010). [1.3, 8.5.4]
11. Amara, P., Hsu, D., Straub, J.E.: Global energy minimum searches using an approximate solution of the imaginary time Schroedinger equation. J. Phys. Chem. **97**(25), 6715–6721 (1993). [8.1, 9.2]
12. Amico, L., Fazio, R., Osterloh, A., Vedral, V.: Entanglement in many-body systems. Rev. Mod. Phys. **80**, 517–576 (2008). [2.2.1]
13. Amit, D.J., Gutfreund, H., Sompolinsky, H.: Storing infinite numbers of patterns in a spin-glass model of neural networks. Phys. Rev. Lett. **55**, 1530–1533 (1985). [9.1.1]
14. Ancona-Torres, C., Silevitch, D.M., Aeppli, G., Rosenbaum, T.F.: Quantum and classical glass transitions in $LiHo_xY_{1-x}F_4$. Phys. Rev. Lett. **101**, 057201 (2008). [1.1, 1.3]
15. Anderson, P.W.: Absence of diffusion in certain random lattices. Phys. Rev. **109**, 1492–1505 (1958). [8.5.4]

[1] Note that numbers in the square brackets in each item show the sections where the reference is cited.

S. Suzuki et al., *Quantum Ising Phases and Transitions in Transverse Ising Models*, Lecture Notes in Physics 862, DOI 10.1007/978-3-642-33039-1,

16. Anderson, P.W.: Random-phase approximation in the theory of superconductivity. Phys. Rev. **112**, 1900–1916 (1958). [1.1, 1.3, 10.1, 10.1.1]
17. Apolloni, B., Cesa-Bianchi, N., de Falco, D.: In: Albeverio, S., Casati, G., Cattaneo, U., Merlini, D., Moresi, R. (eds.) Stochastic Processes, Physics and Geometry. World Scientific, Singapore (1988). [8.1]
18. Arizmendi, C.M., Rizzo, A.H., Epele, L.N., García Canal, C.A.: Phase diagram of the ANNNI model in the Hamiltonian limit. Z. Phys. B, Condens. Matter **83**, 273–276 (1991). [1.3, 4.3, 4.3.7]
19. Ash, R.B.: Information Theory. Dover, New York (1965). [9.2.2, 9.2.4]
20. Auerbach, A.: Interacting Fermions and Quantum Magnetism. Springer, New York (1994). [1.3, 4.1, 4.3]
21. Banerjee, V., Dattagupta, S.: Model calculation for the susceptibility of a quantum spin glass. Phys. Rev. B **50**, 9942–9947 (1994). [7.1.3]
22. Banerjee, V., Dattagupta, S., Sen, P.: Hysteresis in a quantum spin model. Phys. Rev. E **52**, 1436–1446 (1995). [1.1, 1.3, 7.2.3]
23. Barahona, F.: On the computational complexity of Ising spin glass models. J. Phys. A, Math. Gen. **15**(10), 3241 (1982). [8.8]
24. Barber, M.N., Duxbury, P.M.: A quantum Hamiltonian approach to the two-dimensional axial next-nearest-neighbour Ising model. J. Phys. A, Math. Gen. **14**(7), L251 (1981). [1.1, 4.3]
25. Barber, M.N., Duxbury, P.M.: Hamiltonian studies of the two-dimensional axial next-nearest-neighbor Ising (ANNNI) model. J. Stat. Phys. **29**, 427–432 (1982). [3.A.2, 4.3]
26. Bardeen, J., Cooper, L.N., Schrieffer, J.R.: Theory of superconductivity. Phys. Rev. **108**, 1175–1204 (1957). [10.1.1]
27. Barouch, E., McCoy, B.M.: Statistical mechanics of the xy model. ii. Spin-correlation functions. Phys. Rev. A **3**, 786–804 (1971). [10.1.2]
28. Barouch, E., McCoy, B.M., Dresden, M.: Statistical mechanics of the XY model. i. Phys. Rev. A **2**, 1075–1092 (1970). [1.1, 1.3, 7.2.2, 7.2.2.1]
29. Baskaran, G., Mandal, S., Shankar, R.: Exact results for spin dynamics and fractionalization in the Kitaev model. Phys. Rev. Lett. **98**, 247201 (2007). [10.2.2]
30. Battaglia, D.A., Santoro, G.E., Tosatti, E.: Optimization by quantum annealing: lessons from hard satisfiability problems. Phys. Rev. E **71**, 066707 (2005). [8.4.1]
31. Beccaria, M., Campostrini, M., Feo, A.: Density-matrix renormalization-group study of the disorder line in the quantum axial next-nearest-neighbor Ising model. Phys. Rev. B **73**, 052402 (2006). [1.3, 4.3.7]
32. Beccaria, M., Campostrini, M., Feo, A.: Evidence for a floating phase of the transverse ANNNI model at high frustration. Phys. Rev. B **76**, 094410 (2007). [1.3, 4.3.7]
33. Belanger, D., Young, A.: The random field Ising model. J. Magn. Magn. Mater. **100**(1–3), 272–291 (1991). [1.3, 6.7.1, 6.7.2]
34. Benyoussef, A., Ez-Zahraouy, H.: The bond-diluted spin-1 transverse Ising model with random longitudinal field. Phys. Status Solidi (b) **179**(2), 521–530 (1993). [6.7.2]
35. Benyoussef, A., Ez-Zahraouy, H., Saber, M.: Magnetic properties of a transverse spin-1 Ising model with random longitudinal field. Phys. A, Stat. Mech. Appl. **198**(3–4), 593–605 (1993). [6.7.2]
36. Bernardi, L.W., Campbell, I.A.: Critical exponents in Ising spin glasses. Phys. Rev. B **56**, 5271–5275 (1997). [6.1]
37. Bhattacharya, S., Ray, P.: A diluted quantum transverse Ising model in two dimensions. Phys. Lett. A **101**(7), 346–348 (1984). [1.3, 5.2]
38. Bhattacharyya, S., Das, A., Dasgupta, S.: Transverse Ising chain under periodic instantaneous quenches: dynamical many-body freezing and emergence of slow solitary oscillations. Phys. Rev. B **86**(5), 054410 (2012). doi:10.1103/PhysRevB.86.054410. [1.1, 1.3, 7.2.3.1]
39. Binder, K., Young, A.P.: Spin glasses: experimental facts, theoretical concepts, and open questions. Rev. Mod. Phys. **58**, 801–976 (1986). [6.1, 6.4, 6.5, 6.A.3]

40. Blinc, R.: On the isotopic effects in the ferroelectric behaviour of crystals with short hydrogen bonds. J. Phys. Chem. Solids **13**(3–4), 204–211 (1960). [1.1, 1.2, 1.3]
41. Blinc, R., Svetina, S.: Cluster approximations for order-disorder-type hydrogen-bonded ferroelectrics. i. Small clusters. Phys. Rev. **147**, 423–429 (1966). [1.3]
42. Blinc, R., Svetina, S.: Cluster approximations for order-disorder-type hydrogen-bonded ferroelectrics. ii. Application to KH_2PO_4. Phys. Rev. **147**, 430–438 (1966). [1.3]
43. Born, M., Fock, V.: Beweis des Adiabatensatzes. Z. Phys. **51**, 165–180 (1928). [8.3.2]
44. Bray, A.J., Moore, M.A.: Replica theory of quantum spin glasses. J. Phys. C, Solid State Phys. **13**(24), L655 (1980). [6.2, 6.3, 6.A.1]
45. Bray, A.J., Moore, M.A.: Scaling theory of the random-field Ising model. J. Phys. C, Solid State Phys. **18**(28), L927 (1985). [6.7.1]
46. Bricmont, J., Kupiainen, A.: Lower critical dimension for the random-field Ising model. Phys. Rev. Lett. **59**, 1829–1832 (1987). [6.7.1]
47. Brooke, J., Bitko, D., Rosenbaum, F.T., Aeppli, G.: Quantum annealing of a disordered magnet. Science **284**(5415), 779–781 (1999). [8.4.2]
48. Brout, R., Müller, K., Thomas, H.: Tunnelling and collective excitations in a microscopic model of ferroelectricity. Solid State Commun. **4**(10), 507–510 (1966). [1.1, 1.2, 4.5, 6.7.2, 7.1.1]
49. Büttner, G., Usadel, K.D.: Replica-symmetry breaking for the Ising spin glass in a transverse field. Phys. Rev. B **42**, 6385–6395 (1990). [6.3, 6.5, 8.1]
50. Büttner, G., Usadel, K.D.: Stability analysis of an Ising spin glass with transverse field. Phys. Rev. B **41**, 428–431 (1990). [6.2, 6.5, 6.3]
51. Büttner, G., Usadel, K.D.: The exact phase diagram of the quantum XY spin glass model in a transverse field. Z. Phys. B, Condens. Matter **83**, 131–134 (1991). [1.3, 10.1.4]
52. Büttner, G., Kopeć, T., Usadel, K.: Phase diagrams of the quantum XY spin glass model in a transverse field. Phys. Lett. A **149**(5–6), 248–252 (1990). [10.1.4]
53. Buyers, W.J.L., Cowley, R.A., Paul, G.L., Cochran, W.: In: Neutron Inelastic Scattering, vol. 1, p. 269. International Atomic Energy Agency, Vienna (1968). [1.3]
54. Calabrese, P., Cardy, J.: Entanglement entropy and quantum field theory. J. Stat. Mech. Theory Exp. **2004**(06), P06002 (2004). [2.2.1]
55. Calabrese, P., Cardy, J.: Evolution of entanglement entropy in one-dimensional systems. J. Stat. Mech. Theory Exp. **2005**(04), P04010 (2005). [7.2.2.1]
56. Calabrese, P., Cardy, J.: Time dependence of correlation functions following a quantum quench. Phys. Rev. Lett. **96**, 136801 (2006). [1.1, 7.2.2.1]
57. Calabrese, P., Cardy, J.: Quantum quenches in extended systems. J. Stat. Mech. Theory Exp. **2007**(06), P06008 (2007). [7.2.2.1]
58. Caneva, T., Fazio, R., Santoro, G.E.: Adiabatic quantum dynamics of a random Ising chain across its quantum critical point. Phys. Rev. B **76**, 144427 (2007). [1.3, 8.5.2, 8.6]
59. Cardy, J.L.: Random-field effects in site-disordered Ising antiferromagnets. Phys. Rev. B **29**, 505–507 (1984). [6.7.1]
60. Černý, V.: Thermodynamical approach to the traveling salesman problem: an efficient simulation algorithm. J. Optim. Theory Appl. **45**, 41–51 (1985). [8.1]
61. Cesare, L.D., Lukierska-Walasek, K., Rabuffo, I., Walasek, K.: On the p-spin interaction transverse Ising spin-glass model without replicas. Phys. A, Stat. Mech. Appl. **214**(4), 499–510 (1995). [1.3, 6.6]
62. Chakrabarti, B.K.: Critical behavior of the Ising spin-glass models in a transverse field. Phys. Rev. B **24**, 4062–4064 (1981). [1.1, 1.3, 6.2, 6.4, 6.8]
63. Chakrabarti, B.K., Acharyya, M.: Dynamic transitions and hysteresis. Rev. Mod. Phys. **71**, 847–859 (1999). [7.2.3.1]
64. Chakrabarti, B.K., Dasgupta, P.K.: Modelling neural networks. Phys. A, Stat. Mech. Appl. **186**, 33–48 (1992). [9.1.1]
65. Chakrabarti, B.K., Dutta, A., Sen, P.: Quantum Ising Phases and Transitions in Transverse Ising Models. Springer, Berlin (1995). [9.2]

66. Chandra, A.K., Dasgupta, S.: Floating phase in a 2D ANNNI model. J. Phys. A, Math. Theor. **40**(24), 6251 (2007). [1.1, 4.3.7]
67. Chandra, A.K., Dasgupta, S.: Floating phase in the one-dimensional transverse axial next-nearest-neighbor Ising model. Phys. Rev. E **75**, 021105 (2007). [1.1, 1.3, 4.3.7]
68. Chandra, A.K., Dasgupta, S.: Spin-spin correlation in some excited states of the transverse Ising model. J. Phys. A, Math. Theor. **40**(20), 5231 (2007). [1.1, 4.3.7]
69. Chandra, A.K., Das, A., Chakrabarti, B.K.: Quantum Quenching, Annealing and Computation. Lecture Notes in Physics, vol. 802. Springer, Berlin (2010). [7.2.2, 8.1]
70. Chandra, A.K., Inoue, J.i., Chakrabarti, B.K.: Quantum phase transition in a disordered long-range transverse Ising antiferromagnet. Phys. Rev. E **81**, 021101 (2010). [6.3]
71. Chandra, P., Doucot, B.: Possible spin-liquid state at large s for the frustrated square Heisenberg lattice. Phys. Rev. B **38**, 9335–9338 (1988). [4.3]
72. Chayes, L., Crawford, N., Ioffe, D., Levit, A.: The phase diagram of the quantum Curie-Weiss model. J. Stat. Phys. **133**, 131–149 (2008). [3.4.1]
73. Chen, H.D., Nussinov, Z.: Exact results of the Kitaev model on a hexagonal lattice: spin states, string and brane correlators, and anyonic excitations. J. Phys. A, Math. Theor. **41**(7), 075001 (2008). [10.2.1]
74. Cherng, R.W., Levitov, L.S.: Entropy and correlation functions of a driven quantum spin chain. Phys. Rev. A **73**, 043614 (2006). [10.1.2]
75. Choi, V.: Different adiabatic quantum optimization algorithms for the NP-complete exact cover problem. Proc. Natl. Acad. Sci. **108**(7), 19–20 (2011). [8.5.4]
76. Chowdhury, D.: Spin Glass and Other Frustrated Systems. World Scientific, Singapore (1986). [6.1, 6.5, 6.A.3]
77. Christe, P., Henkel, M.: Introduction to Conformal Invariance and Applications to Critical Phenomena. Lecture Notes in Physics Monographs, vol. M 16, pp. 122–136. Springer, Heidelberg (1993). Chapter 10. [2.A.2]
78. Cincio, L., Dziarmaga, J., Rams, M.M., Zurek, W.H.: Entropy of entanglement and correlations induced by a quench: dynamics of a quantum phase transition in the quantum Ising model. Phys. Rev. A **75**, 052321 (2007). [7.2.2]
79. Clay Mathematics Institute: Millennium problems. http://www.claymath.org/millennium/. [8.2]
80. Cochran, W.: Dynamical, scattering and dielectric properties of ferroelectric crystals. Adv. Phys. **18**(72), 157–192 (1969). [1.3]
81. Coldea, R., Tennant, D.A., Wheeler, E.M., Wawrzynska, E., Prabhakaran, D., Telling, M., Habicht, K., Smeibidl, P., Kiefer, K.: Quantum criticality in an Ising chain: experimental evidence for emergent E8 symmetry. Science **327**(5962), 177–180 (2010). [1.3, 2.6]
82. Continentino, M.A.: Quantum scaling in many-body systems. Phys. Rep. **239**(3), 179–213 (1994). [1.1, 1.3, 3.5]
83. Cooke, A., Edmonds, D., Finn, C., Wolf, W.: J. Phys. Soc. Jpn. Suppl. B-1 **17**, 481 (1962). [1.3]
84. Cooke, A., Ellis, C., Gehring, K., Leask, M., Martin, D., Wanklyn, B., Wells, M., White, R.: Observation of a magnetically controllable Jahn Teller distortion in dysprosium vanadate at low temperatures. Solid State Commun. **8**(9), 689–692 (1970). [1.3]
85. Cooke, A., Martin, D., Wells, M.: The specific heat of dysprosium vanadate. Solid State Commun. **9**(9), 519–522 (1971). [1.3]
86. Cooke, A., Swithenby, S., Wells, M.: The properties of thulium vanadate—an example of molecular field behaviour. Solid State Commun. **10**(3), 265–268 (1972). [1.3]
87. Coolen, A.C.C., Ruijgrok, T.W.: Image evolution in Hopfield networks. Phys. Rev. A **38**, 4253–4255 (1988). [9.1.2]
88. Cooper, B.R., Vogt, O.: Singlet ground state magnetism. J. Phys., Colloq. **32**, C1-958–C1-965 (1971). [1.3]
89. Courtens, E.: Vogel-Fulcher scaling of the susceptibility in a mixed-crystal proton glass. Phys. Rev. Lett. **52**, 69–72 (1984). [1.3]

90. Crisanti, A., Rieger, H.: Random-bond Ising chain in a transverse magnetic field: a finite-size scaling analysis. J. Stat. Phys. **77**, 1087–1098 (1994). [6.4]
91. Damski, B., Zurek, W.H.: Adiabatic-impulse approximation for avoided level crossings: from phase-transition dynamics to Landau-Zener evolutions and back again. Phys. Rev. A **73**, 063405 (2006). [7.2.2, 7.A.2]
92. Das, A.: Exotic freezing of response in a quantum many-body system. Phys. Rev. B **82**, 172402 (2010). [1.1, 1.3, 7.2.3.1]
93. Das, A., Chakrabarti, B.K.: Quantum Annealing and Related Optimization Methods. Lecture Notes in Physics, vol. 679. Springer, Berlin (2005). [1.3, 8.1, 9.2]
94. Das, A., Chakrabarti, B.K.: Colloquium: quantum annealing and analog quantum computation. Rev. Mod. Phys. **80**, 1061–1081 (2008). [8.1]
95. Das, A., Chakrabarti, B.K., Stinchcombe, R.B.: Quantum annealing in a kinetically constrained system. Phys. Rev. E **72**(2), 026701 (2005). doi:10.1103/PhysRevE.72.026701. [8.1]
96. Dattagupta, S., Tadić, B., Pirc, R., Blinc, R.: Tunneling in proton glasses: stochastic theory of NMR line shape. Phys. Rev. B **44**, 4387–4396 (1991). [1.3]
97. de Almeida, J.R.L., Thouless, D.J.: Stability of the Sherrington-Kirkpatrick solution of a spin glass model. J. Phys. A, Math. Gen. **11**(5), 983 (1978). [9.1.1]
98. de Gennes, P.G.: Collective motions of hydrogen bonds. Solid State Commun. **1**(6), 132–137 (1963). [1.1, 1.2, 1.3]
99. Deng, S., Ortiz, G., Viola, L.: Dynamical non-ergodic scaling in continuous finite-order quantum phase transitions. Europhys. Lett. **84**(6), 67008 (2008). [10.1.2]
100. Deng, S., Ortiz, G., Viola, L.: Anomalous nonergodic scaling in adiabatic multicritical quantum quenches. Phys. Rev. B **80**, 241109 (2009). [10.1.2]
101. Deng, S., Ortiz, G., Viola, L.: Dynamical critical scaling and effective thermalization in quantum quenches: role of the initial state. Phys. Rev. B **83**, 094304 (2011). [10.1.2]
102. Derian, R., Gendiar, A., Nishino, T.: Modulation of local magnetization in two-dimensional axial-next-nearest-neighbor Ising model. J. Phys. Soc. Jpn. **75**(11), 114001 (2006). [1.1, 4.3.7]
103. Derrida, B.: Random-energy model: limit of a family of disordered models. Phys. Rev. Lett. **45**, 79–82 (1980). [3.4.2, 6.6, 8.5.3.2]
104. Derrida, B.: Random-energy model: an exactly solvable model of disordered systems. Phys. Rev. B **24**, 2613–2626 (1981). [6.6, 8.5.3.2]
105. Dhar, D., Barma, M.: Effect of disorder on relaxation in the one-dimensional Glauber model. J. Stat. Phys. **22**, 259–277 (1980). [8.6]
106. Divakaran, U., Dutta, A., Sen, D.: Quenching along a gapless line: a different exponent for defect density. Phys. Rev. B **78**, 144301 (2008). [10.1.2]
107. Divakaran, U., Mukherjee, V., Dutta, A., Sen, D.: Defect production due to quenching through a multicritical point. J. Stat. Mech. Theory Exp. **2009**(02), P02007 (2009). [10.1.2]
108. Dobrosavljević, V., Stratt, R.M.: Mean-field theory of the proton glass. Phys. Rev. B **36**, 8484–8496 (1987). [6.3]
109. Dobrosavljevic, V., Thirumalai, D.: 1/p expansion for a p-spin interaction spin-glass model in a transverse field. J. Phys. A, Math. Gen. **23**(15), L767 (1990). [6.6, 8.5.3.2]
110. dos Santos, R.R., Stinchcombe, R.B.: Finite size scaling and crossover phenomena: the XY chain in a transverse field at zero temperature. J. Phys. A, Math. Gen. **14**(10), 2741 (1981). [10.1.2]
111. dos Santos, R.R., Sneddon, L., Stinchcombe, R.B.: The 2D transverse Ising model at $t = 0$: a finite-size rescaling transformation approach. J. Phys. A, Math. Gen. **14**(12), 3329 (1981). [3.6.1]
112. dos Santos, R.R., dos Santos, R.Z., Kischinhevsky, M.: Transverse Ising spin-glass model. Phys. Rev. B **31**, 4694–4697 (1985). [6.2]
113. Drell, S.D., Weinstein, M., Yankielowicz, S.: Quantum field theories on a lattice: variational methods for arbitrary coupling strengths and the Ising model in a transverse magnetic field. Phys. Rev. D **16**, 1769–1781 (1977). [2.4]

114. Drzewiński, A., Dekeyser, R.: Renormalization of the anisotropic linear XY model. Phys. Rev. B **51**, 15218–15228 (1995). [2.4.1]
115. Dutta, A., Sen, D.: Gapless line for the anisotropic Heisenberg spin-$\frac{1}{2}$ chain in a magnetic field and the quantum axial next-nearest-neighbor Ising chain. Phys. Rev. B **67**, 094435 (2003). [1.1, 1.3, 4.3.7]
116. Dutta, A., Chakrabarti, B.K., Stinchcombe, R.B.: Phase transitions in the random field Ising model in the presence of a transverse field. J. Phys. A, Math. Gen. **29**(17), 5285 (1996). [6.7.2]
117. Duxbury, P.M., Barber, M.N.: Hamiltonian studies of the two-dimensional axial next-nearest neighbour Ising (ANNNI) model. ii. Finite-lattice mass gap calculations. J. Phys. A, Math. Gen. **15**(10), 3219 (1982). [4.3]
118. Dziarmaga, J.: Dynamics of a quantum phase transition: exact solution of the quantum Ising model. Phys. Rev. Lett. **95**, 245701 (2005). [1.1, 1.3, 7.2.2]
119. Dziarmaga, J.: Dynamics of a quantum phase transition in the random Ising model: logarithmic dependence of the defect density on the transition rate. Phys. Rev. B **74**, 064416 (2006). [1.3, 8.6]
120. Dziarmaga, J.: Dynamics of a quantum phase transition and relaxation to a steady state. Adv. Phys. **59**(6), 1063–1189 (2010). [7.2.2]
121. Edwards, S.F., Anderson, P.W.: Theory of spin glasses. J. Phys. F, Met. Phys. **5**(5), 965 (1975). [6.1]
122. Elliott, R.J.: Phenomenological discussion of magnetic ordering in the heavy rare-earth metals. Phys. Rev. **124**, 346–353 (1961). [4.1]
123. Elliott, R.J.: In: Balkanski, M. (ed.) Proc. of the Second Int. Conf. Light Scattering in Solids. Flammarion Sciences, Paris (1971). [1.3]
124. Elliott, R.J., Parkinson, J.B.: Theory of spin-phonon coupling in concentrated paramagnetic salts and its effect on thermal conductivity. Proc. Phys. Soc. **92**(4), 1024 (1967). [1.3]
125. Elliott, R.J., Wood, C.: The Ising model with a transverse field. i. High temperature expansion. J. Phys. C, Solid State Phys. **4**(15), 2359 (1971). [3.3]
126. Elliott, R.J., Pfeuty, P., Wood, C.: Ising model with a transverse field. Phys. Rev. Lett. **25**, 443–446 (1970). [1.1, 3.1, 3.2]
127. Elliott, R.J., Gehring, G.A., Malozemoff, A.P., Smith, S.R.P., Staude, W.S., Tyte, R.N.: Theory of co-operative Jahn-Teller distortions in $DyVO_4$ and $TbVO_4$ (phase transitions). J. Phys. C, Solid State Phys. **4**(9), 179 (1971). [1.3]
128. Ellis, C.J., Gehring, K.A., Leask, M.J.M., White, R.L.: Spectroscopic properties of dysprosium vanadate. J. Phys., Colloq. **32**, C1-1024–C1-1025 (1971). [1.3]
129. Emery, V.J., Noguera, C.: Critical properties of a spin-(1/2) chain with competing interactions. Phys. Rev. Lett. **60**, 631–634 (1988). [4.3]
130. Farhi, E., Goldstone, J., Gutmann, S., Sipser, M.: Quantum computation by adiabatic evolution. arXiv:quant-ph/0001106 (2000). [1.1, 1.3, 8.1]
131. Farhi, E., Goldstone, J., Gutmann, S., Lapan, J., Lundgren, A., Preda, D.: A quantum adiabatic evolution algorithm applied to random instances of an NP-complete problem. Science **292**(5516), 472–475 (2001). [8.4.1]
132. Farhi, E., Goldstone, J., Gosset, D., Gutmann, S., Meyer, H.B., Shor, P.: Quantum adiabatic algorithms, small gaps, and different paths. arXiv:0909.4766 [quant-ph] (2009). [8.5.4]
133. Fedorov, Y.V., Shender, E.F.: Quantum spin glasses in the Ising model with a transverse field. JETP Lett. **43**, 681 (1986). [6.3]
134. Feynman, R.: Simulating physics with computers. Int. J. Theor. Phys. **21**, 467–488 (1982). [8.1]
135. Finnila, A., Gomez, M., Sebenik, C., Stenson, C., Doll, J.: Quantum annealing: a new method for minimizing multidimensional functions. Chem. Phys. Lett. **219**(5–6), 343–348 (1994). [8.1, 9.2]
136. Fischer, K.H., Hertz, J.A.: Spin Glasses. Cambridge University Press, Cambridge (1991). [6.1, 6.5, 6.A.3]

137. Fisher, D.S.: Scaling and critical slowing down in random-field Ising systems. Phys. Rev. Lett. **56**, 416–419 (1986). [6.7.1]
138. Fisher, D.S.: Random transverse field Ising spin chains. Phys. Rev. Lett. **69**, 534–537 (1992). [1.3, 5.3, 8.5.2]
139. Fisher, D.S.: Critical behavior of random transverse-field Ising spin chains. Phys. Rev. B **51**, 6411–6461 (1995). [1.3, 5.3, 8.5.2]
140. Fisher, D.S.: Phase transitions and singularities in random quantum systems. Phys. A, Stat. Mech. Appl. **263**(1–4), 222–233 (1999). [1.3, 5.3]
141. Fisher, D.S., Young, A.P.: Distributions of gaps and end-to-end correlations in random transverse-field Ising spin chains. Phys. Rev. B **58**, 9131–9141 (1998). [1.3, 5.3]
142. Fisher, M.E.: Perpendicular susceptibility of the Ising model. J. Math. Phys. **4**(1), 124–135 (1963). [1.1]
143. Fisher, M.E., Barber, M.N.: Scaling theory for finite-size effects in the critical region. Phys. Rev. Lett. **28**, 1516–1519 (1972). See also Barber, M.N.: In: Domb, C., Lebowitz, J.L. (eds.) Phase Transition and Critical Phenomena, vol. 8, p. 146. Academic Press, San Diego (1983). [2.3.1]
144. Fisher, M.E., Hartwig, R.E.: Toeplitz determinants: some applications, theorems, and conjectures. Adv. Chem. Phys. **15**, 333–353 (1968). [2.A.3]
145. Fisher, M.E., Selke, W.: Infinitely many commensurate phases in a simple Ising model. Phys. Rev. Lett. **44**, 1502–1505 (1980). [4.2]
146. Fisher, M.P.A., Weichman, P.B., Grinstein, G., Fisher, D.S.: Boson localization and the superfluid-insulator transition. Phys. Rev. B **40**, 546–570 (1989). [1.1]
147. Fishman, S., Aharony, A.: Random field effects in disordered anisotropic antiferromagnets. J. Phys. C, Solid State Phys. **12**(18), 729 (1979). [6.7.1]
148. Fletcher, J., Sheard, F.: The anomalous Schottky heat capacity of cerium ethylsulphate. Solid State Commun. **9**(16), 1403–1406 (1971). [1.3]
149. Fradkin, E., Susskind, L.: Order and disorder in gauge systems and magnets. Phys. Rev. D **17**, 2637–2658 (1978). [5.2]
150. Friedman, Z.: Critical exponents for the three-dimensional Ising model from the real-space renormalization group in two dimensions. Phys. Rev. Lett. **36**, 1326–1328 (1976). [3.6.1]
151. Friedman, Z.: Ising model with a transverse field in two dimensions: phase diagram and critical properties from a real-space renormalization group. Phys. Rev. B **17**, 1429–1432 (1978). [3.6.1]
152. Garel, T., Pfeuty, P.: Commensurability effects on the critical behaviour of systems with helical ordering. J. Phys. C, Solid State Phys. **9**(10), L245 (1976). [4.3]
153. Garey, M.R., Johnson, D.S.: Computers and Intractability: A Guide to the Theory of NP-Completeness. Freeman, New York (1979). [8.2]
154. Gehring, K., Malozemoff, A., Staude, W., Tyte, R.: Observation of magnetically controllable distortion in $TbVO_4$ by optical spectroscopy. Solid State Commun. **9**(9), 511–514 (1971). [1.3]
155. Geman, S., Geman, D.: Stochastic relaxation, Gibbs distributions, and the Bayesian restoration of images. IEEE Trans. Pattern Anal. Mach. Intell. **PAMI-6**(6), 721–741 (1984). [8.1, 8.7.2, 8.A.3, 9.2, 9.2.1]
156. Glauber, R.J.: Time-dependent statistics of the Ising model. J. Math. Phys. **4**(2), 294–307 (1963). [7.1.3, 8.6]
157. Gofman, M., Adler, J., Aharony, A., Harris, A.B., Schwartz, M.: Evidence for two exponent scaling in the random field Ising model. Phys. Rev. Lett. **71**, 1569–1572 (1993). [6.7.1]
158. Goldschmidt, Y.Y.: Solvable model of the quantum spin glass in a transverse field. Phys. Rev. B **41**, 4858–4861 (1990). [1.3, 6.6, 8.5.3.2, 9.2]
159. Goldschmidt, Y.Y., Lai, P.Y.: Ising spin glass in a transverse field: replica-symmetry-breaking solution. Phys. Rev. Lett. **64**, 2467–2470 (1990). [1.3, 6.2, 6.3, 8.1]
160. Greiner, M., Mandel, O., Esslinger, T., Hänsch, T.W., Bloch, I.: Quantum phase transition from a superfluid to a Mott insulator in a gas of ultracold atoms. Nature **415**, 39–44 (2002). [7.2.2, 7.2.2.2]

161. Griffiths, R.B.: Nonanalytic behavior above the critical point in a random Ising ferromagnet. Phys. Rev. Lett. **23**, 17–19 (1969). [5.3]
162. Gross, D., Mezard, M.: The simplest spin glass. Nucl. Phys. B **240**(4), 431–452 (1984). [6.6]
163. Guo, M., Bhatt, R.N., Huse, D.A.: Quantum critical behavior of a three-dimensional Ising spin glass in a transverse magnetic field. Phys. Rev. Lett. **72**, 4137–4140 (1994). [1.3, 6.2, 6.5, 6.4]
164. Guo, M., Bhatt, R.N., Huse, D.A.: Quantum Griffiths singularities in the transverse-field Ising spin glass. Phys. Rev. B **54**, 3336–3342 (1996). [6.2]
165. Hallberg, K., Gagliano, E., Balseiro, C.: Finite-size study of a spin-1/2 Heisenberg chain with competing interactions: phase diagram and critical behavior. Phys. Rev. B **41**, 9474–9479 (1990). [4.3]
166. Hamer, C.J., Barber, M.N.: Finite-lattice methods in quantum Hamiltonian field theory. i. O(2) and O(3) Heisenberg models. J. Phys. A, Math. Gen. **14**(1), 259 (1981). [1.3, 4.3]
167. Hamer, C.J., Barber, M.N.: Finite-lattice methods in quantum Hamiltonian field theory. i. The Ising model. J. Phys. A, Math. Gen. **14**(1), 241 (1981). [1.3, 2.3.1, 4.3]
168. Harley, R., Hayes, W., Smith, S.: Raman study of phase transitions in rare earth vanadates. Solid State Commun. **9**(9), 515–517 (1971). [1.3]
169. Harris, A.B.: Effect of random defects on the critical behaviour of Ising models. J. Phys. C, Solid State Phys. **7**(9), 1671 (1974). [1.3, 5.2]
170. Harris, A.B.: Upper bounds for the transition temperatures of generalized Ising models. J. Phys. C, Solid State Phys. **7**(17), 3082 (1974). [1.3, 5.2]
171. Harris, A.B., Micheletti, C., Yeomans, J.M.: Quantum fluctuations in the axial next-nearest-neighbor Ising model. Phys. Rev. Lett. **74**, 3045–3048 (1995). [1.3, 4.4]
172. Heims, S.P.: Master equation for Ising model. Phys. Rev. **138**, A587–A590 (1965). [7.1.3]
173. Hertz, J.A.: Quantum critical phenomena. Phys. Rev. B **14**, 1165–1184 (1976). [1.1, 3.5]
174. Hikichi, T., Suzuki, S., Sengupta, K.: Slow quench dynamics of the Kitaev model: anisotropic critical point and effect of disorder. Phys. Rev. B **82**, 174305 (2010). [1.3, 10.2.3]
175. Hirsch, J.E., Mazenko, G.F.: Renormalization-group transformation for quantum lattice systems at zero temperature. Phys. Rev. B **19**, 2656–2663 (1979). [2.4.1]
176. Holzhey, C., Larsen, F., Wilczek, F.: Geometric and renormalized entropy in conformal field theory. Nucl. Phys. B **424**(3), 443–467 (1994). [2.2.1]
177. Hopf, E.: An inequality for positive linear integral operators. Indiana Univ. Math. J. **12**, 683–692 (1963). [8.7.1, 8.A.1]
178. Hopfield, J.: Neural networks and physical systems with emergent collective computational abilities. Proc. Natl. Acad. Sci. USA **79**(8), 2554–2558 (1982). [9.1]
179. Hopfield, J.J., Tank, D.W.: "Neural" computation of decisions in optimization problems. Biol. Cybern. **52**, 141–152 (1985). [8.2]
180. Horn, R.A., Johnson, C.R.: Matrix Analysis. Cambridge University Press, Cambridge (1985). [8.A.1, 8.A.2]
181. Hornreich, R.M., Liebmann, R., Schuster, H.G., Selke, W.: Lifshitz points in Ising systems. Z. Phys. B, Condens. Matter **35**, 91–97 (1979). [4.2]
182. Houdayer, J., Hartmann, A.K.: Low-temperature behavior of two-dimensional Gaussian Ising spin glasses. Phys. Rev. B **70**, 014418 (2004). [6.1]
183. Hu, B.: The classical Ising model: a quantum renormalization group approach. Phys. Lett. A **71**(1), 83–86 (1979). [2.4.1, 4.3]
184. Hu, B.: Introduction to real-space renormalization-group methods in critical and chaotic phenomena. Phys. Rep. **91**(5), 233–295 (1982). [2.4.1]
185. Huse, D.A., Fisher, D.S.: Residual energies after slow cooling of disordered systems. Phys. Rev. Lett. **57**, 2203–2206 (1986). [8.8]
186. Husimi, K.: Proc. Int. Conf. Theor. Phys., 531 (1953). [3.4.1]
187. Igarashi, J.i., Tonegawa, T.: Excitation spectrum of a spin-1/2 chain with competing interactions. Phys. Rev. B **40**, 756–759 (1989). [4.3]
188. Iglói, F., Monthus, C.: Strong disorder RG approach of random systems. Phys. Rep. **412**(5–6), 277–431 (2005). [1.3, 5.1, 8.5.2]

189. Ikegami, T., Miyashita, S., Rieger, H.: Griffiths-McCoy singularities in the transverse field Ising model on the randomly diluted square lattice. J. Phys. Soc. Jpn. **67**(8), 2671–2677 (1998). [9]
190. Inoue, J.: Application of the quantum spin glass theory to image restoration. Phys. Rev. E **63**, 046114 (2001). [1.3, 9.2, 9.2.4]
191. Inoue, J.: In: Das, A., Chakrabarti B.K. (eds.) Quantum Annealing and Related Optimization Methods, p. 259. Springer, Berlin (2005). [1.3, 9.2]
192. Inoue, J.: Pattern-recalling processes in quantum Hopfield networks far from saturation. J. Phys. Conf. Ser. **297**(1), 012012 (2011). [1.1, 1.3, 9.1.2]
193. Inoue, J., Tanaka, K.: Dynamics of the maximum marginal likelihood hyperparameter estimation in image restoration: gradient descent versus expectation and maximization algorithm. Phys. Rev. E **65**, 016125 (2001). [9.2, 9.2.4]
194. Inoue, J., Saika, Y., Okada, M.: Quantum mean-field decoding algorithm for error-correcting codes. J. Phys. Conf. Ser. **143**(1), 012019 (2009). [9.2.5]
195. Ishii, H., Yamamoto, T.: Effect of a transverse field on the spin glass freezing in the Sherrington-Kirkpatrick model. J. Phys. C, Solid State Phys. **18**(33), 6225 (1985). [1.3, 6.2, 6.3, 9.2.5]
196. Ishii, H., Yamamoto, T.K.: In: Suzuki, M. (ed.) Quantum Monte Carlo Methods. Springer, Heidelberg (1986). [6.3]
197. Ishizuka, H., Motome, Y., Furukawa, N., Suzuki, S.: Quantum Monte Carlo study of molecular polarization and antiferroelectric ordering in squaric acid crystals. Phys. Rev. B **84**, 064120 (2011). [3.2]
198. Ishizuka, H., Motome, Y., Furukawa, N., Suzuki, S.: Quantum Monte Carlo study of the transverse-field Ising model on a frustrated checkerboard lattice. J. Phys. Conf. Ser. **320**(1), 012054 (2011). [3.2]
199. Itoh, J., Yamagata, Y.: Nuclear magnetic resonance experiments on ammonium halides. ii. Halogen nuclear magnetic resonance. J. Phys. Soc. Jpn. **17**(3), 481–507 (1962). [1.3]
200. Its, A.R., Jin, B.Q., Korepin, V.E.: Entanglement in the XY spin chain. J. Phys. A, Math. Gen. **38**(13), 2975 (2005). [2.2.1]
201. Johnson, M.W., Amin, M.H.S., Gildert, S., Lanting, T., Hamze, F., Dickson, N., Harris, R., Berkley, A.J., Johansson, J., Bunyk, P., Chapple, E.M., Enderud, C., Hilton, J.P., Karimi, K., Ladizinsky, E., Ladizinsky, N., Oh, T., Perminov, I., Rich, C., Thom, M.C., Tolkacheva, E., Truncik, C.J.S., Uchaikin, S., Wang, J., Wilson, B., Rose, G.: Quantum annealing with manufactured spins. Nature **473**(7346), 194–198 (2011). [8.4.2]
202. Jona, F., Shirane, G.: Ferroelectric Crystals. Pergamon, Oxford (1962). [1.3]
203. Jordan, M.: Learning in Graphical Models. MIT Press, Cambridge (1998). [9.2.5]
204. Jörg, T., Krzakala, F., Kurchan, J., Maggs, A.C.: Simple glass models and their quantum annealing. Phys. Rev. Lett. **101**, 147204 (2008). [1.3, 8.5.3.2]
205. Jörg, T., Krzakala, F., Kurchan, J., Maggs, A.C., Pujos, J.: Energy gaps in quantum first-order mean-field-like transitions: the problems that quantum annealing cannot solve. Europhys. Lett. **89**(4), 40004 (2010). [1.3, 3.4.2, 8.5.3.1]
206. Jörg, T., Krzakala, F., Semerjian, G., Zamponi, F.: First-order transitions and the performance of quantum algorithms in random optimization problems. Phys. Rev. Lett. **104**, 207206 (2010). [1.3, 8.5.4]
207. Jullien, R., Fields, J.N., Doniach, S.: Zero-temperature real-space renormalization-group method for a Kondo-lattice model Hamiltonian. Phys. Rev. B **16**, 4889–4900 (1977). [2.4.1]
208. Jullien, R., Pfeuty, P., Fields, J.N., Doniach, S.: Zero-temperature renormalization method for quantum systems. i. Ising model in a transverse field in one dimension. Phys. Rev. B **18**, 3568–3578 (1978). [2.4, 3.5]
209. Jullien, R., Pfeuty, P., Fields, J.N., Penson, K.A.: In: Brukhardt, T.W., van Leeuween, J.M.J. (eds.) Real Space Renormalisation, p. 119. Springer, Berlin (1982). [2.4]
210. Kabashima, Y., Saad, D.: Statistical mechanics of error-correcting codes. Europhys. Lett. **45**(1), 97 (1999). [9.2]

211. Kadowaki, T., Nishimori, H.: Quantum annealing in the transverse Ising model. Phys. Rev. E **58**, 5355–5363 (1998). [1.1, 1.3, 8.1, 8.4.1, 9.2, 9.2.6]
212. Kaminow, I.P., Damen, T.C.: Temperature dependence of the ferroelectric mode in KH_2PO_4. Phys. Rev. Lett. **20**, 1105–1108 (1968). [1.3]
213. Kanzig, W.: In: Seitz, F., Turnbull, D. (eds.) Solid State Physics, p. 1. Academic Press, New York (1957). [1.3]
214. Karevski, D., Lin, Y.C., Rieger, H., Kawashima, N., Igli, F.: Random quantum magnets with broad disorder distribution. Eur. Phys. J. B, Condens. Matter Complex Syst. **20**, 267–276 (2001). [5.3]
215. Kato, T.: On the adiabatic theorem of quantum mechanics. J. Phys. Soc. Jpn. **5**(6), 435–439 (1950). [8.3.2]
216. Katsura, S.: Statistical mechanics of the anisotropic linear Heisenberg model. Phys. Rev. **127**, 1508–1518 (1962). [1.1, 1.3, 2.1.2, 10.1.2]
217. Katzgraber, H.G., Lee, L.W., Young, A.P.: Correlation length of the two-dimensional Ising spin glass with Gaussian interactions. Phys. Rev. B **70**, 014417 (2004). [6.1]
218. Katzgraber, H.G., Körner, M., Young, A.P.: Universality in three-dimensional Ising spin glasses: a Monte Carlo study. Phys. Rev. B **73**, 224432 (2006). [6.1]
219. Kawasaki, K.: Diffusion constants near the critical point for time-dependent Ising models. i. Phys. Rev. **145**, 224–230 (1966). [7.1.3]
220. Kawasaki, K.: Diffusion constants near the critical point for time-dependent Ising models. ii. Phys. Rev. **148**, 375–381 (1966). [7.1.3]
221. Kawasaki, K.: Diffusion constants near the critical point for time-dependent Ising models. iii. Self-diffusion constant. Phys. Rev. **150**, 285–290 (1966). [7.1.3]
222. Kawashima, N., Harada, K.: Recent developments of world-line Monte Carlo methods. J. Phys. Soc. Jpn. **73**(6), 1379–1414 (2004). [3.2]
223. Kibble, T.W.B.: Some implications of a cosmological phase transition. Phys. Rep. **67**(1), 183–199 (1980). [7.2.2.2]
224. Kim, D.H., Kim, J.J.: Infinite-range Ising spin glass with a transverse field under the static approximation. Phys. Rev. B **66**, 054432 (2002). [8.1]
225. Kim, K., Chang, M.S., Korenblit, S., Islam, R., Edwards, E.E., Freericks, J.K., Lin, G.D., Duan, L.M., Monroe, C.: Quantum simulation of frustrated Ising spins with trapped ions. Nature **465**, 590–593 (2010). [8.4.2]
226. Kimball, J.C.: The kinetic Ising model: exact susceptibilities of two simple examples. J. Stat. Phys. **21**, 289–300 (1979). [4.3]
227. Kinoshita, T., Wenger, T., Weiss, D.S.: A quantum Newton's cradle. Nature **440**, 900–903 (2006). [7.2.2]
228. Kirkpatrick, S., Gelatt, C.D., Vecchi, M.P.: Optimization by simulated annealing. Science **220**(4598), 671–680 (1983). [8.1, 9.2, 9.2.1]
229. Kitaev, A.: Anyons in an exactly solved model and beyond. Ann. Phys. **321**(1), 2–111 (2006). [1.3, 10.2.1]
230. Kobayashi, K.K.: Dynamical theory of the phase transition in KH_2PO_4-type ferroelectric crystals. J. Phys. Soc. Jpn. **24**(3), 497–508 (1968). [1.3]
231. Kogut, J.B.: An introduction to lattice gauge theory and spin systems. Rev. Mod. Phys. **51**, 659–713 (1979). [1.1, 2.1.1, 3.A.2]
232. Kopec, T.K.: A dynamic theory of transverse freezing in the Sherrington-Kirkpatrick Ising model. J. Phys. C, Solid State Phys. **21**(36), 6053 (1988). [6.2]
233. Kopec, T.K.: Transverse freezing in the quantum Ising spin glass: a thermofield dynamic approach. J. Phys. C, Solid State Phys. **21**(2), 297 (1988). [6.2, 6.3]
234. Kopeć, T.K., Tadić, B., Pirc, R., Blinc, R.: Random fields and quantum effects in proton glasses. Z. Phys. B, Condens. Matter **78**, 493–499 (1990). [7.1.3]
235. Kovács, I.A., Iglói, F.: Renormalization group study of the two-dimensional random transverse-field Ising model. Phys. Rev. B **82**, 054437 (2010). [1.3, 5.3, 8.5.2]
236. Kovács, I.A., Iglói, F.: Infinite-disorder scaling of random quantum magnets in three and higher dimensions. Phys. Rev. B **83**, 174207 (2011). [1.3, 5.3, 8.8]

237. Kramers, H.A., Wannier, G.H.: Statistics of the two-dimensional ferromagnet. Part i. Phys. Rev. **60**, 252–262 (1941). [2.1.1, 10.1.2]
238. Krzakala, F., Rosso, A., Semerjian, G., Zamponi, F.: Path-integral representation for quantum spin models: application to the quantum cavity method and Monte Carlo simulations. Phys. Rev. B **78**, 134428 (2008). [3.4.1]
239. Lage, E.J.S., Stinchcombe, R.B.: Transverse Ising model with substitutional disorder: an effective-medium theory. J. Phys. C, Solid State Phys. **9**(17), 3295 (1976). [7.1.2]
240. Lai, P.Y., Goldschmidt, Y.Y.: Monte Carlo studies of the Ising spin-glass in a transverse field. Europhys. Lett. **13**(4), 289 (1990). [6.2, 6.3, 6.5, 8.1]
241. Landau, D.P., Binder, K.: A Guide to Monte Carlo Simulations in Statistical Physics. Cambridge University Press, Cambridge (2000). [3.2, 5.3, 6.3]
242. Landau, L.D.: On the theory of transfer of energy at collisions ii. Phys. Z. Sowjetunion **2**, 46 (1932). [7.A.2]
243. Landau, L.D., Lifshitz, E.M.: Quantum Mechanics (Non-relativistic Theory). Butterworth-Heineman, Oxford (1958). [7.A.2]
244. Lee, P.A., Ramakrishnan, T.V.: Disordered electronic systems. Rev. Mod. Phys. **57**, 287–337 (1985). [1.1]
245. Lieb, E., Schultz, T., Mattis, D.: Two soluble models of an antiferromagnetic chain. Ann. Phys. **16**(3), 407–466 (1961). [2.2, 2.A.2, 10.1.2]
246. Lieb, E.H.: Flux phase of the half-filled band. Phys. Rev. Lett. **73**, 2158–2161 (1994). [10.2.1]
247. Liebmann, R.: Statistical Mechanics of Periodic Frustrated Ising Systems. Lecture Notes in Physics, vol. 251. Springer, Berlin (1986). [4.2, 4.A.3]
248. Lin, Y.C., Kawashima, N., Iglói, F., Rieger, H.: Numerical renormalization group study of random transverse Ising models in one and two space dimensions. Prog. Theor. Phys. Suppl. **138**, 479–488 (2000). [5.3]
249. Lubensky, T.C.: In: Balian, R., Maynard, R., Toulouse, G. (eds.) Ill-Condensed Matter. North-Holland, Amsterdam (1979). [5.2]
250. Ma, Y.q., Gong, C.d.: Statics in the random quantum asymmetric Sherrington-Kirkpatrick model. Phys. Rev. B **45**, 793–796 (1992). [1.3, 9.1.1]
251. Ma, Y.q., Gong, C.d.: Hopfield spin-glass model in a transverse field. Phys. Rev. B **48**, 12778–12782 (1993). [1.3,9.1.1]
252. Ma, Y.q., Li, Z.y.: Phase diagrams of the quantum Sherrington-Kirkpatrick Ising spin glass in a transverse field. Phys. Lett. A **148**(1–2), 134–138 (1990). [6.3]
253. Ma, Y.q., Zhang, Y.m., Ma, Y.g., Gong, C.d.: Statistical mechanics of a Hopfield neural-network model in a transverse field. Phys. Rev. E **47**, 3985–3987 (1993). [1.3, 9.1.1]
254. MacKay, D.J.C.: Information Theory, Inference, and Learning Algorithms. Cambridge University Press, Cambridge (2003). [8.1, 9.2.1, 9.2.4]
255. Majumdar, C.K., Ghosh, D.K.: On next-nearest-neighbor interaction in linear chain. i. J. Math. Phys. **10**(8), 1388–1398 (1969). [1.3, 4.3]
256. Majumdar, C.K., Ghosh, D.K.: On next-nearest-neighbor interaction in linear chain. ii. J. Math. Phys. **10**(8), 1399–1402 (1969). [1.3, 4.3]
257. Mangum, B.W., Lee, J.N., Moos, H.W.: Magnetically controllable cooperative Jahn-Teller distortion in $TmAsO_4$. Phys. Rev. Lett. **27**, 1517–1520 (1971). [1.3]
258. Mari, P.O., Campbell, I.A.: Ising spin glasses: interaction distribution dependence of the critical exponents. arXiv:cond-mat/0111174 (2001). [6.1]
259. Martoňák, R., Santoro, G.E., Tosatti, E.: Quantum annealing by the path-integral Monte Carlo method: the two-dimensional random Ising model. Phys. Rev. B **66**, 094203 (2002). [8.7.2]
260. Martoňák, R., Santoro, G.E., Tosatti, E.: Quantum annealing of the traveling-salesman problem. Phys. Rev. E **70**, 057701 (2004). [8.4.1]
261. Matsuda, Y., Nishimori, H., Katzgraber, H.G.: Ground-state statistics from annealing algorithms: quantum versus classical approaches. New J. Phys. **11**(7), 073021 (2009). [8.4.1]

262. Mattis, D.C.: Solvable spin systems with random interactions. Phys. Lett. A **56**(5), 421–422 (1976). [6.8]
263. Mattis, D.C.: Encyclopedia of Magnetism in One Dimension. World Scientific, Singapore (1994). [1.3, 4.1]
264. McCoy, B.: In: Domb, C., Green, M.S. (eds.) Phase Transitions and Critical Phenomena, vol. II. Academic Press, London (1972). [5.2, 5.3]
265. McCoy, B.M.: Spin correlation functions of the X–Y model. Phys. Rev. **173**, 531–541 (1968). [2.2.1, 2.A.3]
266. McCoy, B.M.: Theory of a two-dimensional Ising model with random impurities. iii. Boundary effects. Phys. Rev. **188**, 1014–1031 (1969). [6.4]
267. McCoy, B.M.: Theory of a two-dimensional Ising model with random impurities. iv. Generalizations. Phys. Rev. B **2**, 2795–2803 (1970). [6.4]
268. McCoy, B.M.: Advanced Statistical Mechanics. Oxford University Press, Oxford (2010). [2.A.3]
269. McCoy, B.M., Wu, T.T.: Theory of a two-dimensional Ising model with random impurities. i. Thermodynamics. Phys. Rev. **176**, 631–643 (1968). [5.2, 5.3, 6.4]
270. McCoy, B.M., Wu, T.T.: Theory of a two-dimensional Ising model with random impurities. ii. Spin correlation functions. Phys. Rev. **188**, 982–1013 (1969). [5.2, 6.4]
271. Messiah, A.: Quantum Mechanics, vol. 2. North-Holland, Amsterdam (1962). [8.3.2]
272. Mézard, M., Monasson, R.: Glassy transition in the three-dimensional random-field Ising model. Phys. Rev. B **50**, 7199–7202 (1994). [6.7.1]
273. Mézard, M., Montanari, A.: Information, Physics, and Computation. Oxford University Press, Oxford (2009). [1.3, 8.1]
274. Mézard, M., Young, A.P.: Replica symmetry breaking in the random field Ising model. Europhys. Lett. **18**(7), 653 (1992). [6.7.1]
275. Mézard, M., Parisi, G., Virasoro, M.A.: Spin Glass Theory and Beyond. World Scientific, Singapore (1987). [1.3, 3.4.2, 6.1, 6.5, 6.A.3, 8.1]
276. Miyashita, S.: Dynamics of the magnetization with an inversion of the magnetic field. J. Phys. Soc. Jpn. **64**(9), 3207–3214 (1995). [9.2]
277. Miyashita, S.: Observation of the energy gap due to the quantum tunneling making use of the Landau-Zener mechanism. J. Phys. Soc. Jpn. **65**(8), 2734–2735 (1996). [9.2]
278. Mizel, A., Lidar, D.A., Mitchell, M.: Simple proof of equivalence between adiabatic quantum computation and the circuit model. Phys. Rev. Lett. **99**, 070502 (2007). [8.8]
279. Monasson, R., Zecchina, R., Kirkpatrick, S., Selman, B., Troyansky, L.: Determining computational complexity from characteristic 'phase transitions'. Nature **400**, 133–137 (1999). [8.4.1]
280. Morita, S., Nishimori, H.: Convergence theorems for quantum annealing. J. Phys. A, Math. Gen. **39**(45), 13903 (2006). [1.3, 8.7.2]
281. Morita, S., Nishimori, H.: Convergence of quantum annealing with real-time Schrödinger dynamics. J. Phys. Soc. Jpn. **76**(6), 064002 (2007). [1.3, 8.7.1]
282. Morita, S., Nishimori, H.: Mathematical foundation of quantum annealing. J. Math. Phys. **49**(12), 125210 (2008). [8.8, 9.2.6]
283. Morita, S., Suzuki, S., Nakamura, T.: Quantum-thermal annealing with a cluster-flip algorithm. Phys. Rev. E **79**, 065701 (2009). [3.2]
284. Moruzzi, V., Teaney, D.: Specific heat of EuS. Solid State Commun. **1**(6), 127–131 (1963). [1.3]
285. Motrunich, O., Mau, S.C., Huse, D.A., Fisher, D.S.: Infinite-randomness quantum Ising critical fixed points. Phys. Rev. B **61**, 1160–1172 (2000). [5.3]
286. Mühlschlegel, B., Zittartz, H.: Gaussian average method in the statistical theory of the Ising model. Z. Phys. A, Hadrons Nucl. **175**, 553–573 (1963). [3.6.2]
287. Mukherjee, V., Divakaran, U., Dutta, A., Sen, D.: Quenching dynamics of a quantum XY spin-$\frac{1}{2}$ chain in a transverse field. Phys. Rev. B **76**, 174303 (2007). [10.1.2]

288. Müller-Hartmann, E., Zittartz, J.: Interface free energy and transition temperature of the square-lattice Ising antiferromagnet at finite magnetic field. Z. Phys. B, Condens. Matter **27**, 261–266 (1977). [4.2]
289. Mussardo, G.: Statistical Field Theory. Oxford University Press, Oxford (2010). [2.6]
290. Nagai, O., Yamada, Y., Miyatake, Y.: In: Suzuki, M. (ed.) Quantum Monte Carlo Methods, p. 95. Springer, Heidelberg (1986). [1.3, 3.2]
291. Nagy, A.: Exploring phase transitions by finite-entanglement scaling of MPS in the 1D ANNNI model. New J. Phys. **13**(2), 023015 (2011). [1.1, 1.3, 4.3.7]
292. Nakamura, T., Ito, Y.: A quantum Monte Carlo algorithm realizing an intrinsic relaxation. J. Phys. Soc. Jpn. **72**(10), 2405–2408 (2003). [3.2]
293. Nakano, K.: Associatron—a model of associative memory. IEEE Trans. Syst. Man Cybern. **2**(3), 380–388 (1972). [9.1]
294. Narath, A., Schirber, J.E.: Effect of hydrostatic pressure on the metamagnetic transitions in $FeCl_2{\cdot}2H_2O$, $CoCl_2{\cdot}2H_2O$, $FeCl_2$, and $FeBr_2$. J. Appl. Phys. **37**(3), 1124–1125 (1966). [1.3]
295. Nielsen, M.A., Chuang, I.L.: Quantum Computation and Quantum Information. Cambridge University Press, Cambridge (2000). [2.2.1, 8.1, 8.8]
296. Nishimori, H.: Optimum decoding temperature for error-correcting codes. J. Phys. Soc. Jpn. **62**(9), 2973–2975 (1993). [9.2, 9.2.1]
297. Nishimori, H.: Statistical Physics of Spin Glasses and Information Processing: An Introduction. Oxford University Press, Oxford (2001). [1.3, 8.1, 9.2, 9.2.2]
298. Nishimori, H., Nonomura, Y.: Quantum effects in neural networks. J. Phys. Soc. Jpn. **65**(12), 3780–3796 (1996). [1.1, 1.3, 9.1.1, 9.A]
299. Nishimori, H., Wong, K.Y.M.: Statistical mechanics of image restoration and error-correcting codes. Phys. Rev. E **60**, 132–144 (1999). [9.2, 9.2.1, 9.2.4]
300. Obuchi, T., Nishimori, H., Sherrington, D.: Phase diagram of the p-spin-interacting spin glass with ferromagnetic bias and a transverse field in the infinite-p limit. J. Phys. Soc. Jpn. **76**, 054002 (2007). [6.6]
301. Oitmaa, J., Plischke, M.: Critical behaviour of the Ising model in a transverse field. Physica B+C **86–88**(Part 2), 577–578 (1977). [3.3]
302. Opper, M., Saad, D.: Advanced Mean Field Methods: Theory and Practice. MIT Press, Cambridge (2001). [9.2.5]
303. Osterloh, A., Amico, L., Falci, G., Fazio, R.: Scaling of entanglement close to a quantum phase transition. Nature **416**, 608–610 (2002). [2.2.1]
304. Parisi, G.: Magnetic properties of spin glasses in a new mean field theory. J. Phys. A, Math. Gen. **13**(5), 1887 (1980). [6.1]
305. Parisi, G.: The order parameter for spin glasses: a function on the interval 0–1. J. Phys. A, Math. Gen. **13**(3), 1101 (1980). [6.1]
306. Parisi, G.: A sequence of approximated solutions to the S-K model for spin glasses. J. Phys. A, Math. Gen. **13**(4), 115 (1980). [6.1]
307. Penrose, R.: Shadows of the Mind. Oxford University Press, Oxford (1994). [9.1.1]
308. Penson, K.A., Jullien, R., Pfeuty, P.: Zero-temperature renormalization-group method for quantum systems. iii. Ising model in a transverse field in two dimensions. Phys. Rev. B **19**, 4653–4660 (1979). [3.6.1]
309. Perdomo-Ortiz, A., Dickson, N., Drew-Brook, M., Rose, G., Aspuru-Guzik, A.: Finding low-energy conformations of lattice protein models by quantum annealing. Sci. Rep. **2**, 571 (2012). [8.4.2]
310. Peschel, I.: On the entanglement entropy for an XY spin chain. J. Stat. Mech. Theory Exp. **2004**(12), P12005 (2004). [2.2.1]
311. Peschel, I., Emery, V.J.: Calculation of spin correlations in two-dimensional Ising systems from one-dimensional kinetic models. Z. Phys. B, Condens. Matter **43**, 241–249 (1981). [1.3, 4.3, 4.3.7]
312. Pfeuty, P.: The one-dimensional Ising model with a transverse field. Ann. Phys. **57**(1), 79–90 (1970). [1.1, 1.3, 2.2, 2.2.1, 2.A.3, 4.3, 5.2, 10.1.2]

313. Pfeuty, P., Elliott, R.J.: The Ising model with a transverse field. ii. Ground state properties. J. Phys. C, Solid State Phys. **4**(15), 2370 (1971). [3.6.1]
314. Pfeuty, P., Jullien, R., Penson, K.A.: In: Real Space Renormalisation. Topics in Current Physics, vol. 30, p. 119. Springer, Heidelberg (1982) [1.3, 4.3]
315. Pich, C., Young, A.P., Rieger, H., Kawashima, N.: Critical behavior and Griffiths-McCoy singularities in the two-dimensional random quantum Ising ferromagnet. Phys. Rev. Lett. **81**, 5916–5919 (1998). [1.3, 5.3]
316. Pirc, R., Tadić, B., Blinc, R.: Tunneling model of proton glasses. Z. Phys. B, Condens. Matter **61**, 69–74 (1985). [1.3, 6.2, 6.3]
317. Pirc, R., Tadić, B., Blinc, R.: Random-field smearing of the proton-glass transition. Phys. Rev. B **36**, 8607–8615 (1987). [1.3]
318. Polkovnikov, A.: Universal adiabatic dynamics in the vicinity of a quantum critical point. Phys. Rev. B **72**, 161201 (2005). [1.3, 7.2.2.2, 8.6]
319. Polkovnikov, A., Sengupta, K., Silva, A., Vengalattore, M.: Colloquium: nonequilibrium dynamics of closed interacting quantum systems. Rev. Mod. Phys. **83**, 863–883 (2011). [7.2.2]
320. Pryce, J.M., Bruce, A.D.: Statistical mechanics of image restoration. J. Phys. A, Math. Gen. **28**(3), 511 (1995). [9.2]
321. Quilliam, J.A., Meng, S., Mugford, C.G.A., Kycia, J.B.: Evidence of spin glass dynamics in dilute $LiHo_xY_{1-x}F_4$. Phys. Rev. Lett. **101**, 187204 (2008). [1.1, 1.3]
322. Ray, P., Chakrabarti, B.K.: Exact ground-state excitations of the XY model in a transverse field in one dimension. Phys. Lett. A **98**(8–9), 431–432 (1983). [1.3, 10.1.2]
323. Ray, P., Chakrabarti, B.K., Chakrabarti, A.: Sherrington-Kirkpatrick model in a transverse field: absence of replica symmetry breaking due to quantum fluctuations. Phys. Rev. B **39**, 11828–11832 (1989). [1.3, 6.2, 6.3, 6.5, 8.1]
324. Raymond, J., Sportiello, A., Zdeborová, L.: Phase diagram of the 1-in-3 satisfiability problem. Phys. Rev. E **76**, 011101 (2007). [8.4.1]
325. Read, N., Sachdev, S., Ye, J.: Landau theory of quantum spin glasses of rotors and Ising spins. Phys. Rev. B **52**, 384–410 (1995). [1.3, 6.2, 6.4, 6.5]
326. Reinelt, G.: TSPLIB. http://www.iwr.uni-heidelberg.de/groups/comopt/software/TSPLIB95/. [8.4.1]
327. Rieger, H.: In: Stauffer, D. (ed.) Annual Review of Computational Physics, vol. 2, p. 925. World Scientific, Singapore (1995). [6.2]
328. Rieger, H.: Critical behavior of the three-dimensional random-field Ising model: two-exponent scaling and discontinuous transition. Phys. Rev. B **52**, 6659–6667 (1995). [6.7.1]
329. Rieger, H., Kawashima, N.: Application of a continuous time cluster algorithm to the two-dimensional random quantum Ising ferromagnet. Eur. Phys. J. B, Condens. Matter Complex Syst. **9**, 233–236 (1999). [5.3]
330. Rieger, H., Uimin, G.: The one-dimensional ANNNI model in a transverse field: analytic and numerical study of effective Hamiltonians. Z. Phys. B, Condens. Matter **101**, 597–611 (1996). [4.3.7]
331. Rieger, H., Young, A.P.: Critical exponents of the three-dimensional random field Ising model. J. Phys. A, Math. Gen. **26**(20), 5279 (1993). [6.7.1]
332. Rieger, H., Young, A.P.: Zero-temperature quantum phase transition of a two-dimensional Ising spin glass. Phys. Rev. Lett. **72**, 4141–4144 (1994). [1.3, 6.2, 6.5, 6.4]
333. Rieger, H., Young, A.P.: Griffiths singularities in the disordered phase of a quantum Ising spin glass. Phys. Rev. B **54**, 3328–3335 (1996). [5.3, 6.2, 8.5.2]
334. Rossini, D., Silva, A., Mussardo, G., Santoro, G.E.: Effective thermal dynamics following a quantum quench in a spin chain. Phys. Rev. Lett. **102**, 127204 (2009). [1.1, 1.3, 7.2.2.1]
335. Rossini, D., Suzuki, S., Mussardo, G., Santoro, G.E., Silva, A.: Long time dynamics following a quench in an integrable quantum spin chain: local versus nonlocal operators and effective thermal behavior. Phys. Rev. B **82**, 144302 (2010). [1.1, 1.3, 7.2.2.1]
336. Rujstraightán, P.: Critical behavior of two-dimensional models with spatially modulated phases: analytic results. Phys. Rev. B **24**, 6620–6631 (1981). [1.1, 4.3]

337. Ruján, P.: Finite temperature error-correcting codes. Phys. Rev. Lett. **70**, 2968–2971 (1993). [9.2, 9.2.1]
338. Sachdev, S.: Universal, finite-temperature, crossover functions of the quantum transition in the Ising chain in a transverse field. Nucl. Phys. B **464**(3), 576–595 (1996). [2.A.3]
339. Sachdev, S.: Quantum Phase Transitions. Cambridge University Press, Cambridge (1999). [1.1, 2.A.3]
340. Samara, G.A.: Vanishing of the ferroelectric and antiferroelectric states in KH_2PO_2-type crystals at high pressure. Phys. Rev. Lett. **27**, 103–106 (1971). [1.3]
341. Santoro, G.E., Tosatti, E.: Optimization using quantum mechanics: quantum annealing through adiabatic evolution. J. Phys. A, Math. Gen. **39**(36), 393 (2006). [8.1]
342. Santoro, G.E., Martoňák, R., Tosatti, E., Car, R.: Theory of quantum annealing of an Ising spin glass. Science **295**(5564), 2427–2430 (2002). [8.4.1, 9.2]
343. Sarjala, M., Petäjä, V., Alava, M.: Optimization in random field Ising models by quantum annealing. J. Stat. Mech. Theory Exp. **2006**(01), P01008 (2006). [8.4.1]
344. Satija, I.I.: Symmetry breaking and stabilization of critical phase. Phys. Rev. B **48**, 3511–3514 (1993). [1.3, 10.1.3]
345. Satija, I.I.: Spectral and magnetic interplay in quantum spin chains: stabilization of the critical phase due to long-range order. Phys. Rev. B **49**, 3391–3399 (1994). [1.3, 10.1.3]
346. Satija, I.I., Chaves, J.C.: XY-to-Ising crossover and quadrupling of the butterfly spectrum. Phys. Rev. B **49**, 13239–13242 (1994). [1.3, 10.1.3]
347. Schiff, L.I.: Quantum Mechanics. McGraw-Hill, London (1968). [4.A.3]
348. Schneider, T., Pytte, E.: Random-field instability of the ferromagnetic state. Phys. Rev. B **15**, 1519–1522 (1977). [6.7.1, 6.7.2]
349. Schultz, T.D., Mattis, D.C., Lieb, E.H.: Two-dimensional Ising model as a soluble problem of many fermions. Rev. Mod. Phys. **36**, 856–871 (1964). [1.1, 1.3, 3.1, 3.A.2, 10.1.2]
350. Schwartz, M., Soffer, A.: Critical correlation susceptibility relation in random-field systems. Phys. Rev. B **33**, 2059–2061 (1986). [6.7.1]
351. Seki, Y., Nishimori, H.: Quantum annealing with antiferromagnetic fluctuations. Phys. Rev. E **85**, 051112 (2012). [8.8]
352. Selke, W.: The ANNNI model—theoretical analysis and experimental application. Phys. Rep. **170**(4), 213–264 (1988). [4.1, 4.2]
353. Selke, W.: In: Domb, C., Lebowitz, J.L. (eds.) Phase Transitions and Critical Phenomena, vol. 15. Academic Press, New York (1992). [4.1, 4.2]
354. Selke, W., Duxbury, P.M.: The mean field theory of the three-dimensional ANNNI model. Z. Phys. B, Condens. Matter **57**, 49–58 (1984). [4.2]
355. Sen, D.: Large-S analysis of a quantum axial next-nearest-neighbor Ising model in one dimension. Phys. Rev. B **43**, 5939–5943 (1991). [1.3, 4.4]
356. Sen, D., Chakrabarti, B.K.: Large-S analysis of one-dimensional quantum-spin models in a transverse magnetic field. Phys. Rev. B **41**, 4713–4722 (1990). [1.3, 4.3, 4.4]
357. Sen, P.: Ground state properties of a one dimensional frustrated quantum XY model. Phys. A, Stat. Mech. Appl. **186**(1–2), 306–313 (1992). [4.6]
358. Sen, P.: Order disorder transitions in Ising models in transverse fields with second neighbour interactions. Z. Phys. B, Condens. Matter **98**, 251–254 (1995). [3.3, 4.5]
359. Sen, P., Chakrabarti, B.K.: Ising models with competing axial interactions in transverse fields. Phys. Rev. B **40**, 760–762 (1989). [1.1, 1.3, 4.3]
360. Sen, P., Chakrabarti, B.K.: Critical properties of a one-dimensional frustrated quantum magnetic model. Phys. Rev. B **43**, 13559–13565 (1991). [1.1, 1.3, 4.3]
361. Sen, P., Chakrabarti, B.K.: Frustrated transverse Ising models: a class of frustrated quantum systems. Int. J. Mod. Phys. B **6**, 2439–2469 (1992). [1.1, 1.3, 4.6, 6.2, 6.3]
362. Sen, P., Chakraborty, S., Dasgupta, S., Chakrabarti, B.K.: Numerical estimate of the phase diagram of finite ANNNI chains in transverse field. Z. Phys. B, Condens. Matter **88**, 333–338 (1992). [1.1, 1.3, 4.3]
363. Sen, P., Acharyya, M., Chakrabarti, B.K. (1992, unpublished). [6.5]

364. Sengupta, K., Sen, D.: Entanglement production due to quench dynamics of an anisotropic xy chain in a transverse field. Phys. Rev. A **80**, 032304 (2009). [10.1.2]
365. Sengupta, K., Powell, S., Sachdev, S.: Quench dynamics across quantum critical points. Phys. Rev. A **69**, 053616 (2004). [1.1, 1.3, 7.2.2.1]
366. Sengupta, K., Sen, D., Mondal, S.: Exact results for quench dynamics and defect production in a two-dimensional model. Phys. Rev. Lett. **100**, 077204 (2008). [1.3, 10.2.3]
367. Shankar, R.: In: Pati, J., Shafi, Q., Lu, Yu (eds.) Current Topics in Condensed Matter and Particle Physics. World Scientific, Singapore (1993). [2.2.1]
368. Shankar, R., Murthy, G.: Nearest-neighbor frustrated random-bond model in $d = 2$: some exact results. Phys. Rev. B **36**, 536–545 (1987). [5.3, 6.4]
369. Sherrington, D., Kirkpatrick, S.: Solvable model of a spin-glass. Phys. Rev. Lett. **35**, 1792–1796 (1975). [6.1, 9.1.1, 9.1.2, 9.2, 9.2.4, 9.2.5]
370. Shor, P.W.: Algorithms for quantum computation: discrete logarithms and factoring. In: Proceedings of the 35th Annual Symposium on Foundations of Computer Science, pp. 124–134 (1994). [8.1]
371. Skalyo, J., Frazer, B.C., Shirane, G.: Ferroelectric-mode motion in KD_2PO_4. Phys. Rev. B **1**, 278–286 (1970). [1.3]
372. Smelyanskiy, V.N., Rieffel, E.G., Knysh, S.I., Williams, C.P., Johnson, M.W., Thom, M.C., Macready, W.G., Pudenz, K.L.: A near-term quantum computing approach for hard computational problems in space exploration. arXiv:1204.2821 (2012). [8.1]
373. Sokoloff, J.: Unusual band structure, wave functions and electrical conductance in crystals with incommensurate periodic potentials. Phys. Rep. **126**(4), 189–244 (1985). [10.1.3]
374. Sourlas, N.: Spin-glass models as error-correcting codes. Nature **339**, 693–695 (1989). [9.2, 9.2.2, 9.2.4]
375. Stauffer, D., Aharony, A.: Introduction to Percolation Theory. Taylor & Francis, London (1992). [5.1, 5.2]
376. Steffen, M., van Dam, W., Hogg, T., Breyta, G., Chuang, I.: Experimental implementation of an adiabatic quantum optimization algorithm. Phys. Rev. Lett. **90**, 067903 (2003). [8.4.2]
377. Stella, A.L., Vanderzande, C., Dekeyser, R.: Unified renormalization-group approach to the thermodynamic and ground-state properties of quantum lattice systems. Phys. Rev. B **27**, 1812–1831 (1983). [2.4.1]
378. Stevens, K.W.H., van Eekelen, H.A.M.: Thermodynamic effects of spin-phonon coupling. Proc. Phys. Soc. **92**(3), 680 (1967). [1.3]
379. Stinchcombe, R.B.: Ising model in a transverse field. i. Basic theory. J. Phys. C, Solid State Phys. **6**(15), 2459 (1973). [1.1, 1.2, 1.3, 3.6.2, 6.7.2]
380. Stinchcombe, R.B.: Diluted quantum transverse Ising model. J. Phys. C, Solid State Phys. **14**(10), 263 (1981). [1.3, 5.2]
381. Stinchcombe, R.B.: Exact scalings of pure and dilute quantum transverse Ising chains. J. Phys. C, Solid State Phys. **14**(16), 2193 (1981). [1.3, 5.2]
382. Stinchcombe, R.B.: In: Domb, C., Lebowitz, J.L. (eds.) Phase Transition and Critical Phenomena, vol. VII, p. 151. Academic Press, New York (1983). [1.3, 5.1, 5.2, 6.7.2]
383. Stout, J.W., Chisholm, R.C.: Heat capacity and entropy of $CuCl_2$ and $CrCl_2$ from 11° to 300°K. magnetic ordering in linear chain crystals. J. Chem. Phys. **36**(4), 979–991 (1962). [1.3]
384. Stratt, R.M.: Path-integral methods for treating quantal behavior in solids: mean-field theory and the effects of fluctuations. Phys. Rev. B **33**, 1921–1930 (1986). [3.3, 3.4.1, 4.5]
385. Suzuki, M.: Relationship among exactly soluble models of critical phenomena. i. Prog. Theor. Phys. **46**(5), 1337–1359 (1971). [1.1, 1.3, 3.1, 10.1.2]
386. Suzuki, M.: Relationship between d-dimensional quantal spin systems and $(d + 1)$-dimensional Ising systems. Prog. Theor. Phys. **56**(5), 1454–1469 (1976). [1.1, 1.3, 3.1, 5.2, 8.7.2, 9.1.2, 9.2, 9.2.4, 9.2.5, 9.2.6]
387. Suzuki, M.: In: Suzuki, M. (ed.) Quantum Monte Carlo Methods, p. 1. Springer, Heidelberg (1986). [1.1, 1.3, 3.1, 4.3, 6.5, 6.A.2]

388. Suzuki, S.: In: Das, A., Chakrabarti, B.K. (eds.) Quantum Annealing and Related Optimization Method, p. 207. Springer, Berlin (2005). [7.A.2]
389. Suzuki, S.: Cooling dynamics of pure and random Ising chains. J. Stat. Mech. Theory Exp. **2009**(03), P03032 (2009). [1.3, 8.6]
390. Suzuki, S., Okada, M.: Residual energies after slow quantum annealing. J. Phys. Soc. Jpn. **74**(6), 1649–1652 (2005). [8.6]
391. Swendsen, R.H., Wang, J.S.: Nonuniversal critical dynamics in Monte Carlo simulations. Phys. Rev. Lett. **58**, 86–88 (1987). [1.3, 3.2]
392. Syljuåsen, O.F.: Entanglement and spontaneous symmetry breaking in quantum spin models. Phys. Rev. A **68**, 060301 (2003). [2.2.1]
393. Szegö, G.: On certain Hermitian forms associated with the Fourier series of a positive function. Comm. Sém. Math. Univ. Lund (Medd. Lunds Univ. Mat. Sem.) **1952**(Tome Supplementaire), 228–238 (1952). [2.A.3]
394. Takahashi, K., Matsuda, Y.: Effect of random fluctuations on quantum spin-glass transitions at zero temperature. J. Phys. Soc. Jpn. **76**, 043712 (2010). [6.3]
395. Takahashi, K., Takeda, K.: Dynamical correlations in the Sherrington-Kirkpatrick model in a transverse field. Phys. Rev. B **78**, 174415 (2007). [6.3]
396. Tanaka, K.: Statistical-mechanical approach to image processing. J. Phys. A, Math. Gen. **35**(37), 81 (2002). [9.2]
397. Tanaka, K., Horiguchi, T.: Quantum statistical-mechanical iterative method in image restoration. Electron. Commun. Jpn. **83**(3), 84 (2000). [1.3, 9.2, 9.2.5]
398. Temperley, H.N.V.: Proc. Phys. Soc. **67**, 233 (1954). [3.4.1]
399. Tentrup, T., Siems, R.: Structure and free energy of domain walls in ANNNI systems. J. Phys. C, Solid State Phys. **19**(18), 3443 (1986). [4.5]
400. Thill, M.J., Huse, D.A.: Equilibrium behaviour of quantum Ising spin glass. Phys. A, Stat. Mech. Appl. **214**(3), 321–355 (1995). [6.2]
401. Thirumalai, D., Li, Q., Kirkpatrick, T.R.: Infinite-range Ising spin glass in a transverse field. J. Phys. A, Math. Gen. **22**(16), 3339 (1989). [1.3, 6.2, 6.5, 8.1]
402. Trammell, G.T.: Magnetic ordering properties of rare-earth ions in strong cubic crystal fields. Phys. Rev. **131**, 932–948 (1963). [1.3]
403. Trotter, H.F.: On the product of semi-groups of operators. Proc. Am. Math. Soc. **10**, 545–551 (1959). [3.1, 9.2.4]
404. Tsallis, C., Stariolo, D.A.: Generalized simulated annealing. Phys. A, Stat. Mech. Appl. **233**(1–2), 395–406 (1996). [8.8]
405. Tucker, J.W., Saber, M., Ez-Zahraouy, H.: A study of the quenched diluted spin 32 transverse Ising model. J. Magn. Magn. Mater. **139**(1–2), 83–94 (1995). [5.2]
406. Usadel, K.: Spin glass transition in an Ising spin system with transverse field. Solid State Commun. **58**(9), 629–630 (1986). [6.3]
407. Usadel, K.: Frustrated quantum spin systems. Nucl. Phys. B, Proc. Suppl. **5**(1), 91–96 (1988). [10.1.4]
408. Usadel, K., Schmitz, B.: Quantum fluctuations in an Ising spin glass with transverse field. Solid State Commun. **64**(6), 975–977 (1987). [6.3]
409. Uzelac, K., Jullien, R., Pfeuty, P.: Renormalisation group study of the random Ising model in a transverse field in one dimension. J. Phys. A, Math. Gen. **13**(12), 3735 (1980). [5.2]
410. Vidal, G., Latorre, J.I., Rico, E., Kitaev, A.: Entanglement in quantum critical phenomena. Phys. Rev. Lett. **90**, 227902 (2003). [2.2.1]
411. Villain, J.: Equilibrium critical properties of random field systems: new conjectures. J. Phys. Fr. **46**(11), 1843–1852 (1985). [6.7.1]
412. Villain, J., Bak, P.: Two-dimensional Ising model with competing interactions: floating phase, walls and dislocations. J. Phys. Fr. **42**(5), 657–668 (1981). [1.1, 4.2, 4.A.3]
413. Vitanov, N.V., Garraway, B.M.: Landau-Zener model: effects of finite coupling duration. Phys. Rev. A **53**, 4288–4304 (1996). [7.A.2]
414. Vitiello, G.: Coherence and dissipative dynamics in the quantum brain model. Neural Netw. World **5**, 717 (1995). [9.1.1]

415. Vojta, T.: Rare region effects at classical, quantum and nonequilibrium phase transitions. J. Phys. A, Math. Gen. **39**(22), 143 (2006). [1.3, 5.1, 8.5.2]
416. von Neumann, J., Wigner, E.: Phys. Z. **30**, 467–470 (1929). [8.3.1]
417. Walasek, K., Lukierska-Walasek, K.: Quantum transverse Ising spin-glass model in the mean-field approximation. Phys. Rev. B **34**, 4962–4965 (1986). [1.3, 6.2]
418. Walasek, K., Lukierska-Walasek, K.: Cluster-expansion method for the infinite-range quantum transverse Ising spin-glass model. Phys. Rev. B **38**, 725–727 (1988). [1.3, 6.3]
419. Wang, Y.L., Cooper, B.R.: Collective excitations and magnetic ordering in materials with singlet crystal-field ground state. Phys. Rev. **172**, 539–551 (1968). [1.3]
420. Wielinga, R., Huiskamp, W.: The spontaneous magnetization of the B.C.C. Heisenberg ferromagnet $Cu(NH_4)_2Br_4.2H_2O$. Physica **40**(4), 602–624 (1969). [1.3]
421. Wiesler, A.: A note on the Monte Carlo simulation of one dimensional quantum spin systems. Phys. Lett. A **89**(7), 359–362 (1982). [1.3, 3.2, 6.3]
422. Winkler, G.: Image Analysis, Random Fields, and Markov Chain Monte Carlo Methods: A Mathematical Introduction. Springer, Berlin (2002). [9.2]
423. Wolf, D., Zittartz, J.: On the one-dimensional spin-1/2-chain and its related fermion models. Z. Phys. B, Condens. Matter **43**, 173–183 (1981). [4.3]
424. Wolf, W.P.: Anisotropic interactions between magnetic ions. J. Phys., Colloq. **32**, C1-26–C1-33 (1971). [1.3]
425. Wootters, W.K.: Entanglement of formation of an arbitrary state of two qubits. Phys. Rev. Lett. **80**, 2245–2248 (1998). [2.2.1]
426. Wu, T.T.: Theory of Toeplitz determinants and the spin correlations of the two-dimensional Ising model. i. Phys. Rev. **149**, 380–401 (1966). [5.2]
427. Wu, W., Ellman, B., Rosenbaum, T.F., Aeppli, G., Reich, D.H.: From classical to quantum glass. Phys. Rev. Lett. **67**, 2076–2079 (1991). [1.1, 1.3, 6.2.1, 7.1.3]
428. Wu, W., Bitko, D., Rosenbaum, T.F., Aeppli, G.: Quenching of the nonlinear susceptibility at a $T = 0$ spin glass transition. Phys. Rev. Lett. **71**, 1919–1922 (1993). [1.1, 1.3, 6.2.1]
429. Yamada, Y., Yamada, T.: Inter-dipolar interaction in $NaNO_2$. J. Phys. Soc. Jpn. **21**(11), 2167–2177 (1966). [1.3]
430. Yamada, Y., Fujii, Y., Hatta, I.: Dielectric relaxation mechanism in $NaNO_2$. J. Phys. Soc. Jpn. **24**(5), 1053–1058 (1968). [1.3]
431. Yamada, Y., Fujii, Y., Terauchi, H.: J. Phys. Soc. Jpn. Suppl. **28**, 274 (1970). [1.3]
432. Yamamoto, T.: Ground-state properties of the Sherrington-Kirkpatrick model with a transverse field. J. Phys. C, Solid State Phys. **21**(23), 4377 (1988). [6.3]
433. Yamamoto, T., Ishii, H.: A perturbation expansion for the Sherrington-Kirkpatrick model with a transverse field. J. Phys. C, Solid State Phys. **20**(35), 6053 (1987). [1.3, 6.2, 6.3, 9.2.5]
434. Yanase, A.: Correlation index of the Ising model with a transverse field. J. Phys. Soc. Jpn. **42**(6), 1816–1818 (1977). [3.6.1]
435. Yeomans, J.: The theory and application of axial Ising models. Solid State Phys. **41**, 151–200 (1988). [4.1, 4.2]
436. Yokoi, C.S.O., Coutinho-Filho, M.D., Salinas, S.R.: Ising model with competing axial interactions in the presence of a field: a mean-field treatment. Phys. Rev. B **24**, 4047–4061 (1981). [4.2]
437. Yokota, T.: Numerical study of the SK spin glass in a transverse field by the pair approximation. J. Phys. Condens. Matter **3**(36), 7039 (1991). [6.3, 6.5]
438. Young, A.P.: Quantum effects in the renormalization group approach to phase transitions. J. Phys. C, Solid State Phys. **8**(15), L309 (1975). [1.1, 3.5, 3.6.2]
439. Young, A.P., Rieger, H.: Numerical study of the random transverse-field Ising spin chain. Phys. Rev. B **53**, 8486–8498 (1996). [1.3, 5.3]
440. Young, A.P., Knysh, S., Smelyanskiy, V.N.: Size dependence of the minimum excitation gap in the quantum adiabatic algorithm. Phys. Rev. Lett. **101**, 170503 (2008). [8.5.3.3]
441. Young, A.P., Knysh, S., Smelyanskiy, V.N.: First-order phase transition in the quantum adiabatic algorithm. Phys. Rev. Lett. **104**, 020502 (2010). [8.5.3.3]

442. Zamolodchikov, A.B.: Integrals of motion and s-matrix of the (scaled) $t = t_c$ Ising model with magnetic field. Int. J. Mod. Phys. A **4**, 4235 (1989). [2.6]
443. Zener, C.: Non-adiabatic crossing of energy levels. Proc. R. Soc. Lond. Ser. A **137**, 696–702 (1932). [7.A.2, 9.2]
444. Zurek, W.H.: Cosmological experiments in superfluid helium? Nature **317**, 505 (1985). [7.2.2.2]
445. Zurek, W.H., Dorner, U., Zoller, P.: Dynamics of a quantum phase transition. Phys. Rev. Lett. **95**, 105701 (2005). [1.1, 1.3, 7.2.2]

Index

S. Suzuki et al., *Quantum Ising Phases and Transitions in Transverse Ising Models*, Lecture Notes in Physics 862, DOI 10.1007/978-3-642-33039-1,